AF508607

Spatio-Temporal Analysis of Urbanization Using GIS and Remote Sensing in Developing Countries

Spatio-Temporal Analysis of Urbanization Using GIS and Remote Sensing in Developing Countries

Editors

Yuji Murayama
Matamyo Simwanda
Manjula Ranagalage

MDPI • Basel • Beijing • Wuhan • Barcelona • Belgrade • Manchester • Tokyo • Cluj • Tianjin

Editors
Yuji Murayama
Faculty of Life & Environmental
Sciences
University of Tsukuba
Tsukuba
Japan

Matamyo Simwanda
Department of Plant and
Environmental Sciences
Copperbelt University
Kitwe
Zambia

Manjula Ranagalage
Department of Environmental
Management
Faculty of Social Sciences and
Humanities
Rajarata University of Sri Lanka
Mihintale
Sri Lanka

Editorial Office
MDPI
St. Alban-Anlage 66
4052 Basel, Switzerland

This is a reprint of articles from the Special Issue published online in the open access journal *Sustainability* (ISSN 2071-1050) (available at: www.mdpi.com/journal/sustainability/special_issues/ Spatio_temporal_Analysis_Urbanization).

For citation purposes, cite each article independently as indicated on the article page online and as indicated below:

LastName, A.A.; LastName, B.B.; LastName, C.C. Article Title. *Journal Name* **Year**, *Volume Number*, Page Range.

ISBN 978-3-0365-2541-9 (Hbk)
ISBN 978-3-0365-2540-2 (PDF)

Contents

About the Editors

Yuji Murayama

Yuji Murayama is Professor Emeritus, Adjunct Researcher of the University of Tsukuba, Japan. He has served as President (2018–2020), Chairperson (2016–2018) and Director (2010–2014) of The Association of Japanese Geographers, and Vice President of The Asian Geographical Association (2019–present). His research interests are urban and transport geography, geographical information science and spatial analysis. He has edited and published many books including "Urban development in Asia and Africa: Geospatial analysis of metropolises" (2017, Springer-Nature), "Progress in geospatial analysis" (2012, Springer), "Spatial analysis and modeling in geographical transformation process: GIS-based applications" (2011, Springer), "Recent advances in Remote Sensing and GIS in Sub-Sahara Africa" (2010, Nova Publishers), and "Japanese urban system" (2000, Springer).

Matamyo Simwanda

Matamyo Simwanda is a Senior Lecturer in the Department of Plant and Environmental Sciences, School of Natural Resources, Copperbelt University, Zambia. He has a Bachelor's degree from Copperbelt University where he was awarded the best graduating student in the School of Natural Resources (2004). He has a Master's degree from Oregon State University, USA (2010), and a Ph.D. from Tsukuba University, Japan (2018). He also received a Ph.D. research excellence recognition award from the Faculty of Life and Environmental Sciences of Tsukuba University. His research is focused on the spatiotemporal analysis and modeling of landscapes; remote sensing; ecological assessments; environmental impact assessments; and the management and governance of natural resources. He has a vast track record of peer-reviewed journal publications. He is also an environmental and social compliance specialist and has executed over 30 consultations (as lead) for local and international development projects in Zambia.

Manjula Ranagalage

Manjula Ranagalage is a professor in the Department of Environmental Management, Faculty of Social Science and Humanities of Rajarata University of Sri Lanka, Sri Lanka. He completed a Ph.D. in Geo-Environmental Sciences at the University of Tsukuba, Japan, in 2019. After completing his Ph.D., he was selected as a JSPS postdoctoral research fellow at the University of Tsukuba, Japan. He has published more than 35 index journal articles published in peer-reviewed journals. His research disciplines are urban geography, urban climate, urban heat island, disaster management, sustainability, GIS, and remote sensing. He was recognized as the most outstanding young researcher of Sri Lanka in the field of social sciences and humanities in 2018 and was awarded the University of Tsukuba presidential awards in 2019 for his research excellence.

Editorial

Spatiotemporal Analysis of Urbanization Using GIS and Remote Sensing in Developing Countries

Yuji Murayama [1], Matamyo Simwanda [2] and Manjula Ranagalage [3,*]

1   Faculty of Life and Environmental Sciences, University of Tsukuba, 1-1-1, Tennodai,
    Tsukuba 305-8572, Ibaraki, Japan; mura@geoenv.tsukuba.ac.jp
2   Department of Plant and Environmental Sciences, School of Natural Resources, Copperbelt University,
    P.O. Box 21692, Kitwe 10101, Zambia; matamyo@gmail.com
3   Department of Environmental Management, Faculty of Social Sciences and Humanities,
    Rajarata University of Sri Lanka, Mihintale 50300, Sri Lanka
*   Correspondence: manjularanagalage@gmail.com

Citation: Murayama, Y.; Simwanda, M.; Ranagalage, M. Spatiotemporal Analysis of Urbanization Using GIS and Remote Sensing in Developing Countries. *Sustainability* **2021**, *13*, 3681. https://doi.org/10.3390/su13073681

Received: 15 March 2021
Accepted: 23 March 2021
Published: 26 March 2021

Publisher's Note: MDPI stays neutral with regard to jurisdictional claims in published maps and institutional affiliations.

The international statistics show that the global urban population will increase by up to 68% by 2050 [1]. The United Nations (UN) projection shows that urbanization will be faster in Asian and African countries than in other continents [1]. In the future, rapid urbanization will bring severe environmental and socio-economic problems, such as land degradation, loss of urban ecosystem services, urban heat islands, air pollution, flooding, health, urban poverty, crimes and violence, and traffic congestion [2,3]. Thus, sustainable urban development has become a widely discussed concept in various disciplines, such as geography, engineering, economics, politics, and sociology. Sustainable urban development is viewed as a way of preventing, reducing, and mitigating the environmental and socio-economic negative impacts of urbanization, notwithstanding its positive results (i.e., social and economic improvement of livelihoods). Hence, capturing the spatial-temporal variation of urbanization patterns will help introduce proper sustainable urban planning in developing countries.

During the last two decades, many researchers have focused on studying urbanization patterns. However, the scarcity of spatial data has been an obstacle to study urbanization quantitatively, especially in developing countries of Asia and Africa [4]. The use of remote sensing data and geographical information systems (GIS) techniques can overcome the limitations faced in these developing countries [4]. The data, such as land use and land cover, land surface temperature, population density, and energy consumption, can be extracted based on remote sensing in various spatial and temporal resolutions. GIS techniques can be used to analyze the urbanization patterns and predict future patterns [5]. Thus, the link between urbanization and sustainable urban development has increasingly become a principal issue in designing and developing sustainable cities at the local, regional, and global levels.

This Special Issue discusses the usefulness of the spatiotemporal analysis of urbanization using GIS and remote sensing in developing countries, with a particular focus on future urban sustainability in Asia and Africa. We contribute to this theme through 16 articles to help achieve sustainability in metropolitan cities in Asia and Africa.

The first article in the Special Issue focuses on the simulation of land use and land cover (LULC) changes to forecast strategies for urban sustainability. *"A Cellular Automata Model Constrained by Spatiotemporal Heterogeneity of the Urban Development Strategy for Simulating Land-use Change: A Case Study in Nanjing City, China"* [6] considered the insufficient research on the spatiotemporal heterogeneity of urban development strategies and its application to constraining cellular automata models (CA) models that have become increasingly popular in land-use and land-cover change (LULC) simulations. The authors propose a zoning transition rule and planning influence that consists of a development grade coefficient and traffic facility coefficient in the CA model to reflect the top-down and heterogeneous characteristics of spatial layout and the dynamic and heterogeneous external interferen of traffic facilities on land-use development. They contend that spatial layout planning important for urban green, humanistic, and sustainable development.

The following two articles included in the Special Issue analyze the spatial-tempo dynamics of LULC and landscape pattern changes along with their driving forces for s tainable urban development. *"Remote Sensing-Based Analysis of Landscape Pattern Evoluti in Industrial Rural Areas: A Case of Southern Jiangsu, China"* [7] analyzed the damage on ru landscapes and the ecological environment caused by the rapid economic developme of rural industrial areas using Landsat data. The authors used landscape pattern indic to capture the variation, progress, characteristics, and driving forces of landscape patte evolution over the last 37 years. The authors suggest that their study is important understanding the evolutionary dynamics of the urban–rural industry during urbanizati that can lead to better strategies for improving the landscape pattern and promoting t development of the ecological environment. It can also be used as a reference for oth developing countries for the sustainability of urban and rural development during ind trialization, which helps achieve regional sustainability. *"Spatial-Temporal Dynamic Analy of Land Use and Landscape Pattern in Guangzhou, China: Exploring the Driving Forces from Urban Sustainability Perspective"* [8] focused on (i) analyzing the spatial-temporal dynami of LULC and landscape pattern changes; (ii) figuring out the driving forces of the LUl changes; (iii) assessing the completion and value of green space systems constructic and (iv) forecasting the trend of land use and putting forward proposals about improvi environmental quality. The authors use the above four areas of focus to discuss LUl changes and their causes and propose potential future trends of urban development fro the perspective of sustainability.

The Special Issue also contains three articles addressing the dynamics of LULC chang and the rate of urban expansion towards achieving sustainable urban growth. *"Quan tative Influence of Land-Use Changes and Urban Expansion Intensity on Landscape Pattern Qingdao, China: Implications for Urban Sustainability"* [9] used land-use change and urb expansion intensity (UEI) as inducement factors for changes in landscape patterns and ste wise regression and geographically weighted regression (GWR) were applied to quanti magnitude effects on landscape patterns, respectively. The study suggests that land-us have different contributions to changes in the urban landscape patterns at different urb development zones (downtown, plain suburban area, and mountainous suburban area The authors recommend a compact city and protection policy that should be adapted different regions in the study area to achieve strong urban sustainability. *"An Analys of Urban Land Use/Land Cover Changes in Blantyre City, Southern Malawi (1994–2018)"* [1 addresses how the spatial and temporal LULC changes by rapid and unplanned urb growth could have adverse environmental and social consequences. The findings of t study reveal the pressure of human activities on the land and natural environment ar provide a basis for sustainable urban planning and development in the study area. *"S tiotemporal Patterns and Driving Forces of Urban Expansion in Coastal Areas: A Study on Urb Agglomeration in the Pearl River Delta, China"* [11] focused on capturing the spatiotempor patterns of urban land expansion and further analyze the dynamic driving forces of urb agglomeration. The study applied the urban-land expansion intensity index, urban-lar expansion difference index, fractal dimension, driving force analysis, and driving facto to facilitate their analysis. The authors show how understanding the mechanisms of ban agglomeration could provide useful information for coastal urban planning and al offers new knowledge regarding the interactions between different drivers of urban lar expansion.

The next four articles in the Special Issue address a critical environmental impact of banization that threatens sustainability; the urban heat island (UHI) effect. *"Spatiotempor Patterns and Drivers of the Surface Urban Heat Island in 36 Major Cities in China: A Comparis of Two Different Methods for Delineating Rural Areas"* [12] used the administrative borde (AB) method and an optimized simplified urban-extent (OSUE) algorithm to calcula and map the spatiotemporal patterns of the surface urban heat island intensity (SUHI) in 36 major cities in mainland China and explored whether administrative borders represent an appropriate standard range for the rural extent in SUHI intensity calculations. The study not only explores the standardization of the calculation of urban heat intensity but also provides insights into the relationship between urban development and the SUHI as an important issue of urban sustainability. *"The Impacts of the Expansion of Urban Impervious Surfaces on Urban Heat Islands in a Coastal City in China"* [13] used a combination of remote sensing data and spatial statistical methods to assess the effects of rapid urbanization on the UHI. The results of the study are presented as a useful proxy indicator for implementing sustainable planning of urban areas and for the mitigation of the effects of UHIs. *"The Impacts of Landscape Changes on Annual Mean Land Surface Temperature in the Tropical Mountain City of Sri Lanka: A Case Study of Nuwara Eliya (1996–2017)"* [14] addressed the UHI issue by investigating the impacts of changes in the urban landscape on Land surface temperature (LST) intensity in the tropical mountain city of Nuwara Eliya, Sri Lanka. The results of the study are presented as a useful indicator for improved future landscape and urban planning that can help minimize the negative impacts of LST on urban sustainability. *"Impact of Landscape Structure on the Variation of Land Surface Temperature in Sub-Saharan Region: A Case Study of Addis Ababa using Landsat Data (1986–2016)"* [15] focused on the UHI effect in the African region and assessed the impact of landscape structure on the variation in LST as a geospatial approach in Addis Ababa, Ethiopia. The authors used LULC maps and LST derived from Landsat data and employed geospatial techniques including gradient and intensity analysis, as well as multidirectional and multitemporal LST profiles to comprehend the variations of LST in the study area. The provides insights on how understanding LST variations can help introduce appropriate mitigation techniques to overcome the negative impacts of the UHI effect.

Achieving urban sustainability is also focused on improving the quality of urban environments. Therefore, the Special Issue also added three articles addressing environmental and socio-economic quality issues related to urbanization. *"Analysis of Life Quality in a Tropical Mountain City Using a Multi-Criteria Geospatial Technique: A Case Study of Kandy City, Sri Lanka"* [16] created a life quality index (LQI) and identified the spatial distribution pattern of the LQI in Kandy City, Sri Lanka. The authors used the analytic hierarchy process (AHP) to create the LQI using 13 environmental and socio-economic factors, and employed gradient analysis to examine the spatial distribution pattern of the LQI from the city center to the periphery. The study guides residents and the respective administrative bodies to make the study area a more livable and sustainable city. *"Spatiotemporal Analysis of the Nonlinear Negative Relationship between Urbanization and Habitat Quality in Metropolitan Areas"* [17] focused on examining the spatiotemporal variations and relationship between urbanization intensity (UI) and Habitat Quality (HQ) in the Yangtze River Delta Urban Agglomeration. The study further quantifies and analyses the direct and indirect impacts of urbanization on HQ. The authors demonstrate how the increasing demand for urban land has exacerbated the threat to ecological areas, thus revealing that urbanization might lead to habitat degradation. *"Dynamic Monitoring and Analysis of Ecological Quality of Pingtan Comprehensive Experimental Zone, a New Type of Sea Island City, Based on RSEI"* [18] emphasized the significance of monitoring and evaluating island ecology to curb the increasingly prominent environmental problems with rapid urbanization. The authors used a remote sensing-based ecological index (RSEI) to explore the ecological status and space–time changes in the Pingtan Comprehensive Experimental Zone (PZ) in the east sea of Fujian Province of China. The authors concluded that the increase in large area bare soil can lead to regional ecological degradation, while the implementation of scientific ecological planning can enhance ecological restoration and construction.

The last three articles in the Special Issue are distinct, but they still all address issues that are important for urban sustainability. *"Role of Urban Public Space and the Surrounding Environment in Promoting Sustainable Development from the Lens of Social Media"* [19] proposed a framework that integrates the spatial-temporal distribution for different activity

categories, the urban public spaces'(UPS) check-in time and the UPSs surrounding b
environment. The authors utilized a check-in database collected from Instagram in 2(
and 2017 in two central districts of Ho Chi Minh City (HCMC) to analyze the city dynam
and activities over the course of the day. By quantifying the popularity of contempor;
UPSs, the authors attempt to comprehend the many attractive features spreading o
the two central districts. The results contribute to enhancing the predictability of UI
on socio-economic performance and understanding the role of urban facilities in urb
sustainability.

*"Comparison on Multi-Scale Urban Expansion Derived from Nightlight Imagery betwe
China and India"* [20] takes a global perspective of urban sustainability. The authors cond
a multi-scale comparative analysis of urban development differences between China a
India as a way of capturing global multi-polarization in the 21st century. The auth
used night light data for China and India, and employ several approaches including
Gini coefficient, Getis–Ord Gi* index, and the Urban Expansion Intensity Index (UI
to compare the two countries. The study reveals that understanding the similarities a
differences of urban development between China and India can provide insight into engi
of global economic growth and sustainability.

The final article, "Impact of COVID-19 Induced Lockdown on Environmental Qua
in Four Indian Megacities Using Landsat 8 OLI and TIRS-Derived Data and Mamd;
Fuzzy Logic Modeling Approach" [21] focused on the deadly COVID-19 virus that h
caused a global pandemic health emergency and is a threat to global urban sustainabil
The authors take four megacities (Mumbai, Delhi, Kolkata, and Chennai) of India fo
comprehensive assessment of the dynamicity of environmental quality resulting fro
the COVID-19 induced lockdown situation. The authors create an environmental qu
ity index using remotely sensed biophysical parameters like Particulate Matters (PM
concentration, Land Surface Temperature (LST), Normalized Different Moisture Ind
(NDMI), Normalized Difference Vegetation Index (NDVI), and Normalized Differer
Water Index (NDWI). The results indicated that lockdown is not only capable of controlli
COVID-19 spread but also helpful in minimizing environmental degradation. The findir
of this study can be utilized for assessing and analyzing the impacts of COVID-19 induc
lockdown situation on the overall environmental quality of other megacities of the wo!

The Spatiotemporal analysis of urbanization using GIS and Remote Sensing in «
veloping countries proffers guidance for achieving urban sustainability at local, regior
and global levels. The studies presented in this special issue provide a range of use!
information on various remote sensing spatial data, new geospatial methodologies,
well as other newly developed data types that can be used to capture urbanization and
related problems. Most of the remote sensing data used in the studies are freely availal
and can be easily accessed. We believe that readers can get a wide range of knowled
related to remote sensing data and GIS applications. The research findings can also be us
as a valuable source of information that can be used to implement sustainable devel«
ment goals and thus achieving urban sustainability. The publications also identify furth
research needs.

Funding: This research received no external funding.

Conflicts of Interest: The authors declare no conflict of interest.

References

1. United Nations. *World Urbanization Prospects*; United Nations: New York, NY, USA, 2018.
2. Son, N.-T.; Chen, C.-F.; Chen, C.-R.; Thanh, B.-X.; Vuong, T.-H. Assessment of urbanization and urban heat islands in Ho C Minh City, Vietnam using Landsat data. *Sustain. Cities Soc.* **2017**, *30*, 150–161. [CrossRef]
3. Ranagalage, M.; Ratnayake, S.S.; Dissanayake, D.; Kumar, L.; Wickremasinghe, H.; Vidanagama, J.; Cho, H.; Udagedara, S.; J K.K.; Simwanda, M.; et al. Spatiotemporal variation of urban heat islands for implementing nature-based solutions: A case stu of Kurunegala, Sri Lanka. *ISPRS Int. J. Geo-Inf.* **2020**, *9*, 461. [CrossRef]

Ndzabandzaba, C. Data Sharing for Sustainable Development in Less Developed and Developing Countries. 2015. Available online: https://sustainabledevelopment.un.org/content/documents/615860-Ndzabandzaba-Data%20sharing%20for%20sd%20in%20less%20developed%20and%20developing%20countries.pdf (accessed on 5 March 2021).

Simwanda, M.; Murayama, Y.; Ranagalage, M. Modeling the drivers of urban land use changes in Lusaka, Zambia using multi-criteria evaluation: An analytic network process approach. *Land Use Policy* **2020**, *92*, 104441. [CrossRef]

Yang, J.; Shi, F.; Sun, Y.; Zhu, J. A cellular automata model constrained by spatiotemporal heterogeneity of the urban development strategy for simulating land-use change: A case study in Nanjing City, China. *Sustainability* **2019**, *11*, 4012. [CrossRef]

Zhu, Y.; Wang, C.; Sakai, T. Remote sensing-based analysis of landscape pattern evolution in industrial rural areas: A case of Southern Jiangsu, China. *Sustainability* **2019**, *11*, 4994. [CrossRef]

Liu, S.; Yu, Q.; Wei, C. Spatial-temporal dynamic analysis of land use and landscape pattern in Guangzhou, China: Exploring the driving forces from an urban sustainability perspective. *Sustainability* **2019**, *11*, 6675. [CrossRef]

Yang, J.; Li, S.; Lu, H. Quantitative influence of land-use changes and urban expansion intensity on landscape pattern in Qingdao, China: Implications for urban sustainability. *Sustainability* **2019**, *11*, 6174. [CrossRef]

Mawenda, J.; Watanabe, T.; Avtar, R. An analysis of urban land use/land cover changes in Blantyre City, Southern Malawi (1994-2018). *Sustainability* **2020**, *12*, 2377. [CrossRef]

Yan, Y.; Ju, H.; Zhang, S.; Jiang, W. Spatiotemporal patterns and driving forces of urban expansion in coastal areas: A study on urban agglomeration in the pearl river delta, China. *Sustainability* **2020**, *12*, 191. [CrossRef]

Niu, L.; Tang, R.; Jiang, Y.; Zhou, X. Spatiotemporal patterns and drivers of the surface urban heat island in 36 major cities in China: A comparison of two different methods for delineating rural areas. *Sustainability* **2020**, *12*, 478. [CrossRef]

Hua, L.; Zhang, X.; Nie, Q.; Sun, F.; Tang, L. The impacts of the expansion of urban impervious surfaces on urban heat islands in a coastal city in China. *Sustainability* **2020**, *12*, 475. [CrossRef]

Ranagalage, M.; Murayama, Y.; Dissanayake, D.; Simwanda, M. The Impacts of Landscape Changes on Annual Mean Land Surface Temperature in the Tropical Mountain City of Sri Lanka: A Case Study of Nuwara Eliya (1996–2017). *Sustainability* **2019**, *11*, 5517. [CrossRef]

Dissanayake, D.; Morimoto, T.; Murayama, Y.; Ranagalage, M. Impact of landscape structure on the variation of land surface temperature in sub-saharan region: A case study of Addis Ababa using Landsat Data (1986–2016). *Sustainability* **2019**, *11*, 2257. [CrossRef]

Dissanayake, D.M.S.L.B.; Morimoto, T.; Murayama, Y.; Ranagalage, M.; Perera, E.N.C. Analysis of life quality in a tropical mountain city using a multi-criteria geospatial technique: A case study of Kandy city, Sri Lanka. *Sustainability* **2020**, *12*, 2918. [CrossRef]

Zhu, J.; Ding, N.; Li, D.; Sun, W.; Xie, Y.; Wang, X. Spatiotemporal analysis of the nonlinear negative relationship between urbanization and habitat quality in metropolitan areas. *Sustainability* **2020**, *12*, 669. [CrossRef]

Wen, X.; Ming, Y.; Gao, Y.; Hu, X. Dynamic monitoring and analysis of ecological quality of pingtan comprehensive experimental zone, a new type of sea island city, based on RSEI. *Sustainability* **2020**, *12*, 21. [CrossRef]

Nguyen, T.V.T.; Han, H.; Sahito, N. Role of urban public space and the surrounding environment in promoting sustainable development from the lens of social media. *Sustainability* **2019**, *11*, 5967. [CrossRef]

Zhou, L.; Sun, Q.; Dang, X.; Wang, S. Comparison on multi-scale urban expansion derived from nightlight imagery between China and India. *Sustainability* **2019**, *11*, 4509. [CrossRef]

Ghosh, S.; Das, A.; Hembram, T.K.; Saha, S.; Pradhan, B.; Alamri, A.M. Impact of COVID-19 induced lockdown on environmental quality in four Indian megacities Using Landsat 8 OLI and TIRS-derived data and Mamdani fuzzy logic modelling approach. *Sustainability* **2020**, *12*, 5464. [CrossRef]

 *sustainability* 

Article

A Cellular Automata Model Constrained by Spatiotemporal Heterogeneity of the Urban Development Strategy for Simulating Land-use Change: A Case Study in Nanjing City, China

Jing Yang [1,2,3], Feng Shi [1,2,3], Yizhong Sun [1,2,3,*] and Jie Zhu [4]

[1]   Key Laboratory of Virtual Geographic Environment, Nanjing Normal University, Nanjing 210023, China
[2]   School of Geography, Nanjing Normal University, Nanjing 210023, China
[3]   Jiangsu Center for Collaborative Innovation in Geographical Information Resource Development and Application, Nanjing 210023, China
[4]   College of Civil Engineering, Nanjing Forestry University, Nanjing 210037, China
*   Correspondence: sunyizhong_cz@163.com; Tel.: +86-137-7082-7090

Received: 12 June 2019; Accepted: 22 July 2019; Published: 24 July 2019

Abstract: While cellular automata (CA) has become increasingly popular in land-use and land-cover change (LUCC) simulations, insufficient research has considered the spatiotemporal heterogeneity of urban development strategies and applied it to constrain CA models. Consequently, we proposed to add a zoning transition rule and planning influence that consists of a development grade coefficient and traffic facility coefficient in the CA model to reflect the top-down and heterogeneous characteristics of spatial layout and the dynamic and heterogeneous external interference of traffic facilities on land-use development. Testing the method using Nanjing city as a case study, we show that the optimal combinations of development grade coefficients are different in different districts, and the simulation accuracies are improved by adding the grade coefficients into the model. Moreover, the integration of the traffic facility coefficient does not improve the model accuracy as expected because the deployment of the optimal spatial layout has considered the effect of the subway on land use. Therefore, spatial layout planning is important for urban green, humanistic and sustainable development.

Keywords: cellular automata; spatial layout; transportation infrastructure; LUCC

1. Introduction

Reports on world urbanization prospects reveal that the global urban population exceeded the rural population in 2011, and it is predicted to reach 76% by 2050 [1,2]. Rapid urbanization population growth is a crucial signal that land-use and land-cover change (LUCC) and urban expansion will be serious, which creates great challenges to the sustainable development of cities. Therefore, it seems necessary to develop proper models for simulating and predicting urban growth to guide urban scientific and organic development [3–5]. The cellular automata (CA) model, as a common LUCC simulation model, simplifies the complicated urban development process by calibrating the characteristic parameters from the environmental space to design different transition rules, and it is a heuristic approach to understanding the realistic urban growth dynamics [6,7]. Moreover, the CA model has the advantages of simplification, flexibility, intuitiveness and nonlinearity [8–11].

Wu [6] noted that urban land development consists of two interrelated processes: self-organized growth and spontaneous growth. The former represents the future state of land use that is dependent on the current state of land use in the neighbourhood, which is the result of local land-use interactions.

This process corresponds to the basic principle of bottom-up CA; which states that complex global patterns are generated from the local evolutions by using transition rules [12,13]. Hagoort et al. [14] and Hansen [15,16] conducted research concerning the influence of neighbourhood rules on urban dynamics simulations. Moreno et al. [17,18] implemented a dynamic neighbourhood to generate a more realistic presentation of land-use change. Batty [19] began with models based on CA, simulating urban dynamics through the local actions of automata. The latter reflects that land conversions are affected by the supply-demand relationship, which determines the development propensity. Previous studies were mainly from the perspective of demand subject citizens. Pinto and Antunes [20] applied the CA model to simulate land-use dynamics by considering the evolution of population and employment densities over time. Cohen [21] found that a very significant share of urban land-use dynamics is taking place in those areas where the population is below 500,000, and Ferrás [22] explained this phenomenon as a counter-urbanization trend. Augustijn-Becker et al. [23] developed an agent-based housing model for the simulation of informal settlement dynamics. Although the role of supply subject-urban planning has been mentioned in several studies, strictly speaking, most of them were scenario simulations that were to analyse what the city would be like under different urban planning constraints or events, seldom exploring the essential role of urban development strategy in the urban development process. For example, Sohl et al. [24] developed a location model to stochastically allocate the projected proportions of future land use. Rafiee [4] designed three scenarios (historical, environmentally oriented and specific compound urban growth) to simulate the spatial pattern of urban growth under different conditions. Kok and Winograd [25] simulated urban development under different natural hazards. However, the spatial layout and spatial development control the granularity of land use and the planning implementation, which are controlled and affected by urban development strategy and have the characteristics of spatiotemporal heterogeneity.

Due to large regional differences, it is impossible to achieve comprehensive and all-round urbanization, which will lead to idle and wasted resources. Therefore, we will gradually achieve the urban development goals by levels and regions under the guidance of the development strategy. Liu [7] indicated that urban growth simulation had been affected by a set of primary and secondary transition rules, which are also referred to as local and global transition rules in CA models [26]. The first rule is applied to local areas (neighbourhoods), while the second rule reflects the influence of environmental and institutional factors on global urban growth. Because urban development is neither the purely local process nor the purely global process, the commonly used CA model combines the above two transition rules [27]. However, the implementation of spatial layout is top-down and multi-level [19,28,29]. The allocation of land use will be delivered through multiple levels, and it is impossible to reach the lowest level—the cellular—directly. Moreover, according to the planning of development strategy, to enhance the connotation, quality and sustainable development degree of urban spatial development, the grade and zoning of land-use development are inhomogeneous. At different levels, the same land grades have different control granularity in different control units. Although some scholars have proposed adding an "intermediate level" between the global and local levels, the division of the control unit at the new level does not match with the urban planning system. Yang [30] differentiated the transition probabilities of land use in different spatial regions through a different landscape pattern. Onsted and Clarke [31] divided the study area into several policy zones, determined by farmland security zones, to simulate urban growth and land-use changes. Ward et al. [32] integrated a regional optimization model and CA to discuss the possible growth scenarios in southeast Queensland, Australia. In China and Britain, the spatial planning systems are generally divided into three levels (national level, zoning level and local level), and the strategic requirements at the national level are implemented to the local level through the zoning level to complete target refinement and process control [33,34]. Furthermore, a study also found that zoning is the most effective and universal management and control tool in the planning system of different cities [6,35–37]. Hence, we suggest adding a "zoning transition rule" between the proposed two transition rules and considering the heterogeneity of the control granularity of different zoning units in the process of simulation.

Transportation infrastructure is an important part of the development strategy, and its implementation has a dynamic and heterogeneous external intervention influence on the LUCC process. Conventional CA is a typical spatiotemporal model used to simulate land use dynamics. Arsanjani et al. [38] and Guan et al. [39] used environmental and socio-economic variables to address urban sprawl based on an improved hybrid model. Jantz et al. [40] developed methods that expand the capability of SLEUTH (slope, land-use, excluded, urban, transportation, hill shade) to incorporate economic, cultural and policy information, opening up new avenues for land-use dynamics simulation. In those models, the characteristic of dynamics is represented by land-use attribution change by setting different time steps, but their driving factors are static in the simulation process. Nevertheless, the implementation of transportation infrastructure has a dynamic influence on land use change [3], and the dynamic characteristics are reflected in two aspects. First, the implementation time of transportation facilities is random. At present, most CA are inertia simulation models that drive their transition rules largely from empirical data by assuming that the historical trend of urban development will continue into the future [8,16]. However, there is a case where the facility implementation appears in the simulation process, which is not contained in the initial simulation data. For instance, the construction of a road begins at some point during the simulation. Second, the effect of traffic facilities on the development of urban land use is dynamic and inhomogeneous in the process of spatial evolution, which is reflected in space and time. Murakami and Cervero [41], Willigers and Wee [42] and Garmendia et al. [43] proved that the railway has a spatial spill over effect and spatial agglomeration effect on regional development. Gutiérrez [44], Vickerman [45] and Ureña et al. [46] certified that the two effects coexist. These findings suggest that the construction of transportation facilities (e.g., road planning) will produce inhomogeneous effects on urban land in space and that the spatial effect law changes dynamically with time. Therefore, quantifying the dynamic and heterogeneous external intervention influence of transportation facilities on planning influence is significant for simulating the LUCC in a more realistic and objective way.

It is necessary to pay more attention to the adaptation of rules to real-world planning contexts [47]. Considering the spatiotemporal continuity and heterogeneity of urban development strategy, this paper proposes an urban development strategy-constrained CA model, which is combined with three transition rules. First, according to the slope and location (whether it is within the scope of the ecological reserve) of the central cell, the direction of its transformation can be judged. Second, considering the top-down characteristics and spatial heterogeneity of the implementation of the spatial layout, a special "zoning transition rule" is proposed by constructing evaluating functions to measure the suitability of different land types in different districts. Third, by adding the land development grade coefficient, the traffic facility coefficient and the time variable into the local conversion function, the dynamic and heterogeneous external intervention influence can be reflected in this model. Finally, whether the land use changes or not is dependent on the comprehensive probability of the above rules. Our experiment shows that the proposed model can effectively simulate the LUCC and that the control granularity of land use development is heterogeneous in different districts, which is reflected in the difference in the development grade coefficient. Moreover, the simulation results confirm that the design of the spatial layout is a comprehensive and all-round process which considered the spatial difference of the subway effect. Therefore, the addition of the traffic facility coefficient has not produced the expected results in this experiment. These findings suggest that we need to pay more attention to the guiding mechanism of the urban development strategy for urban sustainable development.

The remainder of this paper is organized as follows. In Section 2, we describe the study area, driving factors, the design of transition rules, the determination of parameters and the verification of the model. In Section 3, we present the case study experiment conducted in Nanjing. Section 4 discusses the experimental results. Section 5 presents our conclusions on the experimental results and proposes future work.

2. Materials and Methods

2.1. Study Area and Data Sources

The provincial capital—Nanjing city—is located in the southwest of Jiangsu Province. It is the central city in the middle and lower reaches of the Yangtze River. Its geographic coordinates are 118°22′–119°14′ E, 31°14′–32°37′ N. The projection of land-use dynamics was supplemented by analyses of land-use changes for the years 1995, 2000 and 2005, the Nanjing City Master Plan (1991–2010) (development grades of land use), the Nanjing city urban planning functional zoning (ecological reserve) and several driving factors (Figure 1 and Table 1). The classified maps that were obtained from the Geographical Information Monitoring Cloud Platform [48], include six thematic land-use classes (river, forest land, agricultural land, urban and rural residential land, grassland and unused land), which were mapped at a spatial resolution of 30 × 30 m. The overall classification accuracy of the agricultural land and urban and rural residential land are more than 85%, and that of other types of land-use are more than 75%. According to the zoning policy, the main study areas are the urban areas of Nanjing city, which are divided into Qixia District, Main District, Jiangning District and Pukou District. The Nanjing Metro Line 1 began construction in 2000 and officially began operation on 3 September, 2005. It runs through the Qixia District, Main District and Jiangning District.

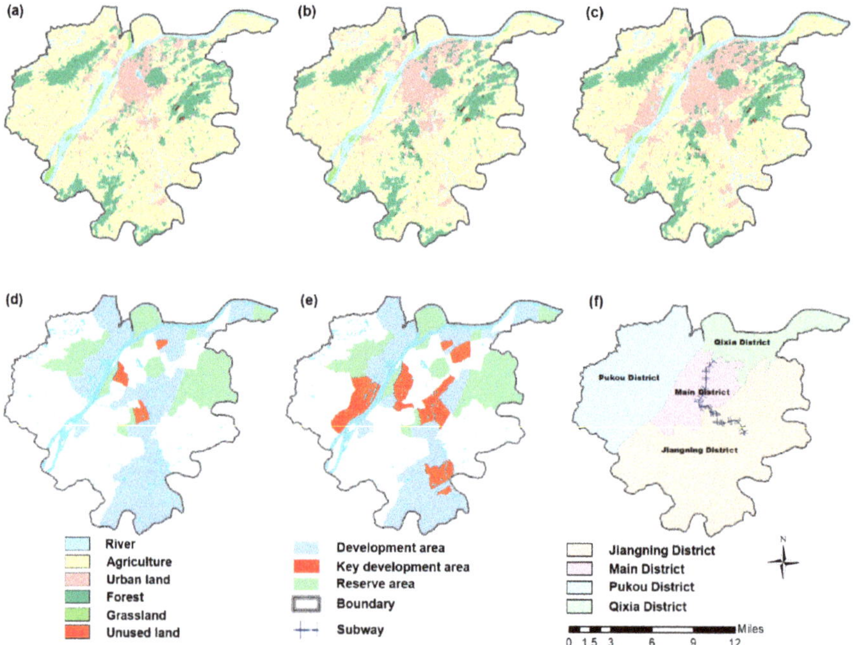

Figure 1. Illustration of the study area (Nanjing, China): (**a**) 1995 satellite-derived classified map; (**b**) 2000 satellite-derived classified map; (**c**) 2005 satellite-derived classified map; (**d**) 1995 urban land planning functional zoning; (**e**) 2000 urban land planning functional zoning; and (**f**) division maps of different districts and metro spatial layout.

Table 1. LUCC driving factors adopted for different transition rules.

Types	Transition Rules	Driving Factors
Natural factors	Global transition rule	Slope
	Zoning transition rule	Distance to river (DtoR) Elevation
Traffic factors	Zoning transition rule	Distance to the national highway (DtoNH) Distance to the provincial highway (DtoPH) Distance to the first-class highway (DtoFH) Distance to the railway (DtoRl) Distance to the motorway (DtoM)
	Local transition rule	Distance to the subway
Urban factors	Zoning transition rule	Distance to the city centre (DtoCC) Distance to the county centre (DtoCoC) Distance to the town centre (DtoTC)
Planning factors	Global transition rule	Ecological reserve
	Local transition rule	Grade of land-use development

2.2. LUCC Driving Factors

There are two main forms of LUCC: the transition of land-use types and the change of land-use intensity. Generally, the driving force and cause, which result in those land-use dynamics, are called te LUCC driving factors. In this paper, the LUCC driving factors are divided into four types: natural factors, traffic factors, urban factors, and planning factors (Table 1).

2.3. Transition Rules

Transition rules are the key of the CA model, which is used to determine the state of the cell at the next moment based on the current state of the target cell and its neighbourhood cells. Liu [49] found that the simulation result would be better when the iteration number is an integer multiple of the observation, so we set the iteration number from 1995 to 2005 as 40. The design of the rules should penetrate three levels of the urban planning system (global, zoning and local) and follow a series of principles of land-use change. In this paper, we proposed a spatiotemporal difference-constrained CA model whose whole process is shown in Figure 2.

2.3.1. Global Transition Rule

As the first rule among all transition rules, the global transition rule remaining stable in a specific period is a constant variable that is not affected by the region and time, and its probability is commonly defined as the set of {0, 1}. When the probability value is 0, it means that it is impossible to convert the land-use type into a specific land-use type, so the other rules will not work; otherwise, land use meets the primary conditions for the conversion to a specific type. In this paper, there are the following two global constraints.

Slope Constraints

Slope is the ratio of the vertical height to horizontal distance of each cell, indicating the inclination degree. Different types of land use have different slope requirements. Here, there are two main kinds of slope constraints: the planning mandatory constraint and the analytical constraint.

(1) Planning mandatory constraint: in the code for vertical planning on urban field [50], the slope of urban and rural residential land must be less than 25°.

(2) Analytical constraint: taking classified maps for the years 1995 and 2000 as examples, land slope maps and classified maps are overlapped to analyse the slope constraints of other types of land use that are without mandatory planning constraints. According to the specifications for base-map production of the second national land investigation [51], the slopes of urban land use are divided into five

levels for statistics, namely ≤2°, 2~6°, 6~15°, 15~25°, and >25°. Then, the slope distribution index (Equation (1)) is used to analyse the dominant slopes of different types of land use.

$$P = \frac{A_{ij} \times A}{A_i \times A_j}$$ (1)

where A_{ij} is the total area of the ith type of land use with the jth level of slope, A is the total area, A_i is the total area of the ith type of land use, A_j is the total area of the jth level of slope, and P is the slope distribution index. With the increase of P, the dominance of the ith type of land use on the jth level of slope also increases. When P is greater than 1, it indicates that the ith type of land use is suitable to generate on the jth level of slope. The specific dominance is shown in Tables 2 and 3. The dominant slopes of river, agricultural land, forest land, and unused land, with slopes of P greater than 1, are less than 2°, less than 15°, greater than 6° and greater than 6°, respectively. The dominance of grassland with different slopes is greater than or close to 1; in other words, the grassland has no global slope constraints.

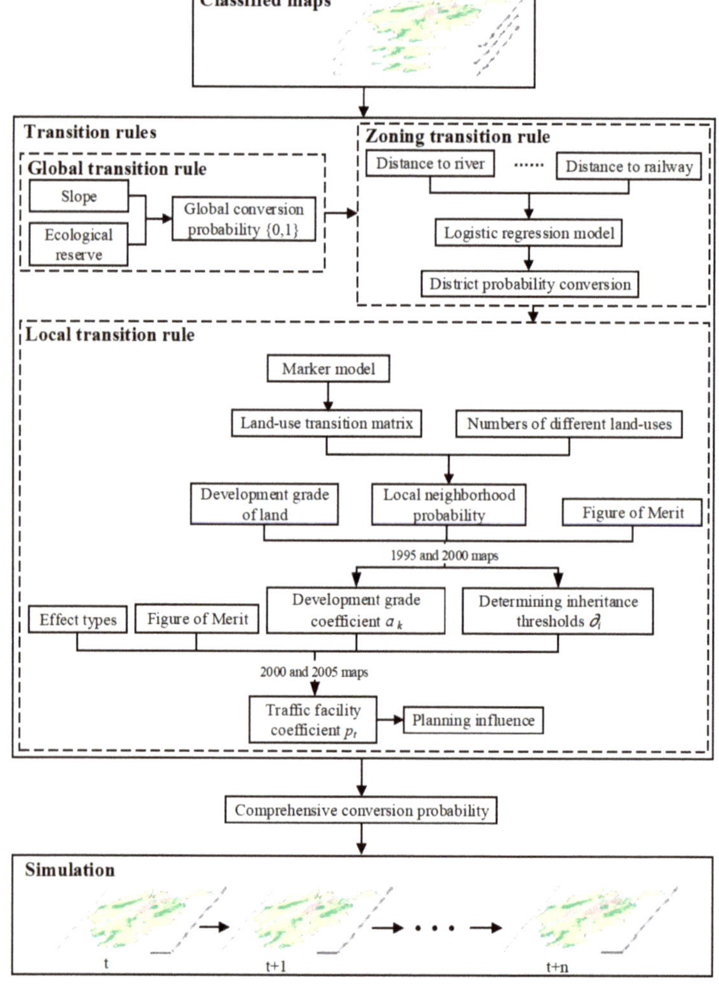

Figure 2. Flowchart of the spatiotemporal difference-constrained CA model.

Table 2. The slope dominance of different land-use types in 1995.

Slope	The Types of Land Use					
	River	Agricultural Land	Residential Land	Forest Land	Grassland	Unused Land
0–2	3.749715	0.831182	0.635183	0.33193	1.666305	0.194276
2–6	0.531526	1.157626	1.068633	0.490533	0.889708	0.510358
6–15	0.4847	1.004208	1.144227	1.155114	0.854234	1.297419
15–25	0.529827	0.468231	0.671574	4.302906	0.992741	4.15173
>25	0.423168	0.171718	0.266865	6.362546	0.916602	3.593295

Table 3. The slope dominance of different land-use types in 2000.

Slope	The Types of Land Use					
	River	Agricultural Land	Residential Land	Forest Land	Grassland	Unused Land
0–2	3.849988	0.882763	0.672931	0.252777	1.665558	0.159312
2–6	0.562731	1.225256	1.108991	0.38004	0.905429	0.418503
6–15	0.511673	0.950711	1.088816	1.314651	0.867801	1.406244
15–25	0.535038	0.396583	0.619102	3.61699	0.960287	3.324652
>25	0.442164	0.143669	0.304322	4.852855	0.905072	2.946072

Ecological Reserve Constraints

In the process of urban construction, forest, grassland, agricultural land and river area are constantly transformed into urban and rural residential land for maximizing economic benefit, which leads to the destruction of urban ecosystems and the erosion of urban ecological function areas. To maintain regional ecological security and sustainable development of the economy and society, the state has delimited ecological reserve (Figure 1d,e), where the type of land used for forest, grassland or river remains unchanged. Otherwise, the type of land use will be converted when the transition probability exceeds the threshold.

In summary, the global transition rules are as follows (see Figure 3):

1	**If** $S_{i,j}^t \in U$
2	$S_{i,j}^{t+1} = S_{i,j}^t,\ P_u = 0$
3	$Pglobal_{river} = Pglobal_{urban} = Pglobal_{agriculture} = Pglobal_{forest} = Pglobal_{grassland} = Pglobal_{unused} = 0$
4	**else**
5	$P_u = 1$
6	**If** $Slope_{i,j} \le 2$
7	$Pglobal_{river} = Pglobal_{urban} = Pglobal_{agriculture} = Pglobal_{grassland} = 1$
8	$Pglobal_{forest} = Pglobal_{unused} = 0$
9	**elif** $2 < Slope_{i,j} \le 6$
10	$Pglobal_{urban} = Pglobal_{agriculture} = Pglobal_{grassland} = 1$
11	$Pglobal_{river} = Pglobal_{forest} = Pglobal_{unused} = 0$
12	**elif** $6 < Slope_{i,j} \le 15$
13	$Pglobal_{urban} = Pglobal_{agriculture} = Pglobal_{forest} = Pglobal_{unused} = Pglobal_{grassland} = 1$
14	$Pglobal_{river} = 0$
15	**elif** $15 < Slope_{i,j} \le 25$
16	$Pglobal_{urban} = Pglobal_{forest} = Pglobal_{unused} = Pglobal_{grassland} = 1$
17	$Pglobal_{river} = Pglobal_{agriculture} = 0$
18	**else**
19	$Pglobal_{forest} = Pglobal_{unused} = Pglobal_{grassland} = 1$
20	$Pglobal_{urban} = Pglobal_{river} = Pglobal_{agriculture} = 0$

Figure 3. Main steps of the global transition rules.

Where $S_{i,j}^{t}$ is the land-use type of the grid in row i and column j at time t, and $S_{i,j}^{t+1}$ is the land-use type at time t + 1. U is a set of ecological reserves, including forest, grassland or river areas. $Slope_{i,j}$ is the slope of the grid in row i and column j. P_u is used to determine whether the grid is within the scope of the ecological reserve. When the grid is within the scope, its state remains unchanged with P_u being 0, and its probability of being converted to other types of land-use ($Pglobal_{river}$, $Pglobal_{urban}$, $Pglobal_{agriculture}$, $Pglobal_{forest}$, $Pglobal_{grassland}$, $Pglobal_{unused}$) is all 0. Otherwise, P_u is 1, and the probability of being converted is determined according to the value of $Slope_{i,j}$. The probability is 1, which means it is possible to be converted into this type of land, or the probability is 0, which means it is impossible to be converted into this type of land.

2.3.2. Zoning Transition Rule

After meeting the first transition rule (P_u = 1), this paper notes that a new zoning transition rule needs to be considered in the CA model. As the management and control unit in zoning planning, districts are set as study sub-units to analyse the suitability of different land-use types by using a logistic regression model (Equation (2)). Therefore, we consider 10 factors (nature, traffic and urban factors) (in Table 1) to construct the zoning transition rule.

$$Pmiddle_j = \frac{\exp(\beta_0 + \beta_1 X_1 + \beta_2 X_2 + \ldots + \beta_n X_n)}{1 + \exp(\beta_0 + \beta_1 X_1 + \beta_2 X_2 + \ldots + \beta_n X_n)} \tag{2}$$

where X_1, X_2, $\ldots$, X_n are the driving factors; β_1, β_2, $\ldots$, β_n are the regression coefficients; and β_0 is the constant term. The regression coefficients and the constant term are different in different districts. $Pmiddle_j$ is the probability of being the land-use type j, which is also defined as the suitability of the land-use type j by some scholars.

2.3.3. Local Transition Rule

Neighbourhood Constraint

In urban development, the distribution of planning land is affected by the types of land uses in the neighbourhood of the candidate areas where the land may be developed. Therefore, the CA model always follows the principle that the types of central cells are affected by spatial neighbouring cells.

This model chooses the commonly used extended Moore neighbourhood of the 5 × 5 window; each cell represents an area of 50 m × 50 m on the ground for the simulation. In the neighbourhood window, land types of the other cells except the central cell are recorded and set as $DevelopNum$ = {$Land_{river}$, $Land_{agriculture}$, $Land_{urban}$, $Land_{forest}$, $Land_{grassland}$, $Land_{unused}$}, where $Land_{river}$, $Land_{agriculture}$, $Land_{urban}$, $Land_{forest}$, $Land_{grassland}$, and $Land_{unused}$ represent the numbers of the different types of land use. In general, the greater the number of certain land-use types in the neighbourhood, the greater the probability that the central cell will be converted into this land-use type; that is, the more likely the central cell will be assimilated. However, the number of land-use types in the neighbourhood is not the only criterion of local constraints. According to the construction demand and the change of national conditions, the transfer rates among different land-use types are different. They are obtained using the Markov model and saved as a land-use transition matrix $DevelopExtent$, where R_r, A_a, RU_{ru}, F_f, G_g, and U_u are the inheritance degree of the original land-use types, and the other parameters indicate the evolution degree in the transformation direction of the different land-use types. For instance, R_{ru} indicates the evolution probability of river area being converted into urban area (Equation (3)).

$$\text{DevelopExtent} = \begin{array}{c} \\ river \\ agri \\ urban \\ forest \\ grass \\ unused \end{array} \begin{array}{cccccc} river & agri & urban & forest & grass & unused \\ \left[\begin{array}{cccccc} R_r & R_a & R_{ru} & R_f & R_g & R_u \\ A_r & A_a & A_{ru} & A_f & A_g & A_u \\ RU_r & RU_a & RU_{ru} & RU_f & RU_g & RU_u \\ F_r & F_a & F_{ru} & F_f & F_g & F_u \\ G_r & G_a & G_{ru} & G_f & G_g & G_u \\ U_r & U_a & U_{ru} & U_f & U_g & U_u \end{array} \right] \end{array} \qquad (3)$$

In addition, the land-use transition matrix that was obtained based on the classified maps for the years 1995 and 2000 is the initial matrix. In the iteration process, the matrix will change dynamically according to the evolution state of the land use in the last two moments. When the land type of the central cell is i and the numbers of the different land-use types within the neighbourhood of the $n*n$ window are recorded in the matrix *DevelopNum*, then its local neighbourhood probability of being converted into the other land-use types ($Plocaln_{i\to j}$) is as follows:

$$DevelopExtent_i = \{i_r, i_a, i_{ru}, i_f, i_g, i_u\}, \qquad (4)$$

$$Plocaln_{i\to j} = \frac{i_j Land_j}{n*n-1}. \qquad (5)$$

Planning Dynamic Constraint

Metro planning will produce urban land spill-over and aggregation effects that can be quantified into the planning influence to represent the influence degree of metro planning policy, which is related to space and time. The planning influence is determined by the types of intraregional effects; when the spill-over effect occurs in the region, it indicates that the policy has a spatial damping effect on the development of the land use, so the traffic facility coefficient (TFC, p_l) is a negative value. When the aggregation effect is generated in the region, it signifies that the policy exerts a spatial pull on the development of land use, so the traffic facility coefficient (p_l) is a positive value. In the experiment, we divided the development of the study area into two stages; from 1995 to 2000 and from 2000 to 2005. If the subway did not produce any effects in the second stage, the growth rates of residential land in this stage should be equal to those in the first stage. That is, the relative growth rate (RGR) of the two stages is equal to 0. If the sign of RGR is negative with some ranges, it means that the subway generates the spill-over effect on the land use with these ranges. Reviewing the reference on the influence of a subway on urban land [52–54], this paper divides the research range of the traffic facility coefficients into six levels: 0–200 m, 200–500 m, 500–800 m, 800–1600 m, 1600–2400 m, and 2400–3200 m. The experiment studies the dynamic change of the planning policy effect in the stage of subway construction. It is assumed that the policy effect has a linear relationship with time in the whole evolution process.

In addition, according to the Nanjing City Master Plan (1991–2010) and 2000 planning content adjustment text, the urban land is divided into three development grades (Figure 1d,e). The key development area, development area and other land belong to the first, second and third grades, respectively, and correspond to different development grade coefficients $\alpha_k(\alpha_1, \alpha_2, \alpha_3)$. The specific equation is shown as follows:

$$Plocalp_{k,l,t} = \alpha_k\left(1 + p_l\frac{t-t_s}{t_e-t_s}\right) \qquad (6)$$

where $Plocalp_{k,l,t}$ is the planning influence, $p_l(p_1, p_2, p_3, p_4, p_5, p_6)$ is the traffic facility coefficient (when the region is affected by the planning spill-over effect, it is a negative value, or not a positive value), $\alpha_k(\alpha_1, \alpha_2, \alpha_3)$ is the development grade coefficient, which is decided by the development grade, t_e and t_s are the end time and start time of the planning policy, respectively, and t is the current time of planning implementation.

In summary, the calculation of the local dynamic conversion probability ($Plocal_{i \to j,k,l,t}$) is as follows:

$$Plocal_{i \to j,k,l,t} = \begin{cases} Plocaln_{i \to j} \times Plocalp_{k,l,t} & j = urban, \ t \geq t_s \\ Plocaln_{i \to j} \times Plocalp_{k,l,t} \ and \ p_l = 0 & j = urban, \ t \leq t_s \\ Plocaln_{i \to j} & j = others \end{cases} \quad (7)$$

2.3.4. Comprehensive Transition Rule

According to Wu [55], the comprehensive conversion probability ($Pcomp_{i \to j, \ k,l,t}$) can be obtained by combining the above transition rules (Equation (8)) as follows:

$$Pcomp_{i \to j, \ k,l,t} = P_u \times Pglobal_j \times Pmiddle_j \times Plocal_{i \to j,k,l,t}. \quad (8)$$

Each cell has six conversion probabilities of being converted to different types of land use. There are six inheritance thresholds—∂_{river}, $\partial_{agriculture}$, ∂_{urban}, ∂_{forest}, $\partial_{grassland}$, and ∂_{unused}—in the model. When the inheritance value of a cell whose land-use type is i $\left(Pcomp_{i \to i,k,l,t}\right)$ and is greater than the threshold ∂_i, the land-use type remains unchanged. Otherwise, the cell will be transited to the land type j with the greatest conversion probability ($Max(Pcomp_{i \to j,k,l,t})$).

2.4. Model Parameter Estimation and Accuracy Verification

The figure of merit (FOM)—which consists of misses, hits, wrong hits and false alarms—is a popular metric for model validation by using three-map comparison, whose range is from 0 to 1. The closer the value is to 1, the more similar the simulated change is to the actual changes [56,57]. In this experiment, we used FOM to determine parameters and judge the simulation results.

There are three kinds of parameters that need to be set up in the model, including the inheritance thresholds (∂_i), development grade coefficients (α_k) and traffic facility coefficient (p_l). First, based on the reference map of 1995, we simulated the land-use changes in 2000 by calculating the six conversion probabilities of each cell without considering the planning influence. Comparing the simulation results with the actual map, the minimum conversion probabilities of each type of land-use without changes were what were defined as inheritance thresholds by us. Second, combined with the above obtained inheritance thresholds, the iteration method was used to calculate the FOM under different combinations of development grade coefficients, and the combination with the maximum FOM was the optimal combination of development grade coefficients. The traffic facility coefficient is set as 0 in the process of determining the development grade coefficient because the planning policy had not been implemented during the period. Third, the sign of the traffic facility coefficient (p_l) can be determined by judging the effect types based on the relative growth comparison of urban land in different regions before and after the planning policy. Then, we applied the iteration method to test the optimal traffic facility coefficients of the different districts with a step size of 0.1. The FOM of the simulated result was the largest, as we used this optimal traffic facility coefficient.

3. Results

The global transition probability of every cell ($P_u \times Pglobal_j$) can be obtained based on its location and slope, which has been introduced above. In the following, we explain how to determine the middle transition probability ($Pmiddle_j$) based on a logistic regression model and calculate the local transition probability ($Plocal_{i \to j,k,l,t}$) by using the Markov model and planning influence.

3.1. Logical Regression Coefficient

According to the introduction in Section 2.3.2, logistic regression models of four districts were built based on the 10 LUCC driving factors of the zoning transition rules in Table 1, and the regression coefficients of each district are shown in Table 4. Since there is no change in the unused land, we only selected five types of land use to construct regression models.

Table 4. Logistic regression factors and their coefficient of different districts.

District	Factors	Land-Use Types				
		River	Agricultural Land	Residential Land	Forest Land	Grassland
Main District	DtoFH	−0.945	1.948	0.569	—	−4.036
	DtoNH	0.386	−0.185	−0.249	−0.903	4.556
	DtoPH	—	0.529	−0.203	−1.303	−1.911
	DtoRl	—	0.573	—	−0.438	−1.182
	DtoM	−0.717	—	0.745	—	5.421
	DtoCC	2.030	—	−2.267	−2.056	—
	DtoCoC	−1.249	—	—	2.028	—
	DtoTC	0.267	—	—	0.346	3.369
	DtoR	—	−0.218	0.810	−0.218	—
	Elevation	−3.372	−0.518	−0.898	2.040	1.287
	Constant	−1.520	−0.509	−0.101	−1.707	−5.297
Qixia District	DtoFH	5.751	−3.080	4.934	—	21.998
	DtoNH	—	1.993	−0.755	—	−19.279
	DtoPH	−4.766	4.104	—	—	−11.763
	DtoRl	−1.705	−2.169	—	−1.361	12.676
	DtoM	11.319	−5.574	1.677	—	39.650
	DtoCC	−4.282	3.193	−6.822	—	−17.897
	DtoCoC	−3.973	−0.962	0.578	—	−11.817
	DtoTC	0.657	—	—	—	0.853
	DtoR	—	0.839	0.585	0.454	−1.693
	Elevation	−1.756	−0.774	−0.132	1.985	0.820
	Constant	−1.545	0.080	−0.558	−2.15	−10.201
Pukou District	DtoFH	0.642	—	−0.663	−3.552	0.642
	DtoNH	—	−0.142	—	2.299	—
	DtoPH	0.761	−0.242	−0.471	—	−1.911
	DtoRl	−0.408	0.442	—	−0.653	0.761
	DtoM	—	—	—	—	—
	DtoCC	−2.738	3.082	—	2.889	−2.738
	DtoCoC	−0.274	—	0.734	—	−0.274
	DtoTC	2.309	−0.608	−0.125	−2.700	2.309
	DtoR	—	—	—	—	—
	Elevation	−3.433	−1.225	−0.433	4.303	−3.433
	Constant	−1.656	−0.036	−0.124	−1.707	−1.656
Jiangning District	DtoFH	−1.766	1.165	—	1.249	−4.511
	DtoNH	−0.760	—	−0.387	1.214	−1.984
	DtoPH	−0.335	—	−0.321	0.459	—
	DtoRl	1.795	0.952	—	−2.199	4.154
	DtoM	0.736	−0.638	—	—	—
	DtoCC	—	0.603	—	—	—
	DtoCoC	—	—	—	—	—
	DtoTC	0.349	−0.584	—	—	5.843
	DtoR	—	0.316	0.170	—	−0.453
	Elevation	−0.637	−1.609	−0.395	3.894	0.760
	Constant	0.041	0.184	−0.009	0.888	−2.320

3.2. Land-use Transition Matrices

Based on classified maps for the years 1995 and 2000, we obtained the initial land use transition matrices of four districts by using the Markov model (Table 5). In the following table, every element represents the evolution probability between the different land uses, and the inheritance degrees of the original land-use types are all close to 1. Combining Equations (4) and (5), the neighbourhood conditions and the following transition matrices, we can obtain the local neighbourhood probability of every cell ($Plocaln_{i \rightarrow j}$).

Table 5. Land-use transition matrices of different districts.

District	Land-Use Types	Land-Use Types					
		River	Agricult-Ural LAND	Resident-Ial LAND	Forest Land	Grassland	Unused Land
Main District	River	0.999	0.740	0.801	0	0	0
	Agricult-ural land	0.604	0.992	0.881	0.878	0	0
	Resident-ial land	0.575	0.777	0.999	0	0	0
	Forest land	0	0	0.833	0.998	0	0
	Grassland	0	0	0	0	1	0
	Unused land	0	0	0	0	0	1
Qixia District	River	0.999	0.620	0.620	0	0	0
	Agricult-ural land	0.732	0.994	0.889	0	0.773	0
	Resident-ial land	0	0.782	0.999	0.601	0	0
	Forest land	0	0.746	0.842	0.998	0	0
	Grassland	0	0.686	0.886	0	0.995	0
	Unused land	0	0	0	0	0	1
Pukou District	River	0.999	0.800	0.595	0	0	0
	Agricult-ural land	0.711	0.997	0.860	0.558	0	0
	Resident-ial land	0	0.654	0.999	0.583	0	0
	Forest land	0	0.620	0.743	0.999	0	0
	Grassland	0.643	0	0.764	0	0.999	0
	Unused land	0	0	0	0	0	1
Jiangning District	River	0.999	0.678	0.718	0.613	0.592	0
	Agricult-ural land	0.767	0.997	0.836	0.794	0	0
	Resident-ial land	0.743	0.681	0.999	0	0	0
	Forest land	0.659	0.632	0.797	0.999	0	0
	Grassland	0.648	0	0	0	0.999	0
	Unused land	0	0	0	0	0	1

3.3. Determination of Planning Influence

Planning influence ($Plocalp_{k,l,t}$) consists of the development grade coefficient (α_k) and the traffic facility coefficient (∂_i), which can be trained from the classified maps of 1995 and 2000 and of 2000 and 2005, respectively.

3.3.1. Development Grade Coefficient and Inheritance Thresholds

According to the planning text, there are $\alpha_1 > \alpha_2 > \alpha_3$. Based on the iterative method, the model accuracies (FOM) of the different districts are calculated under different combinations of grade coefficients. As shown in Figure 4, each point displays a combination of grade coefficients, and its colour represents the level of the FOM. The optimal combinations of Main District, Qixia District, Pukou District and Jiangning District are (2.5, 2.4, 2.2), (2.7, 2.6, 1), (-, 2.7, 2.5) and (2.3, 2.2, 1.9), respectively. That is, model accuracies are the largest when using those combinations. Based on the optimal combinations, we simulated the land use in 2000 by using the proposed model (Figure 5a).

By analysing Table 6, it can be found that for Qixia District, the α_1 value is close to the α_2 value and quite far from the α_3 value. However, for Main District and Jiangning District, their grade coefficients are very close. Since there was no key development area in Pukou District in 1995, it only had two almost equal grade coefficients (α_2 and α_3). For the four districts, their inheritance thresholds of river (∂_1) are all very small, which implies that it is hard for the river to be changed into other land. Almost all the inheritance thresholds of agricultural land (∂_2) and grassland (∂_5) are greater than 0.3, which may be due to the occupation of farmland and grassland for urban expansion. The inheritance threshold of forest land (∂_4) is obviously higher than that in other districts because a large number of forest lands adjacent to Main District in Jiangning District were cut down for district development. In addition, the inheritance thresholds of residential lands (∂_3) in Main District and Qixia District are clearly less than those in Pukou District and Jiangning District. Comparing the simulated result

(Figure 5a) with the actual result (Figure 1a,b), we calculated the model accuracies (FOM), which are all greater than 0.19.

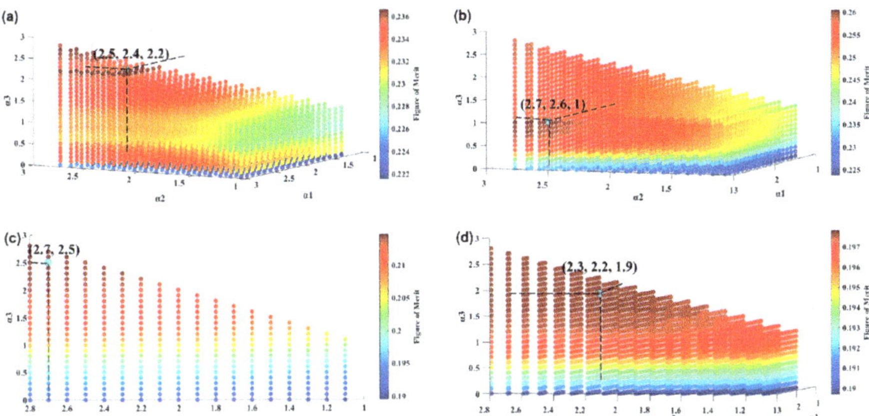

Figure 4. The development grade coefficient combinations of different districts: (**a**) Main District; (**b**) Qixia District; (**c**) Pukou District; and (**d**) Jiangning District.

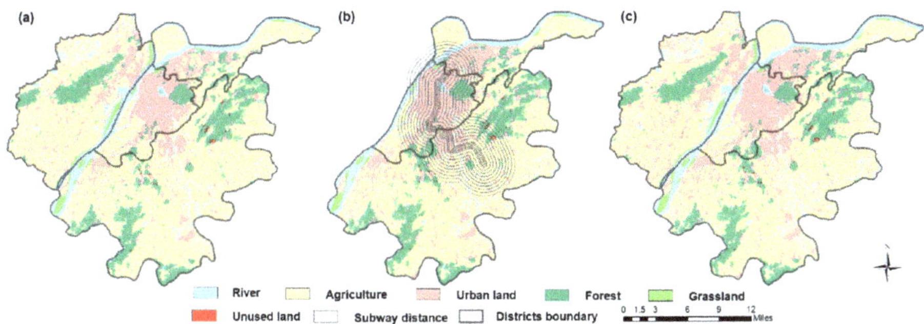

Figure 5. Simulation result of the LUCC: (**a**) in 2000 with the development grade coefficient; (**b**) in 2005 with the traffic facility coefficient; and (**c**) in 2005 without the traffic facility coefficient.

Table 6. The optimal development grade coefficient combinations and their inheritance thresholds of different districts.

District	α_k			∂_i						FOM
	α_1	α_2	α_3	∂_1	∂_2	∂_3	∂_4	∂_5	∂_6	
Main District	2.5	2.4	2.2	0.001	0.452	0.049	0.191	0.345	0	0.2365
Qixia District	2.7	2.6	1	0.002	0.402	0.070	0.154	0.292	0	0.2605
Pukou District	—	2.7	2.5	0.005	0.450	0.219	0.113	0.190	0	0.2147
Jiangning District	2.3	2.2	1.9	0	0.377	0.145	0.449	0.500	0	0.1978

3.3.2. Traffic Facility Coefficient

The Nanjing Metro Line 1 began construction in 2000, so we should consider the traffic facility coefficient (p_l) in the simulation process from the 21st iteration. As shown in Table 6, the development grade coefficients and inheritance thresholds of the different districts were trained by using the data of 1995 and 2000. In 2000, Pukou District began to plan key development areas, and planning text specifies that α_1 must be larger than α_2, so we set α_1 to 2.8 in the next experiment. Through the signs of the relative growth rate of land use obtained by using the classified map of 1995, 2000 and 2005, it can be judged whether the subway has a spatial spill-over effect or spatial agglomeration effect on the land use within the different ranges. Then, we used the iterative method to simulate the LUCC and test the optimal TFC combinations. As shown in Table 7, the subway produced a spill-over effect within 500 to 800 m of the Main District, but an aggregation effect within 800 to 3200 m. Jiangning District creates an aggregation effect within 200 to 2400 m. For Qixia District, the subway generates a spill-over effect within 200 to 1600 m, but an aggregation effect within 2400 to 3200 m. In summary, the aggregation effect coefficients or spill-over effect coefficients are all very small, less than or equal to 0.2. Comparing the simulated result (Figure 5b) with the actual result (Figure 1b,c), we found that the simulated accuracies (FOM) of Main District and Qixia District are both greater than 0.21, excepting the Jiangning District.

Table 7. The relative growth rate (RGR), optimal traffic facility coefficient (TFC) combinations and FOM of different districts.

| Level | Distance (m) | District | | | | | | | | |
| | | Main District | | | Qixia District | | | Jiangning District | | |
		RGR	TFC (p_l)	FOM	RGR	TFC (p_l)	FOM	RGR	TFC (p_l)	FOM
1	0–200	−0.406	0		0	0		3.787	0	
2	200–500	0.308	0		−2.040	−0.1		2.651	0.1	
3	500–800	−3.775	−0.1	0.2106	−2.670	−0.2	0.2371	2.081	0.2	0.1890
4	800–1600	0.797	0.1		−2.008	−0.1		2.252	0.2	
5	1600–2400	2.921	0.2		−2.881	0		0.603	0.1	
6	2400–3200	1.300	0.1		3.794	0.2		0.710	0	

3.4. Simulation of LUCC

Combining the classified map and land development grade data of 2000 with the above obtained grade coefficients and thresholds, we used our proposed model to simulate what the land use will be like in 2005 without the traffic facility coefficient (subway effect) (Figure 5c).

We compared the simulation result (Figure 5c) with the actual result (Figure 1b,c) and calculated the FOM of the simulation result. As shown in Table 8, the accuracies of the different districts are all above 0.18. Comparing the accuracies in Tables 7 and 8, we found that they are almost the equal in Main District, Qixia District and Jiangning District. The result reflects that the subway effect on land use is negligible after considering the land development grade. This is because the planning process of land development is interactive and comprehensive—when classifying the spatial land development grade, the planning departments have considered the role of subway planning in land-use development. Therefore, the planning influence can be expressed only by the development grade, without a subway effect.

Table 8. The accuracy of the simulation results without the traffic facility coefficient in 2005.

| | District | | | |
	Main District	Qixia District	Pukou District	Jiangning District
FOM	0.2296	0.2366	0.2145	0.1810

4. Discussion

In fact, the differences in the grade coefficients of the districts—which are the land grade being different in one district and the coefficient of same land grade being different in different districts—reflect the differences in spatial control granularity of the land use. The form is due to the great differences in regional development and resource distribution, so the granularity of spatial development is heterogeneous. The latter is due to the characteristic of the Chinese planning system. where the management and control powers of different land-use grades are heterogeneous in different districts. As shown in Figure 6a,b, we also simulated what the urban land will be like without considering the development grade coefficients; in other words, we assume that the planning granularity of land development is homogeneous in the whole study area. Compared with Figure 5a,c, it is obvious that the residential lands of Figure 6a,b in the first and second development areas are larger, which confirms that the division of land grades has different driving forces for land expansion.

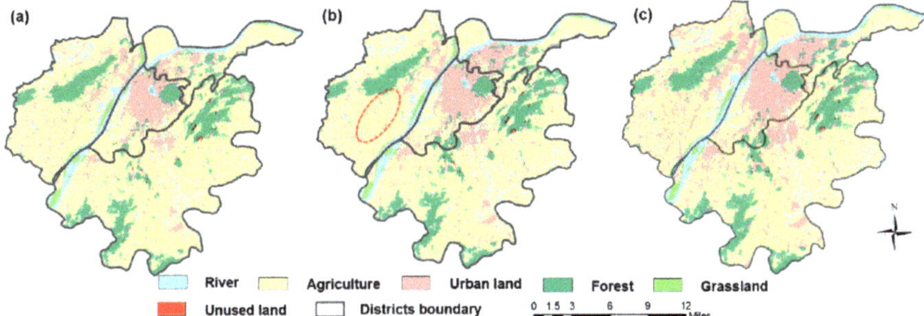

Figure 6. Simulation results: (**a**) without planning influence in 2000, (**b**) without planning influence in 2005, and (**c**) with planning influence in 2020.

The simulation results (Figure 6a,b) are compared with the actual results (Figure 1b,c) to obtain the accuracies of the different districts. By comparing Table 9 with Tables 6 and 8, it can be found that the simulation accuracies considering the differences in grade coefficients are almost higher. This conclusion further confirms that the differences in zoning control granularity are inevitable.

Table 9. The accuracies of the simulation results without planning influence.

	Year	District			
		Main District	**Qixia District**	**Pukou District**	**Jiangning District**
FOM	2000	0.2243	0.2315	0.2001	0.1956
	2005	0.2197	0.2034	0.2109	0.1711

Further, comparing Figure 6b or Figure 5c with Figure 1c, it can be found that neither of the two simulation methods can simulate the residential land in the southwest of Pukou District (the scope circled in Figure 6b). This is due to the limitation of the CA model itself, which pays too much attention to the role of the neighbourhood. Although the planning department had designated this scope as a key development area in 2000, the original number of the residential land in this scope was too small to simulate the real development status.

In addition, we simulated the land-use changes of Nanjing in 2020 based on the data of 2005 (Figure 6c). In the general land use planning of Nanjing (2006–2020), there are clear control indicators for the area of agricultural land, residential land and forest land in 2020. By comparing the simulated results with the planning indicators (Table 10), we found that the overall results of the simulation were good, because the simulation results of each type of land-use reached 80% of the planning indicator. The simulation results of residential land in the four districts are less than the planning indicators, while the other two types of land-use are opposite. This finding indicates that the growth rate of residential land in the model is slightly smaller than that of planning. In other words—the development of residential land will accelerate between 2005 and 2020.

Table 10. The comparison of simulated results and planning indicators in 2020.

District	Land-Use Types					
	Agricultural Land		Residential Land		Forest Land	
	Simulated	Planning	Simulated	Planning	Simulated	Planning
Main District	1731.7	1480.5	20,710.7	25,281.1	2611	2270.8
Qixia District	16,189	14,803.7	10,504.7	12,943.8	4190.2	4207.7
Pukou District	58,058.25	57,611.7	14,914.5	17,438.5	20,603.5	20,488.5
Jiangning District	120,668.7	111,404.7	26,119.2	30,146.9	24,363.5	28,226.1

5. Conclusions

The urban development strategy plays an absolute guiding role in the process of urban land-use development. To simulate and discuss the development process of urban land use in a more real and objective way, this paper proposes a CA model under the constraint of the development strategy of urban planning. First, a new zoning transition rule is added to the two traditional rules to show the multi-level characteristics of the planning system. Second, the development grade coefficient is put out in the model to reflect the heterogeneity of the spatial control granularity in different districts. Third, we explored the spatial effects of subways by setting a traffic facility coefficient in the local transition rule. Through the above research, some significant results are summarized as follows:

- By comparing the simulation results of this proposed model and a model without considering the planning influence, it can be found that the heterogeneity of the spatial control granularity does exist and affects the simulation accuracies.
- Other contents of regulatory plans (e.g., road planning) have been taken into account in the design process of land-use spatial layouts, so the effect of subways on the simulation results in this model is very small.

People often define a city as a community that maintains a small scale for the good life. With the concepts of humanism and sustainable development becoming more and more popular, people are beginning to attach great importance to the living environment and comprehensive development of community. The results of this study suggest that planners can solve the problems of urban landscape fragmentation and achieve the goal of harmonious coexistence between man and nature by designating ecological reserves to establish continuous green grids and increase the continuity of urban green spaces. Based on the demand of people and the spatial distribution of infrastructure, planners can also realize intensive land use and develop compact cities by classifying the development level of land-use and controlling the granularity of different levels, thus reducing the use of resources and promoting urban sustainable development. Besides, we can simulate the land-use development under different planning strategies by using the model, and analysing the green, humanistic and sustainability of cities under different scenarios, so as to provide scientific guidance for the development of cities.

In cases where the residential land is too sparse to simulate the real development status, the proposed method revealed some limitations that must be resolved in future works. Apart from this, there are still some potential factors that need to be considered to better understand the driving mechanism of sustainable development. First, limited by the availability of data, this paper did not consider the influence of economic elements on the model, which need to be further studied. Second, the urban planning system is multi-level [19,28,29], and the system of each city is slightly different. Although this paper has considered the general level-zoning in order to simulate the process of land-use allocation more objectively—which is from the top to bottom—more levels need to be added. Third, urban population is the basis of land-use planning for planners, so we need to combine the population prediction model with the proposed model to improve the accuracy of urban growth simulation in the following work.

Author Contributions: Conceptualization, Y.S. and J.Y.; methodology, J.Y.; validation, F.S.; formal analysis, J.Z.; writing-original draft preparation, J.Y.; writing-review and editing, J.Y. and F.S.; visualization, J.Y. and F.S.; supervision, Y.S.

Funding: This research was funded by the National Natural Science Foundation of China grant number 41671392 and 41871297.

Acknowledgments: The authors would like to thank the editors and the anonymous reviewers for their constructive comments and suggestions, which greatly helped to improve the quality of the manuscript.

Conflicts of Interest: The authors declare no conflicts of interest.

References

1. Heilig, G.K. *World Urbanization Prospects: The 2011 Revision*; United Nations, Department of Economic and Social Affairs (DESA), Population Division, Population Estimates and Projections Section: New York, NY, USA, 2012; p. 14.
2. Li, X.; Liu, X.; Yu, L. A systematic sensitivity analysis of constrained cellular automata model for urban growth simulation based on different transition rules. *Int. J. Geogr. Inf. Sci.* **2014**, *28*, 1317–1335.
3. Dahal, K.R.; Chow, T.E. An agent-integrated irregular automata model of urban land-use dynamics. *Int. J. Geogr. Inf. Sci.* **2014**, *28*, 2281–2303.
4. Rafiee, R.; Mahiny, A.S.; Khorasani, N.; Darvishsefat, A.A.; Danekar, A. Simulating urban growth in Mashad City, Iran through the SLEUTH model (UGM). *Cities* **2009**, *26*, 19–26.
5. Heistermann, M.; Müller, C.; Ronneberger, K. Land in sight?: Achievements, deficits and potentials of continental to global scale land-use modeling. *Agric. Ecosyst. Environ.* **2006**, *114*, 141–158.
6. Wu, F. Calibration of stochastic cellular automata: The application to rural-urban land conversions. *Int. J. Geogr. Inf. Sci.* **2002**, *16*, 795–818.
7. Liu, Y. Modelling sustainable urban growth in a rapidly urbanising region using a fuzzy-constrained cellular automata approach. *Int. J. Geogr. Inf. Sci.* **2012**, *26*, 151–167.
8. Liu, X.; Li, X.; Shi, X.; Zhang, X.; Chen, Y. Simulating land-use dynamics under planning policies by integrating artificial immune systems with cellular automata. *Int. J. Geogr. Inf. Sci.* **2010**, *24*, 783–802.
9. Lu, Y.; Cao, M.; Zhang, L. A vector-based Cellular Automata model for simulating urban land use change. *Chin. Geogr. Sci.* **2015**, *25*, 74–84.
10. Shi, W.; Pang, M.Y.C. Development of Voronoi-based cellular automata-an integrated dynamic model for Geographical Information Systems. *Int. J. Geogr. Inf. Sci.* **2000**, *14*, 455–474.
11. Feng, Y.; Liu, Y. A heuristic cellular automata approach for modelling urban land-use change based on simulated annealing. *Int. J. Geogr. Inf. Sci.* **2013**, *27*, 449–466.
12. Couclelis, H. Cellular worlds: A framework for modeling micro—macro dynamics. *Environ. Plan. A* **1985**, *17*, 585–596.
13. He, C.; Okada, N.; Zhang, Q.; Shi, P.; Zhang, J. Modeling urban expansion scenarios by coupling cellular automata model and system dynamic model in Beijing, China. *Appl. Geogr.* **2006**, *26*, 323–345.
14. Hagoort, M.; Geertman, S.; Ottens, H. Spatial externalities, neighbourhood rules and CA land-use modelling. *Ann. Reg. Sci.* **2008**, *42*, 39–56.
15. Hansen, H.S. *Quantifying and Analysing Neighbourhood Characteristics Supporting Urban Land-Use Modelling*; Springer: Berlin/Heidelberg, Germany, 2008; pp. 283–299.
16. Hansen, H.S. Empirically derived neighbourhood rules for urban land-use modelling. *Environ. Plan. B Plan. Des.* **2012**, *39*, 213–228.
17. Moreno, N.; Ménard, A.; Marceau, D.J. VecGCA: A vector-based geographic cellular automata model allowing geometric transformations of objects. *Environ. Plan. B Plan. Des.* **2008**, *35*, 647–665.
18. Moreno, N.; Wang, F.; Marceau, D.J. A geographic object-based approach in cellular automata modeling. *Photogramm. Eng. Remote Sens.* **2010**, *76*, 183–191.
19. Batty, M. *Cities and Complexity: Understanding Cities with Cellular Automata, Agent-Based Models, and Fractals*; The MIT press: Cambridge, MA, USA, 2007.
20. Pinto, N.N.; Antunes, A.P. A cellular automata model based on irregular cells: Application to small urban areas. *Environ. Plan. B Plan. Des.* **2010**, *37*, 1095–1114.

21. Cohen, B. Urban growth in developing countries: A review of current trends and a caution regarding existing forecasts. *World Dev.* **2004**, *32*, 23–51.
22. Ferrás, C. El enigma de la contraurbanización: Fenómeno empírico y concepto caótico. *Eure (Santiago)* **2007**, *33*, 5–25.
23. Augustijn-Beckers, E.W.; Flacke, J.; Retsios, B. Simulating informal settlement growth in Dar es Salaam, Tanzania: An agent-based housing model. *Comput. Environ. Urban Syst.* **2011**, *35*, 93–103.
24. Sohl, T.L.; Sayler, K.L.; Drummond, M.A.; Loveland, T.R. The FORE-SCE model: A practical approach for projecting land cover change using scenario-based modeling. *J. Land Use Sci.* **2007**, *2*, 103–126.
25. Kok, K.; Winograd, M. Modelling land-use change for Central America, with special reference to the impact of hurricane Mitch. *Ecol. Model.* **2002**, *149*, 53–69.
26. El Yacoubi, S. *Cellular Automata: 7th International Conference on Cellular Automata for Research and Industry, ACRI 2006, Perpignan, France, 20–23 September 2006, Proceedings*; Science Business Media: Berlin, Germany, 2006.
27. Couclelis, H. From cellular automata to urban models: New principles for model development and implementation. *Environ. Plan. B Plan. Des.* **1997**, *24*, 165–174.
28. Liu, X.; Liang, X.; Li, X.; Xu, X.; Ou, J.; Chen, Y.; Li, Y.; Wang, S.; Pei, F. A future land use simulation model (FLUS) for simulating multiple land use scenarios by coupling human and natural effects. *Landsc. Urban Plan.* **2017**, *168*, 94–116.
29. Verburg, P.H.; Overmars, K.P. Combining top-down and bottom-up dynamics in land use modeling: Exploring the future of abandoned farmlands in Europe with the Dyna-CLUE model. *Landsc. Ecol.* **2009**, *24*, 1167.
30. Yang, X.; Chen, R.; Zheng, X.Q. Simulating land use change by integrating ANN-CA model and landscape pattern indices. *Geomat. Nat. Hazards Risk* **2016**, *7*, 918–932.
31. Onsted, J.A.; Clarke, K.C. Forecasting enrollment in differential assessment programs using cellular automata. *Environ. Plan. B Plan. Des.* **2011**, *38*, 829–849.
32. Ward, D.P.; Murray, A.T.; Phinn, S.R. Integrating spatial optimization and cellular automata for evaluating urban change. *Ann. Reg. Sci.* **2003**, *37*, 131–148.
33. Ministry of Housing, Communities & Local Government. Plain English Guide to the Planning System [EB/OL]. Available online: https://www.gov.uk/government/publications/plain-english-guide-to-the-planning-system (accessed on 10 May 2019).
34. Greater London Authority. The London Plan: The Spatial Development Strategy for London Consolidated with Alterations Since 2011 [R/OL]. Available online: https://www.london.gov.uk/what-we-do/planning/london-plan/current-london-plan/london-plan-2016-pdf (accessed on 10 June 2018).
35. Owens, D.W. *Introduction to Zoning*; Institute of Government University of North Carolina: Chapel Hill, NC, USA, 2001.
36. CoE [Council of Europe]. European Regional/Spatial Planning Charter–Torremolinos Charter. In Proceedings of the 6th Conference of the Council of Europe of Ministers Responsible for Spatial Planning (CEMAT), Torremolinos, Spain, 19–20 May 1983; Council of Europe: Strasbourg, France, 1983.
37. Faludi, A. A turning point in the development of European spatial planning? The 'Territorial Agenda of the European Union'and the 'First Action Programme'. *Prog. Plan.* **2009**, *71*, 1–42.
38. Arsanjani, J.J.; Helbich, M.; Kainz, W.; Bolooranic, A.D. Integration of logistic regression, Markov chain and cellular automata models to simulate urban expansion. *Int. J. Appl. Earth Obs. Geoinf.* **2013**, *21*, 265–275.
39. Guan, D.J.; Li, H.F.; Inohae, T.; Su, W.; Nagaie, T.; Hokao, K. Modeling urban land use change by the integration of cellular automaton and Markov model. *Ecol. Model.* **2011**, *222*, 3761–3772.
40. Jantz, C.A.; Goetz, S.J.; Donato, D.; Claggett, P. Designing and implementing a regional urban modeling system using the SLEUTH cellular urban model. *Comput. Environ. Urban Syst.* **2010**, *34*, 1–16.
41. Murakami, J.; Cervero, R. *High-Speed Rail and Economic Development: Business Agglomerations and Policy Implications*; UC Berkeley, University of California Transportation Center: Berkeley, CA, USA, 2012.
42. Willigers, J.; Van Wee, B. High-speed rail and office location choices. A stated choice experiment for the Netherlands. *J. Transp. Geogr.* **2011**, *19*, 745–754.
43. Garmendia, M.; Romero, V.; Ureña, J.M.; Coronado, J.M. High-speed rail opportunities around metropolitan regions: Madrid and London. *J. Inf. Syst.* **2012**, *18*, 305–313.
44. Gutiérrez, J. Location, economic potential and daily accessibility: An analysis of the accessibility impact of the high-speed line Madrid–Barcelona–French border. *J. Transp. Geogr.* **2001**, *9*, 229–242.

45. Vickerman, R. High-speed rail in Europe: Experience and issues for future development. *Ann. Reg. Sci.* **1997**, *31*, 21–38.
46. Ureña, J.M.; Menerault, P.; Garmendia, M. The high-speed rail challenge for big intermediate cities: A national, regional and local perspective. *Cities* **2009**, *26*, 266–279.
47. Torrens, P.M.; O'Sullivan, D. Cellular automata and urban simulation: Where do we gofrom here? *Environ. Plan. B Plan. Des.* **2001**, *28*, 163–168. [CrossRef]
48. Beijing Digital View Technology Co., Ltd. Geographical Information Monitoring Cloud platform [DS/OL]. Available online: http://www.dsac.cn/ (accessed on 1 October 2012).
49. Liu, X.; Li, X.; Ye, J.; He, J.; Tao, J. Using ant colony intelligence to mine the transformation rules of geographic cellular automata. *Chin. Sci. (Ser. D Geosci.)* **2007**, *6*, 824–834.
50. Ministry of Housing and Urban-Rural Development of the People's Republic of China (MOHURD). Announcement of the Ministry of Housing and Urban-Rural Development on the Publication of the Industry Standard "Regulations for Vertical Planning of Urban and Rural Construction Land" [EB/OL]. Available online: http://www.mohurd.gov.cn/wjfb/201607/t20160722_228283.html (accessed on 1 August 2016).
51. Ministry of Natural Resources of the People's Republic of China. Notice on the Review of the Second National Land Survey Slope Map and the Arable Land Field Coefficient [EB/OL]. Available online: http://www.mnr.gov.cn/gk/tzgg/200903/t20090302_1989991.html (accessed on 25 February 2009).
52. Cervero, R.; Kockelman, K. Travel demand and the 3Ds: Density, diversity, and design. *Transp. Res. Part D Transp. Environ.* **1997**, *2*, 199–219.
53. Atkinson-Palombo, C.; Kuby, M.J. The geography of advance transit-oriented development in metropolitan Phoenix, Arizona, 2000–2007. *J. Transp. Geogr.* **2011**, *19*, 189–199.
54. Bowes, D.R.; Ihlanfeldt, K.R. Identifying the impacts of rail transit stations on residential property values. *J. Urban Econ.* **2001**, *50*, 1–25.
55. Wu, F.; Webster, C.J. Simulation of land development through the integration of cellular automata and multicriteria evaluation. *Environ. Plan. B* **1998**, *5*, 103–126.
56. Pontius, R.G.; Boersma, W.; Castella, J.C.; Clarke, K.C.; de Nijs, T.; Dietzel, C.; Duan, Z.; Fotsing, E.; Goldstein, N.; Kok, K.; et al. Comparing the input, output, and validation maps for several models of land change. *Ann. Reg. Sci.* **2008**, *42*, 11–37.
57. Varga, O.G.; Pontius, R.G., Jr.; Singh, S.K.; Szabó, S. Intensity Analysis and the Figure of Merit's components for assessment of a Cellular Automata–Markov simulation model. *Ecol. Indic.* **2019**, *101*, 933–942.

 *sustainability*

Article

Remote Sensing-Based Analysis of Landscape Pattern Evolution in Industrial Rural Areas: A Case of Southern Jiangsu, China

Yifan Zhu [1], Chengkang Wang [1,*] and Takeru Sakai [2]

[1] College of Landscape Architecture, Nanjing Forestry University, Nanjing 210037, China; zhuyifan1997@outlook.com

[2] Campus Planning Office, Kyushu University, Fukuoka 819-0383, Japan; sakai@kyudai.jp

* Correspondence: chengkang.wang@njfu.edu.cn; Tel.: +86-18951841571

Received: 9 August 2019; Accepted: 10 September 2019; Published: 12 September 2019

Abstract: With the rapid economic development of industrial rural areas in Southern Jiangsu, the rural landscape and ecological environment of these industrial rural areas are getting damaged. Based on GIS and RS techniques, Landsat Satellite remote sensing images from 1981, 1991, 2001, 2011 and 2018 were collected for Jiangyin, Zhangjiagang, Changshu and Kunshan, to extract landscape pattern indexes and spatial distribution data. Landscape pattern indexes of the patch-class level and landscape level from each year were calculated by FRAGSTATS. After analyzing and comparing landscape pattern variation of five years, progress, characteristics and driving forces of landscape pattern evolution were explored. At the patch-class level, construction land had continuously encroached on green and cultivated land, exhibiting trends of expansion and centralization. At the landscape level, the number of small patches and degree of landscape fragmentation generally increased. The direct cause of landscape pattern evolution in industrial rural areas of Southern Jiangsu was the encroachment and segmentation of green and cultivated land by construction land, and the dominant factors driving the changes in construction land in the industrial rural areas of Southern Jiangsu were the effects of land and population aggregation exerted by the development of township enterprises and rural industries.

Keywords: landscape pattern; industrial rural area; rural landscape; landscape ecology; southern Jiangsu

1. Introduction

Industrial rural areas are clusters of rural land with high industrial development, a large number of industrial areas and industrial enterprises based on raw material collection and processing–manufacturing industries, where the industrial output value accounts for the highest proportion of the total rural community assets [1]. During the early period of economic reform in China during the 1980s, the location of the rural areas in Southern Jiangsu Province facilitated the rapid development of township enterprises based on manufacture, which led to the establishment of a unique industry-dominated rural economic development pathway referred to as the "Southern Jiangsu Model" and the formation of large industrial rural areas. The Southern Jiangsu Model is one of the most representative models of the urbanization process in China and has certain research value. However, as development in Southern Jiangsu occurred during the early period of urbanization, there was a lack of preceding examples that could provide a reference and theoretical guidance for sustainable development. Consequently, the original, natural appearance of these rural landscapes was destroyed, diminishing their unique characteristics and rurality [2]. Additionally, the rapid urbanization of peripheral areas also caused the continuous extension of urban landscapes into rural areas [3]. Under the joint influence of the aforementioned internal and external factors, the landscape patterns of the industrial rural areas in Southern Jiangsu gradually evolved to their present state, which is characterized by the widespread loss of traditional rural landscapes. Additionally, the encroachment of urban construction on rural land and the commingling of residential and industrial zones in rural construction areas have caused the continuous deterioration of the environments surrounding human settlements in these areas [4]. With the proposal of the ecological civilization construction concept and rural revitalization strategy, strategies to mitigate the destruction of rural landscapes and assimilation effects of urban-rural integration, enhance the environment surrounding rural settlements and promote the sustainable development of rural areas have become the focus of research on Chinese rural areas [5].

Industrial rural development has resulted in the evolution of rural functions and the modification of industrial rural structures on a global scale. Although industrial rural and manufacturing models had previously promoted economic development and increased the income of rural residents, they significantly impacted land use patterns and rural environments, which is unfavorable for sustainable development [6]. Recent major economic and social reforms have driven the development of tourism, which has, in turn, promoted the development of the national economy through industrial, agricultural, construction, transportation and trade activities [7,8]. The development of the tourism sector has profoundly influenced the transformation of various economic, social and cultural factors, leading to the establishment of the rural tourism market and the consolidation of the tourism sector while also promoting sustainable development and environmental conservation [9,10]. Rural tourism aims to sustain rural architecture and landscapes, thereby promoting creative activities related to the natural and historical backgrounds of rural areas for the appropriate development of these rural landscapes and their cultural heritage. This is also a major pathway in the sustainable development of rural areas [11,12].

After sorting out the relevant research results, it is found that researchers have studied the spatial and temporal distribution characteristics of carbon emissions [13], industrial pollution [14], ecological sensitivity [15] and the natural landscape level [16] in the industrial rural areas in Southern Jiangsu by using the methods of composite analysis, buffer analysis and data weighted overlay analysis in the GIS platform. These studies have achieved effective results, but generally there are some shortcomings in data and methods, such as small research scope, low accuracy of data and a lack of macro-control of the development of urban and rural areas for the region as a whole.

Previous studies indicated that landscape pattern indexes calculated based on land-use categories directly reflect the characteristics of changes to a certain region from the perspective of landscape ecology. The widespread application of geospatial techniques such as geographic information systems (GIS) and remote sensing (RS) in landscape ecology research, has demonstrated the necessity of assessing the progress of research in this field [17]. With the prevalence of Landsat imagery-based landscape indexes and application methods, land-use change maps constructed using multi-temporal data can be used to calculate landscape metrics and analyze the urban-to-rural gradient characteristics. Consequently, the spatiotemporal characteristics of landscape changes in metropolitan areas can be monitored, facilitating the investigation of the influence of urban expansion on both urban and rural landscapes [18,19]. Land-use change maps can also reflect the influence of land consolidation on local landscape patterns, which is vital for optimizing land consolidation models and accelerating the sustainable development of local communities [20,21]. The evolution of landscape patterns also leads to changes in the spatial and ecosystem structures of landscapes, which ultimately influences ecological security [22–24]. In previous studies related to agriculture and forestry, landscape pattern indexes and land-use change maps have not only reflected the influence of urbanization on cultivated land and forests, but have also been used to determine the service functions of cultivated lands and forests in urban ecosystems, thereby providing a reference for promoting sustainable development in urban and rural regions [25]. In this study, GIS and RS techniques were used for a case study that will help achieve sustainability in industrial rural areas, Southern Jiangsu, China.

To reach this goal, we aimed to determine the actual influence of industrial rural development in Southern Jiangsu during 1981–2018 on local rural landscapes and the environments surrounding human settlements by using landscape pattern indexes. Four representative regions, i.e., the cities of Jiangyin, Zhangjiagang, Changshu, and Kunshan, were selected as study areas. In the County Economy Top 100 (2019) List of China by the government, Kunshan ranked No. 1, Jiangyin ranked No. 2, Zhangjiag ranked No. 3 and Changshu ranked No. 4, which means the four counties are the counties with the most developed economy in China. The study area is so peculiar, and representative of the most successful rural economic development model and one of the most representative models of the urbanization process in China since the economic reform of China. The economic achievement was based on industrialization, with damage to rural landscapes and natural environments, so it is necessary to sort out this typical kind of fast urbanization in regions driven by industry development. The specific aims of this case study are as follows: (1) Use GIS and RS techniques to establish land-use change maps from remotely sensed image data acquired by Landsat satellites during five periods in 1981, 1991, 2001, 2011, and 2018. (2) Calculate and analyze eight landscape pattern indexes at two different levels with ENVI and FRAGSTATS software. (3) Compare the landscape pattern evolution processes of the four study areas to analyze the landscape pattern evolution characteristics of industrial rural areas in Southern Jiangsu. (4) Explore factors driving landscape changes in this region. The results of this study concluded the landscape pattern and evolution dynamics of industrial rural areas in Southern Jiangsu, China, providing relevant fields with methods to investigate the evolution dynamic of the urban–rural industry during urbanization. It could also support and serve as a reference for other developing countries in Asia for sustainability of urban and rural development during industrialization.

The structure of this paper is organized as follows: In Section 2, we introduced and elaborated on the study area, data collection and pre-processing, as well as on the landscape pattern indexes chosen for this study; in Section 3, calculation results of the case study conducted in Southern Jiangsu were presented; in Section 4, experimental results and factors driving landscape pattern changes were discussed, and strategies for improving the landscape pattern and promoting the development of ecological environment and tourism were proposed, which is helpful to achieve sustainability; and in Section 5, conclusions on the experimental results and future work were drawn.

2. Materials and Methods

2.1. Study Area

Jiangsu Province is divided into three major regions by the Yangtze and Huai Rivers, i.e., Southern, Central and Northern Jiangsu. Specifically, Southern Jiangsu contains the cities of Nanjing, Suzhou, Wuxi, Changzhou and Zhenjiang. According to statistics from 2017, the total population of Southern Jiangsu is approximately 33.4752 million, and the gross domestic product (GDP) of the region is approximately 5.0175 trillion yuan, accounting for 6% of China's GDP. The per-capita GDP of the region is 150,200 yuan, which is approximately three times the average per-capita GDP of the country. Industrial rural areas are rural locations dominated by industry, where the industrial output value accounts for the highest proportion of the total community assets, which have mainly been implemented in county-level cities under the jurisdiction of the three prefectural-level cities of Suzhou, Wuxi and Changzhou. Based on governmental statistics, we selected the following county-level cities as representative regions for this study, which have similar areas, similar population sizes, and respective GDPs of over 200 billion yuan: Jiangyin, Zhangjiagang, Changshu and Kunshan. The geographic coordinates of the spatial extent shown in Figure 1 are Jiangyin: 120°15′36″ E, 31°54′36″ N; Zhangjiagang: 120°33′00″ E, 31°52′12″ N; Changshu: 120°44′24″ E, 31°54′36″ N; and Kunshan: 120°57′00″ E, 31°23′24″ N (Figure 1 and Table 1).

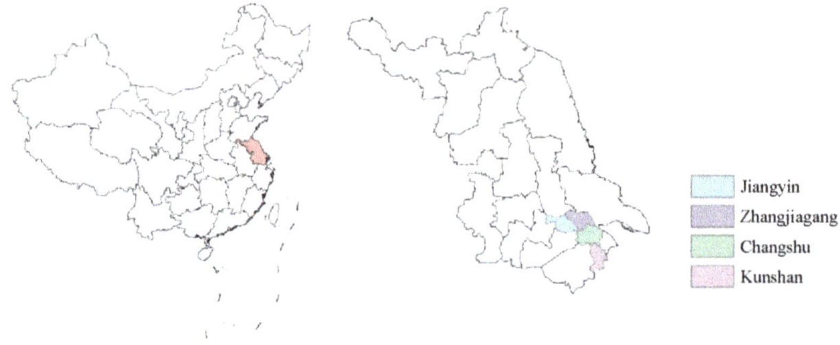

Figure 1. Study area.

Table 1. Statistical data of study area [26–29].

Prefectural-Level	County-Level	Year	Population/Million	GDP/Billion	Area/km^2
Wuxi	Jiangyin	1990	1.10	3.67	987.53
		2001	1.15	36.50	
		2011	1.21	233.59	
		2017	1.25	348.83	
Suzhou	Zhangjiagang	1990	0.83	2.78	999.00
		2001	0.85	30.68	
		2011	0.90	186.03	
		2017	0.92	260.61	
	Changshu	1990	1.03	3.63	1264.00
		2001	1.04	30.30	
		2011	1.06	171.05	
		2017	1.07	227.96	
	Kunshan	1990	0.56	2.01	927.68
		2001	0.60	23.08	
		2011	0.72	243.23	
		2017	0.86	352.03	

2.2. Data Collection and Pre-Processing

Based on GIS and RS techniques, remotely sensed image data of the four subject cities acquired by Landsat satellites during five periods in 1981, 1991, 2001, 2011 and 2018 were utilized to extract the spatial distribution of various landscape categories. The satellite images were all captured in fall included the dates of 7 December 1981, 12 November 1991, 6 November 2001, 11 November 2011 and 24 December 2018. The resolution of the image in 1981 was 60 m × 60 m, whereas the others were all 30 m × 30 m. These were the highest resolution of images accessible to us. Resolution has been unified during land classification and resampling. FRAGSTATS 4.2 was used to calculate the landscape pattern indexes of each city during the different periods. By quantitatively and qualitatively analyzing the landscape pattern indexes, the influence of the rural industry on the evolution of landscape patterns was investigated. The remotely-sensed images were geometrically corrected in ERDAS 9.2 using the corresponding topographic maps as a reference. The images were then vectorized using ENVI 5.0. By following a human–computer interaction interpretation method, the landscapes were classified into four categories based on the characteristics of the studied areas: Construction land, cultivated land, green land and water bodies. Ground Truth ROI in ENVI 5.0 was used to assess the accuracy. And the kappa coefficients were all over 0.8648, which was reliable. The classification maps obtained from visual interpretation were converted into the ArcGrid format using ArcGIS 10.2 and subsequently imported into FRAGSTATS 4.2 for calculating the landscape pattern indexes.

2.3. Landscape Pattern Indexes

In this study, we utilized landscape pattern indexes, including patch-class-level and landscape-level indexes, to portray the characteristics of the spatiotemporal changes in the various landscape categories. The patch-class-level indexes included the percentage of landscape (PLAND), number of patches (NP), patch density (PD) and largest patch index (LPI), while the landscape-level indexes included the contagion index (CONTAG), splitting index (SPLIT), Shannon's diversity index (SHDI) and Shannon's evenness index (SHEI). The evolution process of the landscape patterns was determined by investigating the basic characteristics, morphological changes and spatiotemporal evolution of the various landscape categories, as shown in Table 2. The annual changing ratio of landscape pattern indexes in the patch-class level was calculated to show the annual change of four kinds of lands clearly. The formula was $((year2 - year1)/year1) \times 100\%$.

Table 2. Landscape pattern index used in this study.

Name	Calculation Formula	Notes
Proportion of landscape types (PLAND) [30]	$PLAND = P_i = \frac{\sum_{j=1}^{n} a_{ij}}{A}\,(100)$	a_{ij} represents the area of patches numbered ij, and A represents the total area of all patches.
Number of patches (NP) [30]	$NP = n_i$	n_i represents the total number of patches contained in type I of the entire landscape.
Patch density (PD) [30]	$PD = \frac{1}{A}\sum_{j=1}^{M} N_i$	A represents the total area of all patches M represents the total number of landscape element types at a spatial resolution within the scope of the study
Largest patch index (LPI) [30]	$LPI = \frac{\max(a_{ij})}{A}(100)$	a_{ij} represents the area of patches numbered ij, and A represents the total area of all patches.
Contagion index (CONTAG) [30]	$CONTAG = $ $\left[1 + \frac{\sum_{i=1}^{m}\sum_{k=1}^{m}\left[(P_i)\frac{g_{ik}}{\sum_{k=1}^{m} g_{ik}}\right]\left[\ln(P_i)\frac{g_{ik}}{\sum_{k=1}^{m} g_{ik}}\right]}{2\ln(m)}\right](100)$	P_i represents the percentage of area occupied by type I; g_{ik} represents the number of IK adjacent to the plaque type. M represents the total number of patch types in the landscape
Splitting Index (SPLIT) [30]	$C_i = \frac{N_i}{A_i}$	N_i represents the number of patches. A_i represents the total area of all patches.
Shannon's diversity index (SHDI) [30]	$SHDI = -\sum_{k=1}^{m} P_k \ln(P_k)$ $SHDI_{max} = \ln(m)$	P_k represents the probability of patch type K appearing in landscape. m represents the total number of patch types in the landscape.
Shannon's evenness index (SHEI) [30]	$SHEI = \frac{SHDI}{SHDI_{max}} = \frac{-\sum_{k=1}^{m} P_k \ln(P_k)}{\ln(m)}$	P_k represents the probability of patch type K appearing in landscape. m represents the total number of patch types in the landscape.

3. Results

3.1. Land Classification Results

The characteristics of the land-use categories of the four study areas during the five periods described above extracted using ENVI 5.0 are shown in Figure 2, as well as the patch-class-level and landscape-level landscape pattern indexes of each study area calculated using FRAGSTATS 4.2.

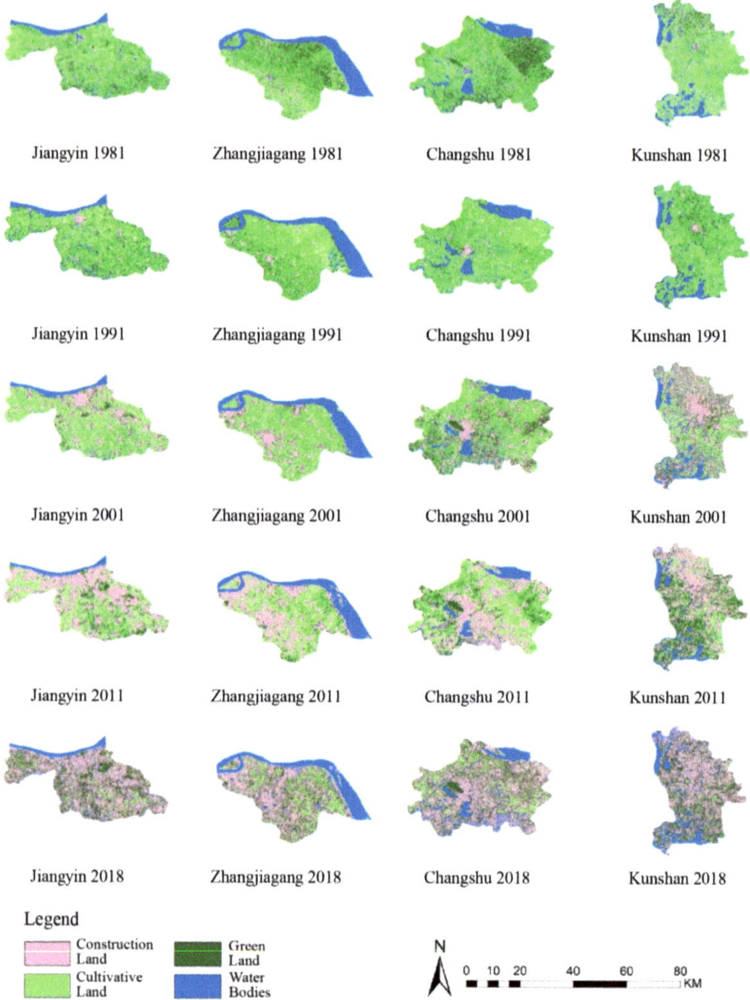

Figure 2. Land classification results.

3.2. Patch-Class-Level Landscape Pattern Results

Tables 3–10 show the results of the calculation and analysis of the patch-class-level landscape pattern indexes selected in this study (PLAND, NP, PD and LPI).

In all four study areas, the PLAND of construction land exhibited a rapid, continuous increase. The annual increase during 1981–2018 exceeded 400% in all areas, with the highest annual increase of 637.03% being observed in Changshu City. In contrast, the PLAND of cultivated land in all four study areas decreased continuously over time. Figure 3 shows that the annual decrease during 1981–2018 exceeded 50% in all areas, with Kunshan City exhibiting the highest annual decrease of 80.17%. The PLAND of green land decreased in the cities of Jiangyin, Zhangjiagang and Changshu, while that in Kunshan increased. No significant changes in the PLAND of water bodies were observed in all four study areas.

Table 3. Calculation results from the patch-class-level in Jiangyin.

City	Category	Year	PLAND	NP	PD	LPI
Jiangyin	Construction Land	1981	9.59	2186.00	2.24	0.20
		1991	17.56	1269.00	1.30	1.23
		2001	28.90	1897.00	1.95	6.42
		2011	46.76	923.00	0.95	32.83
		2018	57.61	228.00	0.23	56.50
	Green Land	1981	23.27	2201.00	2.26	1.97
		1991	29.23	797.00	0.82	1.23
		2001	23.04	2836.00	2.91	0.44
		2011	10.11	956.00	0.98	0.64
		2018	18.71	2234.00	2.29	0.95
	Water Bodies	1981	9.18	1056.00	1.08	5.31
		1991	10.75	873.00	0.90	6.09
		2001	10.97	480.00	0.49	5.20
		2011	10.60	315.00	0.32	4.45
		2018	10.71	1441.00	1.48	4.71
	Cultivated Land	1981	57.96	245.00	0.25	55.75
		1991	42.46	690.00	0.71	30.91
		2001	37.09	359.00	0.37	47.25
		2011	32.53	1010.00	1.04	19.91
		2018	12.98	2365.00	2.42	0.52

Table 4. Calculation results from the patch-class-level in Zhangjiagang.

City	Category	Year	PLAND	NP	PD	LPI
Zhangjiagang	Construction Land	1981	8.86	1615.00	1.61	0.31
		1991	10.33	884.00	0.45	0.48
		2001	21.24	1446.00	1.44	4.54
		2011	36.69	739.00	0.73	22.18
		2018	45.40	411.00	0.41	38.06
	Green Land	1981	29.08	1140.00	1.13	16.94
		1991	15.98	1250.00	1.24	2.34
		2001	12.71	2414.00	2.40	0.11
		2011	6.98	869.00	0.86	0.54
		2018	13.68	2033.00	2.02	0.17
	Water Bodies	1981	21.80	592.00	0.59	20.16
		1991	22.69	459.00	0.46	20.55
		2001	21.03	281.00	0.28	20.09
		2011	21.22	182.00	0.18	17.53
		2018	21.09	943.00	0.94	17.40
	Cultivated Land	1981	40.25	456.00	0.45	28.14
		1991	51.01	161.00	0.16	49.08
		2001	45.02	226.00	0.22	41.77
		2011	35.11	608.00	0.60	31.55
		2018	19.83	1430.00	1.42	1.92

Table 5. Calculation results from the patch-class-level in Changshu.

City	Category	Year	PLAND	NP	PD	LPI
Changshu	Construction Land	1981	7.15	1987.00	1.64	0.28
		1991	16.11	1258.00	1.04	0.83
		2001	24.26	2110.00	1.74	4.72
		2011	36.69	964.00	0.80	25.08
		2018	52.70	412.00	0.34	48.41
	Green Land	1981	34.12	1655.00	1.37	12.42
		1991	19.98	2627.00	0.86	0.64
		2001	22.79	2293.00	0.70	2.07
		2011	13.68	996.00	0.82	2.83
		2018	15.29	2510.00	2.07	1.13
	Water Bodies	1981	13.36	1349.00	1.11	5.52
		1991	12.88	1038.00	0.86	5.30
		2001	11.61	850.00	0.70	5.02
		2011	12.79	484.00	0.40	3.88
		2018	13.62	1414.00	1.17	3.84
	Cultivated Land	1981	45.36	811.00	0.67	37.50
		1991	51.04	143.00	0.12	59.43
		2001	41.34	862.00	0.71	28.91
		2011	36.85	919.00	0.76	26.93
		2018	18.40	2008.00	1.66	0.97

Table 6. Calculation results from the patch-class-level in Kunshan.

City	Category	Year	PLAND	NP	PD	LPI
Kunshan	Construction Land	1981	9.87	2475.00	2.60	0.25
		1991	15.37	862.00	0.90	0.66
		2001	42.79	1118.00	1.17	24.39
		2011	41.40	932.00	0.98	26.52
		2018	53.96	312.00	0.33	50.99
	Green Land	1981	12.03	2201.00	2.31	0.70
		1991	30.47	1965.00	2.06	6.04
		2001	12.07	2787.00	2.92	0.25
		2011	22.13	1236.00	1.30	3.60
		2018	18.87	2243.00	2.35	0.69
	Water Bodies	1981	15.58	1510.00	1.58	3.48
		1991	13.43	944.00	0.99	3.02
		2001	13.69	1472.00	1.54	1.86
		2011	13.62	773.00	0.81	1.87
		2018	14.77	1727.00	1.81	1.92
	Cultivated Land	1981	62.52	190.00	0.20	60.98
		1991	40.72	352.00	0.37	46.08
		2001	31.45	1270.00	1.33	5.27
		2011	22.85	1942.00	2.04	1.82
		2018	12.39	2086.00	2.19	0.20

Table 7. Changes of landscape indexes in Jiangyin.

City	Category	Year	PLAND	NP	PD	LPI
Jiangyin	Construction Land	1981–1991	83.22%	−41.95%	−41.95%	518.57%
		1991–2001	64.55%	49.49%	49.49%	423.80%
		2001–2011	61.80%	−51.34%	−51.34%	411.29%
		2011–2018	23.19%	−75.30%	−75.30%	72.09%
		1981–2018	500.94%	−89.57%	−89.57%	28,407.87%
	Green Land	1981–1991	32.51%	73.44%	73.45%	60.33%
		1991–2001	1.13%	−57.99%	−57.98%	−80.91%
		2001–2011	−50.73%	207.26%	207.25%	−98.65%
		2011–2018	−82.45%	319.30%	319.25%	−98.86%
		1981–2018	−19.62%	1.50%	1.50%	−51.44%
	Water Bodies	1981–1991	17.11%	−17.33%	−17.33%	14.70%
		1991–2001	2.07%	−45.02%	−45.01%	−14.63%
		2001–2011	−3.45%	−34.38%	−34.38%	−14.41%
		2011–2018	1.03%	357.46%	357.43%	5.98%
		1981–2018	16.61%	36.46%	36.46%	−11.17%
	Cultivated	1981–1991	26.71%	−64.69%	−64.70%	74.39%
		1991–2001	−26.75%	181.63%	181.65%	−44.55%
		2001–2011	−12.65%	−47.97%	−47.97%	52.85%
		2011–2018	−12.28%	181.34%	181.34%	−57.87%
		1981–2018	−77.60%	865.31%	865.33%	−99.07%

Table 8. Changes of landscape indexes in Zhangjiagang.

City	Category	Year	PLAND	NP	PD	LPI
Zhangjiagang	Construction Land	1981–1991	16.49%	−45.26%	−71.76%	54.54%
		1991–2001	105.65%	63.57%	217.10%	844.20%
		2001–2011	72.75%	−48.89%	−48.89%	389.01%
		2011–2018	23.75%	−44.38%	−44.38%	71.64%
		1981–2018	412.16%	−74.55%	−74.55%	12,147.14%
	Green Land	1981–1991	−45.06%	9.65%	9.65%	−86.21%
		1991–2001	−20.41%	93.12%	93.12%	−95.25%
		2001–2011	−45.08%	−64.00%	−64.00%	389.10%
		2011–2018	95.95%	133.95%	133.95%	−69.15%
		1981–2018	−52.95%	78.33%	78.34%	−99.01%
	Water Bodies	1981–1991	4.09%	−22.47%	−22.46%	1.93%
		1991–2001	−7.33%	−38.78%	−38.79%	−2.21%
		2001–2011	0.93%	−35.23%	−35.23%	−12.77%
		2011–2018	−0.63%	418.13%	418.19%	−0.70%
		1981–2018	−3.26%	59.29%	59.29%	−13.66%
	Cultivated Land	1981–1991	26.71%	−64.69%	−64.70%	74.39%
		1991–2001	−11.74%	40.37%	40.44%	−14.89%
		2001–2011	−22.02%	169.03%	168.98%	−24.46%
		2011–2018	−43.52%	135.20%	135.19%	−93.92%
		1981–2018	−50.74%	213.60%	213.59%	−93.18%

Table 9. Changes of landscape indexes in Changshu.

City	Category	Year	PLAND	NP	PD	LPI
Changshu	Construction Land	1981–1991	125.25%	−36.69%	−36.69%	193.36%
		1991–2001	50.61%	67.73%	67.73%	465.91%
		2001–2011	51.24%	−54.31%	−54.31%	430.88%
		2011–2018	43.65%	−57.26%	−57.26%	92.99%
		1981–2018	637.03%	−79.27%	−79.27%	16,908.96%
	Green Land	1981–1991	−41.45%	58.73%	−37.28%	−94.87%
		1991–2001	14.09%	−12.71%	−18.11%	225.15%
		2001–2011	−40.01%	−56.56%	17.17%	36.67%
		2011–2018	11.78%	152.01%	152.01%	−60.26%
		1981–2018	−55.20%	51.66%	51.66%	−90.94%
	Water Bodies	1981–1991	−3.65%	−23.05%	−23.05%	−3.86%
		1991–2001	−9.87%	−18.11%	−18.11%	−5.29%
		2001–2011	10.20%	−43.06%	−43.06%	−22.78%
		2011–2018	6.47%	192.15%	192.17%	−0.97%
		1981–2018	1.89%	4.82%	4.82%	−30.37%
	Cultivated Land	1981–1991	12.51%	−82.37%	−82.37%	58.49%
		1991–2001	−19.00%	502.80%	502.96%	−51.36%
		2001–2011	−10.87%	6.61%	6.61%	−6.83%
		2011–2018	−50.08%	118.50%	118.49%	−96.40%
		1981–2018	−59.45%	147.60%	147.62%	−97.41%

Table 10. Changes of landscape indexes in Kunshan.

City	Category	Year	PLAND	NP	PD	LPI
Kunshan	Construction Land	1981–1991	55.85%	−65.17%	−65.17%	159.74%
		1991–2001	178.30%	29.70%	29.71%	3587.65%
		2001–2011	−3.23%	−16.64%	−16.64%	8.73%
		2011–2018	30.34%	−66.52%	−66.52%	92.29%
		1981–2018	447.01%	−87.39%	−87.39%	19,926.20%
	Green Land	1981–1991	153.33%	−10.72%	−10.72%	761.93%
		1991–2001	−60.39%	41.83%	41.83%	−95.84%
		2001–2011	83.36%	−55.65%	−55.65%	1334.49%
		2011–2018	−14.74%	81.47%	81.48%	−80.76%
		1981–2018	56.86%	1.91%	1.91%	−1.00%
	Water Bodies	1981–1991	−13.84%	−37.48%	−37.49%	−13.26%
		1991–2001	1.96%	55.93%	55.94%	−38.29%
		2001–2011	−0.52%	−47.49%	−47.48%	0.28%
		2011–2018	8.46%	123.42%	123.41%	2.74%
		1981–2018	−5.21%	14.37%	14.37%	−44.84%
	Cultivated Land	1981–1991	−34.86%	85.26%	85.25%	−24.43%
		1991–2001	−22.77%	260.80%	260.83%	−88.57%
		2001–2011	−27.37%	52.91%	52.91%	−65.41%
		2011–2018	−45.74%	7.42%	7.42%	−89.22%
		1981–2018	−80.17%	997.89%	997.94%	−99.68%

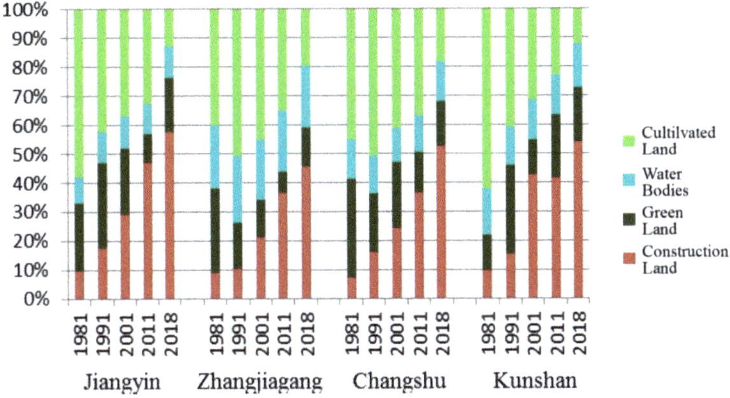

Figure 3. Changes in percentage of landscape (PLAND).

NP and PD reflect the degree of fragmentation of a certain patch class. Figures 4 and 5 show that the NP and PD of construction land decreased in all four study areas, while those of cultivated land, green land and water bodies increased. Therefore, the degree of fragmentation of construction land decreased in all study areas, while that of cultivated land, green land and water bodies increased.

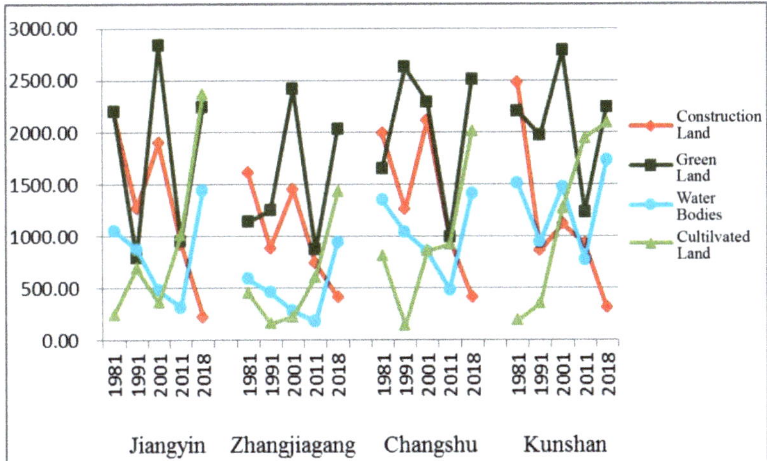

Figure 4. Changes in number of patches (NP).

LPI is a measure of landscape dominance, and the dynamic characteristics of the calculated LPI values indicate that the dominant landscape type in Jiangyin, Zhangjiagang and Changshu was cultivated land during 1981–2011, and construction land during 2011–2018. Additionally, the LPI of water bodies in all four study areas gradually decreased during 1981–2018, i.e., the landscape dominance of the "water body" patch class decreased continuously. The dominant landscape category in Kunshan City was cultivated land during 1981–2001, and construction land during 2001–2018 (Figure 6).

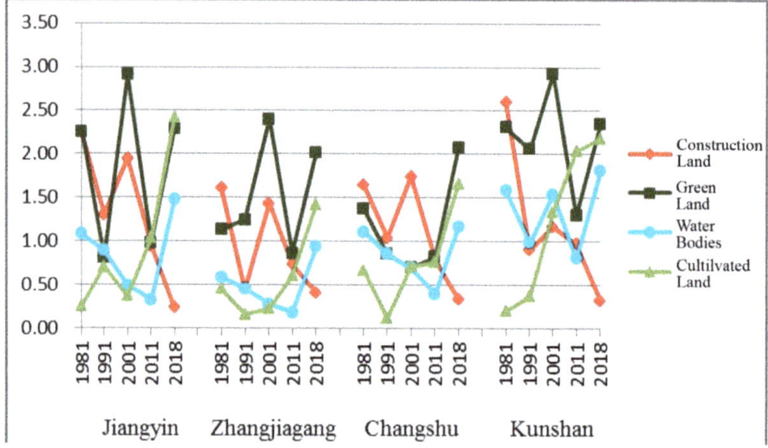

Figure 5. Changes in patch density (PD).

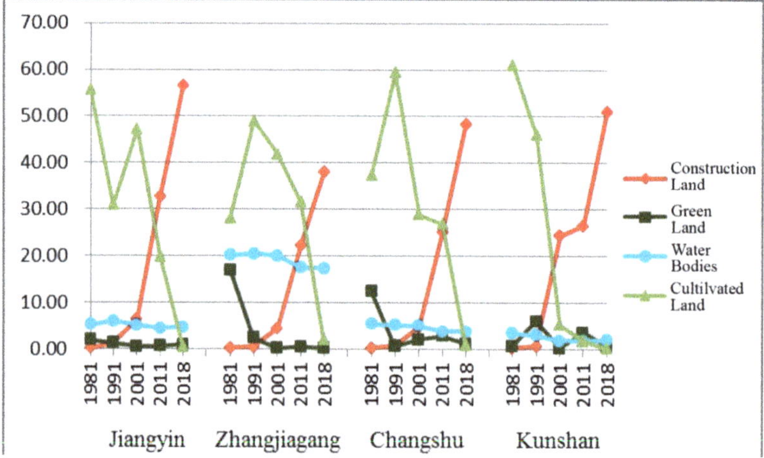

Figure 6. Changes in largest patch index (LPI).

In all four study areas, construction land continuously encroached on green land, exhibiting trends of expansion, centralization and the continuous consolidation of small patches into large patches during its evolution and ultimately replaced cultivated land as the dominant landscape category. Concurrently, the green and cultivated land patches, which are high-quality landscape resources, were continuously segmented into a larger number of small patches, which led to fragmentation during the evolution of green and cultivated land. Additionally, the water bodies in the four study areas should receive special attention; although the changes in PLAND during 1981–2018 were not significant, the dynamic changes in NP, PD and LPI indicate that the water bodies also faced fragmentation and reductions in dominance. Therefore, the effects of human disturbance on water bodies should not be neglected, even though they are less severe than those experienced by green and cultivated land.

3.3. Landscape-Level Landscape Pattern Results

Table 11 shows the results of the calculation and analysis of the landscape-level landscape pattern indexes, including CONTAG, SPLIT, SHDI and SHEI.

Table 11. Calculation results from the landscape-level.

	Year	CONTAG	SPLIT	SHDI	SHEI
Jiangyin	1981	25.93	3.18	1.10	0.79
	1991	22.88	8.02	1.17	0.84
	2001	24.24	4.33	1.15	0.83
	2011	28.84	6.53	1.12	0.81
	2018	22.25	3.11	1.14	0.82
Zhangjiagang	1981	22.75	2.56	1.27	0.92
	1991	31.06	3.52	1.16	0.84
	2001	24.29	4.59	1.28	0.92
	2011	28.37	5.50	1.23	0.89
	2018	18.72	5.67	1.28	0.92
Changshu	1981	20.15	2.25	1.18	0.85
	1991	29.92	2.80	1.06	0.76
	2001	13.46	11.13	1.30	0.93
	2011	24.21	16.98	1.20	0.87
	2018	17.31	14.23	1.21	0.87
Kunshan	1981	28.57	2.68	1.07	0.77
	1991	26.55	4.56	1.11	0.80
	2001	15.20	14.90	1.25	0.90
	2011	16.25	13.43	1.29	0.93
	2018	19.99	13.83	1.18	0.85

The calculated values of CONTAG, SPLIT, SHDI and SHEI indicate that the number of small patches and degree of landscape fragmentation in all four study areas generally increased, with the landscape patches exhibiting an even distribution and the landscape patterns exhibiting increased fragmentation and uniformity during evolutions (Figure 7).

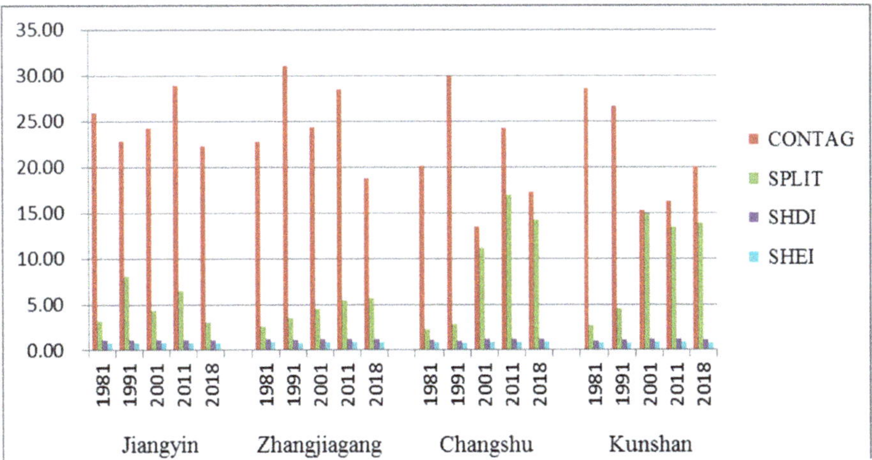

Figure 7. Calculation results from the landscape-level.

The CONTAG values of Jiangyin, Zhangjiagang and Changshu exhibited certain fluctuations, but remained relatively stable during 1981–2018, indicating that there was no dominant patch class that exhibited a high degree of contagion in these three study areas. The CONTAG values of Kunshan decreased continuously during 1981–2001 and increased continuously during 2001–2018, indicating a continuous decrease in the degree of contagion of the dominant patch class in Kunshan during 1981–2001, and a continuous increase during 2001–2018.

The SPLIT values of Jiangyin fluctuated during 1981–2018, while those of Zhangjiagang, Changshu and Kunshan continuously increased. The SPLIT values of Changshu and Kunshan during 2001–2018 were approximately double those of Jiangyin and Zhangjiagang, indicating that the rates of landscape fragmentation in Zhangjiagang, Changshu and Kunshan were higher than that of Jiangyin, and the degree of landscape fragmentation in Changshu and Kunshan was higher than that in Jiangyin and Zhangjiagang.

The SHDI and SHEI values of the four study areas during 1981–2018 were relatively stable and exhibited a slight overall increase without significant changes, indicating an increase in the number of patch classes and landscape diversity. SHEI exceeded 0.77 in all four study areas during 1981–2018. SHEI ranges from 0 to 1, with values closer to 1 indicating a higher degree of evenness in the distribution of the various landscape patch classes. Thus, the landscape patches of the four study areas were relatively evenly distributed. Cross-sectional comparisons were conducted and indicated that the SHDI and SHEI values of Kunshan exhibited the greatest rates of increase, indicating that Kunshan experienced the most significant changes in landscape diversity among the four study areas.

4. Discussion

During the analysis on landscape pattern indexes of industrial rural areas at the patch-class level and landscape-class level, this study discovered that:

(1) At the patch-class level, the landscape pattern evolution characteristics of industrial rural areas in Southern Jiangsu were that construction land had continuously encroached on green and cultivated land in all four study areas, exhibiting trends of expansion, centralization and the continuous consolidation of small patches into large patches during evolution and ultimately replacing cultivated land as the dominant landscape category. The result was in accordance with those in the studies of Yang, Sun et al., Ma et al., Xu et al. and Chuai et al. [31–35]. The reasons might be as follows: (1) Relevant studies showed that under the background of rapid socio-economic development, urban master planning is constantly updated, and relevant policies and regulations constantly promote human beings to expand the scope of urban built-up areas through deforestation, farmland reclamation and civil construction. Therefore, this study inferred that these human activities were likely to be the main reason for the changes of the area with urban landscape patches and the PLAND index [31]; and (2) in recent years, the process of urbanization development had been observed to show a strong aggregation (that is, all kinds of land shrink to the city center), and all kinds of land in the urban area were forced to "squeeze" out of the core area of the city. This phenomenon reflected in the landscape pattern in that the number of urban land patches had been reduced and the complexity increased [32].

(2) At the landscape-class level, the landscape patches were evenly distributed, and the landscape patterns exhibited increased fragmentation and uniformity during evolution. The evolution of landscape patterns in the four study areas was most intense during 1981–2001. Among the four study areas, Kunshan exhibited the most significant landscape pattern evolution characteristics. This result was also in accordance with the results of the studies by Yang, Sun et al., Ma et al., Xu et al. and Chuai et al. [31–35]. The reason for the results might be as follows: (1) The gradual improvement of a traffic network would cause disorderly cutting of the original patch landscape, which seriously affects the circulation of the ecological function of the regional landscape. From the perspective of landscape pattern, it is embodied in the increase of the SPILT index and the decrease of the CONTAG index [33,34]; and (2) for the rapid urbanization in China, the area of urban construction land had increased sharply, and the landscape tends to be homogeneous and fragmented. These might mainly explain the decline of the SHDI index and the SHEI index [35].

In order to help achieve sustainability for this region by improving the landscape pattern and promoting the development of ecological environment and tourism, the factors driving landscape pattern changes in the four study areas were analyzed to investigate the forces driving landscape pattern evolution in the industrial rural areas of Southern Jiangsu.

During the urbanization of the Southern Jiangsu region, triggered by China's economic reform, a series of economic support policies that were implemented by the central and local governments to meet the needs of economic development promoted the development of township enterprises and industries in Southern Jiangsu. This then boosted investments and economic prosperity, which attracted labor to the region and increased the demand for land utilization [36]. Consequently, the continuous aggregation of construction land patches led to the encroachment and segmentation of patches of other landscape categories, driving the evolution of landscape patterns in the industrial rural areas of Southern Jiangsu during 1981–2001. During the period of 1981–2001, a sort of significant event occurred, which pushed the industry and economic development hardly. Since the cold war in 1981, China improved the relationship with the West, especially the USA and Japan, the first and second developed countries in the world. China's economic reform began in 1978 and its market economy had begun to be gradually established since 1991. Pudong New District was set up in Shanghai in 1992. Hong Kong's return in 1997 has enhanced China's overall financial environment. Under this background, foreign capital and private capital flowed into Southern Jiangsu and accelerated the development of industry and enterprises in the study area. With the increase in the number and expansion of township enterprises in Southern Jiangsu, the urban and rural areas between industrial and agricultural populations became increasingly less distinct. This overcame the urban–rural economic divide, which drove the progress of urban–rural integration in Southern Jiangsu, resulting in the continuous aggregation and development of industrial rural areas and towns. However, such industrial rural development has been accompanied by the undesirable consequence of "smoke billowing from every village, factories emerging at every corner" [37]. To resolve this, the local governments in Jiangsu Province requested the establishment of industrial parks in townships, which led to further modifications to urban–rural structures in Southern Jiangsu. With the implementation of these development plans, the originally fragmented construction land patches aggregated further, thereby driving the decrease in the degree of fragmentation of construction land patches [38]. Since the onset of the 21st Century, "coordinated urban–rural development planning" has become China's national strategy, and there is a consensus that the urban–rural divide should be overcome among various circles of society. Consequently, the planning of residential, industrial and agricultural spaces in the industrial rural areas of Southern Jiangsu has changed. Plans for residential, industrial, agricultural, ecological water systems and transportation projects in small towns and villages have been matched with the development of urban areas, which integrated resource allocation, public service facilities and industrial layouts in both urban and rural areas. The influence of these strategies is most significant in Kunshan owing to its proximity to Shanghai, China's largest city. This is consistent with the comparative analysis results described above, which indicates that changes in landscape diversity were more significant in Kunshan than those in the other three study areas.

Despite the adverse effects of urbanization in the industrial rural areas of Southern Jiangsu, the industrial development that has occurred over the past four decades should not be wholly negated, and overcorrection measures to restore the rural appearances of these areas, such as the reconstruction of old villages or the reversion of industrial land to cultivated land, are strongly discouraged. The industrial rural areas of Southern Jiangsu should strive to achieve a balance between economic development and ecological sustainability, and develop a landscape construction pathway that is best suited to its circumstances.

Based on the discussions above, the landscape pattern evolution of industrial rural areas in Southern Jiangsu was dominated by the increase of construction land, change of policies and overall plan, resulting in fragmentation and homogenization. We propose the following strategies for improving the landscape pattern and promoting the development of the ecological environment and tourism, which are helpful to achieving sustainability for this region.

(1) Conserve existing cultivated land patches, appropriately expand grassland and forest land patches

The planning of cultivated landscapes and surrounding spaces in the industrial rural areas of Southern Jiangsu should focus on conservation. Agricultural tourism can be developed and landscapes can be enhanced by developing modern circular agriculture and establishing ecological farmsteads [39,40]. Additionally, the construction of facilities such as rural greenways and parks, grassland, and forest land patches can be expanded or even organically merged, thereby increasing the degree of landscape aggregation of these patches and reducing the dominance of construction land patches.

(2) Respect the four decades of historical progress, reasonably modify industrial landscapes

The industrial development of the rural areas of Southern Jiangsu is a collective memory that is unforgettable for the local residents. Instead of demolishing factories that have been closed due to pollution problems, the land where they are located can be modified to create post-industrial landscapes that integrate with the natural and rural landscapes. Through such modifications, the history of the area can be respected and the unique characteristics of industrial rural areas can be fully manifested.

(3) Capitalize on the economic and positional advantage, optimize urban–rural landscape spaces

As the industrial rural areas of Southern Jiangsu are located in the Yangtze River Delta Economic Region, which exhibits the highest degree of urbanization in China, the urban–rural integration progress in these areas has been relatively rapid. Therefore, industrial rural areas possess economic and positional advantages, despite the severe rural landscape encroachment by urban landscapes. These favorable conditions of the industrial rural areas enable the construction of landscape connectors between urban and rural areas, such as urban–rural greenways and urban forest parks, to establish appropriate transition spaces between urban and rural landscapes. Consequently, urban and rural landscape spaces can be reasonably delineated through multi-plan integration at city, county, town and village levels, to alleviate the encroachment on rural landscapes by urban landscapes.

5. Conclusions

Four county-level cities in Southern Jiangsu with large numbers of industrial rural areas, i.e., Jiangyin, Zhangjiagang, Changshu and Kunshan, were selected as study areas for this work, which employed GIS and RS techniques. Remotely-sensed image data of the four study areas acquired during 1981, 1991, 2001, 2011 and 2018, as well as eight landscape pattern indexes at the patch-class and landscape levels, were selected to investigate landscape pattern evolution in the industrial rural areas of Southern Jiangsu from a provincial perspective, which involved analyzing the evolution characteristics and determining the factors driving evolution. The following conclusions were drawn:

(1) At the patch-class level, construction land had continuously encroached on green and cultivated land in all four study areas, exhibiting trends of expansion, centralization and the continuous consolidation of small patches into large patches during evolution and ultimately replacing cultivated land as the dominant landscape category. Concurrently, green and cultivated land patches, which are high-quality landscape resources, were continuously segmented into a large number of small patches, which led to fragmentation during the evolution of green and cultivated land. The effects of human disturbance on water bodies should receive attention, even though they are less severe than those experienced by green and cultivated land. Urban master planning was being constantly updated, and relevant policies and regulations were constantly promoting human beings to expand the scope of urban built-up areas through deforestation, farmland reclamation and civil construction. The processes of urbanization development had been observed to show strong aggregation. And all kinds of lands in the urban area were forced to "squeeze" out of the core area of the city.

(2) At the landscape level, the number of small patches and degree of landscape fragmentation generally increased in all four study areas. The landscape patches were evenly distributed

and the landscape patterns exhibited increased fragmentation and uniformity during evolution. The evolution of landscape patterns in the four study areas was most intense during 1981–2001. Among the four study areas, Kunshan exhibited the most significant landscape pattern evolution characteristics. The gradual improvement of traffic network would cause disorderly cutting of the original patch landscape, which seriously affected the circulation of the ecological function of the regional landscape. For rapid urbanization in China, the areas of urban construction land had increased sharply, and the landscape tends to be homogeneous and fragmented.

(3)　The direct cause of landscape pattern evolution in the industrial rural areas of Southern Jiangsu was the encroachment and segmentation of green and cultivated land patches by construction land patches, and the dominant factors driving the changes in construction land patches in the industrial rural areas of Southern Jiangsu were the effects of land and population aggregation exerted by the development of township enterprises and rural industries.

(4)　This study concluded the landscape pattern and evolution dynamic of industrial rural areas, providing relevant fields with methods to investigate the evolution dynamic of urban–rural industry during urbanization and propose strategies for improving the landscape pattern and promoting the development of the ecological environment and tourism. It would also serve as a reference for other developing countries in Asia for sustainability of urban and rural development during industrialization, which is helpful to achieve sustainability for this region.

For the long period of 37 years and great dynamic changes of administrative division in this region, this study was limited in the scale of county-level cities. In the future we will try to focus more on the relationship between urban and rural areas with multivariate analysis based on GIS and RS techniques, which can provide more cases for the sustainability of similar regions in developing countries of Asia.

Author Contributions: Conceptualization, C.W.; methodology, C.W.; software, Y.Z.; validation, Y.Z.; formal analysis, Y.Z.; investigation, Y.Z.; resources, Y.Z.; data curation, Y.Z.; writing—original draft preparation, Y.Z.; C.W.; writing—review and editing, C.W.; T.S.; visualization, T.S.; supervision, T.S.; project administration, C.W.; funding acquisition, C.W.

Funding: This research was funded by the Humanity and Social Sciences Youth Foundation of the Ministry of Education of China, grant number 18YJCZH167.

Acknowledgments: Special thanks to all the team members for their helpful comments.

Conflicts of Interest: The authors declare no conflict of interest.

References

1.　Rongtian, Z.; Huafu, J.; Xiaolin, Z. Rural Development Types and Rurality in the Yangtze River Delta. *J. Nanjing Norm. Univ.* **2014**, *37*, 132–136.

2.　Zhu, J.; Zhan, Y.; Han, C. *The Way to Urbanization in Southern Jiangsu: Change and Innovation of Hudai Town*; China Social Sciences Press: Beijing, China, 2008.

3.　Zhou, X. Study on the Change of Rural Landscape in Urbanization Process: A Case Study of Southern Jiangsu Province. Ph.D. Thesis, Nanjing Normal University, Nanjing, China, 2006.

4.　Huo, G. Analysis on Planning and Design of Rural Landscape in Sunan Area. Master's Thesis, Suzhou University, Suzhou, China, 2012.

5.　Zhao, Q.; Guo, X.; Li, G. The Transformation of the Rural Space in Southern Jiangsu under the Dominance of Development Zones: A Case study of Suzhou industrial park. *Mod. Urban Res.* **2014**, *5*, 9–14.

6.　Cucari, N.; Wankowicz, E.; De Falco, S.E. Rural tourism and Albergo Diffuso: A case study for sustainable land-use planning. *Land Use Policy* **2019**, *82*, 105–119. [CrossRef]

7.　Cattaneo, T.; Giorgi, E.; Ni, M. Landscape, Architecture and Environmental Regeneration: A Research by Design Approach for Inclusive Tourism in a Rural Village in China. *Sustainability* **2019**, *11*, 128. [CrossRef]

8.　Balan, M.; Burghelea, C. Rural tourism and its implication in the development of the Fundata Village. *Procedia Soc. Behav. Sci.* **2015**, *188*, 276–281. [CrossRef]

9. Dolejs, M.; Nadvornik, J.; Raska, P.; Riezner, J. Frozen Histories or Narratives of Change? Contextualizing Land-Use Dynamics for Conservation of Historical Rural Landscapes. *Environ. Manag.* **2019**, *63*, 352–365. [CrossRef] [PubMed]

10. Prestia, G.; Scavone, V. Enhancing the Endogenous Potential of Agricultural Landscapes: Strategies and Projects for a Inland Rural Region of Sicily. In *Smart and Sustainable Planning for Cities and Regions, Sspcr 2017*; Bisello, A., Vettorato, D., Laconte, P., Costa, S., Eds.; Springer: Cham, Switzerland, 2018; pp. 635–648.

11. Gonzalez Alvarez, D. Rethinking tourism narratives on the cultural landscapes of Asturias (Northern Spain) from the perspective of Landscape Archaeology: Do archaeologists have anything to say? *Landsc. Res.* **2019**, *44*, 117–133. [CrossRef]

12. Gonzalez Diaz, J.A.; Celaya, R.; Fernandez Garcia, F.; Osoro, K.; Rosa Garcia, R. Dynamics of rural landscapes in marginal areas of northern Spain: Past, present, and future. *Land Degrad. Dev.* **2019**, *30*, 141–150. [CrossRef]

13. Qi, W.; Tang, C. Study on the Influencing Factors of Carbon Emission in Industrial Villages: A Case Study of Fenghuang Village, Xiaoshan City, Hangzhou. *Archit. Cult.* **2018**, *42*, 174–178.

14. Li, H.; Wu, J.; Zhang, X.; Li, C. The Distribution Characteristics and Mechanism of the Rural Industrial Land with the "Southern Jiangsu Pattern"—A Case Study of Changshu. *Econ. Geogr.* **2018**, *38*, 152–159.

15. Liu, L.; Tang, X.; Xiong, X.; Du, J.; Wang, J. Ecological sensitivity analysis of rural natural landscape in the south of Jiangsu Province based on GIS. *J. Nanjing For. Univ.* **2018**, *42*, 159–164.

16. Zhao, Y.; Wu, X.; Huang, X. Research on the performance assessment of rural water landscape in South of Jiangsu. *J. Nanjing For. Univ.* **2018**, *42*, 174–178.

17. Dadashpoor, H.; Azizi, P.; Moghadasi, M. Land use change, urbanization, and change in landscape pattern in a metropolitan area. *Sci. Total Environ.* **2019**, *655*, 707–719. [CrossRef] [PubMed]

18. Guo, L.; Di, L.; Tian, Q. Detecting spatio-temporal changes of arable land and construction land in the Beijing-Tianjin corridor during 2000–2015. *J. Geogr. Sci.* **2019**, *29*, 702–718. [CrossRef]

19. Jiang, G.; Ma, W.; Zhou, D.; Zhao, Q.; Zhang, R. Agglomeration or dispersion? Industrial land-use pattern and its impacts in rural areas from China's township and village enterprises perspective. *J. Clean. Prod.* **2017**, *159*, 207–219. [CrossRef]

20. Van der Sluis, T.; Pedroli, B.; Frederiksen, P.; Kristensen, S.B.; Busck, A.G.; Pavlis, V.; Cosor, G.L. The impact of European landscape transitions on the provision of landscape services: An explorative study using six cases of rural land change. *Landsc. Ecol.* **2019**, *34*, 307–323. [CrossRef]

21. Stokes, E.C.; Seto, K.C. Characterizing and measuring urban landscapes for sustainability. *Environ. Res. Lett.* **2019**, *14*. [CrossRef]

22. Xi, Y.; Nguyen Xuan, T.; Li, C. Spatio-Temporal Variation Analysis of Landscape Pattern Response to Land Use Change from 1985 to 2015 in Xuzhou City, China. *Sustainability* **2018**, *10*, 4287. [CrossRef]

23. Wang, S.; Yang, K.; Yuan, D.; Yu, K.; Su, Y. Temporal-spatial changes about the landscape pattern of water system and their relationship with food and energy in a mega city in China. *Ecol. Model.* **2019**, *401*, 75–84. [CrossRef]

24. Ma, L.; Bo, J.; Li, X.; Fang, F.; Cheng, W. Identifying key landscape pattern indexes influencing the ecological security of inland river basin: The middle and lower reaches of Shule River Basin as an example. *Sci. Total Environ.* **2019**, *674*, 424–438. [CrossRef]

25. Li, Y.; Xue, C.; Shao, H.; Shi, G.; Jiang, N. Study of the Spatiotemporal Variation Characteristics of Forest Landscape Patterns in Shanghai from 2004 to 2014 Based on Multisource Remote Sensing Data. *Sustainability* **2018**, *10*, 4397. [CrossRef]

26. Jiangsu Statistical Bureau. *Statistical Yearbook of Jiangsu 1991*; China Statistical Press: Beijing, China, 1991.

27. Jiangsu Statistical Bureau. *Statistical Yearbook of Jiangsu 2002*; China Statistical Press: Beijing, China, 2002.

28. Jiangsu Statistical Bureau. *Statistical Yearbook of Jiangsu 2012*; China Statistical Press: Beijing, China, 2012.

29. Jiangsu Statistical Bureau. *Statistical Yearbook of Jiangsu 2018*; China Statistical Press: Beijing, China, 2018.

30. O'neill, R.V.; Krummel, J.R.; Gardner, R.E.; Sugihara, G.; Jackson, B.; DeAngelis, D.L.; Milne, B.T.; Turner, M.G.; Zygmunt, B.; Christensen, S.W.; et al. Indices of Landscape Pattern. *Landsc. Ecol.* **1988**, *1*, 153–162. [CrossRef]

31. Yang, W. Spatiotemporal change and driving forces of urban landscape in Beijing. *Acta Ecol. Sin.* **2015**, *35*, 4357–4366.

32. Sun, K.; Yang, Y.; Zhao, P.; Zhang, Z. Spatial Temporal Evolution of Landscape Pattern in Xi'an Based on 3S Technology. *J. Northwest For. Univ.* **2015**, *30*, 180–185.

33. Ma, S.; Zhang, Y.; Sun, C. Optimization and Application of Integrated Land Use and Transportation Model in Small- and Medium-Sized Cities in China. *Sustainability* **2019**, *11*, 2555. [CrossRef]
34. Xu, L.; Chen, S.S.; Xu, Y.; Li, G.; Su, W. Impacts of Land-Use Change on Habitat Quality during 1985–2015 in the Taihu Lake Basin. *Sustainability* **2019**, *11*, 3513. [CrossRef]
35. Chuai, X.; Wen, J.; Zhuang, D.; Guo, X.; Yuan, Y.; Lu, Y.; Zhang, M.; Li, J. Intersection of Physical and Anthropogenic Effects on Land-Use/Land-Cover Changes in Coastal China of Jiangsu Province. *Sustainability* **2019**, *11*, 2370. [CrossRef]
36. Hu, M. Development Process and Enlightenment of Township Enterprises in China since Reform and Opening: The Case of Township Enterprises in Jiangsu from 1978 to 1992. *Lit. Chin. Communist Party* **2008**, *4*, 24–29.
37. Yu, G.; Li, Y. Present Main Problems of Township Enterprises: Investigation and Research of Township Enterprises in Jiangsu Province from 1978–1992. *Issues Agric. Econ.* **1989**, *10*, 24–29.
38. Sun, X. Research on the Development Model of Urbanization of Village Area in Suzhou. Master's Thesis, Suzhou University of Science and Technology, Suzhou, China, 2017.
39. Zhong, Z.; Chen, Z.; Xu, Y.; Ren, C.; Yang, G.; Han, X.; Ren, G.; Feng, Y. Relationship between Soil Organic Carbon Stocks and Clay Content under Different Climatic Conditions in Central China. *Forests* **2018**, *9*, 598. [CrossRef]
40. Dax, T.; Zhang, D.; Chen, Y. Agritourism Initiatives in the Context of Continuous Out-Migration: Comparative Perspectives for the Alps and Chinese Mountain Regions. *Sustainability* **2019**, *11*, 4418. [CrossRef]

 *sustainability* 

Article

Spatial-Temporal Dynamic Analysis of Land Use and Landscape Pattern in Guangzhou, China: Exploring the Driving Forces from an Urban Sustainability Perspective

Siqi Liu [1,2,3,*], Qing Yu [2,4] and Chen Wei [1]

[1]   Shaanxi Provincial Land Engineering Construction Group Co., Ltd., Xi'an 710075, China; 2015226043@chd.edu.cn

[2]   The Center for Housing Innovations of the Chinese University of Hong Kong, Hong Kong 999077, China; kjc@aku.edu.cn

[3]   Institute of Land Engineering and Technology, Shaanxi Provincial Land Engineering Construction Group Co., Ltd., Xi'an 710024, China

[4]   Natural Resources Monitoring Center of Zhejiang Province, Zhejiang 310000, China

*   Correspondence: 4102090209@chd.edu.cn

Received: 23 October 2019; Accepted: 22 November 2019; Published: 26 November 2019

Abstract: Rapid urbanization is one of the most important factors causing land-use change, which mainly results from the orientation of government policies, adjustment of industrial structure, and migration of the rural population. Land use and land cover change (LUCC) is the natural foundation of urban development that is significantly influenced by human activities. By analyzing the LUCC and its inner driving force, as well as landscape pattern change, human activity and urban sustainable development can be better understood. This research adopted a geographic information system (GIS) and remote sensing (RS) technology to comprehensively analyze land use of Guangzhou, respectively, in 1995, 2005, and 2015. Fragmentation Statistics (FRAGSTATS) is the most authoritative software to calculate landscape metrics. Landscape pattern change was analyzed by FRAGSTATS. The results showed that urban land significantly increased from 16.33% in 1995 to 36.05% in 2015. Farmland greatly decreased from 45.16% in 1995 to 27.82% in 2005 and then slightly decreased to 25.10% in 2015. In the first decade, the non-agricultural conversion of rural land and the expansion of urban land was the dominant factor that led to the change. In the second decade, urban land had been supplemented through the redevelopment of low-efficiency land. The fragmentation of landscape patterns significantly increased from 1995 to 2005 and slightly decreased from 2005 to 2015. It indicated that the change in land use in the second decade was different from that in the first. This difference mainly resulted from three aspects: (1) urban development area and ecological conservation area were clearly defined in Guangzhou; (2) many small towns had developed into urban centers, and the scattered urban land gathered into these centers; (3) the establishment of greenway improved the connection of fragmented patches. After that, this study discussed land-use change and its causes and proposed the trend of urban development from the perspective of sustainability.

Keywords: LUCC; landscape pattern; fragmentation; non-agricultural conversion of rural land; urban green space

---

1. Introduction

In the past few decades, many parts of the world have experienced rapid urbanization. However, urbanization in China is considered to be the fastest and the most extensive one [1]. The area of land-use change in China from 1980 to 2015 accounted for 17% of the total land area of China. In particular, the net increment of construction land area is the largest worldwide. Rapid urbanization is the major cause of land-use change [2]. China's urbanization is divided into two stages. The urbanization process from 1949 to 1978 was slow and tortuous; after which, China's urbanization entered a period of rapid development with the urbanization rate increased from 17.9% in 1978 to 59.58% in 2018. Especially after 1995, China's urbanization increment was as much as 1.5% per year [3]. The focus of urbanization had gradually changed from the expansion of metropolitan areas to the overall construction of small cities and towns in the central and western regions.

Fast economic growth supports rapid urbanization. Conversely, rapid urbanization provides the material basis for economic development, such as land resources and labor force. However, this economy-oriented mode neglected the development of society and the environment. In this context, urban sustainability is proposed to mitigate the negative impact of rapid urbanization on society and the environment. Over time, urban sustainability has been developed by a variety of approaches, including efficient land use that focuses on compact, the transformation of industrial structure, and a healthy ecological environment that focuses on human living quality improvements [4]. Particularly, environmental issues have received widespread attention with environmental problems, including land degradation, contamination, and the decline in green space, becoming evident [5]. On the level of urban construction, increasing urban green space is considered to be an effective way to alleviate these problems and to improve living quality. Urban green space, called "the lung of the city", is an essential part of the urban ecosystem. It can offer multiple ecosystem services, including climate dampening, water supply, and nutrient cycle [6]. Urban green space plays an active role in maintaining ecosystem stability and forming the urban landscape, meeting urban ecological service functions, and improving urban environmental quality. What's more, it is also an important way to achieve urban sustainable development [7,8].

The fundamental components of urban green space are landscape patches with various sizes and shapes [9]. Every patch has a regular structure, specific function, and relative independence. The spatial distribution of landscape patches is called landscape pattern, which is generated under the influence of many factors [10]. Landscape pattern is not only the arrangement and combination of landscape elements with different sizes and shapes in space but also the spatial manifestation of landscape heterogeneity [11]. It can embody urban development characteristics, including land resources, urban morphology, and urban management [12].

Land use and land cover change (LUCC) is the natural foundation that significantly influences human existence development. It is a dynamic process that can lead to changes in landscape patterns [13,14]. Urban development policies directly affect human activities, which directly affects the change in land use. Conversely, by studying the optimization of land use, we can better formulate urban development policies. With the continuous improvement of geographic information system (GIS) technology and the establishment of a remote sensing (RS) platform, China has entered a period of remote sensing and big data. RS technology is widely used in LUCC monitoring, which has the characteristics of large area coverage, high precision, and timeliness [15,16]. RS data has become the most important and stable data source for LUCC analysis [17].

Many researchers [18–20] attempted to analyze LUCC and landscape pattern change, respectively, in China, but they seldom integrate both of them to discuss the implication for urban sustainability. From the perspective of urban planning and sustainability, this research analyzed the internal driving force of the change by using GIS and RS technologies. It aimed at (1) analyzing spatial-temporal dynamics of LUCC and landscape pattern from 1995 to 2015; (2) figuring out the internal driving force of LUCC and landscape pattern from 1995 to 2015; (3) assessing the completion and the value of green space system construction; (4) forecasting the trend of land use and putting forward proposals about improving environmental quality.

2. Study Area

Guangzhou is the capital and largest city of Guangdong province in the south of China (from 112°57′ to 114°03′ E longitude, 22°26′ to 23°56′ N latitude), which is located on the Pearl River Delta, adjacent to Hong Kong and the Macao Special Administrative Region. Moreover, Guangzhou is one of the fastest-growing cities in China in terms of urbanization. It has been playing a crucial role in China's economic reform and development since 1978. The urban population grew from 6,122,000 in 1992 to 14,043,000 in 2016. In recent years, it has continued to push forward the adjustment of urban spatial layout, realizing the transformation from single-center to multi-center development of urban spatial structure, expanding the city to its suburbs, and making great changes in land use.

In this research, we selected eight districts of Guangzhou, including Yuexiu district, Liwan district, Haizhu district, Tianhe district, Baiyun district, Huangpu district, Panyu district, Huadu district, as study areas (Figure 1). They form the core areas of Guangzhou, which have a relatively higher urbanization proportion and faster urbanization speed. Nansha district, Zengcheng district, and Conghua district belong to the Guangzhou eco-tourism area and ecological agriculture area, whose land cover types are mainly forestland and farmland. They are less affected by rapid urbanization. Therefore, these areas were not in the scope of this research.

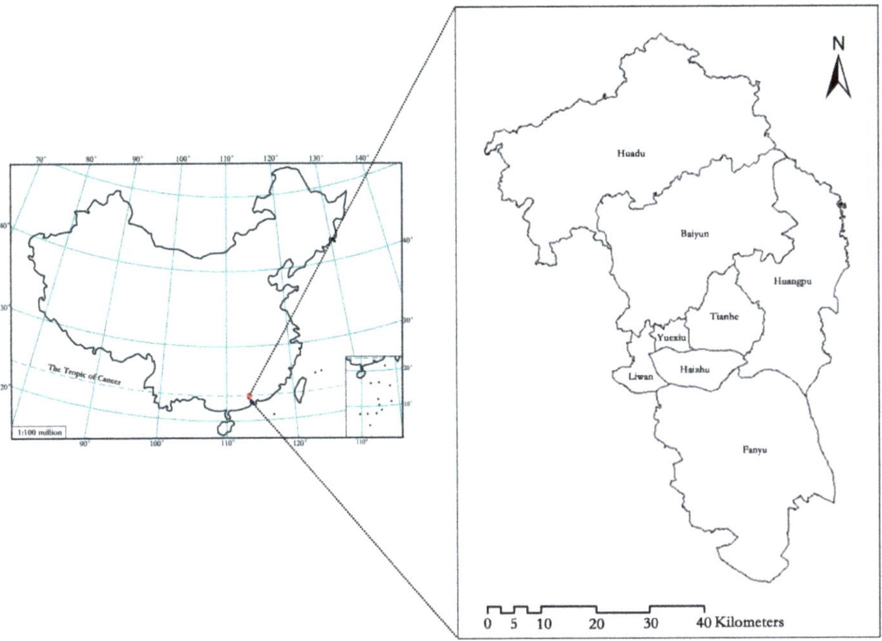

Figure 1. Study areas (Yuexiu district, Liwan district, Haizhu district, Tianhe district, Baiyun district, Huangpu district, Panyu district, Huadu district).

3. Materials and Methods

3.1. Data Source

The research data was used to dynamically analyze LUCC and landscape pattern from 1995 to 2015. Three years' satellite image data (1995, 2005, and 2015) was selected, coming from Landsat5 and Landsat8. We used the Landsat5 image as the data source of 1995 and 2005, while Landsat8 image as 2015 image data, as it just launched in 2013. Both the spatial resolution of Landsat5 and Landsat8 are 30 × 30 m. As a new satellite, the Landsat8 had been greatly improved in the aspects of image resolution, number of wave band, and scanning mode compared with other previous satellites. In addition, the time of images acquired in different years should be relatively close, and in order to minimize the weather's negative influence on analysis, the images were selected with low cloud cover. As a result, the acquisition time of the image was 19 January 2015, 23 November 2005, 30 December 1995, respectively.

3.2. Analysis Method for LUCC

The Environment for Visualizing Image (ENVI) is complete remote sensing (RS) image processing platform, which is developed by Interactive Data Language (IDL). It is powerful in remote sensing (RS) image display, processing, and analysis, including image enhancement, correction, classification, and analysis of multiple features [21]. It is an effective approach to research LUCC.

Image data were processed by ENVI, and land cover involved six classes in this study: water, forestland, grassland, urban land, farmland, and bare land (including unused land). Thus, we obtained the green land information from classification results, including farmland, forestland, and grassland, as well as the changes in urban land and bare land, which enabled us to analyze the changes in the landscape pattern afterward.

The classification of land cover leveraged a collection of various technologies and approaches. The green land information extraction should choose suitable bands. In this study, six original bands were used except band 6 (thermal infrared band). Band 1, band 2, and band 3 are visible light ones, and band 1 has the greatest penetration of water. Band 4 is the near-infrared band, which reflects the high-reflection area of plants and reflects a large amount of plant information. Besides, band 5 is the short-wave infrared band (SWIR), which is sensitive to plant and soil moisture content. Band 7, the middle infrared band, has the longest wavelength in those six bands [22]. Due to the characteristics of each band of Landsat 4-5 thematic mapper (TM) image, there is redundant information between each band. In order to reduce the interference between the bands, we should select several bands with a large amount of information and less correlation to integrate and analyze afterward. We first analyzed the correlation of six original bands, as Table 1 shows below.

Table 1. Landsat 4-5 thematic mapper (TM) original bands correlation matrix in 2015.

Correlation	Band 1	Band 2	Band 3	Band 4	Band 5	Band 7
Band 1	1					
Band 2	0.970924	1				
Band 3	0.956297	0.983961	1			
Band 4	0.62476	0.709536	0.662028	1		
Band 5	0.818552	0.870241	0.861311	0.904781	1	
Band 7	0.907242	0.923027	0.925918	0.779751	0.960959	1

Band 1 contains much more information than band 2 and band 3, but a reflection of surface features except water is generally lower than band 4. In addition, band 1 is greatly disturbed by the atmosphere [23]. According to the correlation between bands and the amount of information contained, TM bands 2, 3, 4 were selected for combination. When extracting information, we used six original bands' data. Moreover, it may also include various kinds of "derivative" band data, like the principal components, NDVI (normalized difference vegetation index), and other ratio indices [24]. The principal component transform, also known as K–L transform, is a multidimensional orthogonal linear transformation based on the statistical characteristics. It can be compressed correlated multi-band to fewer fully independent bands data, to make image interpretation easier [25]. Normalized difference vegetation index (NDVI) is a simple graphical indicator, which is calculated from the visible (VIS) and near-infrared (NIR) light reflected by vegetation [26,27]. The formula is:

$$NDVI = (NIR - VIS)/(NIR + VIS) \tag{1}$$

Figure 2a–c below are the results of principal component transformation by using principal components tool in ENVI, together with NDVI results (Figure 2d), which we could compare directly.

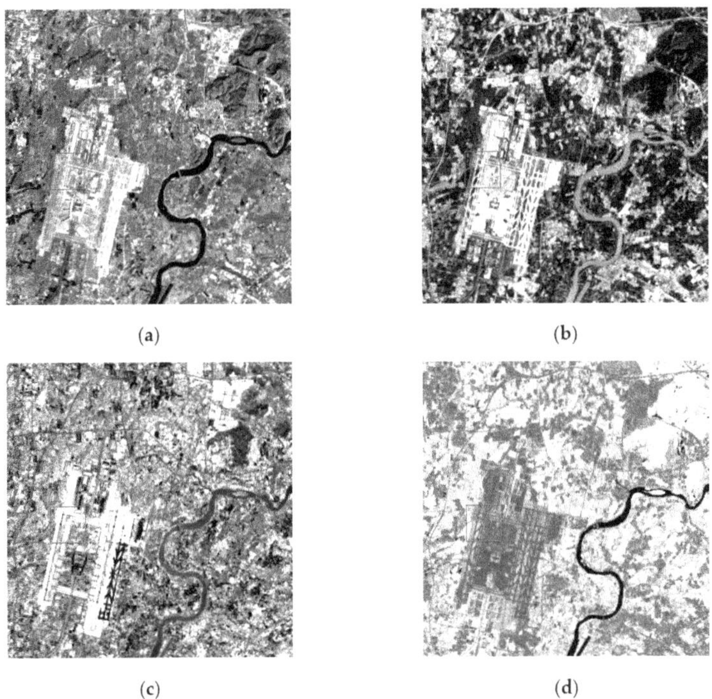

(a)

(b)

(c)

(d)

Figure 2. The (**a**) shows PC1; the (**b**) shows PC2; the (**c**) shows PC3; the (**d**) shows NDVI results. PC, principal component; NDVI, normalized difference vegetation index.

The first principal component (PC1) contains the maximum variance percentage, the second principal component (PC2) contains the second large variance, and so on. The last band's principal component, due to the less variance, most of which is caused by the original spectrum noise, it appears as noise [28–30]. From the four figures above, it is obvious that they showed different characteristics of different objects. For example, water was very clear in PC1 and NDVI results.

According to the correlation coefficient of each band, the optimal band combination was selected. In this study, we selected the first principal component band (PC1) and NDVI, as well as TM bands 2, 3, 4 for correlation analysis of each band. After the analysis and comparison of the correlation of images, we found the combination of band 3, PCI, and NDVI had a better visual interpretation result for classification reference, as shown in Table 2.

Table 2. TM bands 2, 3, 4, PC1, and NDVI correlation matrix in 2015.

Correlation	PC1	NDVI	Band 2	Band 3	Band 4
PC1	1				
NDVI	0.338435	1			
Band 2	0.912088	0.101084	1		
Band 3	0.888872	0.062101	0.983961	1	
Band 4	0.924823	0.507719	0.709536	0.662028	1

[1] PC1 = the first principal component band; NDVI = normalized difference vegetation index

In this study, we started with a preliminary interpretation in the research area by using the supervised classification, and then used a combination of supervised classification result (maximum likelihood classification) and decision tree classification. In supervised classification, a region of interest (ROI) was selected as samples, which was acquired on the basis of prior knowledge experience and higher resolution images. According to the rules of knowledge and successive comparison, decision tree classification extracted various types of surface objects from the image. [31,32].

TM image has some limitations, such as the same type of objects have different spectrums, and different types of objects may have a similar spectrum. Therefore, as a consequence, there might be certain omission errors. The step of visual interpretation, combined with a large number of field research, for correcting is necessary, which could minimize the errors. Thus, the image classification result could reflect the objects' information on the ground as far as possible. After classification, accuracy evaluation is indispensable. In this study, we firstly selected some parts of classification results and compared them with the Google Earth image (Figures 3 and 4), following the inspection standards.

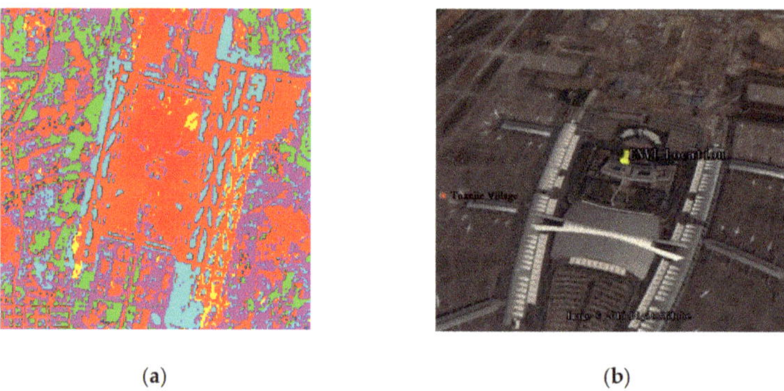

(**a**) (**b**)

Figure 3. Classification results, where the red area represents urban land (**a**); the real image of the airport by Google Earth with the same location (**b**).

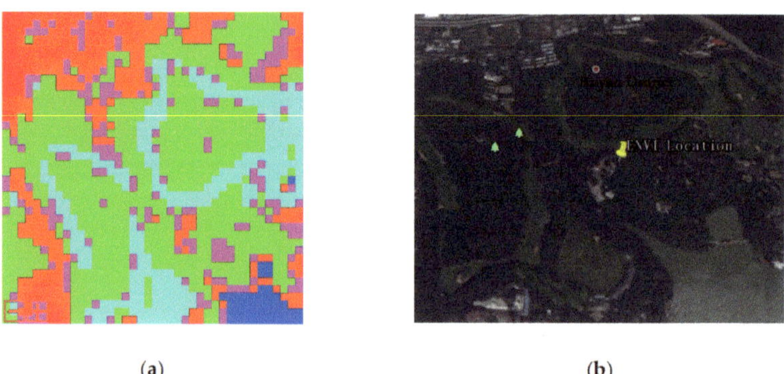

(**a**) (**b**)

Figure 4. Classification results, where the cyan area represents grassland, and the green area is forestland (**a**); the real image of the park by Google Earth with the same location (**b**).

Then, the image classification accuracy was also evaluated using the Kappa coefficient index [33]. It is an important and also widely used index to measure the accuracy of image classification. The classification results were assessed by using a randomly scattered diagram and generating the confusion matrix in ENVI [34] (Tables 3–5).

According to the accuracy assessment, the overall accuracy and kappa coefficient were both above 0.9; as a result, the classification results were approximately acceptable.

Table 3. Confusion matrix of the classification results in 2015.

	Water	Urban Land	Forestland	Farmland	Grassland	Bare Land	Total
Water	148	1	0	0	0	1	150
Urban land	6	149	1	0	0	0	156
Forestland	2	0	149	0	0	0	152
Farmland	0	7	8	136	0	0	151
Grassland	0	3	2	1	46	1	53
Bare land	0	3	0	0	0	55	58
Total	156	163	160	137	47	57	720

Overall Accuracy = (683/720) 94.8611%. Kappa Coefficient = 0.9364.

Table 4. Confusion matrix of the classification results in 2005.

	Water	Urban Land	Forestland	Farmland	Grassland	Bare Land	Total
Water	149	1	0	1	0	1	152
Urban land	5	144	1	0	0	0	150
Forestland	2	0	148	1	0	0	151
Farmland	3	4	1	142	0	0	150
Grassland	0	1	2	3	51	0	57
Bare land	0	1	0	0	0	59	60
Total	159	151	152	147	51	60	720

Overall Accuracy = (693/720) 96.2500%. Kappa Coefficient = 0.9537.

Table 5. Confusion matrix of the classification results in 1995.

	Water	Urban Land	Forestland	Farmland	Grassland	Bare Land	Total
Water	147	0	2	0	0	0	149
Urban land	9	146	0	0	0	0	155
Forestland	0	2	148	1	0	0	151
Farmland	1	6	3	141	0	0	151
Grassland	0	0	4	2	50	0	56
Bare land	0	2	0	0	0	56	58
Total	157	156	157	144	50	56	720

Overall Accuracy = (688/720) 95.5556%. Kappa Coefficient = 0.9451.

3.3. Landscape Pattern Analysis

Fragmentation statistics (FRAGSTATS) is a computer software program designed to compute a wide variety of landscape metrics for categorical map patterns. FRAGSTATS 4.2 is the most authoritative software to calculate landscape metrics. It can help users to quantify the structure of landscapes. The landscape subject to analysis is user-defined and can represent any spatial phenomenon [35]. Based on the features of the study area, we emphasized on the quantity, shape features, aggregation, and diversity to analyze the spatial pattern of the green landscape [36]. This paper conducted patch level, class level, and landscape-level research on spatial scale, which is ranging from microscope to macroscope. Through landscape pattern metrics calculation and analysis, the changes in landscape pattern from 1995 to 2005 and from 2005 to 2015 were studied, respectively.

According to different ecological significance, the indices were classified into area-edge, shape, core area, aggregation, and diversity. Eleven indices were selected to analyze, including number of patches (NP), patch density (PD), mean patch size (MPS), fractal dimension index (FRAC), Euclidean nearest neighbor (ENN), largest patch index (LPI), Shannon's evenness index (SHEI), Shannon's diversity index (SHDI), aggregation index (AI), contagion index (CONTAG), and interspersion juxtaposition index (IJI).

4. Results

4.1. Land Use Dynamics

The land use/cover image results are shown in Figures 5–7. The area and percentage of each land use classification are listed in Table 6. The results showed that land use structure had changed significantly from 1995 to 2015. The major manifestations were the continuous decrease of farmland (from 45.16% in 1995 to 25.10% in 2015) and the notable increase of urban land (from 16.33% in 1995 to 36.05% in 2015). It could be divided into two stages: namely 1995–2005 and 2005–2015. In the first stage (1995–2005), urban land increased by 289.64 km^2, farmland decreased by 559.59 km^2, and other land types, including water, bare land, forestland, and grassland, increased slightly; in the second stage (2005–2015), urban land increased by 346.76 km^2, while other land types were relatively stable and slightly reduced.

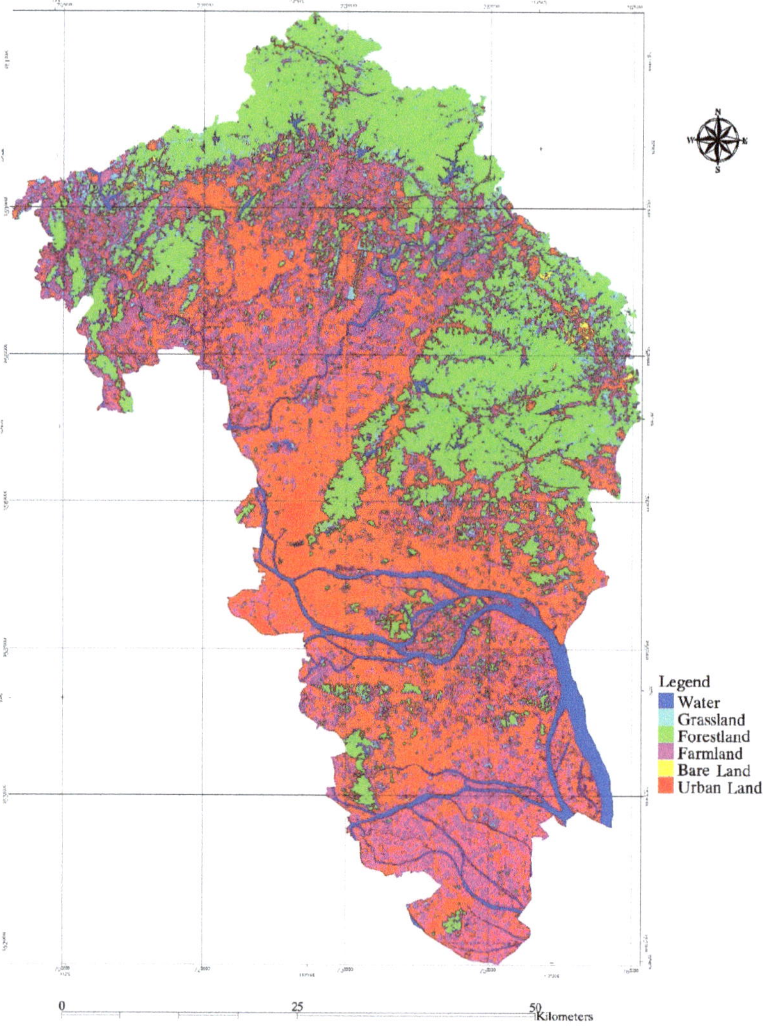

Figure 5. Land use/cover map of the central area of Guangzhou in 2015.

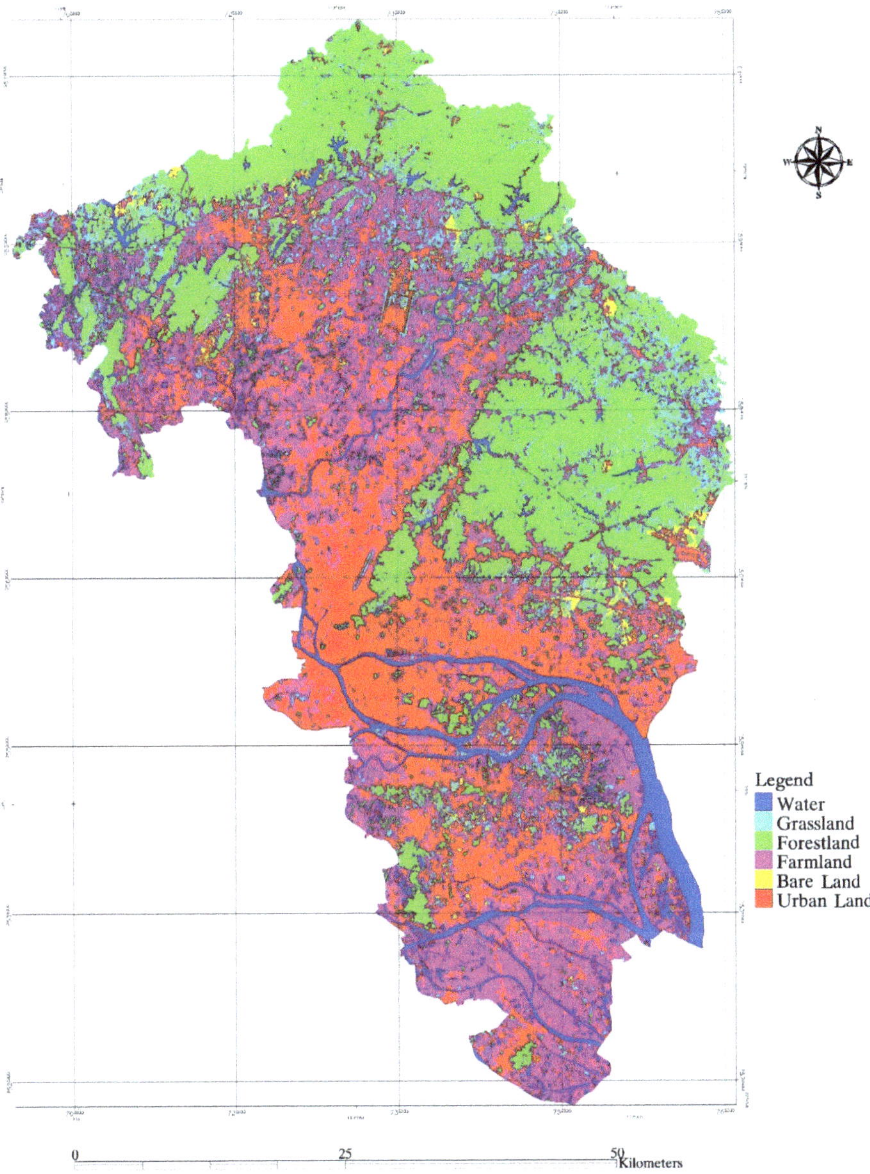

Figure 6. Land use/cover map of the central area of Guangzhou in 2005.

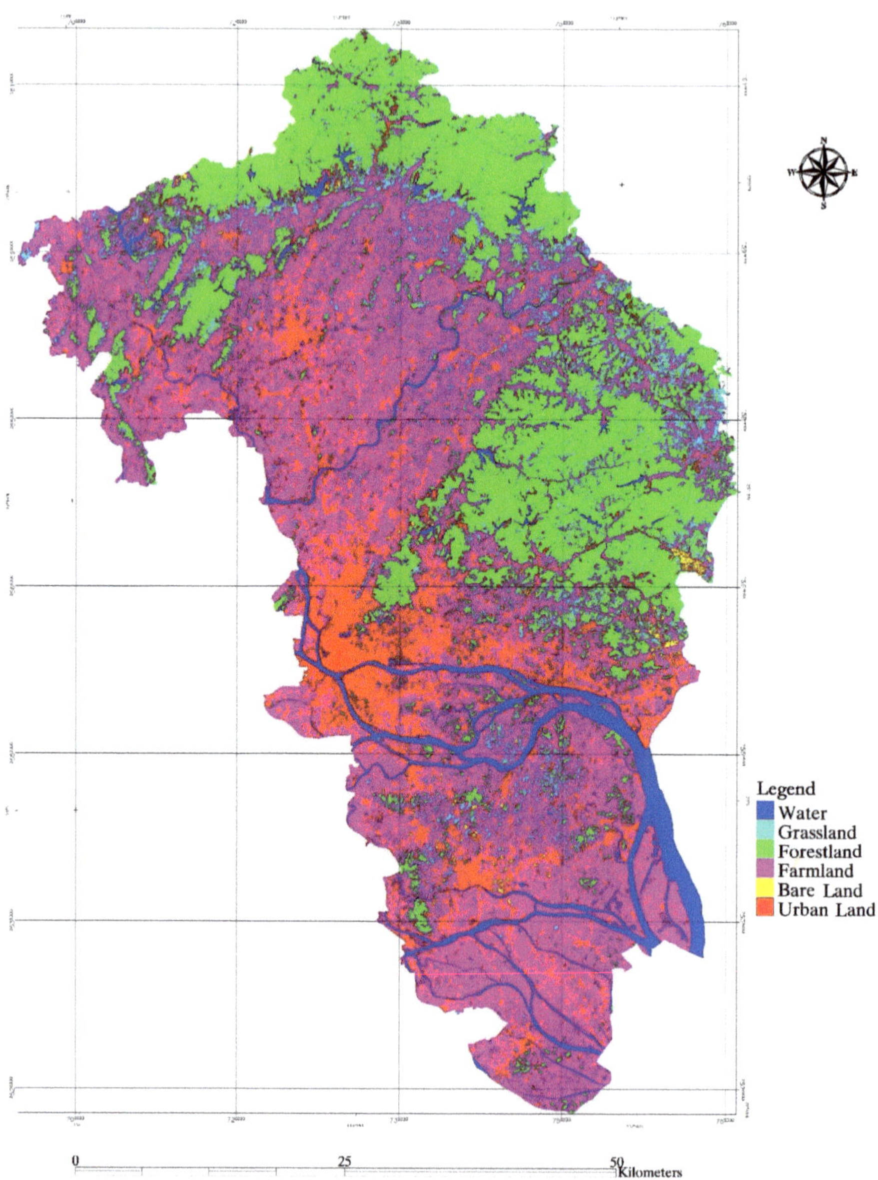

Figure 7. Land use/cover map of the central area of Guangzhou in 1995.

Table 6. Land-use change statistics.

Area (km²) Year	Water	Urban Land	Bare Land	Forestland	Farmland	Grassland
1995	192.47	526.96	23.28	857.95	1457.29	169.36
	5.96%	16.33%	0.72%	26.58%	45.16%	5.25%
2005	239.13	816.60	60.22	915.98	897.70	297.67
	7.41%	25.30%	1.87%	28.38%	27.82%	9.22%
2015	160.60	1163.36	39.55	865.47	809.99	188.23
	4.98%	36.05%	1.23%	26.82%	25.10%	5.83%

4.2. Landscape Pattern Analysis

4.2.1. General Landscape Pattern Metrics Analysis

In order to illustrate the overall trend of landscape pattern change from 1995 to 2015, six indicators were selected for analysis, as shown in Table 7. NP and MPS reflected the spatial landscape pattern, which was used to describe the heterogeneity of the entire landscape. While the value of NP had a good positive correlation with landscape fragmentation, MPS and the landscape fragmentation had a negative correlation. From 1995 to 2005, the value of NP rocketed from 3400 to 4962, while MPS rapidly decreased from 73.69 to 43.64, reflecting the further fragmentation of the landscape. To some extent, it reflected that human disturbance to landscape had increased significantly in this period. From 2005 to 2015, the value of NP remained stable, and MPS slightly decreased. SHEI and SHDI are significant indices of landscape diversity. They are inversely proportional to landscape dominance. The larger values of SHEI ($0 \leq$ SHEI ≤ 1) and SHDI (SHDI ≥ 0) were, the lower the landscape dominance was. As a result, various patches became distributed evenly; the higher the diversity was and the richer the land use was. According to Figure 8, it can be seen that SHEI went up from 0.71 in 1995 to 0.75 in 2005, then slightly went down to 0.72 in 2015; SHDI rapidly went up from 1.37 in 1995 to 1.47 in 2005, then slightly went down to 1.42 in 2015. It indirectly reflected the state of land-use types from a rapid increase to a slow decrease. The change in this period was closely related to the development of the economy and the adjustment of industrial structure. IJI and CONTAG are widely used for ecological risk assessment, reflecting the condition of landscape contagion and interspersion. IJI and CONTAG describe the degree of agglomeration and extension trend of different patches. When IJI becomes higher, it represents more patch types that are adjacent. While when CONTAG becomes lower, it represents a more intensive landscape pattern with multiple landscape elements. According to Figure 9, the landscape richness increased from 1995 to 2015. To be specific, it rapidly increased from 1995 to 2005 and relatively slightly decreased from 2005 to 2015. The results of these indicators showed strong consistency.

Table 7. Related metrics of overall characteristics of the landscape.

Year	NP	MPS	SHDI	SHEI	IJI	CONTAG
1995	3400	73.69	1.37	0.71	60.09	54.29
2005	4962	43.64	1.47	0.75	69.15	51.09
2015	4942	39.37	1.41	0.72	61.68	53.38

[1] NP = Number of Patches; MPS = Mean Patch Size; SHDI = Shannon's Diversity Index; SHEI = Shannon's Evenness Index; IJI = Interspersion Juxtaposition Index; CONTAG = Contagion Index.

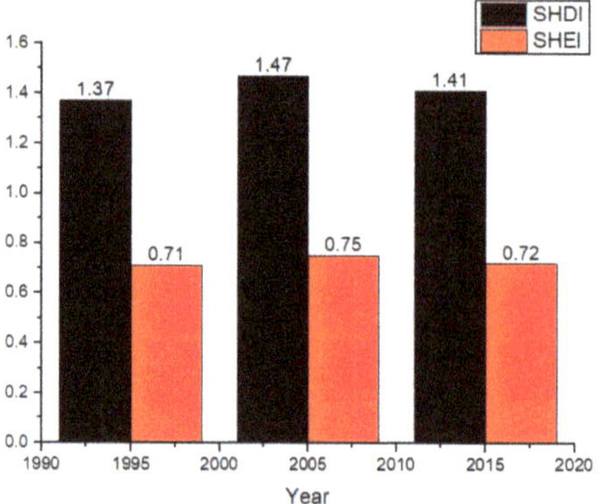

Figure 8. Column chart of Shannon's diversity index (SHDI) and Shannon's evenness index (SHEI).

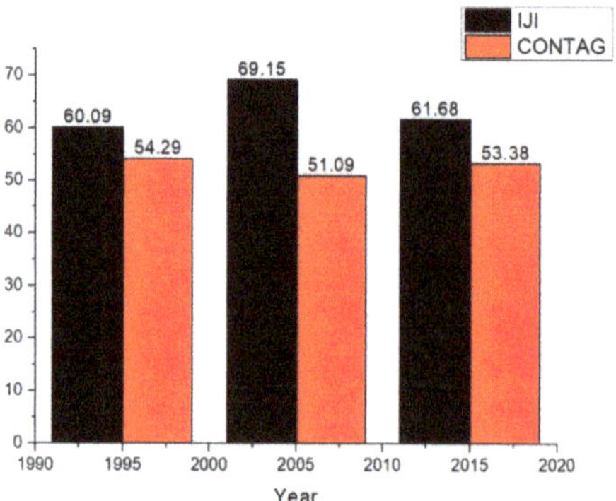

Figure 9. Column chart of interspersion juxtaposition index (IJI) and contagion index (CONTAG).

4.2.2. Landscape Pattern Metrics Analysis of Different Land Use Types

The changes in the landscape pattern are the results of changes in various land-use types [37,38]. The individual analysis of landscape pattern metrics of each land-use type can help us better understand the process of urban development. Seven representative indices were selected from three different scale levels: patch, class, and landscape (Table 8).

From the level of patch metrics, ENN reflects the distance between the nearest patches. It is the simplest measure to quantify patch context, which has been used extensively to assess the condition of patch isolation. From 1995 to 2005, the distance among various types of patch significantly decreased, suggesting that the patches became more compact. From 2005 to 2015, the distance of forestland and farmland between patches reduced, while the distance between patches of bare land and grassland slightly increased. Forestland, farmland, and grassland's ENN values were relatively small, indicating that their distribution was more concentrated. The bare land's ENN value was much larger, and its distribution was more scattered.

From the level of class metrics, FRAC is a measurement index of landscape shape. It is mainly used to analyze the complexity of patches. The value of FRAC is between 1 and 2. When FRAC is closer to 1, the shape of the patch becomes simpler, implying rectangular patches and obvious artificial influence [39]. From 1995 to 2005, the FRAC value of bare land and grassland increased, and the patch's shape became more complex and irregular. From 2005 to 2015, FRAC values of farmland, bare land and grassland decreased, which means that the shape of the patch tended to be simple and regular and looked like a square. Among them, FRAC values of forestland and farmland were relatively larger than the value of bare land and grassland, implying the shape of patches were relatively complex, suggesting human activities had fewer impacts on these type of landscape. However, bare land and grassland's FRAC values were smaller, and the patches' shapes were relatively simple, indicating that these two types were greatly affected by human activities.

From the level of landscape metrics, NP, PD, MPS, and AI, respectively, represents the total number, density, size, and overall aggregation. They form a whole to measure landscape aggregation and fragmentation. According to the results, the changes in the landscape pattern of forestland was the most stable from 1995 to 2015. NP decreased by 47 (4.20%), and MPS increased by 4.49 (5.87%). AI slightly decreased from 65.80 in 1995 to 65.59 in 2015. It showed a relatively stable condition in terms of aggregation. Landscape fragmentation of farmland had been increasing sharply with a rapid increase in NP and a sharp decline in MPS. Especially from 1995 to 2005, NP increased by 713 (89.91%), while MPS decreased by 124.18 (67.56%). Meanwhile, the AI of farmland significantly decreased by 16.85 (27.60%). Differently, AI of Bare land and grassland had experienced an increase from 1995 to 2005 and a great decrease from 2005 to 2015. In other words, the aggregation of bare land and grassland increased first and then decreased. The NP of grassland increased by 28.96%. The MPS became smaller in 2015 compared with 1995, and the shape of its patch was very close to rectangle, which indicated that the man-made urban green space had increased during this period. It can be found that the LPI of forestland was relatively large, and the change was the most stable. It showed that the forestland maintained its natural growth condition, and there was almost no human interference. From the quantitative point of view, the change of farmland was continuous and the most drastic, sharply decreased from 45.68 to 10.94, showing that the change of farmland during this period was directly affected by human activities.

Table 8. Landscape pattern metrics of each land-use type.

Land Type	Year	NP	PD	MPS	AI	LPI	FRAC	ENN
Forestland	1995	1118	1.31	76.50	65.80	38.38	1.219	705.2
	2005	1160	1.28	78.36	65.20	34.08	1.212	676.6
	2015	1071	1.23	80.99	65.59	36.46	1.215	555.3
Farmland	1995	793	0.54	183.82	61.04	45.68	1.285	607.2
	2005	1506	1.57	59.64	44.19	24.19	1.209	600.8
	2015	1836	2.28	43.89	37.92	10.94	1.164	546.9
Bare land	1995	232	9.95	10.05	14.88	8.79	1.026	1522.9
	2005	451	7.79	12.84	21.42	3.14	1.039	1144.8
	2015	414	10.98	9.11	8.57	3.35	1.018	1240.2
Grassland	1995	1257	7.42	13.47	16.22	1.55	1.047	769.5
	2005	1847	6.14	16.28	20.16	1.93	1.064	681.1
	2015	1621	8.76	11.42	12.49	0.80	1.033	685.6

[1] NP = Number of Patches; PD = Patch Density; MPS = Mean Patch Size; AI = Aggregation Index; LPI = Largest Patch Index; FRAC = Fractal Dimension Index; ENN = Euclidean Nearest Neighbor.

5. Discussion

From the beginning of the 1990s, as a regional center, Guangzhou entered the fast urban construction period. With the promulgation of the 15th Round Master Plan in 1996, the northern group of the city was developed. Consequently, Guangzhou had developed from a single-center city to a multi-center city. The multi-center development mode stimulated the development of the suburbs, and the township industrial enterprises developed rapidly. At the same time, many service-oriented villages were formed. From 1995 to 2005, the value of NP greatly increased, and the value of MPS greatly decreased. The fragmentation of the landscape increased significantly as a result. Particularly, the increase of NP and the decrease of MPS of farmland in quantity was the largest among various land-use types. During the first decade, the urbanization in Guangzhou was a spontaneous process in which the rural land was transformed into urban land. The main manifestation was that the rural population was transformed into an urban population in rural areas [40]. The representation in land-use change was the non-agricultural conversion of rural land and the expansion of urban land. Non-agricultural conversion of rural land generally refers to the process in which agricultural land was transformed into urban construction land. The driving force of this process was the development of the rural non-agricultural industry. Under the influence of rapid urbanization and industrialization in Guangzhou, the development of rural land conversion was an inevitable trend. As indicated in Figure 10, the proportion of the agricultural industry in Guangzhou had been dropping, accounting for only 1.64% of the total GDP in 2005. Great changes had taken place in the industrial structure. This process had been promoting the differentiation and reorganization of the productive factors in the rural areas through the expansion of urban land, the evolution of industrial structure, the migration of population, and the construction of infrastructure, and so on, which had profoundly changed the traditional spatial structure of rural land in Guangzhou.

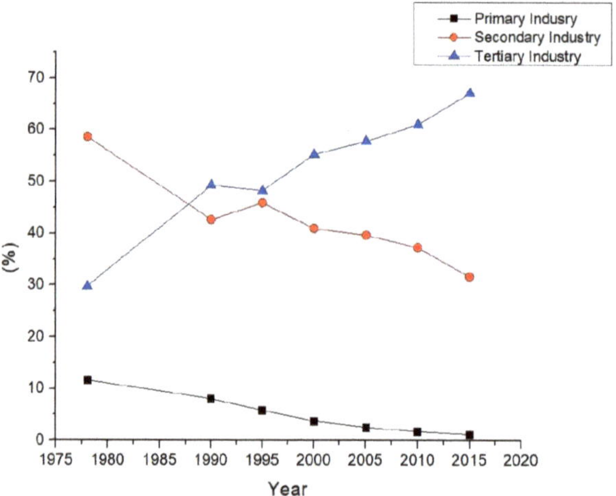

Figure 10. Composition of gross domestic product (GDP) of Guangzhou in main years. Source: Guangzhou Statistical Yearbook.

In general, different metrics of landscape showed the same trend of change. In 2005, the landscape diversity reached the top. Particularly, the value of FRAC of farmland obviously decreased by 0.076 from 1995 to 2005. It showed that human activities directly affected the change in farmland. Therefore, the farmland became more fragmented and scattered. This change was closely related to the extensive and spontaneous construction in Guangzhou. The dominant factors of rural land conversion were complex and diversified. The dominant factors showed a trend of transition from natural and location factors to production and industrial factors, and eventually to life and personal factors. As a result, the spatial manifestation of this transformation was the increase in patch fragmentation. Obviously, it was a development mode that lacked professional planning and guidance. The rapid increase in landscape fragmentation was not conducive to urban sustainability. The city was developed in a low-efficiency approach, and the land resources were wasted.

The 11th Five-Year Plan, promulgated in 2006, emphasizes the optimization of land-use layout and the adjustment of land-use structure, so as to control urban development scale and to realize high-efficiency urban land-use. From 2005 to 2015, urban land was still growing rapidly, while other land-use types slightly decreased (Figure 11). Additionally, the government promulgated a series of policies to protect farmland, which had reduced the loss of farmland to some extent. However, it was still an obvious contradiction between the supply and demand of urban land. From 2005 to 2015, urban land had been greatly supplemented through the redevelopment of low-efficiency land. During this period, the improvement of urban infrastructure was one of the reasons for the decline in

MPS. For example, the increase in urban road density made the landscape patches smaller. However, the fragmentation of landscape decreased. This contrast mainly resulted from three aspects: (1) urban development area and ecological conservation area were clearly defined in Guangzhou. Specifically, the natural land, including forestland, grassland, and farmland, was zoned; (2) many small towns had developed into urban centers, and the scattered urban land gathered into these centers. As a result, the agglomeration of urban land obviously increased. The multi-center structure of Guangzhou became more apparent; (3) the establishment of greenway improved the connection of fragmented patches. Generally, greenway refers to ecological corridors of different sizes. It includes regional greenway, urban greenway, and community greenway in Guangzhou.

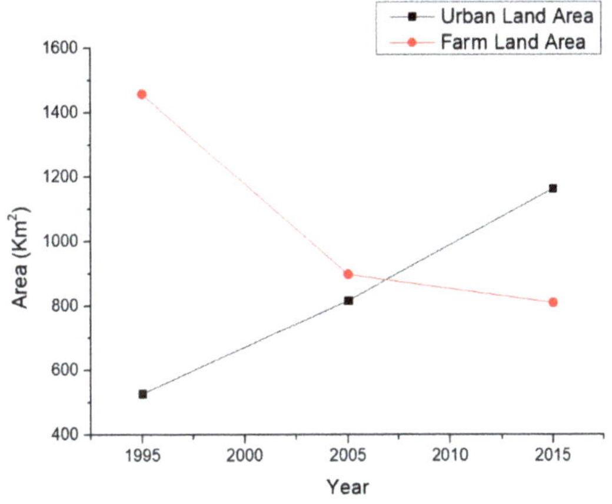

Figure 11. The changes in urban land and farmland in 1995, 2005, 2015.

The outline of the Pearl River Delta (PRD) Reform and Development Plan, promulgated in 2008, stressed the importance of improving the quality of the urban environment. In 2010, Urban Green System Planning of Guangzhou (2001–2020) was officially promulgated, and the development objective of Guangzhou was pushed forward: improving the structure of urban ecological space system, forest conservation, and the establishment of the greenway. We should optimize the construction of green space in central urban areas and strengthen the construction of street green space and community parks. Besides, it was required by the plan to reach 41.5% green coverage rate in the administrative region of Guangzhou and 40% in the central area by 2015.

From 1995 to 2015, the proportion of grassland and forestland increased slightly. Particularly, the landscape pattern aggregation of forestland increased as well, indicating that the forestland system has a better continuity while maintaining a stable natural growth state. The main forms of urban green space are forestland and grassland. Although the growth of forestland and grassland was only 7.52 km^2 and 18.87 km^2, compared with remote sensing images, the increase of green space in urban core areas (Haizhu, Tianhe, Yuexiu) was obvious (Figure 12). On the basis of protecting the existing green space, a large number of integrated parks had been planned through park reconstruction and expansion (Figure 13). Additionally, small-scale green spaces were encouraged to be established by a variety of forms, such as community parks, village parks, and roadside green space. Community parks exist in residential communities of different scales and provide ecosystem services for residents. Village parks are built in rural villages or suburbs. It provides space for villagers' public activities and still retains rural landscape patches, such as farmland and natural waters. Roadside green space is built outside urban roads. It is relatively small and scattered in the city. The fragmentation of grassland increased at the beginning and then decreased, which was also directly related to these government policies. Urban ecological environment quality had been greatly improved during this period.

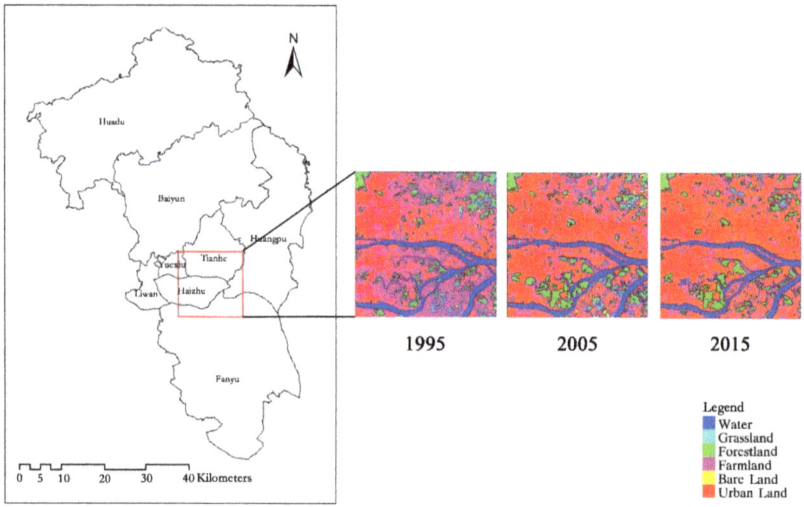

Figure 12. The changes in green space in core areas of Guangzhou in 1995, 2005, 2015.

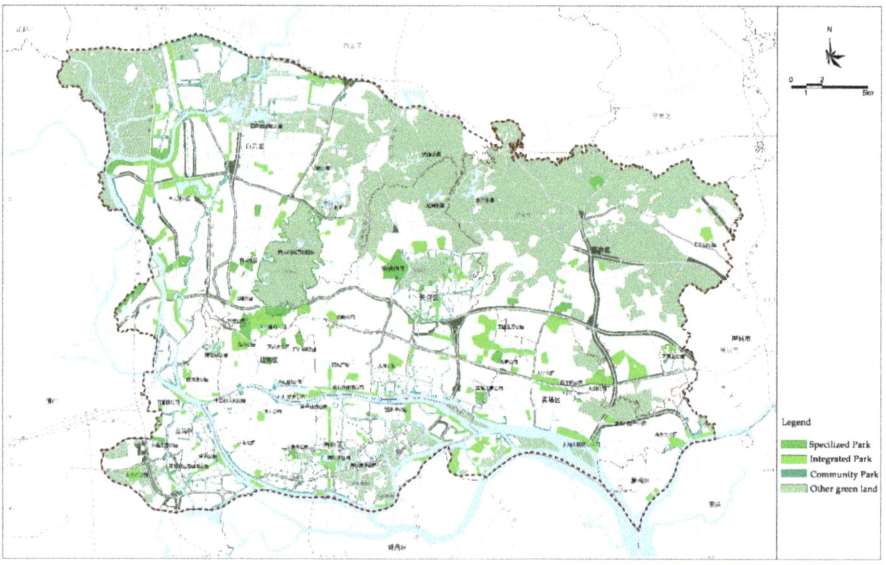

Figure 13. Green system planning of the central area of Guangzhou. Source: Planning Bureau of Guangzhou.

The improvement of the ecological environment benefits from the guidance of planning and policies. Among them, the most important thing is the delineation of basic ecological control lines, including basic farmland protection areas, rural parks, and ecological corridors. This strategy reasonably controls the size of the city and optimizes the urban structure. According to Urban Green System Planning of Guangzhou (2001–2020), a three-level park system, including a specialized park, integrated park, and community park, had been established, which forms a network-based green space system [41]. Additionally, the greenway system had been well established on different scales. Based on actual completion and statistics, the number of parks in Guangzhou reached 246 in 2015, and the green coverage rate in developed areas reached 41.53%, achieving the goal of 41.5% (Figures 14 and 15).

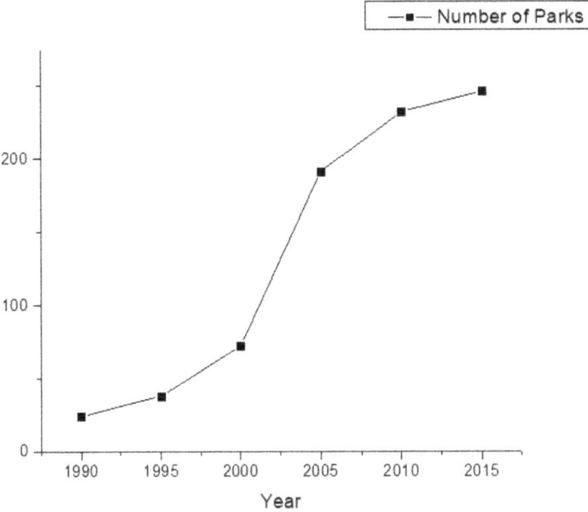

Figure 14. The number of parks in Guangzhou in main years.

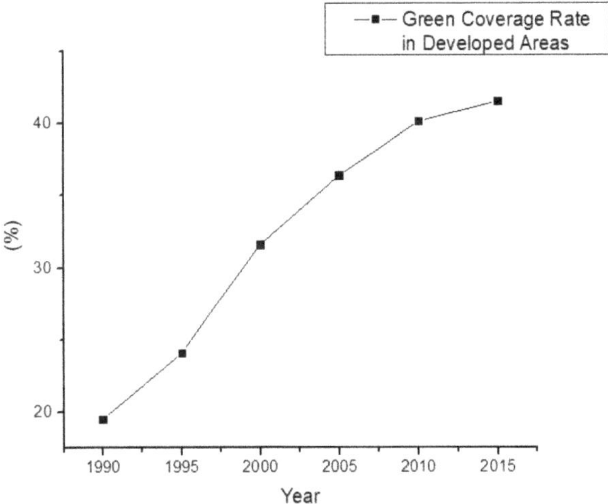

Figure 15. The green coverage rate in developed areas of Guangzhou in main years.

6. Conclusions

Based on the land-use data of 1995, 2005, and 2015, the spatial-temporal dynamics of land-use and landscape pattern in Guangzhou were analyzed through RS and GIS technologies. In the context of rapid urbanization, the driving force, which led to the change, was explored. According to the results, the change in land-use and landscape pattern in the second decade was quite different from that in the first one. The mode of urban land-use was transformed from incremental development to inventory development, which also meant an extensive mode to an efficient one. The focus of urban development had become the improvement of living quality.

(1) From 1995 to 2015, the urban land of Guangzhou greatly increased from 526.96 km^2 (16.33%) to 1163.36 km^2 (36.05%). From 1995 to 2005, the increase of urban land was mainly from the non-agricultural conversion of rural land. The driving force of this process was the development of the

rural non-agricultural industry. From 2005 to 2015, the increase in urban land mainly resulted from the redevelopment of low-efficiency lands, such as bare land.

(2) Farmland greatly decreased from 45.16% in 1995 to 27.82% in 2005. A large number of farmland was transformed into urban land. After 2005, a series of policies were promulgated to protect farmland. The loss of farmland had been obviously controlled. As a result, farmland only decreased by 2.72% until 2015.

(3) Landscape fragmentation is mainly affected by human activity. From 1995 to 2005, human activity in terms of urban construction was extensive and spontaneous. The landscape fragmentation increased greatly. From 2005 to 2015, the efficient land-use that focuses on compact and the establishment of the urban green system had reduced the landscape fragmentation. This was helpful to improve urban sustainability.

Although the importance of agriculture in Guangzhou has been decreasing, it still has great social and environmental significance according to the policy. With the end of the rapid non-agricultural conversion of rural land, farmland will remain in a relatively stable state. On the other hand, the basic ecological control lines will be strictly enforced. As a result, the spatial pattern, including ecological, agricultural, and urban space, will remain stable and continue to be optimized. In order to improve the urban environment, urban green space will become more systematic from a macro perspective; from the micro point of view, with the increase of community parks, roadside green space, and village parks, as well as roof greening, the fragmentation of urban green space will be higher. Eventually, the city and green space will become integrated, and urban environmental quality of Guangzhou will be further improved.

Author Contributions: Conceptualization S.L. and Q.Y.; methodology, S.L. and Q.Y.; software, S.L. and Q.Y.; validation, S.L. and C.W.; formal analysis, S.L. and Q.Y.; investigation, S.L. and C.W.; resources, S.L. and Q.Y.; data curation, Q.Y.; writing—original draft preparation, S.L.; writing—review and editing, S.L. and Q.Y.; visualization, S.L. and C.W.; supervision, Q.Y.; project administration, S.L.

Funding: This research received no external funding.

Conflicts of Interest: The authors declare no conflict of interest.

References

1. Ren, Z.B.; Fu, Y.; Du, Y.X.; Zhao, H.B. Spatiotemporal patterns of urban thermal environment and comfort across 180 cities in summer under China's rapid urbanization. *PeerJ* **2019**, *7*, 7424. [CrossRef] [PubMed]
2. Jin, K. Spatio-Temporal Variations of Vegetation Cover and its Relationships between Climate Change and Human Activities over CHINA. Ph.D. Thesis, Northwest A&F University, Yangling, China, April 2019.
3. Zong, J.F.; Lin, Z.J. 70 years of retrospect and reflection on China's urbanization. *Econom. Probl.* **2019**, *07*, 1–8.
4. Kremer, P.; Haase, A.; Hasse, D. The future of urban sustainability: Smart, efficient, green or just? Introduction to the special issue. *Sustain. Cities Soc.* **2019**, *51*, 101761. [CrossRef]
5. Chan, K.M.; Vu, T.T. A landscape ecological perspective of the impacts of urbanization on urban green spaces in the Klang Valley. *Appl. Geogr.* **2017**, *85*, 89–100. [CrossRef]
6. Costanza, R.; d'Arge, R.; de Groot, R. The value of the world's ecosystem services and natural and natural capital. *Nature* **1997**, *3*, 253–260. [CrossRef]
7. Stessens, P.; Khan, A.Z.; Huysmans, M.; Canters, F. Analysing urban green space accessibility and quality: A GIS-based model as spatial decision support for urban ecosystem services in Brussels. *Ecosyst. Serv.* **2017**, *28*, 328–340. [CrossRef]
8. Lu, Y.; Gan, H.H.; Shi, Z.J.; Liu, Z.L. Soil Fertility Quality Assessment and Managing Measures for Urban Green Space in Shenzhen City. *J. Soil. Water Conserv.* **2015**, *19*, 153–155.
9. Ali, B.; Frederic, C.; Bernard, K. Improving the multi-functionality of urban green spaces: Relations between components of green space and urban services. *Sustain. Cities Soc.* **2018**, *43*, 1–10. [CrossRef]
10. Hashem, D.; Parviz, A.; Mahdis, M. Land use change, urbanization, and change in landscape pattern in a metropolitan area. *Sci. Total Environ.* **2019**, *655*, 707–719. [CrossRef]
11. Turner, M.G. Landscape ecology: The effect of pattern on process. *Annu. Rev. Ecol. Syst.* **1989**, *20*, 171–177. [CrossRef]

12. Wu, Z.; Chen, R.S.; Michael, E.W.; Dhritiraj, S.; Di, X. Changing urban green space in Shanghai: Trends, drivers and policy implications. *Land Use Policy* **2019**, *87*, 104080. [CrossRef]
13. Gillanders, S.N.; Coops, N.C.; Wulder, M.A. Multitemporal remote sensing of landscape dynamics and pattern change: Describing natural and anthropogenic trends. *Prog. Phys. Geogr.* **2008**, *32*, 503–528. [CrossRef]
14. Yang, X.; Zheng, X.-Q.; Chen, R. A land use change model: Integrating landscape pattern indexes and Markov-CA. *Ecol. Model.* **2014**, *283*, 1–7. [CrossRef]
15. Zhou, X.L.; Wang, Y.C. Spatial–temporal dynamics of urban green space in response to rapid urbanization and greening policies. *Landsc. Urban Plan.* **2011**, *100*, 268–277. [CrossRef]
16. Chen, B.M.; Liu, X.W.; Yang, H. Review of most recent progresses of study on land use and land cover change. *Prog. Geogr.* **2003**, *22*, 22–29.
17. Yu, H.P.; Cheng, P.G.; Xia, Y.Q. Application of 3S Technology in Land Dynamic Monitoring. *J. Anhui Agric. Sci.* **2011**, *3*, 1844–1846.
18. Wang, C.D.; Wang, Y.T.; Wang, R.Q.; Zheng, P.M. Modeling and evaluating land-use/land-cover change for urban planning and sustainability: A case study of Dongying city, China. *J. Clean. Prod.* **2018**, *172*, 1529–1534. [CrossRef]
19. Yu, W.H.; Zang, S.Y.; Wu, C.S. Analyzing and modeling land use land cover change (LUCC) in the Daqing City, China. *Appl. Geogr.* **2011**, *31*, 600–608. [CrossRef]
20. Gong, J.Z.; Xia, B.C. Spatio-temporal characteristics of vegetation fraction calculation from TM imagines in Guangzhou from 1990 to 2005. *Ecol. Environ.* **2016**, *15*, 1289–1294.
21. Chen, Y.H. Application of GIS and remote sensing technology in the production of image map. *Electron. Test* **2016**, *11*, 107–108.
22. Dai, C.D.; Hu, D.Y. Information characteristics of TM data. *Remote Sens. Environ.* **1987**, *2*, 7–10.
23. Dai, C.D.; Lei, L.P. TM Image spectral information characteristics and optimum band combination. *Remote Sens. Environ.* **1984**, *4*, 282-2.
24. Bruzzone, L. Detection of changes in remotely-sensed images by the selective use of multi -spectral information. *Int. J. Remote Sens.* **1997**, *18*, 3883–3888. [CrossRef]
25. Zhao, Y.S. *The Principle and Method of Analysis of Remote Sensing Application*, 2nd ed.; Science Press: Beijing, China, 2003.
26. Measuring Vegetation (NDVI and EVI). Available online: http://earthobservatory.nasa.gov/Features/MeasuringVegetation/measuring_vegetation_2.php (accessed on 19 July 2019).
27. Alejandro, C.S.; Jorge, L.B. Delineation of suitable areas for crops using a multi-criteria evaluation approach and land use/cover mapping: A case study in Central Mexico. *Agric. Syst.* **2003**, *77*, 117–136.
28. Byrne, G.F.; Crapper, P.F.; Mayo, K.K. Monitoring land-cover by principal component analysis of multi-temporal Landsat data. *Remote Sens. Environ.* **1980**, *10*, 175–184. [CrossRef]
29. Feng, D.J.; Li, Y.S.; Lan, Y. The automatic detection methods of changing information for dynamic monitoring by principal component transform. *Comput. Eng. Appl.* **2004**, *36*, 202–205.
30. Mo, D.L.; Liu, K.J.; Cao, B.C.; Bao, Y.Q. Remote sensing image change detection based on principal component analysis. *Image Technol.* **2013**, *25*, 53–56.
31. Cihlar, J.; Xiao, Q.H.; Chen, J. Classification by progressive generalization: A new automated methodology for remote sensing multichannel data. *Int. J. Remote. Sens.* **1998**, *19*, 2685–2704. [CrossRef]
32. Li, D.; Luo, Y.F.; Meng, Y.B.; Wang, X.; Li, M.L. Study on spatial change of land use based on GIS. *Mod. Surv. Mapp.* **2019**, *42*, 13–17.
33. Fisher, P.F. Visualization of the reliability in classified remote sensing images. *PE RS* **1994**, *60*, 905–910.
34. Yi, L.; Zhang, G. Object-oriented remote sensing imagery classification accuracy assessment based on confusion matrix. *Proc. Int. Conf. Geoinf.* **2012**, 1–8. [CrossRef]
35. Lamine, S.; Petropoulos, G.P.; Sudhir, K.S. Quantifying land use/land cover spatio-temporal landscape pattern dynamics from Hyperion using SVMs classifier FRAGSTATS. *Geo. Int.* **2018**, *33*, 862–878. [CrossRef]
36. UMass Landscape Ecology Lab. FRAGSTATS: Spatial Pattern Analysis Program for Categorical Maps. Available online: http://www.umass.edu/landeco/research/fragstats/fragstats.html (accessed on 19 July 2019).
37. Li, X.; Lu, L.; Cheng, J.D.; Xiao, H.L. Quantifying landscape structure of the Heihe river basin, north-west China using FRAGSTATS. *J. Arid Environ.* **2001**, *48*, 521–535. [CrossRef]

38. Chen, W.B.; Xiao, D.N.; Li, X.Z. Classification, application and construction of landscape metrics. *Chin. J. Appl. Ecol.* **2002**, *1*, 121–125.
39. Meghann, M.; Paul, B.; Anna, J.; Ebru, E.; Rave, M. Green space spatial characteristics and human health in an urban environment: An epidemiological study using landscape metrics in Sheffield, UK. *Ecol. Indic.* **2019**, *106*, 105464. [CrossRef]
40. Liu, Y.; Zhou, C.S.; Huang, W.L.; Zhu, Q.Q. Spatial features and forming mechanism of rural land non-agricultural degree in metropolitan areas: A case study of Guangzhou City. *Prog. Geogr.* **2018**, *37*, 1119–1130.
41. Liao, Y.T.; Xiao, R.B. Establishing urban green space planning system. *Planner* **2012**, *3*, 46–49.

 *sustainability*

Article

Quantitative Influence of Land-Use Changes and Urban Expansion Intensity on Landscape Pattern in Qingdao, China: Implications for Urban Sustainability

Jinming Yang, **Shimei Li and Huicui Lu ***

School of Landscape Architecture and Forestry, Qingdao Agriculture University, Qingdao 266109, China; 201401038@qau.edu.cn (J.Y.); lism@qau.edu.cn (S.L.)
* Correspondence: huicui.lu@qau.edu.cn

Received: 9 October 2019; Accepted: 1 November 2019; Published: 5 November 2019

Abstract: The spatial structure and configuration of land-use patches, i.e., landscape patterns could affect the flow of energy and materials in inner-urban ecosystems, and hence the sustainable development of urban areas. Studying landscape pattern changes under the process of urbanization would have implicational significance to urban planning and urban sustainability. In this paper, land-use change and urban expansion intensity (UEI) were treated as the inducement factors for changes in landscape patterns, and stepwise regression and geographically weighted regression (GWR) were adapted to quantify their integrated and distributed magnitude effects on landscape patterns, respectively. The findings suggested that land-uses have different contributions to changes in landscape patterns at different urban development zones (downtown, suburban plain area and mountainous suburban areas). Furthermore, the GWR analysis results indicated that the effect of UEI on landscape patterns has spatial and temporal heterogeneity. From 1987 to 2000, the UEI had great explanatory capacity on changes in landscape patterns and helped the landscape assemble faster in the downtown and adjacent areas. However, with the shifting of the center of urban construction from downtown to the suburbs, the high explanatory ability was oriented towards suburban areas during 2000–2016 and the magnitude of influence spatially changed. Therefore, a compact city and protection policy should be adapted to different regions in the study area to achieve strong urban sustainability.

Keywords: urbanization; GIS; urban development zones; urban sustainability; regression analysis; GWR

---

1. Introduction

The concentrations of population and socioeconomic activities in urban areas have led to fast urban expansion all around the world in recent decades [1,2]. If this aggregation phenomenon continues, land transformed into urban areas will nearly triple by the end of 2030 [3]. Meanwhile, other land-use types such as agricultural and forest land around urban areas will be embezzled during the rapid expansion of cities. These tremendous land-use changes and intensive human activities put forward serious challenges to human and natural environment [4,5], such as the loss of biodiversity [6], an increase of the urban heat island effect [7], continuous environmental degradation [8,9], decreased watershed runoff and increased flood potential in urban areas [10], and enhanced CO_2 emissions [11]. Furthermore, changes in the spatial structure and configuration of land-use patches, i.e., landscape patterns, have a direct impact on urban sustainability because they determine energy flow efficiency and air pollution [12,13]. Therefore, the magnitude of influence from variations of land-use on landscape patterns determines their impact on urban sustainability. Thus, the quantitative relationship between land-use and landscape patterns is an important issue to study.

A lot of previous studies have observed land-use change and its impact on landscape patterns in rapidly urbanizing regions [1,12,14–21]. However, most of these studies have only qualitatively analyzed this relationship. For example, the conversion of agricultural land and forest land into built-up land in these areas has led to a prominent, tremendously fragmented landscape, but the magnitude of this influence has not been articulated. At the same time, the few studies using a quantitive analysis of this relationship have done so just from a perspective of their study area [15,16]. However, the effect of land-use changes on landscape patterns has been shown to vary between different urban development zones [1]. Learning the behaviors of land-use changes and the corresponding change in landscape patterns in and around urban core areas will be conductive to urban planning and urban sustainability. In this study, we analyze changes in land-use change and landscape patterns, and we describe quantitative relationships between them at different urban development zones.

Amidst land-use types, changes in built-up land is indicative of urbanization [22–27]. Urban expansion intensity, the growth rate of built-up land in an unit time interval, has been used to investigate urbanization's impact on agricultural landscape patterns [28] and landscape patterns in land utilization [15]. However, these studies were conducted at the block and pixel levels with only several landscape metrics. However, the intensity of urbanization is more related to local government behaviors such as regional urban planning and attracting investment decisions [29]. Thus, spatial relationships between urban expansion and landscape patterns would be better conducted at the administrative division level, a comprehensive study of which could enhance the understanding of the influence of urban expansion on landscape patterns and, ultimately, benefit urban sustainable planning.

It has become easier to characterize a landscape and quantify its structural changes with advanced developments in remote sensing and geographic information science (GIS) techniques. In recent decades, a set of indices has been created to measure landscape patterns from the perspective of area, shape, aggregation and diversity [30–32]. In fact, these landscape indices are algorithms for quantifying specific spatial characteristics of patches, classes of patches, or entire landscape mosaics [23]. FRAGSTATS—which integrates most of the landscape indices in categories of patch, class and landscape levels—is widely used to calculate landscape metrics [33].

Qingdao, like most of the eastern cities of China, has experienced intensified urbanization during the last few decades [34,35]. This urban land expansion has caused great transformation in land-use and landscape patterns. Agricultural land has been converted to urban land, causing the landscape patterns to become fragmented [36]. As a result, the patch size for the remaining habitat has been reduced, and the edge effects and the isolation of patches through the destruction of connecting corridors have increased [37]. Thus, it is necessary to analyze the effect of land-use changes and urbanization on landscape pattern changes in Qingdao and to apply the results in sustainable urban planning and policymaking.

Consequently, the objectives of this study are listed as follows: (1) Identify the spatial–temporal changes of land-use and landscape patterns in three urban development zones in Qingdao; (2) quantify the effect of land-use changes on landscape pattern changes; (3) evaluate the impact of urban expansion intensity on the spatial changes of landscape patterns. Studies of these questions could provide support for urban planning and urban sustainability.

2. Materials and Methods

2.1. Study Area

Qingdao (35°35′ N–37°09′ N and 119°30′ E–121°00′ E) is in the eastern Shandong Peninsula on the east coast of China, encompassing a total land area of 1174.56 km². Surrounded by the Yellow Sea to the east and south, Qingdao has a typical maritime climate with a mean annual temperature of 12.6 °C with a cold winter (an average temperature in December–February of 0.9 °C) and mild summers (an average temperature in June–August of 23.3 °C). The mean annual precipitation is around 662 mm, and it is mostly distributed in summer (http://www.weather.com.cn/; http://www.data.ac.cn/). A natural inlet of the Yellow Sea named Jiaozhou Bay is located on the southern coast of the Shandong Peninsula. The bay is 32 km long and 27 km wide, and it has witnessed a fast development of urbanization. Qingdao City has 6 districts. We excluded Huangdao, so the other 5 districts of Shinan, Shibei, Licang, Laoshan and Chengyang were selected as the study area (Figure 1). These districts are geographically contiguous and have experienced great landscape changes during the last few decades in the process of urbanization. The terrain in this area is high in the east and low in the west, with elevation between 0 and 1133 m. Like most cities in coastal China, Qingdao has had documented high economic growth rates since 1990. Except for the global financial crisis year (2007–2008), the GDP and GDP per capita of Qingdao grew annually by double digits from 1998 to 2009 [38]. Huge economic growth resulted in great urban expansion and landscape changes in the study area. The study of these changes in response to urban sprawl could provide a clear basis for promoting the coordinated development between economics and the urban ecosystem in Qingdao.

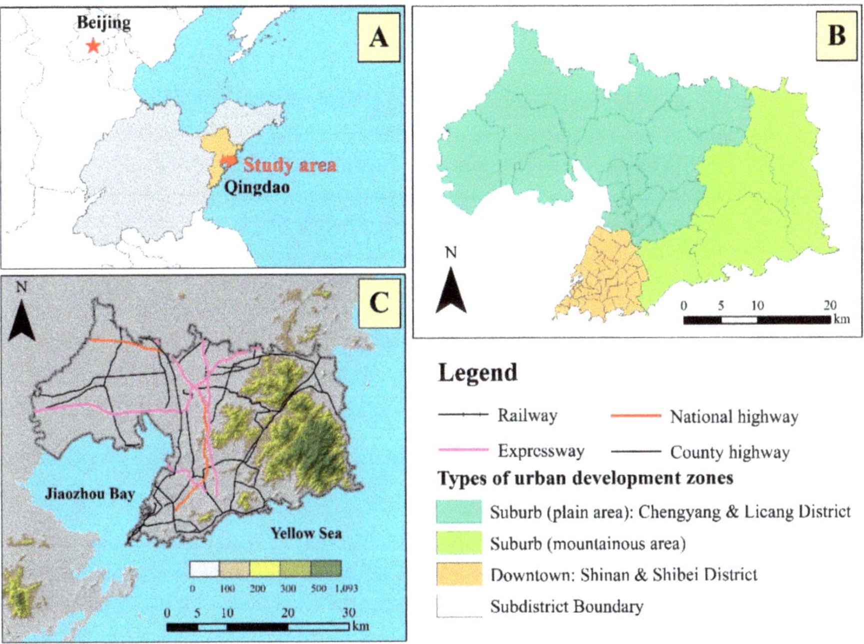

Figure 1. The location (**A**), three types of urban development zones in 2016 (**B**), and the terrain as well as main road (**C**) of the study area.

2.2. Mapping and Analysis of Land-Use/Land Cover

Landsat images downloaded from the Geospatial Data Cloud Platform (http://www.gscloud.cn/) were used to produce the land-use/cover map of the study area. Before images were selected, two criteria were considered: (1) the date acquired must have been in the growing season (typically from around March 18 to around November 25) (https://weatherspark.com/y/136066/Average-Weather-in-Qingdao-China-Year-Round) which could be beneficial for dividing bare land and built-up land from other land types; (2) the cloud coverage had to be less than 1%. Finally, three images of Landsat 5 TM (Thematic Mapper, one sensor type) and Landsat 8 OLI (Operational Land Imager, one sensor type) acquired on 7 October 1987, 26 October 2000, and 16 June 2016 were sought out. The classification system was determined as 6 one-class level land-use types—forest land, agricultural land, built-up land, grassland, bare land and waterbody—according to previous studies and the aim of learning the influence of urbanization on agricultural and forest land [1,15,39]. During the process of classification, the images were first radiometrically calibrated and clipped with the vector boundary of the study area in the ENVI 5.3 software (Exelis Visual Information Solutions, Inc., Boulder, CO 80301 USA). Then, the spectral indices of a normalized difference vegetation index (NDVI) and normalized difference built-up index (NDBUI) [40] were calculated and stacked into multispectral bands to improve classification. Next, regions of interest (ROI), evenly distributed in the study area for both types, were delineated through visual interpretation to train and test the subsequent classification, respectively. In total, the training ROIs were 28,082, 56,484 and 45,751 pixels, and the testing ROIs were 9911, 59,464 and 36,410 pixels, respectively, for images of 1987, 2000 and 2016. Finally, the maximum likelihood classification method was used to classify images into the six land-use categories mentioned above. The overall accuracy was 92.76%, 89.66%, and 93.07% for the years of 1987, 2000 and 2016, respectively. Moreover, the Kappa coefficients were 0.90, 0.86 and 0.90, respectively, and indicated that the accuracy of land-use classification maps was high.

The conversion matrix method in ENVI was used to analyze changes in land-use/cover. The conversion matrix could provide the magnitude and intensity of changes between the different land-use types.

2.3. Analysis of Landscape Patterns

Landscape patterns have been often analyzed through landscape metrics [41]. All the metrics, which are classified into the three levels of patch, class, and landscape, quantify such aspects as the fragmentation, heterogeneity, and connectivity of the landscape pattern [21]. However, it is vital to select appropriate metrics when analyzing landscape patterns. Some methods have been used for metric selection by predecessors. For example, in order to remove surplus or redundant metrics, Schindler et al. [42] tested methods including expert knowledge, decision tree analysis, principal component analysis, and principal component regression for metric selection. On the other hand, Su et al. [39] selected metrics for an agriculture landscape pattern analysis based on the three key criteria: (1) Metrics should have some ecological significance in categories of edge/density/area, shape, and contagion; (2) metrics should be comparable with previous studies; and (3) metrics should be of low redundancy.

To comprehensively examine the urban expansion on landscape patterns, 15 frequently used landscape-level metrics [43–46] were initially selected: total area (TA), patch density (PD), the largest patch index (LPI), total edge (TE), edge density (ED), large shape index (LSI), mean shape index (SHAPE_MN), area-weighted mean shape index (SHAPE_AM), mean fractal dimension index (FRAC_MN), area-weighted fractal dimension index (FRAC_AM), contagion (CONTAG), effective mesh size (MESH), splitting index (SPLIT), Shannon's diversity index (SHDI), and aggregation index (AI). Based on a two-tailed Pearson's correlation analysis, 4 metrics with the least correlation between each other were chosen in this study: the LPI, SHAPE-AM, PD and CONTAG. These metrics were categorized into three groups [47]: area and edge metrics (LPI), shape metrics (SHAPE-AM), and aggregation metrics (PD and CONTAG). The area and edge metrics describe the patches' number, size, and edge, and the shape metrics measure the geometric complexity of the patches. The aggregation metrics calculate landscape texture to describe the tendency of patch types to be spatially aggregated. All the metrics for each sub-district area were computed in FRAGSTATS 4.2 (http://www.umass.edu/landeco/research/fragstats/fragstats.html), and their change ratios were calculated based on the following Formula (1).

$$R_{\Delta t} = \frac{M_{t+\Delta t} - M_t}{M_t} \times 100\% \tag{1}$$

where $R_{\Delta t}$ is the change ratio of landscape metrics from time t to $t + \Delta t$; M_t and $M_{t+\Delta t}$ represent the value of metrics for the target unit in time t and $t + \Delta t$, respectively; and Δt is the time interval for the study period (measured in years).

2.4. Urban Expansion Intensity

In our study, the urban expansion intensity (UEI) was selected to identify urban expansion. UEI indicates the annual spatial changes of urban land area per unit area [24,25] by using Formula (2).

$$UEI_{\Delta t} = \frac{UA_{t+\Delta t} - UA_t}{TA} \times \frac{1}{\Delta t} \times 100\% \tag{2}$$

where $UEI_{\Delta t}$. is the urban expansion intensity during the time t and $t + \Delta t$; UA_t and $UA_{t+\Delta t}$ indicate the urban area of calculated unit at time t and $t + \Delta t$, respectively; Δt is the time interval of the study period (measured in years); and TA is the total land area for the target subdistrict.

2.5. Statistical and Regression Analysis

The two-tailed Pearson's correlation method was used to analyze correlations between change in land-use and landscape metrics' change ratio at different urban development zones and the whole study area. In addition, stepwise linear regressions were performed to quantify and compare the effects of land-use change (a unit of km^2 was designated) on landscape pattern changes. To evaluate the explanatory ability of models, the adjusted R^2 value was calculated for each model. All analyses were performed using IBM SPSS Statistics version 22 software (SPSS Inc., Chicago, IL, USA).

To assess the spatial influence of urban expansion on landscape pattern changes, geographically weighted regression (GWR) was applied to explore the relationship between urban expansion intensity and landscape metrics. Under the assumption that nearer observations exert more significant influences, GWR can accurately describe a spatial relationship by taking neighbor effects into consideration [48]. It expresses the spatial relationships of a variable by generating a group of local coefficients containing a local R^2 value, local residuals, local parameter estimates, and t-test values. The formulation of the GWR model can be expressed as Formula (3).

$$y_j = \beta_0\left(u_j, v_j\right) + \sum_{i}^{k} \beta_i\left(u_j, v_j\right)x_{ij} + \varepsilon_j \tag{3}$$

where u_j and v_j indicate the spatial coordinates of location j; $\beta_0\left(u_j, v_j\right)$ and $\beta_i\left(u_j, v_j\right)$ are the intercept value and the local-specific coefficient of the independent variable x_{ij}, respectively; k is the number of independent variable; and ε_j represents the error. GWR uses kernel bandwidth to calculate the spatial dependency range, and the weight for all observations are computed by a distance–decay function within the spatial range. This distance decay function is given as Formula (4).

$$w_{ij} = \exp\left(-d_{ij}^2/b^2\right) \tag{4}$$

where w_{ij} represents the weight of observation i relative to neighborhood observation j; d_{ij} is the Euclidean distance between locations i and j; and b represents the kernel bandwidth. Coarser kernel bandwidth leads to a slowed weight decrease with distance and generates more global relationships [49]. The weight value equals 1 when the observation points i and j are located at the same place; conversely, the weight value decay is close to 0 when the distance between them increases toward the spatial scope of kernel bandwidth.

The spatial regression model was performed with the GWR tool in the ArcGIS 10.3 software (Environmental Systems Research Institute, Redlands, CA, USA). The spatial map for the changes of each of the metrics during a time interval was taken as a dependent variable and the corresponding urban expansion intensity change was entered as an independent factor. There are two types of kernel function in the spatial regression tool: FIXED and ADAPTIVE. This means that the spatial context (the Gaussian kernel) can be a fixed distance or a function of a specified number of neighbors. In this study, the basic unit for the GWR modeling are sub-districts whose boundary and the area were irregular and unequal, i.e., the spatial distribution of observations was not uniform. Therefore, we adapted the

ADAPTIVE type of kernel matched up with the lowest Akaike information criteria (AICc) to determine the extent of the kernel.

3. Results

3.1. Changes in Land-Use/Cover

The land-use changes during the years 1987 and 2016 were examined by using spatial maps (Figure 2) and a conversion matrix (Appendix A, Tables A1–A4). The results showed that urbanization has led to significant land-use changes over the years 1987–2016. Therefore, the most obvious changes occurred in built-up land, with 14,416.2 ha (growth rate of 98.04%) and 27,635.04 ha (growth rate of 94.90%) increases in the periods of 1987–2000 and 2000–2016, respectively. As a result, the proportion of built-up land in the total land has increased from 13.03% to 50.28%. This rapid urbanization process has mainly occurred around the Jiaozhou Bay, which occupies a flat terrain area near the sea. With the dramatic upsurge in built-up land, other land-use types such as forest land, agricultural land and waterbodies have been bulldozed. For example, the forest land, agricultural land and waterbody had 6160.32, 6967.98, and 1744.74 ha converted to built-up land during 1987–2000, which account for 59.19%, 48.20% and 90.95% in the changed area of the three types, respectively. This trend was more pronounced from 2000 to 2016, with the above-mentioned proportion reached 95.21%, 69.33% and 91.67%, respectively. At the same time, the built-up land had transformed into other land-use types such as waterbody and forest land during this period. This could be due to river regulation, reservoir building, and old villages' reconstruction, all of which would generate urban green space.

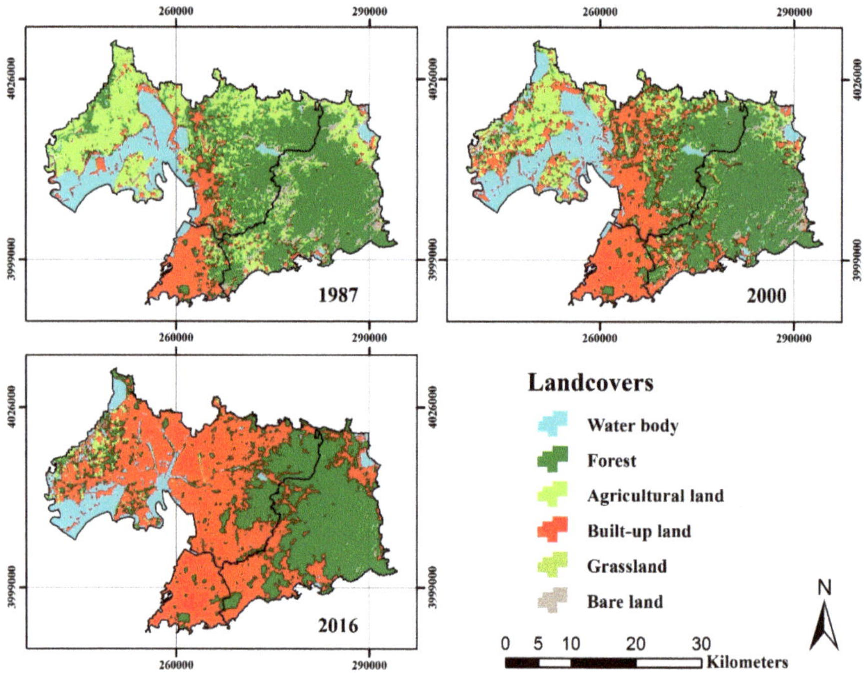

Figure 2. Land-use/cover maps in 1987, 2000, and 2016 in study areas.

Land-use changes exhibited distinct variations among different urban development zones. In the downtown area, the built-up land continuously increased from 6176.97 to 8275.32 ha during the years 1987–2016, which is 88.09% of the total downtown area. Among the augmenter in built-up land, the forest land had the largest contribution, with 49.06% land transformed during the period of 1987–2000 and 54.31% land converted from 2000 to 2016. The sea reclamation in the east of downtown is the most important inducement of the decrease of waterbody, whose area declined 62.26% and 88.69% in the periods of 1987–2000 and 2000–2016, respectively. At the same time, the other three land-use types experienced dramatic decreases because of urbanization progress. The grassland and bare land had disappeared, and the agricultural land almost vanished in 2016. However, there were also 159.75 and 301.41 ha of built-up land transformed into forest land during 1987–2000 and 2000–2016, respectively. This may benefit from the restoration and conservation of several big parks (e.g., Zhongshan Park, Beiling Mountain Forest Park, Jiading Mountain Park, Guanxiang Mountain Park and Shuangshan Park) and the construction of the coastal green belt.

In the suburban plain area, the built-up land area grew 141.83% and 123.48% during the years of 1987–2000 and 2000–2016, respectively, indicating an extreme urban expansion in this region. The increase in built-up land area was stronger across the transportation network around the Jiaozhou Bay. In other land-use types, the agricultural land had the greatest decline, with 7208.19 ha (33.45%) and 11,927.79 ha (83.17%) decreases in the periods of 1987–2000 and 2000–2016, respectively. Among the decrement, 74.39% and 75.77% of agricultural land, respectively, in the periods of 1987–2000 and 2000–2016 were converted to built-up land. Similarly, a large amount of forest land, grassland and bare land were transformed into urban land, resulting in continuous declinations for these types. On the contrary, during the years 1987–2000, the waterbody area increased 1631.25 ha, with the growth rate being 12.98%. This increase was mainly due to the construction of Jihongtan Reservoir in the north of this region. However, this trend turned in the opposite direction, with 35.66% (5063.49 ha) of the waterbody transformed into other land-use types—mainly built-up land.

In the mountainous suburban areas, the increase of built-up land in southwestern and northeastern Laoshan district reached 4254.84 ha (during 1987 and 2000) and 10,622.88 ha (during 2000 and 2016), which is 11.08% and 27.66% of the total area. At the same time, the forest land area increased 2332.71 ha from 1987 to 2016 under the influence of the Grain for Green Project and forest conservation policy in this region. In contrast, the agricultural land and bare land areas significantly declined 99.17% and 93.68%, respectively, in the years of 1987–2016, whereas the grassland completely vanished by the end of 2016. The waterbody area increased 73.17 ha from 1987 to 2016, but a great decrease of 246.42 ha followed because of the encroachment of urban land.

3.2. Spatial and Temporal Change of Landscape Patterns

The landscape patterns of different urban development zones in the study area were analyzed through four landscape metrics during the periods of 1987, 2000 and 2016. As shown in Figure 3, the changes were obviously different across different zones for the metrics: (1) The LPI significantly increased in the downtown and mountainous suburban areas, but it decreased in the suburban plain area from 1987 to 2000. However, the LPI had a huge increase in the suburban plain area when it slightly increased in the downtown and mountainous suburban areas. The significant increase of the LPI in the downtown and mountainous suburban areas was due to the expansion and infilling of built-up land and forest land, respectively (Figure 2). As a result, the largest patch of area increased. The changes of the LPI in the suburban plain area was because the largest patch of agricultural land decreased and was finally replaced by a large patch of built-up land. (2) The area-weighted mean shape index (SHAPE-AM) continually increased in the downtown area, whereas it firstly decreased and then enormously increased in the suburban plain area and exhibited a first increasing and then decreasing tendency in the mountainous suburban areas. (3) The urbanization transformed other land-use types into built-up land and promoted the appearance of green space in the downtown area. As a result, PD significantly decreased in the period of 1987–2000 and increased in the period of 2000–2016. However, there had been consistent, slight growth and a sustained, great decrease in the suburban plain and mountainous areas, respectively. The conversion of a large patch of agricultural land and waterbody into dispersed built-up land in the suburban area, and the infilling of forest land by disappeared bare land in the mountainous suburban areas was responsible for these changes. (4) CONTAG significantly increased in downtown and slightly increased in the other two places from 1987 to 2000. However, during 2000–2016, it increased a lot in suburban plain and mountainous areas but had a slight increase in downtown.

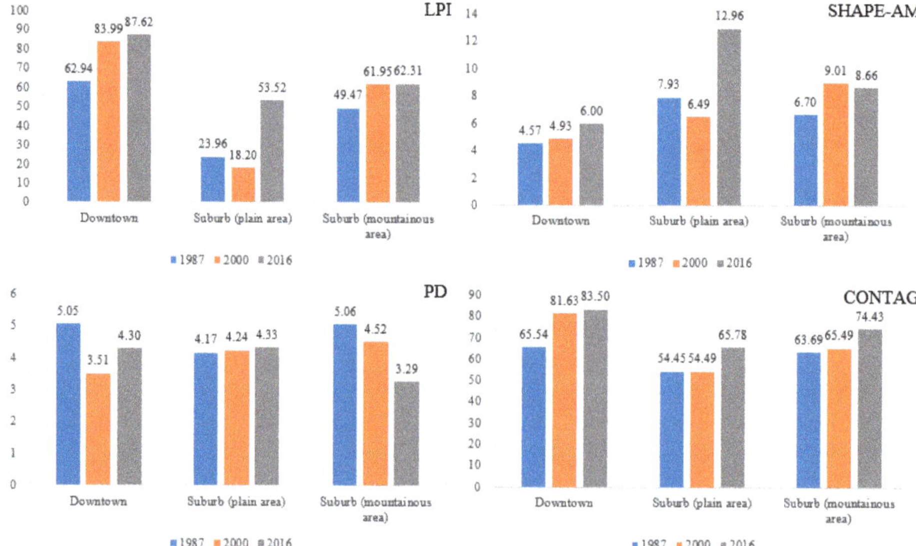

Figure 3. Changes of four landscape metrics in different urban development zones during 1987–2016. LPI: largest patch index; SHAPE-AM: area-weighted mean shape index; PD: patch density; CONTAG: contagion.

3.3. The Influence of Land-Use Change and UEI on Landscape Patterns

A correlation analysis was conducted to identify the relationships between selected landscape metrics and land-use changes and UEI (Tables 1 and 2). The results indicate that there have been great variations across three urban development zones and the whole study area at different time stages. From the perspective of the whole region, none of four metrics were related with land-use changes during 1987–2000. However, in the last period, the LPI was sensitive to waterbody, agricultural and built-up land; SHAPE-AM was not significantly related to grassland and bare land; PD was well-related with forest land and grassland; and CONTAG was only sensitive to built-up land. In terms of UEI, it was strongly related to all the landscape metrics in the two periods other than SHAPE-AM from 1987 to 2000.

In the downtown area, the area-edge metrics (LPI) and aggregation metrics (PD, CONTAG) were significantly related to all the land-use types except waterbody during the period of 1987–2000, whereas CONTAG was no longer significantly related to any land-use types, and the LPI and CONTAG were not sensitive to bare land during the period of 2000–2016. The shape metric (SHAPE-AM) related well to agricultural, grassland and bare land from 1987 to 2000, and it was correlated well with forest and built-up land in 2000 and 2016, respectively. In the suburban plain area, only the metrics of SHAPE-AM and PD were strongly related to forest and agricultural land during the stage of 2000–2016. In the mountainous suburban areas, only SHAPE-AM had a significant relationship with the land use of grassland from 1987 to 2000.

Table 1. Pearson's correlation coefficients between landscape metrics changes and urbanization related to land-use changes and urban expansion intensity in different urban development zones and the whole region from 1987 to 2000.

Urban Development Zones	Landscape Metrics	Waterbody	Forest	Agricultural Land	Built-up Land	Grassland	Bare Land	UEI
Downtown area	LPI	−0.069	−0.673 **	−0.705 **	0.797 **	−0.532 **	−0.776 **	
	SHAPE–AM	−0.073	0.039	−0.416 *	0.281	−0.377 *	−0.440 *	
	PD	−0.182	0.637 **	0.551 **	−0.602 **	0.567 **	0.566 **	
	CONTAG	−0.018	−0.595 **	−0.638 **	0.684 **	−0.563 **	−0.554 **	
Suburb (mountainous area)	SHAPE-AM	0.802	−0.750	0.143	0.415	−0.974 *	0.319	
The whole region	LPI	−0.124	−0.027	0.206	−0.126	0.071	−0.037	0.696 **
	SHAPE-AM	−0.067	0.067	0.023	0.025	−0.080	−0.147	0.068
	PD	0.057	0.119	−0.240	0.161	−0.079	−0.050	−0.691 **
	CONTAG	−0.081	−0.087	0.220	−0.127	0.045	−0.008	0.756 **

The superscripts ** and * indicate a significantly correlation at the 0.01 level (two-tailed) and the 0.05 level (two-tailed), respectively. LPI: largest patch index; SHAPE-AM: area-weighted mean shape index; PD: patch density; CONTAG: contagion; UEI: urban expansion intensity.

Table 2. Pearson's correlation coefficients between landscape metrics changes and urbanization related to land-use changes and urban expansion intensity in different urban development zones and the whole region from 2000 to 2016.

Urban Development Zones	Landscape Metrics	Waterbody	Forest	Agricultural Land	Built-up Land	Grassland	Bare Land	UEI
Downtown area	LPI	−0.391 *	−0.724 **	−0.706 **	0.853 **	−0.463 **	−0.149	
	SHAPE-AM	−0.023	0.695 **	0.304	−0.413 *	0.228	−0.192	
	PD	0.103	0.376 *	0.493 **	−0.407 *	0.354 *	0.141	
	CONTAG	−0.268	−0.193	−0.255	0.315	−0.098	0.316	
Suburb (plain area)	SHAPE-AM	−0.299	0.367	−0.582 **	0.388	0.309	0.013	
	PD	−0.308	0.642 **	−0.267	0.063	0.378	0.332	
The whole region	LPI	−0.341 *	−0.013	−0.343 **	0.392 **	0.108	−0.123	0.630 **
	SHAPE-AM	−0.291 *	0.429 **	−0.639 **	0.404 **	0.081	−0.240	0.319 *
	PD	−0.061	0.307 *	0.060	−0.214	0.323 *	0.246	−0.489 **
	CONTAG	−0.191	−0.154	−0.246	0.367 **	−0.150	−0.251	0.511 **

The superscripts ** and * indicate a significantly correlation at the 0.01 level (two-tailed) and the 0.05 level (two-tailed), respectively. LPI: largest patch index; SHAPE-AM: area-weighted mean shape index; PD: patch density; CONTAG: contagion; UEI: urban expansion intensity.

A stepwise regression model was used to assess which land-use was more important to the change of landscape patterns across different types of urban development zones (Tables 3 and 4). During the period of 1987–2000, the changes of built-up land, grassland, waterbody and bare land played a significant role in predicting the LPI and CONTAG, and forest land and bare land were more important in the prediction of SHAPE-AM and PD in the downtown area. However, grass land was the only land-use type that could effectively predict SHAPE-AM in the mountainous suburban areas. Furthermore, in the period of 2000–2016, built-up land, waterbody and agricultural land played an important role in predicting the LPI in downtown, forest and agricultural land were more important to SHAPE-AM and PD in the downtown and suburban plain areas. These results indicated that the influence of land-use changes on landscape patterns was significantly different across three urban development zones.

Table 3. Stepwise regression models of land-use changes during 1987–2000 and the related landscape metrics in different urban development zones. The regression equations were selected because the probabilities of their F values were less than 0.05.

Urban Development Zones	Dependent Variables	Final Model	Adjusted R^2	p Value
Downtown area	LPI	LPI = 1.925 + 22.801 built-up land + 89.567 grassland – 69.838 bare land	0.718	0.000
	SHAPE-AM	SHAPE-AM = -0.004 – 2.359 bare land + 0.504 forest land	0.285	0.002
	PD	PD = -0.852 + 6.439 forest land	0.387	0.000
	CONTAG	CONTAG = 0.010 + 0.220 built-up land	0.451	0.000
Suburb (mountainous area)	SHAPE-AM	SHAPE-AM = -0.179 – 1.195 grassland	0.922	0.026

LPI: largest patch index; SHAPE-AM: area-weighted mean shape index; PD: patch density; CONTAG: contagion; UEI: urban expansion intensity.

Table 4. Stepwise regression models of land-use changes during 2000–2016 and the related landscape metrics in different urban development zones. The regression equations were selected because of the probabilities of their F values were less than 0.05.

	Dependent Variables	Final Model	Adjusted R^2	p Value
Downtown area	LPI	LPI = -0.783 + 24.866 built-up land + 13.121 waterbody – 25.781 agricultural	0.871	0.000
	SHAPE-AM	SHAPE-AM = 0.138+ 0.784 forest land	0.466	0.000
	PD	PD = 1.474 + 18.293 agricultural	0.219	0.004
Suburb (plain area)	SHAPE-AM	SHAPE-AM = 0.290 – 0.047 agricultural	0.299	0.009
	PD	PD = 0.156 + 0.214 forest land	0.377	0.003

LPI: largest patch index; SHAPE-AM: area-weighted mean shape index; PD: patch density; UEI: urban expansion intensity.

3.4. Spatial Relationships between Urban Expansion Intensity and Landscape Patterns

The Pearson's correlation analysis indicated that the area–edge metrics (LPI) and aggregation metrics (PD and CONTAG) were strongly related to UEI from 1987 to 2016. However, the shape metric (SHAPE-AM) was sensitive to UEI only in the period of 2000–2016.

The spatial relationships between UEI and changes in the landscape metrics at the subdistrict level through GWR are presented in Figure 4. A high explanatory ability of UEI on landscape pattern changes was detected with adjusted R^2 values ranged from 0.377 to 0.622 for area–edge metrics (LPI) and aggregation metrics (PD and CONTAG). However, this explanatory ability exhibited obvious spatial variations. The highest adjusted R^2 values shifted from the core urban area in the first stage to the adjacent area where high urbanization was experienced during 2000–2016. It can be concluded that higher urban expansion intensity would explain more landscape pattern changes. The spatial patterns of coefficients revealed that the LPI and CONTAG were positively related, whereas PD was negatively related to UEI in most of the regions from 1987 to 2016.

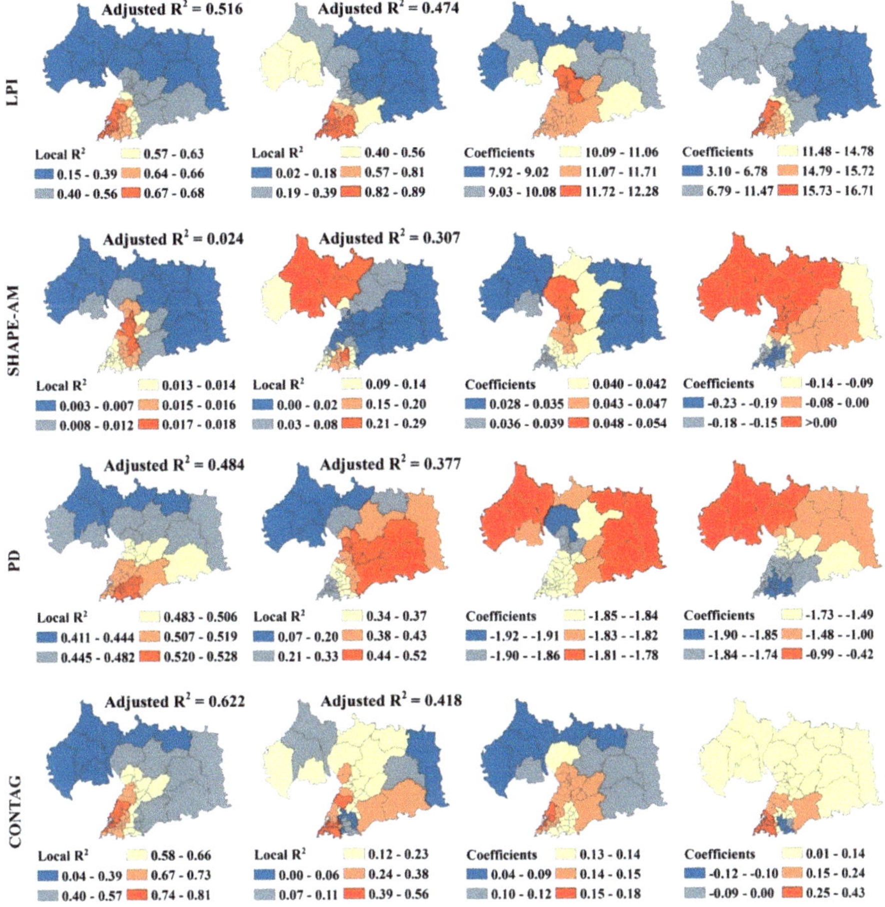

Figure 4. Spatial distribution of local R^2 value and correlation coefficients between urban expansion intensity and landscape metrics during 1987–2000 (the first and third columns) and 2000–2016 (the second and fourth column) acquired from geographically weighted regression (GWR) models. The Jenks Natural Breaks classification method was used, and 0 was set as the bread value when the value of coefficients crossed it.

4. Discussion

4.1. How Much Does Land-Use Change Affect Landscape Patterns?

Changes in land-use could effectively lead to landscape pattern changes [29,50]. The linear regression equations conducted in our study revealed that a strong relationship existed between land-use change and landscape pattern changes (Tables 3 and 4). This result is similar to the result of Dadashpoor et al. (2019 B) who concluded that landscape pattern changes have a strong relationship with changes in agricultural land, garden land, bare land and built-up land [15]. However, the relationship conducted in our study was varied among different urban development zones and time intervals.

In the downtown area, a 1 km^2 increase in built-up land caused a 22.801 increase in the LPI and a 0.220 increase in CONTAG from 1987 to 2000, and a 1 km^2 decrease in forest land was usually accompanied with decreases of 0.504 and 6.439 in SHAPE-AM and PD, respectively (Table 3).

In addition, changes in grass land and bare land also contributed to the landscape pattern changes. The reason for this might be that there was no great difference among the changes in these land-use types. However, the dominant influence on landscape patterns was attributed to agricultural land from 2000 to 2016. A 1 km^2 decrease in agricultural land led to a 25.781 increase in the LPI and caused 18.293 decrease in PD. Meanwhile, the forest land, built-up land and waterbody contributed to the LPI and SHAPE-AM.

In the suburban plain area, the changes in land-use had no significant influence on the four landscape metrics during 1987 and 2000. However, from 2000 to 2016, forest land and agricultural land had a slight influence on landscape pattern change. A 1 km^2 decrease in forest and agricultural land were usually together with a 0.214 decrease in PD and a 0.047 increase in SHAPE-AM. This may be related to the dispersion of change in forest land. This implied that changes of forest land can fragmentate the landscape [3]. The embezzlement of forest land in the west and the encroachment of built-up land in the center reduced the patch numbers and increased the concentration of patches.

In the mountainous suburban areas, only grassland had a significant effect on changes in SHAPE-AM. A 1 km^2 decrease in grassland caused a 1.195 increase in SHAPE-AM. This change in the shape properties of the landscape pattern was mainly due to the spatial distribution change in the grassland in this region.

4.2. How Does Urban Expansion Influence Landscape Pattern Changes?

Over the last three decades, the urban expansion in the area has transformed agricultural land, forest land, bare land and so on into built-up land, a transformation which has then led to changes in landscape patterns. The quantitative relationship between them was conducted through the GWR method (Figure 4). Ours findings show that the effect of urban expansion on landscape patterns has spatial and temporal heterogeneity. During the period of 1987–2000, urban expansion was mainly in the form of infilling and edge expansion from the urban core areas. Therefore, UEI has strong explanatory power for landscape patterns among these areas, and the coefficients indicated that the fragmentation level (reflected by PD and the LPI) was significantly, negatively correlated with the intensity of urban expansion. However, in regions far from downtown, the landscape pattern change was relatively less related with urban expansion than the urban core areas. Furthermore, the urban expansion even led to fragmentation in the northwestern area. This agrees with the result of the study of Dadashpoor et al. [15]. They concluded that the growth of built-up lands in a Tabriz metropolitan area has resulted in an increase in aggregation and integration in central and adjacent urban areas, whereas it was opposite in areas far away from built-up lands. Similarly, a few previous studies have also discovered that the expansion of built-up land has reduced the diversity of landscapes and created fragmentation and heterogeneity [1,14,51,52].

From 2000 to 2016, the high explanatory ability has shifted towards suburban areas along with a shift of urban focus, and the correlation coefficients also spatially changed (Figure 4). It can be said that urban expansion has had a significant effect on landscape pattern changes. Chowdhury et al. [4] and Islam [5] also believed that urbanization directly influences the degree of fragmentation in landscapes. Their results also indicated that the built-up landscape has been aggregating in the suburbs where the fragmentation of landscapes was once increased (Table A3 and Figure 4).

Both regression analyses revealed that urbanization has a strong influence on landscape pattern changes, and the effect was different in the three urban development zones. However, the GWR models could give a precise and detailed relationship at each location, which was suitable for questions with spatial heterogeneity [1,53,54].

4.3. Implications

Qingdao has experienced high economic growth since 1990, with an annual growth rate of GDP greater than 6% [38]. This sustained and rapid economic growth has greatly changed the spatial pattern of land-use types in the study area. As a result, a simplified land-use structure has emerged in the downtown area. Meanwhile, the profound influence from urbanization on landscape patterns has been shifting from the city center to the suburbs (Figure 4). According to weak sustainability theory, this region may still be considered sustainable since the total capital of natural and manufactured has increased [7]. However, this sustainability is not sustainable over the long haul. Thus, weak sustainable regions such as well-planned natural areas, agricultural systems and urban centers should be spatially configurated at a large scale in order to reach long-term sustainability, i.e., strong sustainability [6,7,10]. For this purpose, three recommendations are drawn for long-term sustainable development in Qingdao.

Firstly, priority should be given to protect large forest patches and to plant more trees in the downtown. Urban forests play an important role in improving environmental quality and public health in urban areas [55]. Just 20% more trees could double the benefits of urban forests and make cities more environmentally sustainable [50]. In 1987, there are various land-use types in this area, but almost only the forest and construction land have been retained over three decades of urbanization. As a result, the landscape diversity has decreased, and the AI of built-up land has intensified, both of which are not conducive to the sustainable development of urban landscapes. Thus, decision-makers should concentrate on protecting existing forest plots and exploiting new green spaces.

Secondly, the concept of a compact city should be adapted to guide urban planning in the suburban plain area. A compact city is an idea for sustainable urban development to avoid unorderly urban sprawl and achieve harmony between human beings and the natural environment [8,56]. During the process of urban expansion from 1987 to 2016, a large amount of contiguous agricultural land, forest land and water body patches was encroached on by built-up land, leading to a stronger concentration landscape of construction land. Therefore, rational land development strategies, such as establishing limited development zones and retaining large patches of forest and agricultural land with equal distribution, should be made in this region.

Thirdly, anthropogenic disturbances should be continued to be controlled in the suburban mountainous area to protect the forest landscapes and maintain the ecosystem resilience. Due to forest protection policies on Mountain Lao, the forest land area in this region grew from 1987 to 2016 (Table A4). However, it should be noticed that built-up land also had great increases of 179.05% and 149.67% from 1987 to 2000 and from 2000 to 2016, respectively, indicating an increase in anthropic influence. Therefore, decision-makers should focus on setting up prohibited development zones to protect the integrity of the forest ecosystem in this region.

5. Conclusions

This paper aimed to recognize changes in land-use and landscape patterns and to investigate the quantitative influence of land-use change and UEI on landscape patterns in three urban development zones in Qingdao during 1987–2016. For this purpose, firstly, a land-use transform matrix and the landscape metrics of the LPI, SHAPE-AM, PD and CONTAG were used to identify land-use and landscape pattern changes. The results suggested that the growth of the economy in Qingdao has transformed other land-use types such as agricultural and forest land to built-up land in the study area. Therefore, land-use change has caused aggregation and homogeneity in the landscape. Then, a correlation analysis and a stepwise regression were adapted to investigate the quantitative relationship between land-use change and landscape patterns. It has been observed that the influence magnitude of different land-use types on landscape patterns varied for different urban development zones and periods of time. In the downtown area, all the land-use types significantly influenced landscape patterns, and the change in agricultural and forest land had the greatest contribution, especially during the period 2000–2016. However, the agricultural and forest land, respectively, became the dominant factors of landscape pattern changes during 1987–2000 and 2000–2016 in the suburban plain area. The change in grass land had the biggest impact on landscape change in the mountainous suburban areas. Finally, to evaluate urbanization's impact on landscape pattern changes, GWR regression was used to identify the spatial relationship between UEI and the change in the landscape patterns. The result showed that the effect of UEI on landscape patterns has spatial and temporal heterogeneity.

From 1987 to 2000, the UEI strongly explained the change in the landscape patterns and made the landscape assemble faster in the downtown and adjacent areas. However, the a high explanatory ability shifted towards suburban areas during 2000–2016, and the correlation coefficients also spatially changed. The reason for this is the shifting of the focus of urban construction from downtown to the suburbs. Thus, it can be said that UEI has a significant effect on the landscape pattern changes. From this point of view, a compact city and protection policy should be adapted to different regions in the study area to achieve a strong sustainability of urban development.

Author Contributions: Formal analysis, J.Y.; funding acquisition, H.L.; resources, S.L.; writing—original draft, J.Y.; writing—review and editing, H.L.

Funding: This research was funded by National Natural Science Foundation of China, grant number 31800374, Natural Science Foundation of Shandong Province, grant number ZR2019BC083 and Research Foundation for Advanced Talents of Qingdao Agricultural University, grant number 663/1115007.

Conflicts of Interest: The authors declare no conflict of interest. The funders had no role in the design of the study; in the collection, analyses, or interpretation of data; in the writing of the manuscript, or in the decision to publish the results.

Appendix A

Table A1. Land-use/cover conversion matrix for the whole region during the period 1987–2016 (in ha).

		1987						
		Forest	Agricultural	Built-Up	Grassland	Bare Land	Water	Class Total
2000	Forest	36460.8	5893.11	658.71	923.76	1609.38	117.09	45662.85
	Agricultural	1863.63	13950.63	241.65	459.18	980.91	35.1	17531.1
	Built-up	6160.32	6967.98	12318.57	463.95	1464.39	1744.74	29119.95
	Grassland	172.26	64.8	5.49	29.16	135.72	0.09	407.52
	Bare land	1290.6	872.82	148.95	49.05	2486.79	21.42	4869.63
	Water	920.16	656.46	1330.38	322.02	64.71	11990.61	15284.34
	Class Total	46867.77	28405.8	14703.75	2247.12	6741.9	13909.05	
	Image Difference	−1204.92	−10874.7	14416.2	−1839.6	−1872.27	1375.29	
	Image Difference (%)	−2.57	−38.28	98.04	−81.86	−27.77	9.89	
		2000						
2016	Forest	33714.99	4623.57	1713.33	216.63	2659.50	509.04	43437.06
	Agricultural	248.58	1783.71	194.85	7.38	132.30	101.25	2468.07
	Built-up	11375.28	10918.53	25729.83	164.79	1762.38	6804.18	56754.99
	Grassland	11.07	50.13	10.62	18.63	9.99	6.30	106.74
	Bare land	25.83	11.97	31.05	0.09	233.19	1.89	304.02
	Water	287.10	143.19	1440.27	0.00	72.27	7861.68	9804.51
	Class Total	45662.85	17531.1	29119.95	407.52	4869.63	15284.34	
	Image Difference	−2225.79	−15063.03	27635.04	−300.78	−4565.61	−5479.83	
	Image Difference (%)	−4.87	−85.92	94.90	−73.81	−93.76	−35.85	

Table A2. Land-use/cover conversion matrix for the downtown area during the period 1987–2016 (in ha).

		1987						
		Waterbody	Forest	Agricultural	Built-Up	Grassland	Bare Land	Class Total
2000	Waterbody	190.71	4.14	0	1.08	3.06	0	198.99
	Forest	22.32	812.7	56.52	159.75	49.32	11.97	1112.58
	Agricultural	4.68	19.89	29.7	10.98	13.68	4.59	83.52
	Built-up	305.91	954.72	333.9	5997.24	138.06	213.57	7943.4
	Grassland	0	3.33	0.09	0	0	0	3.42
	Bare land	3.6	13.5	6.93	7.92	3.69	16.92	52.56
	Class Total	527.22	1808.28	427.14	6176.97	207.81	247.05	
	Image Difference	−328.23	−695.7	−343.62	1766.43	−204.39	−194.49	
	Image Difference (%)	−62.26	−38.47	−80.45	28.60	−98.35	−78.72	
		2000						
2016	Waterbody	3.42	8.91	0.09	9.99	0	0.09	22.50
	Forest	19.80	753.12	7.02	301.41	1.80	11.70	1094.85
	Agricultural	0	0.45	0	1.35	0	0	2.00
	Built-up	175.77	350.10	76.41	7630.65	1.62	40.77	8275.32
	Grassland	0	0	0	0	0	0	0
	Bare land	0	0	0	0	0	0	0
	Class Total	198.99	1112.58	83.52	7943.4	3.42	52.56	
	Image Difference	−176.49	−17.73	−81.52	331.92	−3.42	−52.56	
	Image Difference (%)	−88.69	−1.59	−97.61	4.18	−100	−100	

Table A3. Land-use/cover conversion matrix for the suburb (plain area) during the period 1987–2016 (in ha).

		1987						
		Waterbody	Forest	Agricultural	Built-Up	Grassland	Bare Land	Class Total
2000	Waterbody	11104.83	855.63	629.37	1257.75	318.33	34.38	14200.29
	Forest	77.94	13770.72	3206.34	276.3	497.61	342.09	18171.00
	Agricultural	22.23	1488.78	11903.85	159.75	406.62	360.36	14341.59
	Built-up	1355.76	4063.86	5361.84	5208.03	289.35	620.46	16899.3
	Grassland	0	56.07	32.04	5.49	29.07	66.78	189.45
	Bare land	8.28	280.71	416.34	80.64	30.87	278.64	1095.48
	Class Total	12569.04	20515.77	21549.78	6987.96	1571.85	1702.71	
	Image Difference	1631.25	−2344.77	−7208.19	9911.34	−1382.4	−607.23	
	Image Difference (%)	12.98	−11.43	−33.45	141.83	−87.95	−35.66	
		2000						
2016	Waterbody	7325.91	270.36	140.31	1329.39	0.00	70.83	9136.80
	Forest	467.82	10234.71	3335.22	1041.66	74.70	315.54	15469.65
	Agricultural	101.16	236.16	1776.69	189.18	7.29	103.32	2413.80
	Built-up	6299.10	7418.43	9037.62	14326.83	88.47	595.71	37766.16
	Grassland	6.30	11.07	50.13	10.62	18.99	9.99	107.10
	Bare land	0.00	0.27	1.62	1.62	0.00	0.09	3.60
	Class Total	14200.29	18171.00	14341.59	16899.3	189.45	1095.48	
	Image Difference	−5063.49	−2701.35	−11927.79	20866.86	−82.35	−1091.88	
	Image Difference (%)	−35.66	−14.87	−83.17	123.48	−43.47	−99.67	

Table A4. Land-use/cover conversion matrix for the suburb (mountainous area) during the period 1987–2016 (in ha).

		1987						
		Waterbody	Forest	Agricultural	Built-Up	Grassland	Bare Land	Class Total
2000	Waterbody	679.23	59.67	27.09	69.93	0.18	30.15	866.25
	Forest	16.74	21812.76	2621.52	221.85	372.69	1252.35	26297.91
	Agricultural	8.19	352.44	1972.44	70.92	37.26	613.71	3054.96
	Built-up	79.38	1138.14	1271.07	1101.60	34.38	630.27	4254.84
	Grassland	0.09	112.86	32.67	0	0.09	67.95	213.66
	Bare land	9.45	992.79	447.93	60.48	14.49	2186.28	3711.42
	Class Total	793.08	24468.66	6372.72	1524.78	459.09	4780.71	
	Image Difference	73.17	1829.25	−3317.76	2730.06	−245.43	−1069.29	
	Image Difference (%)	9.23	7.48	−52.06	179.05	−53.46	−22.37	
		2000						
2016	Waterbody	517.68	5.04	2.79	93.42	0	0.90	619.83
	Forest	20.70	22669.65	1273.86	369.81	139.41	2327.94	26801.37
	Agricultural	0.09	11.79	7.56	4.32	0.09	28.98	52.83
	Built-up	325.89	3585.87	1760.40	3757.86	74.07	1118.79	10622.88
	Grassland	0	0	0	0	0	0	0
	Bare land	1.89	25.56	10.35	29.43	0.09	234.81	302.13
	Class Total	866.25	26297.91	3054.96	4254.84	213.66	3711.42	
	Image Difference	−246.42	503.46	−3002.13	6368.04	−213.66	−3409.29	
	Image Difference (%)	−28.45	1.91	−98.27	149.67	−100	−91.86	

References

1. Li, H.; Peng, J.; Liu, Y.; Hu, Y. Urbanization impact on landscape patterns in Beijing City, China: A spatial heterogeneity perspective. *Ecol. Indic.* **2017**, *82*, 50–60. [CrossRef]
2. Zhang, Q.; Su, S. Determinants of urban expansion and their relative importance: A comparative analysis of 30 major metropolitans in China. *Habitat Int.* **2016**, *58*, 89–107. [CrossRef]
3. Luederitz, C.; Lang, D.J.; von Wehrden, H. A systematic review of guiding principles for sustainable urban neighborhood development. *Landsc. Urban Plan.* **2013**, *118*, 40–52. [CrossRef]
4. Sahana, M.; Hong, H.; Sajjad, H. Analyzing urban spatial patterns and trend of urban growth using urban sprawl matrix: A study on Kolkata urban agglomeration, India. *Sci. Total Environ.* **2018**, *628–629*, 1557–1566. [CrossRef] [PubMed]
5. Dadashpoor, H.; Azizi, P.; Moghadasi, M. Analyzing spatial patterns, driving forces and predicting future growth scenarios for supporting sustainable urban growth: Evidence from Tabriz metropolitan area, Iran. *Sustain. Cities Soc.* **2019**, *47*, 101502. [CrossRef]
6. Bihamta, N.; Soffianian, A.; Fakheran, S.; Gholamalifard, M. Using the SLEUTH Urban Growth Model to Simulate Future Urban Expansion of the Isfahan Metropolitan Area, Iran. *J. Indian Soc. Remote Sens.* **2014**, *43*, 407–414. [CrossRef]

7. Min, M.; Lin, C.; Duan, X.; Jin, Z.; Zhang, L. Spatial distribution and driving force analysis of urban heat island effect based on raster data: A case study of the Nanjing metropolitan area, China. *Sustain. Cities Soc.* **2019**, *50*, 101637. [CrossRef]

8. Dadashpoor, H.; Salarian, F. Urban sprawl on natural lands: Analyzing and predicting the trend of land use changes and sprawl in Mazandaran city region, Iran. *Environ. Dev. Sustain.* **2018**. [CrossRef]

9. Simwanda, M.; Murayama, Y. Spatiotemporal patterns of urban land use change in the rapidly growing city of Lusaka, Zambia: Implications for sustainable urban development. *Sustain. Cities Soc.* **2018**, *39*, 262–274. [CrossRef]

10. He, Y.; Song, J.; Hu, Y.; Tu, X.; Zhao, Y. Impacts of different weather conditions and landuse change on runoff variations in the Beiluo River Watershed, China. *Sustain. Cities Soc.* **2019**, *50*, 101674. [CrossRef]

11. Ali, R.; Bakhsh, K.; Yasin, M.A. Impact of urbanization on CO2 emissions in emerging economy: Evidence from Pakistan. *Sustain. Cities Soc.* **2019**, *48*, 101553. [CrossRef]

12. Alalouch, C.; Al-Hajri, S.; Naser, A.; Hinai, A. The impact of space syntax spatial attributes on urban land use in Muscat: Implications for urban sustainability. *Sustain. Cities Soc.* **2019**, *46*, 101417. [CrossRef]

13. Khalil, H.A. Energy Efficiency Strategies in Urban Planning of Cites. In Proceedings of the 7th International Energy Conversion Engineering Conference, Denver, CO, USA, 2–5 August 2009.

14. Weng, Y.C. Spatiotemporal changes of landscape pattern in response to urbanization. *Landsc. Urban Plan.* **2007**, *81*, 341–353. [CrossRef]

15. Dadashpoor, H.; Azizi, P.; Moghadasi, M. Land use change, urbanization, and change in landscape pattern in a metropolitan area. *Sci. Total Environ.* **2019**, *655*, 707–719. [CrossRef] [PubMed]

16. Shen, S.; Yue, P.; Fan, C. Quantitative assessment of land use dynamic variation using remote sensing data and landscape pattern in the Yangtze River Delta, China. *Sustain. Comput. Inform. Syst.* **2019**, *23*, 111–119. [CrossRef]

17. Jiao, M.; Hu, M.; Xia, B. Spatiotemporal dynamic simulation of land-use and landscape-pattern in the Pearl River Delta, China. *Sustain. Cities Soc.* **2019**, *49*, 101581. [CrossRef]

18. Hassan, M.M. Monitoring land use/land cover change, urban growth dynamics and landscape pattern analysis in five fastest urbanized cities in Bangladesh. *Remote Sens. Appl. Soc. Environ.* **2017**, *7*, 69–83. [CrossRef]

19. Schulp, C.J.E.; Levers, C.; Kuemmerle, T.; Tieskens, K.F.; Verburg, P.H. Mapping and modelling past and future land use change in Europe's cultural landscapes. *Land Use Policy* **2019**, *80*, 332–344. [CrossRef]

20. Wan, L.; Zhang, Y.; Zhang, X.; Qi, S.; Na, X. Comparison of land use/land cover change and landscape patterns in Honghe National Nature Reserve and the surrounding Jiansanjiang Region, China. *Ecol. Indic.* **2015**, *51*, 205–214. [CrossRef]

21. Cabral, A.I.R.; Costa, F.L. Land cover changes and landscape pattern dynamics in Senegal and Guinea Bissau borderland. *Appl. Geogr.* **2017**, *82*, 115–128. [CrossRef]

22. Qian, J.; Peng, Y.; Luo, C.; Wu, C.; Du, Q. Urban Land Expansion and Sustainable Land Use Policy in Shenzhen: A Case Study of China's Rapid Urbanization. *Sustainability* **2016**, *8*, 16. [CrossRef]

23. Liu, X.; Li, X.; Chen, Y.; Tan, Z.; Li, S.; Ai, B. A new landscape index for quantifying urban expansion using multi-temporal remotely sensed data. *Landsc. Ecol.* **2010**, *25*, 671–682. [CrossRef]

24. Xu, X.; Min, X. Quantifying spatiotemporal patterns of urban expansion in China using remote sensing data. *Cities* **2013**, *35*, 104–113. [CrossRef]

25. Li, S.; Liu, X.; li, Z.; Wu, Z.; Yan, Z.; Chen, Y.; Gao, F. Spatial and Temporal Dynamics of Urban Expansion along the Guangzhou–Foshan Inter-City Rail Transit Corridor, China. *Sustainability* **2018**, *10*, 593. [CrossRef]

26. Ren, Z.; He, X.; Zheng, H.; Wei, H. Spatio-Temporal Patterns of Urban Forest Basal Area under China's Rapid Urban Expansion and Greening: Implications for Urban Green Infrastructure Management. *Forests* **2018**, *9*, 272. [CrossRef]

27. Sun, X.; Crittenden, J.C.; Li, F.; Lu, Z.; Dou, X. Urban expansion simulation and the spatio-temporal changes of ecosystem services, a case study in Atlanta Metropolitan area, USA. *Sci. Total Environ.* **2018**, *622–623*, 974–987. [CrossRef]

28. Su, S.; Jiang, Z.; Zhang, Q.; Zhang, Y. Transformation of agricultural landscapes under rapid urbanization: A threat to sustainability in Hang-Jia-Hu region, China. *Appl. Geogr.* **2011**, *31*, 439–449. [CrossRef]

29. Zhang, W.; Wang, M.Y. Spatial-temporal characteristics and determinants of land urbanization quality in China: Evidence from 285 prefecture-level cities. *Sustain. Cities Soc.* **2018**, *38*, 70–79. [CrossRef]

30. O'Neill, R.V.; KrummelR, J.R.; Gardner, H.; Sugihara, G.; Jackson, B.; DeAngelis, D.L.; Milne, B.T.; Turner, M.G.; Zygmunt, B.; Christensen, S.W.; et al. Indices of landscape pattern. *Landsc. Ecol.* **1988**, *1*, 153–162.
31. Matsushita, B.; Xu, M.; Fukushima, T. Characterizing the changes in landscape structure in the Lake Kasumigaura Basin, Japan using a high-quality GIS dataset. *Landsc. Urban Plan.* **2006**, *78*, 241–250. [CrossRef]
32. Feng, Y.; Liu, Y.; Tong, X. Spatiotemporal variation of landscape patterns and their spatial determinants in Shanghai, China. *Ecol. Indic.* **2018**, *87*, 22–32. [CrossRef]
33. McGarigal, K.; Cushman, S.; Ene, E. Fragstats V4: Spatial Pattern Analysis Program for Categorical and Continuous Maps. 2012. Available online: http://www.umass.edu/landeco/research/fragstats/fragstats.html (accessed on 2 November 2019).
34. Verburg, P.H.; Veldkamp, A.; Fresco, L.O. Simulation of changes in the spatial pattern of land use in China. *Appl. Geogr.* **1999**, *19*, 211–233. [CrossRef]
35. Ma, Q.; He, C.; Wu, J.; Liu, Z.; Zhang, Q.; Sun, Z. Quantifying spatiotemporal patterns of urban impervious surfaces in China: An improved assessment using nighttime light data. *Landsc. Urban Plan.* **2014**, *130*, 36–49. [CrossRef]
36. Tan, M.; Li, X.; Xie, H.; Lu, C. Urban land expansion and arable land loss in China—A case study of Beijing–Tianjin–Hebei region. *Land Use Policy* **2005**, *22*, 187–196. [CrossRef]
37. Botequilha Leitão, A.; Ahern, J. Applying landscape ecological concepts and metrics in sustainable landscape planning. *Landsc. Urban Plan.* **2002**, *59*, 65–93. [CrossRef]
38. Zhang, M.; Rasiah, R. Qingdao. *Cities* **2013**, *31*, 591–600. [CrossRef]
39. Su, S.; Ma, X.; Xiao, R. Agricultural landscape pattern changes in response to urbanization at ecoregional scale. *Ecol. Indic.* **2014**, *40*, 10–18. [CrossRef]
40. Zha, Y.; Gao, J.; Ni, S. Use of normalized difference built-up index in automatically mapping urban areas from TM imagery. *Int. J. Remote Sens.* **2003**, *24*, 583–594. [CrossRef]
41. McGarigal, K.; Marks, B.J. *FRAGSTATS: Spatial Pattern Analysis Program for Quantifying Landscape Structure*; USDA: Washington, DC, USA, 1995.
42. Schindler, S.; Wehrden, H.; Poirazidis, K.; Hochachka, W.M.; Wrbka, T.; Kati, V. Performance of methods to select landscape metrics for modelling species richness. *Ecol. Model.* **2015**, *295*, 107–112. [CrossRef]
43. Schindler, S.; Poirazidis, K.; Wrbka, T. Towards a core set of landscape metrics for biodiversity assessments: A case study from Dadia National Park, Greece. *Ecol. Indic.* **2008**, *8*, 502–514. [CrossRef]
44. Plexida, S.G.; Sfougaris, A.I.; Ispikoudis, I.P.; Papanastasis, V.P. Selecting landscape metrics as indicators of spatial heterogeneity—A comparison among Greek landscapes. *Int. J. Appl. Earth Obs. Geoinf.* **2014**, *26*, 26–35. [CrossRef]
45. Feng, Y.; Liu, Y. Fractal dimension as an indicator for quantifying the effects of changing spatial scales on landscape metrics. *Ecol. Indic.* **2015**, *53*, 18–27. [CrossRef]
46. Niesterowicz, J.; Stepinski, T.F. On using landscape metrics for landscape similarity search. *Ecol. Indic.* **2016**, *64*, 20–30. [CrossRef]
47. McGarigal, K. Fragstats v4: Spatial Pattern Analysis Program for Categorical and Continuous Maps-Help Manual. University of Massachusetts, Amherst. 2014. Available online: http://www.umass.edu/landeco/research/fragstats/fragstats.html (accessed on 2 November 2019).
48. Brunsdon, C.; Fotheringham, A.S.; Charlton, M.E. Geographically Weighted Regression: A Method for Exploring Spatial Nonstationarity. *Geogr. Anal.* **1996**, *28*, 281–298. [CrossRef]
49. Li, S.; Zhao, Z.; Miaomiao, X.; Wang, Y. Investigating spatial non-stationary and scale-dependent relationships between urban surface temperature and environmental factors using geographically weighted regression. *Environ. Model. Softw.* **2010**, *25*, 1789–1800. [CrossRef]
50. Endreny, T.; Santagata, R.; Perna, A.; de Stefano, C.; Rallo, R.F.; Ulgiati, S. Implementing and managing urban forests: A much needed conservation strategy to increase ecosystem services and urban wellbeing. *Ecol. Model.* **2017**, *360*, 328–335. [CrossRef]
51. Herold, M.; Goldstein, N.C.; Clarke, K.C. The spatiotemporal form of urban growth: Measurement, analysis and modeling. *Remote Sens. Environ.* **2003**, *86*, 286–302. [CrossRef]
52. Abdullah, S.A.; Nakagoshi, N. Changes in landscape spatial pattern in the highly developing state of Selangor, peninsular Malaysia. *Landsc. Urban Plan.* **2006**, *77*, 263–275. [CrossRef]
53. Fotheringham, A.S.; Brunsdon, C.; Charlton, M.E. *Geographically Weighted Regression: The Analysis of Spatially Varying Relationships*; Wiley: New York, NY, USA, 2002.

54. Su, S.; Xiao, R.; Zhang, Y. Multi-scale analysis of spatially varying relationships between agricultural landscape patterns and urbanization using geographically weighted regression. *Appl. Geogr.* **2012**, *32*, 360–375. [CrossRef]
55. Gerrish, E.; Watkins, S.L. The relationship between urban forests and income: A meta-analysis. *Landsc. Urban Plan.* **2018**, *170*, 293–308. [CrossRef]
56. Jim, C.Y. Sustainable urban greening strategies for compact cities in developing and developed economies. *Urban Ecosyst.* **2013**, *16*, 741–761. [CrossRef]

 *sustainability*

Article

An Analysis of Urban Land Use/Land Cover Changes in Blantyre City, Southern Malawi (1994–2018)

John Mawenda [1,*] , Teiji Watanabe [1,2] and Ram Avtar [1,2]

1. Graduate School of Environmental Science, Hokkaido University, Sapporo, Hokkaido 060-0810, Japan; teiwata@mac.com (T.W.); ram@ees.hokudai.ac.jp (R.A.)
2. Faculty of Environmental Earth Science, Hokkaido University, Sapporo, Hokkaido 060-0810, Japan
* Correspondence: mawendaj@gmail.com; Tel.: +81-70-4233-1012

Received: 17 February 2020; Accepted: 16 March 2020; Published: 18 March 2020

Abstract: Rapid and unplanned urban growth has adverse environmental and social consequences. This is prominent in sub-Saharan Africa where the urbanisation rate is high and characterised by the proliferation of informal settlements. It is, therefore, crucial that urban land use/land cover (LULC) changes be investigated in order to enhance effective planning and sustainable growth. In this paper, the spatial and temporal LULC changes in Blantyre city were studied using the integration of remotely sensed Landsat imageries of 1994, 2007 and 2018, and a geographic information system (GIS). The supervised classification method using the support vector machine algorithm was applied to generate the LULC maps. The study also analysed the transition matrices derived from the classified map to identify prominent processes of changes for planning prioritisation. The results showed that the built-up class, which included urban structures such as residential, industrial, commercial and public installations, increased in the 24-year study period. On the contrary, bare land, which included vacant lands, open spaces with little or no vegetation, hilly clear-cut areas and other fallow land, declined over the study period. This was also the case with the vegetation class (i.e., forests, parks, permanent tree-covered areas and shrubs). The post-classification results revealed that the LULC changes during the second period (2007–2018) were faster compared to the first period (1994–2007). Furthermore, the results revealed that the increase in built-up areas systematically targeted the bare land and avoided the vegetated areas, and that the vegetated areas were systematically cleared to bare land during the study period (1994–2018). The findings of this study have revealed the pressure of human activities on the land and natural environment in Blantyre and provided the basis for sustainable urban planning and development in Blantyre city.

Keywords: urban planning; LULC change; transition matrix; systematic transition; Blantyre city

---

1. Introduction

Studying land use/land cover (LULC) is vital for enhancing our understanding of global environmental change and sustainability [1]. In recent times, LULC changes have remarkably intensified due to increased anthropogenic processes such as urbanisation [2]. In 2018, 55 per cent of the world's population lived in urban areas, a proportion that is anticipated to reach 68 per cent by 2050. Almost 90 per cent of this expected growth will occur in Asia and Africa, especially in medium and small-sized cities [3]. In Africa alone, the urban population was 42.9 per cent in 2018 and is projected to reach 56 per cent by 2050 [4]. Moreover, in sub-Saharan Africa, where over 200 million people in urban areas reside in informal settlement, a higher urbanisation rate at 4.5 per cent annually has been reported [5]. The growth of urban areas has a significant influence on the global and regional environments, including LULC changes, and has implications for environmental, social and economic sustainability [6]. Rapid and unplanned urbanisation has had dire consequences, such as a reduction in vegetation cover and loss of biodiversity, as habitats for species become fragmented through the conversion of land for infrastructure development [7].

Like other anthropogenic–environmental interactions, urban LULC changes are due to a myriad of factors, as no single factor can account for these changes. The interactions are different in every region, but most scholars agree that most LULC changes are influenced by specific economic, demographic, socio-political and environmental conditions [8,9]. These factors are usually interrelated. For example, the economic and social advantages found in urban areas compared to rural areas attract many people to cities, leading to rapid population growth that contributes to the over-exploitation of natural resources for settlement and livelihoods.

Malawi is among the world's least developed and one of the most densely populated countries in Africa. Like other developing countries in sub-Saharan Africa, Malawi has been experiencing progressive urban growth since it attained its independence on 6th July 1964. Malawi's system of governance changed from a one-party rule dictatorship (1964–1993) to a multiparty democratic system from 1994 to the present. During the democratic rule, the urban population increased 2-fold from 1,095,419 in 1998 to 2,115,867 in 2018 in its four main cities of Lilongwe, Blantyre, Mzuzu and Zomba, where over 70 per cent lives in informal settlements [10,11]. Presently, Malawi is ranked among the top 10 countries in the world projected to have the largest population increase in both its rural and urban areas [4]. The urban population as a percentage of the national population was 16 per cent in 2018 and it is projected to rise to 30 per cent by 2030 and 50 per cent by 2050 [10–12].

Despite that several studies have been done on LULC changes in Malawi [13–18], comprehensive studies on urban LULC change in Malawi's cities remain scarce, as such, the understanding of urbanisation is primarily based on population figures only. This inadequacy of LULC change information constrains effective economic and environmental planning, resulting in uninformed policy decisions [16]. This is particularly prominent in low-income countries like Malawi, which commits most of its resources to address urgent needs such as poverty reduction at the expense of maintaining a vibrant LULC system [16].

Over the past three decades, advances in remote sensing technologies have expedited LULC change studies. By obtaining satellite imagery over a period of time, remote sensing methods can be utilised to analyse historical LULC changes [19]. Regardless of some shortfalls, such as spatial and spectral confusion of the urban areas [20], remote sensing remains a reliable source of data to support LULC studies [21]. Therefore, studies of urban LULC changes using remote sensing data, such as Landsat, are essential for land management and urban land use planning, especially in developing countries where they can provide fundamental and cost-effective information that is not available from other sources [22,23].

Although cities in Malawi actively engage in planning—including the development of master plans that provide an overview of spatial and infrastructural intentions, strategic plans that outline the broad ways in which thematic issues are to be addressed and investment plans that list priority infrastructure—the plans are barely recognised by the public in urban jurisdictions and often lack effective implementation mechanisms. This is usually blamed on a lack of resources [24]. Additionally, the existence of multiple institutions in the land administration in cities causes the disorder, which arises from unclear authority and mandates over land including development control [24]. Furthermore, cities in Malawi neither exercise control over key sectors (such as utility providers) nor have the authority to make other agencies align their plans to citywide coordination, and thus are unable to facilitate cross-sectoral planning [12]. As a result, cities continue to experience various challenges, such as rapid urban growth, which can lead to irreversible LULC changes, the proliferation of unplanned settlements and environmental degradation.

In Blantyre city, the second-largest city in Malawi by population and size, the urban LULC situation is unclear and has not yet been quantified. There is also not enough empirical data on urban LULC changes over time. Therefore, the aim of this study is to analyse urban LULC changes from 1994 to 2018 in Blantyre city using remotely sensed Landsat data in order to support sustainable urban planning. The year of 1994 was chosen as a start date because it is when Malawi changed its governance system from a one-party system to a multiparty system. The year of 2018 was chosen as the end date because it is when the field verification was taken, while the year of 2007 was chosen as an intermediate date to illustrate the rates of change.

The study also identified random and systematic transitions derived from the classified maps to focus on prominent signals of change in the study area. The study period 1994–2018 was chosen in order to understand how the democratic governance systems influenced the LULC changes in urban environments of Blantyre city. The significance of the study is threefold: (1) the study will reveal the underlying human processes in the urban environment and their interactions; (2) the information generated can assist with managing the pressures of human activities and urban developments on the land, and (3) the results of this study can be used as baseline information to determine future urban land use and for setting policy priorities to promote inclusive and equitable urban development. Ultimately, this will help to realise well-balanced urban growth for citizens and the environment in Blantyre city.

2. Materials and Methods

2.1. Study Area

Blantyre is the oldest and second-largest city in Malawi. It was established by the Scottish missionaries in the 1870s and declared a planning area in 1897. It is the main commercial city hosting most of the private sector headquarters in the country. The city lies within the latitude 15°47′10″ S and longitude 35°00′20.10″ E, as shown in Figure 1. It covers an area of 240 km^2 [11].

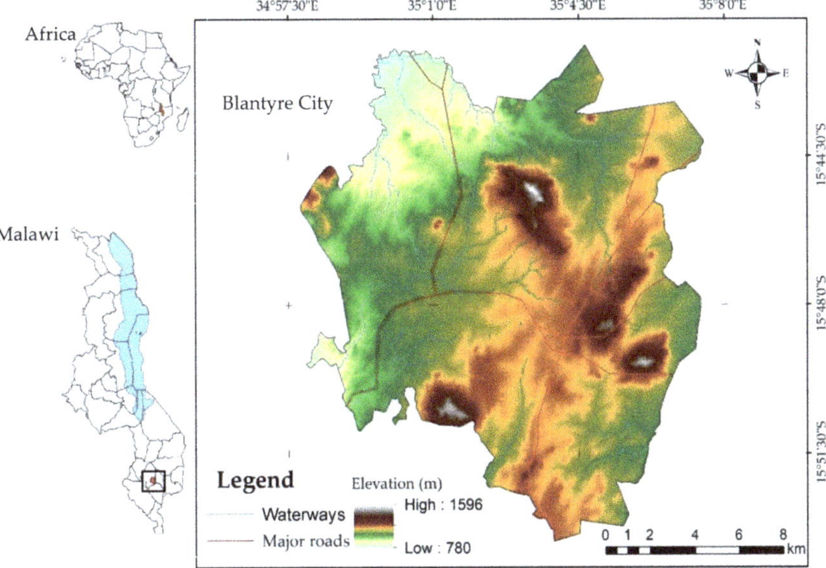

Figure 1. Study area—Blantyre city in southern Malawi, Africa. Data source: Diva GIS website and The Shuttle Radar Topographic Mission (SRTM) Digital Elevation Models (DEM) data.

The city is located in rugged terrain with multiple hills and river systems which have a direct effect on the development of the city [25]. Its topography ranges from an elevation of about 780 to 1612 m above sea level (m.a.s.l), as shown in Figure 1. The main mountains, such as Ndirande, Soche, Bangwe and Michiru, are also designated forest reserves [25].

In 1966, Blantyre was the prime city in Malawi with 109,461 people, seconded by Zomba, the then capital city, with only 19,666 inhabitants. With the transfer of the capital city from Zomba to Lilongwe, as part of a deliberate government policy to redistribute the urban population and implement spatially balanced development across the country [26], Lilongwe started to grow rapidly. The population grew from a mere 19,425 in 1966 to nearly 100,000 in a decade. By the year 2008, Lilongwe had overtaken Blantyre by about 7700 people. Lilongwe city continued to grow ahead of Blantyre city in 2018, registering a population of 989,318 against 800,852 in Blantyre [11]. Blantyre city currently hosts almost 5 per cent of the national population and has the highest population density in the country with 3328 people/km^2 [11]. Over 91 per cent of the city residents are below the age of 45 years, and over 65 per cent reside in the informal settlements [11].

2.2. Remote Sensing Data

The remotely sensed data used for this study were Landsat Thematic Mapper (TM) and Landsat Operational Land Imager (OLI) with 30 m spatial resolution and a 16 day repeat cycle [27]. The cloud-free Landsat images covering the study area (path 167, row 71) were downloaded from the United States Geological Survey (USGS) Earth Explorer https://earthexplorer.usgs.gov/, as shown in Table 1. The acquisition quality of these images was high (09, meaning no quality issues/errors detected) [28]. The data were acquired during the dry season, which starts from May to October, to best distinguish the spectral signature of different land cover types, and near-anniversary dates were chosen for consistency between the two time periods.

Table 1. Landsat data used for the analysis of land use/land cover (LULC) changes in the study area.

Satellite	Sensor	Path/row	Spatial Resolution (m)	Date of Acquisition	Source
Landsat 5	TM	167/71	30	17th September 1994	USGS
Landsat 5	TM	167/71	30	4th August 2007	USGS
Landsat 8	OLI	167/71	30	19th September 2018	USGS

USGS: United States Geological Survey.

2.3. Image Processing

The Landsat images were an L1T product (systematically, geometrically and topographically corrected). The study area was clipped as a vector file onto the Landsat image in ArcGIS Pro software (ESRI, Redland, CA, USA). In order to reduce confusion between spectral features and improve overall image classification, the radiometric and atmospheric corrections were applied using the Fast Line-of-sight Atmospheric Analysis of Spectral Hypercubes (FLAASH) Model in ENVI 5.2 software (Harris Geospatial, Broomfield, CA, USA) prior to image classification [29].

2.4. Image Classification

The support vector machine (SVM) classifier in the supervised classification scheme in ArcGIS Pro software (ESRI, Redland, CA, USA) was employed to derive four major LULC classes, namely bare land, built-up area, vegetation and water, as shown in Table 2.

Table 2. Description of LULC classes.

LULC Classes	Description
Bare land	Vacant lands, open areas with little or no vegetation, exposed rocks, quarry, hilly clear-cut areas and other idle fallow land sometimes illegally used for agriculture.
Built-up area	Urban structures of all types: residential, industrial, commercial, public installations, roads/highways and other similar facilities.
Vegetation	Forest, parks and permanent tree-covered areas, temporary croplands, grassland, shrubs and other idle lands along streams.
Water	Permanent open water, especially manmade dams/ponds.

The classes were adopted from [30] with minor modifications to suit the study area. The training samples for each LULC class were determined by comparing the false colour composites using spectral bands 5, 4, 3, 2 and 1 for Landsat 5 TM and 6, 5, 4, 3 and 2 for Landsat 8 OLI as shown in Figure 2. Reference was also made to the Google Earth archived images and the ground control points collected during the study area field visit in August 2018 when collecting the training samples.

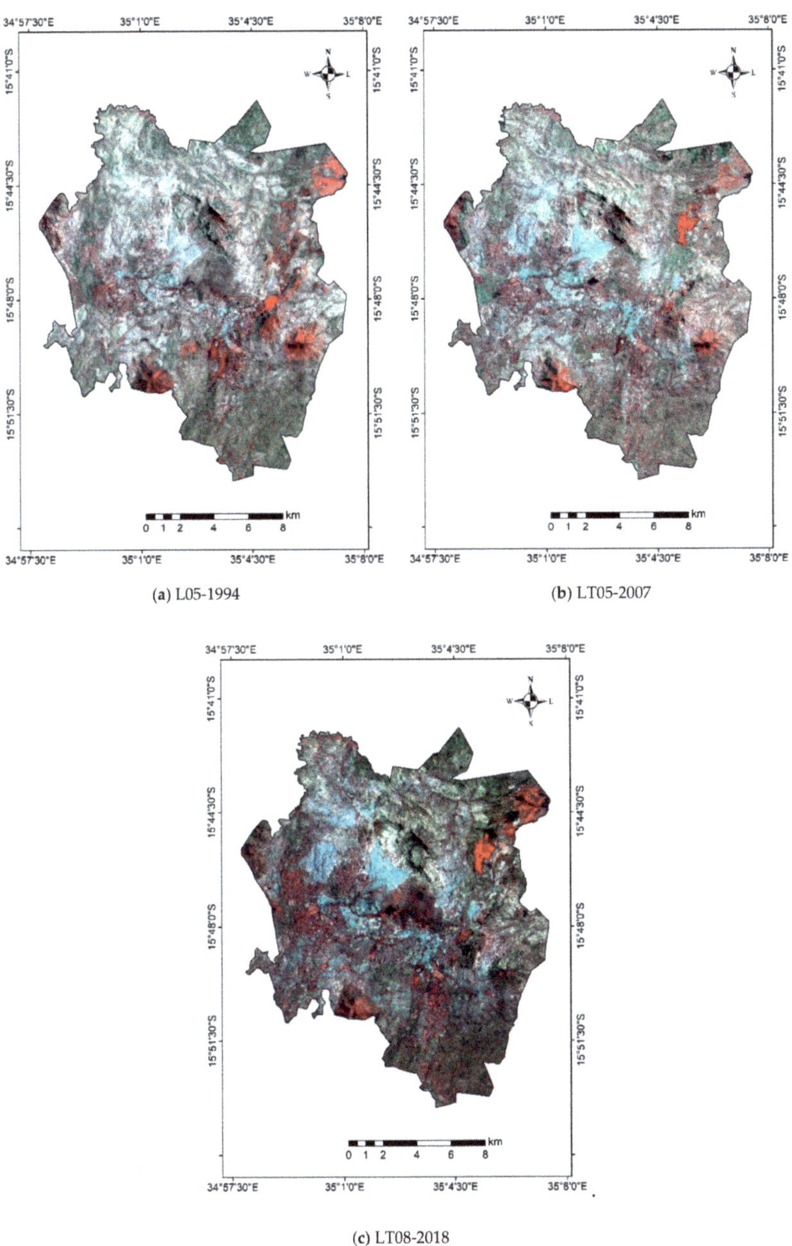

(**a**) L05-1994

(**b**) LT05-2007

(**c**) LT08-2018

Figure 2. False colour composite for (**a**) 1994, (**b**) 2007 and (**c**) 2018.

2.5. Accuracy Assessment

Accuracy assessment is a prerequisite step in the LULC classification process. This aims to quantitatively determine how effectively pixels are grouped into the correct feature classes in the area under investigation [31]. Accuracy assessment, therefore, compares the classified results to geographically referenced data that are presumed to be true. This comparison is typically achieved in a statistically rigorous fashion using error matrices. For this study, the reference data for 1994, 2007 and 2018 maps were collected from Google Earth image archives for the respective years. A set of at least 100 reference points were collected using stratified random sampling based on the sizes of the LULC classes for 1994, 2007 and 2018 classified images. The producer's, user's and overall accuracies, as well as a non-parametric Kappa index, were calculated [32,33]. The Kappa coefficient was computed using Equation (1).

$$k = \frac{N \sum_{i=1}^{r} X_{ii} - \sum_{i=1}^{r} (X_{i+} X_{+1})}{N^2 - \sum_{i=1}^{r} (X_i X_{+i})} \tag{1}$$

where N is the total number of observations in the error matrix, r is the number of rows and columns in the error matrix, X_{ii} is the number of observations in row i and column i (i.e., the diagonal elements), X_{+i} is the marginal total of row I, X_{i+} is the marginal total of column i.

A Kappa value greater than 0.80 indicates excellent agreement, while a value between 0.4 and 0.80 indicates moderate agreement and a value less than or equal to 0.4 indicates poor agreement between classification categories [32].

2.6. Annual Rates of Change

For an improved description of LULC changes, we calculated the annual rate of change. This process measures the amount of LULC change per year and was calculated according to the approach proposed by [34] and described in Equation (2).

$$R = \left\{ \frac{1}{t_2 - t_1} \right\} * \left\{ \ln \frac{A_2}{A_1} \right\} \tag{2}$$

where R is the rate of change, A_1 and A_2 are the area in km^2 at the beginning and end of the analysis period, and t_1 and t_2 correspond to the time in years from start to finish, respectively.

2.7. Land Use/Land Cover Change Detection

LULC change detection was carried out by a post-classification comparison of bi-temporal maps. This is perhaps the most common approach for change detection [35] and has been successfully used by many studies, such as [1,19,36–41]. The cross-tabulation matrix for the LULC changes for the first period (1994–2007), the second period (2007–2018) and the overall period (1994–2018) were generated. From the transitional matrices, the diagonal values in each matrix indicate persistency between LULC classes from initial time t_1, and the later time t_2, while the off-diagonal entries indicate the transition from one LULC class to the other. The gain was also calculated through the difference between the total value of each LULC class for the later period P_{+j}, and persistency P_{jj}, while the loss was the difference between the total for the initial time P_{i+}, and persistency P_{ii}. On the other hand, swapping tendencies were calculated as two times the minimum value of the gains and losses for each LULC class [42]. Furthermore, the concept of systematic and random change was applied to the transition matrix in order to detect significant inter-category transitions [42]. An abrupt or unique process of LULC change is regarded as a random transition while that characterised by a permanent, stable or common process of change is regarded as a systematic transition. In order to determine whether a transition is random or systematic, the expected value was compared with the observed value in the matrix. A transition is said to be random if the difference between the observed and expected values is zero, while it is a systematic transition if it is not near zero. The expected values in terms of gain, G_{ij}, and loss, L_{ij}, were calculated using Equations (3) and (4) for ($\forall i \neq j$).

$$G_{ij} = \left(P_{+j} - P_{jj}\right)\left(\frac{P_i+}{\sum_{i=1,\ i\neq j}^{J} P_{j+}}\right) \tag{3}$$

where $\left(P_{+j} - P_{jj}\right)$ is the observed total gross gain of category j, P_i+ is the size of category i in the start date, 1994 in our case. $\sum_{i=1,\ i\neq j}^{J} P_{j+}$ is the sum of the sizes of all categories, excluding categories of j, in the start date (1994).

$$L_{ij} = (P_{i+} - P_{ii})\left(\frac{P_{+j}}{\sum_{j=1,\ j\neq j}^{J} P_{+j}}\right) \tag{4}$$

where $(P_{i+} - P_{ii})$ is the observed total gross loss of category i, P_{+j} is the size of category j in the last date, 2018 in our case. $\sum_{j=1,\ j\neq j}^{J} P_{+j}$ is the sum of all sizes of all categories, excluding j, in the later date (2018).

Transitions with observed values that were larger than expected values by 1 per cent point under both random processes of gain and losses were identified as systematic transitions [43,44].

3. Results

3.1. Land Use/Land Cover Classification

The spatiotemporal patterns of LULC classes considered in this study for 1994, 2007 and 2018 derived from the Landsat images using the SVM classifier are shown in Figure 3. Table 3 illustrates the temporal changes in the LULC from 1994 to 2018. From the results, it is clear that bare land was the main LULC category in the study area. In 1994, bare land represented 83.90 per cent of the total area followed by vegetation (10.90 per cent) and built-up area (5.10 per cent), respectively.

There have been distinct urban LULC changes in Blantyre city over the 24-year period. The changes are mainly characterised by an increase in a built-up area and a decrease in bare land and vegetation. The water class remained constant (less than 0.5 per cent of the study area) throughout the study period. Overall, the built-up area increased at an annual rate of 5.34 per cent from 1994 to 2018. In the first period (1994–2007), the built-up class increased at a lower annual growth rate of 4.01 per cent compared to almost 7 per cent annual rate during the second period (2007–2018). The increase in the built-up area signifies the urban growth in Blantyre city.

Bare land declined at the annual rate of 0.52 per cent over the study period. However, during the first period (1994–2007), the bare land declined at a lower annual rate of 0.69 per cent compared to the later period at 0.86 per cent. Lastly, with the overall annual declining rate of 1.69 per cent, vegetation registered a lower decline at the annual declining rate of 0.84 per cent in the first period (1994–2007) compared to the sharp decline at an annual rate of 2.69 per cent in the later period (2007–2018).

Table 3. Land use/land cover (LULC) distribution for Blantyre city.

No.	LULC Classes	1994		2007		2018	
		Area (km²)	%	Area (Km²)	%	Area (km²)	%
1	Bare land	199.88	83.90	194.08	81.50	176.63	74.20
2	Built-up	12.19	5.10	20.53	8.60	43.93	18.40
3	Vegetation	25.88	10.90	23.20	9.70	17.26	7.20
4	Water	0.15	0.10	0.28	0.10	0.29	0.10
	Total	238.1	100.0	238.1	100.0	238.1	100.0

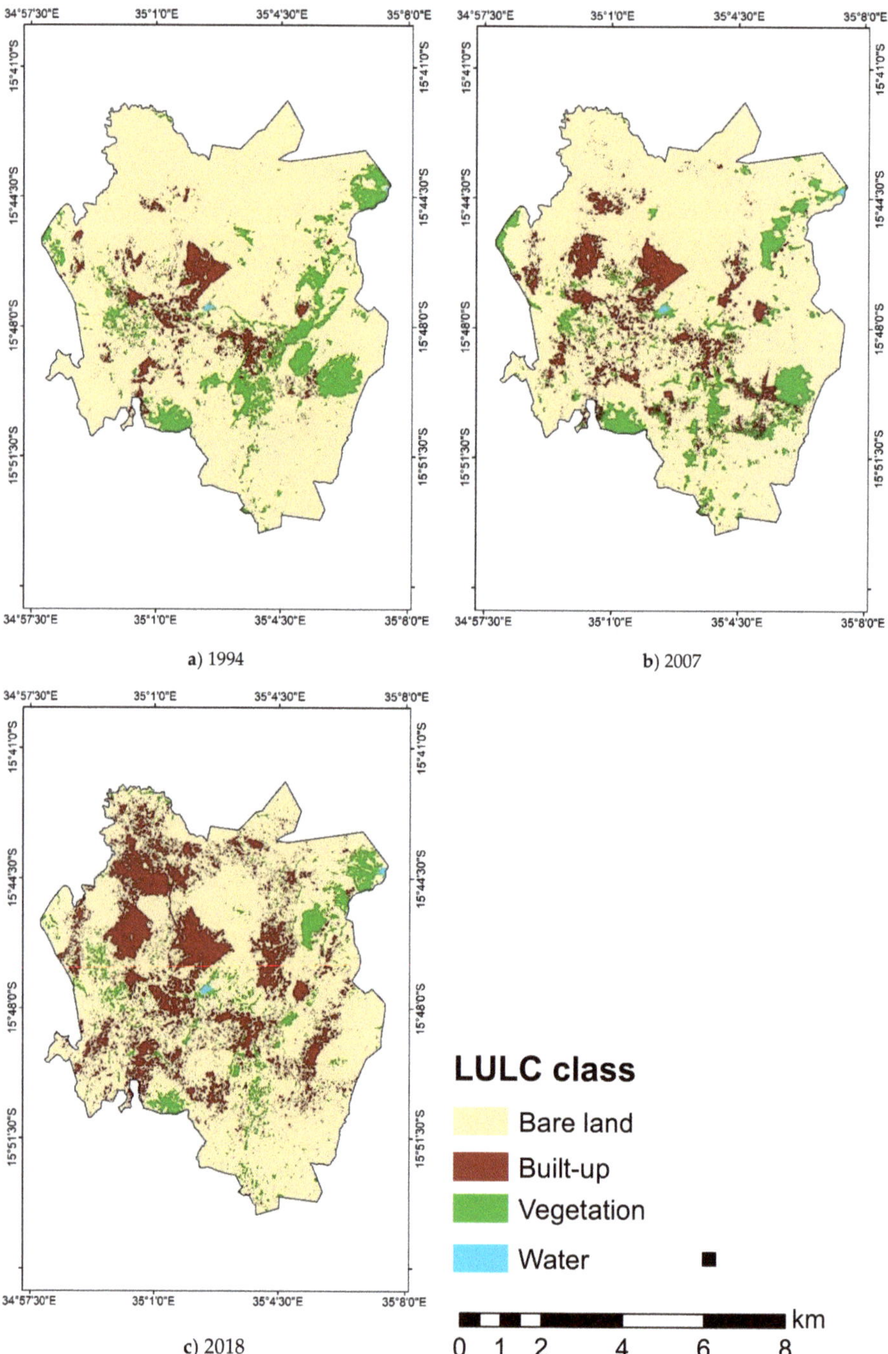

Figure 3. Classified land use/land cover (LULC) maps for Blantyre city in (**a**) 1994, (**b**) 2007 and (**c**) 2018.

3.2. Accuracy Assessment of the LULC Classification

Table 4 shows the results of the accuracy assessment for the 1994, 2007 and 2018 classified LULC maps. The overall accuracies were 89.5, 87.5 and 86.6 per cent for 1994, 2007 and 2018 images, respectively. These accuracies met the minimum overall accuracy set out by USGS (i.e., greater than 85 per cent). Hence, the classified results can be used as a data source for post-classification comparisons and further analyses. The Kappa coefficient of 0.79, 0.76 and 0.75 for the respective maps showed a good agreement between the classified map and reference data [33].

Table 4. Accuracy assessment for 1994, 2007 and 2018 classified maps of Blantyre city.

LULC Class	1994		2006		2018	
	User's Accuracy (%)	Producer's Accuracy (%)	User's Accuracy (%)	Producer's Accuracy (%)	User's Accuracy (%)	Producer's Accuracy (%)
Bare land	100	86	91	96	88	91
Built-up	67	100	76	72	79	71
Vegetation	61	100	100	63	90	75
Water	100	100	70	100	90	100
Overall Accuracy (%)	89.5		87.5		86.6	
Kappa Coefficient	0.79		0.76		0.75	

3.3. Post-Classification Change Detection

Table 5 shows the LULC transition matrix derived from the classified maps for the first period (1994–2007), the second period (2007–2018) and the overall period (1994–2018). In Table 5a–c, the proportions of the LULC classes that were persistent are provided in the diagonal entries of each matrix. During the first period, as shown in Table 5a, only 18 per cent of the landscape changed from one class to the other compared to about 24 per cent in the later period (2007–2018) as shown in Table 5b, affirming that changes were more rapid in the later period than the former. Further analysis of the LULC transition matrix for the entire period from 1994 to 2018 revealed that about 74 per cent of the landscape remained stable, meaning that only 26 per cent of the study area exhibited the transitions from one LULC class to a different LULC class.

The matrices were also analysed to understand the transition budget by calculating the gains, losses, total change, swap and net change as shown in Table 6. These calculations suggest the degree of interactions among the LULC classes in the landscape. The analysis of Table 6 revealed that the higher proportion of bare land and vegetation classes experienced a swap type of change throughout the study period. For example, 87 and 71 per cent of the total changes for the bare land class swapped locations during the first and second intervals, respectively. Similarly, almost 91 and 74 per cent of the total change for the vegetation class also swapped locations during the respective intervals. The swapping tendencies for bare land and vegetation classes could be attributed to the simultaneous reforestation and deforestation activities within the landscape. Unlike the bare land and vegetation classes, the built-up class experienced a larger proportion of the net change during all the intervals, taking up 55 and 70 per cent of the total change in the class during the first and second study interval, respectively.

The losses and the resulting swap proportions for the built-up class would not be expected but it could have resulted from the errors of commission and omission during the classification process. The error of commission and omission are the complement of the user's and the producer's accuracies, respectively, as shown in Table 4. In addition, the Landsat spatial resolution of 30 m and the edge effects in the determination of the changes could also result in some misclassifications along with the difficulties with spectral confusion between the built-up and bare land classes in the urban environment [19–23].

Table 5. Transition matrix of land use/land cover (LULC), expressed in percentage.

(a) 1994–2007

		2007					
		Bare land	Built-up	Vegetation	Water	Total	Losses
1994	Bare land	74.27	4.62	5.42	0.04	84.34	10.07
	Built-up	1.32	3.58	0.05	0.00	4.95	1.37
	Vegetation	6.37	0.16	4.11	0.02	10.66	6.55
	Water	0.00	0.00	0.00	0.06	0.06	0.00
	Total	81.97	8.35	9.57	0.12	100.00	
	Gains	7.70	4.77	5.46	0.06		

(b) 2007–2018

		2018					
		Bare land	Built-up	Vegetation	Water	Total	Losses
2007	Bare land	66.85	11.38	3.72	0.00	81.95	15.10
	Built-up	1.93	6.40	0.04	0.00	8.37	1.97
	Vegetation	6.36	0.00	3.05	0.02	9.42	6.38
	Water	0.01	0.00	0.00	0.26	0.27	0.01
	Total	75.14	17.78	6.81	0.28	100.00	23.45
	Gains	8.29	11.38	3.77	0.02	23.45	

(c) 1994–2018

		2018					
		Bare land	Built-up	Vegetation	Water	Total	Losses
1994	Bare land	66.80	13.66	3.83	0.03	84.33	17.53
	Built-up	1.03	3.90	0.02	0.00	4.95	1.05
	Vegetation	7.31	0.38	2.94	0.02	10.66	7.71
	Water	0.00	0.00	0.00	0.06	0.06	0.00
	Grand Total	75.13	17.95	6.80	0.12	100.00	26.29
	Gains	8.33	14.04	3.85	0.06	26.29	

Table 6. Summary of land use/land cover (LULC) changes, expressed in percentage.

(a) 1994–2007

LULC Class	Persistence	Gain	Loss	Total change	Swap	Net change
Bare land	74.27	7.7	10.07	17.77	15.4	2.37
Built-up	3.58	4.77	1.37	6.14	2.74	3.4
Vegetation	4.11	5.46	6.55	12.01	10.92	1.09
Water	0.06	0.06	0	0.06	0	0.06
Total	82.02	17.99	17.99	17.99	14.53	3.46

(b) 2007–2018

LULC Class	Persistence	Gain	Loss	Total change	Swap	Net change
Bare land	66.85	8.29	15.1	23.39	16.58	6.81
Built-up	6.4	11.38	1.97	13.35	3.94	9.41
Vegetation	3.04	3.77	6.38	10.15	7.54	2.61
Water	0.26	0.02	0.01	0.03	0.02	0.01
Total	76.55	23.46	23.46	23.46	14.04	9.42

(c) 1994–2018

LULC Class	Persistence	Gross gain	Gross loss	Total change	Swap	Net change
Bare land	66.8	8.33	17.53	25.86	16.66	9.2
Built-up	3.9	14.04	1.05	15.09	2.1	12.99
Vegetation	2.94	3.85	7.71	11.56	7.7	3.86
Water	0.06	0.06	0	0.06	0	0.06
Total	73.7	26.28	26.29	26.285	13.23	13.055

3.4. Detection of Random and Systematic Transitions

This study further analysed the transition matrix for 1994 and 2018 in order to identify the most significant changes among the LULC classes. The expected gains under the random process of gain were calculated using Equation (3). The expected gains and the differences between the observed and expected proportions are given in Table 7a,b. In Table 7b, numbers closer to zero indicate a random transition between LULC classes, while numbers further from zero implied systematic transition [42]. In this study, only transitions that were larger than 1 per cent point were identified as systematic transitions [43,44].

Table 7. Inter-category gains for transitions in the landscape.

1994	2018			
	Bare Land	**Built-Up**	**Vegetation**	**Water**
(a) Expected gains under the random process of gain (%)				
Bare land	66.80	12.46	3.64	0.05
Built-up	2.68	3.90	0.21	0.00
Vegetation	5.67	1.57	2.94	0.01
Water	0.03	0.01	0.00	0.06
(b) Difference between observed and expected gains under the random process of gain (%)				
Bare land	0.00	1.20 *	0.20	−0.02
Built-up	−1.61 *	0.00	−0.19	−0.00
Vegetation	1.64 *	−1.19 *	0.00	0.02
Water	−0.03	−0.01	0.00	0.00

* indicates a systematic transition.

The differences between the observed and expected gains under the random processes of gain from the bare land class to the built-up class (1.20) and the vegetation class to the bare land class (1.64) was positive and above 1 per cent point, meaning they occurred systematically rather than randomly.

The differences between the observed and expected gains for vegetation to the built-up area (−1.19) and built-up to bare land (−1.61) were negative and above 1 per cent point. This implies that when vegetation gains, new vegetation systematically avoids gaining from built-up areas. Similarly, when bare land gains, it systematically avoids gaining from the built-up class [42].

The equivalent relationship between the different LULC classes, which examines the actual and expected losses under the random process of loss, is shown in Table 8a, b.

Table 8. Inter-category losses for transitions in the landscape.

1994	2018			
	Bare Land	**Built-Up**	**Vegetation**	**Water**
(a) Expected losses under the random process of loss (%)				
Bare land	66.80	12.65	4.79	0.09
Built-up	0.96	3.90	0.09	0.00
Vegetation	6.22	1.49	2.94	0.01
Water	0.00	0.00	0.00	0.06
(b) Difference between observed and expected losses under the random process of loss (%)				
Bare land	0.00	1.01 *	−0.96	−0.05
Built-up	0.07	0.00	−0.07	−0.00
Vegetation	1.09 *	−1.10 *	0.00	0.01
Water	0.00	0.00	0.00	0.00

* indicates a systematic transition.

The difference between observed and expected losses under the random process of loss for the bare land class to the built-up class and the vegetation class to the bare land class was 1.01 and 1.09 per cent points, respectively. This infers that the bare land class and the vegetation class lost systematically to the built-up and the bare land classes, respectively. Likewise, the difference between the observed and expected losses for vegetation to the built-up area (−1.10) was above 1 per cent point but negative, implying that vegetation systematically avoided losing to the built-up class [42].

In a nutshell, there has been a systematic conversion of the bare land class to the built-up class as well as a systematic degradation of the vegetation class to the bare land class in Blantyre city over the 24-year study period. This was demonstrated by the concurrent incidences of the systematic gains and losses [44]. This means that 13.66 per cent transition of bare land to the built-up class and the degradation of 7.31 per cent of vegetation to bare land were due to a systematic process of change, as shown in Table 5c. The LULC map showing systematic changes from bare land to built-up area and vegetation to bare land, as well as persistence and other random changes, is shown in Figure 4.

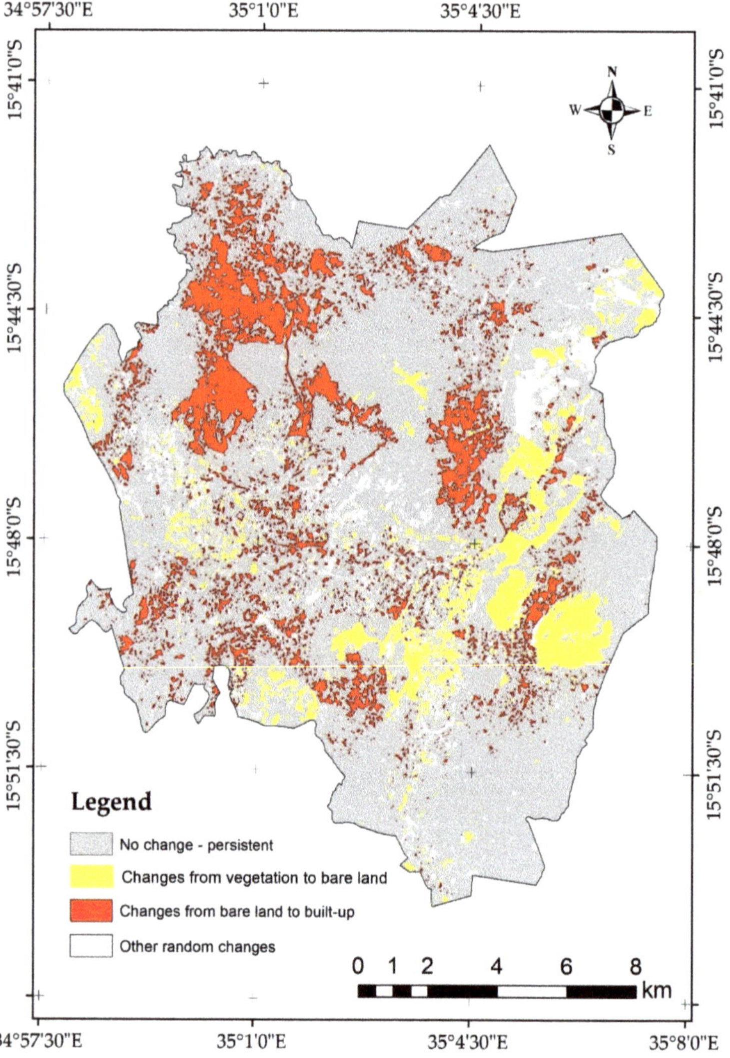

Figure 4. Spatial distribution of random and systematic land use/land cover (LULC) changes and persistence during 1994–2018.

4. Discussion

The results indicated that there were increased anthropogenic-induced urban LULC changes in Blantyre city over the past 24 years. This was substantiated by the observed accelerated increase in built-up area from 4.1 per cent during the initial period to almost 7 per cent in the later period, with an overall annual change at 5.3 per cent. This is similar to the accelerated declining rates observed for the vegetation and bare land classes (Section 3.1). The increase in the built-up class is comparable to other studies conducted in the sub-Saharan African cities, including the Dakar metropolitan area in Senegal, Nairobi city in Kenya, and Harare city in Zimbabwe, which experienced an increase in built-up areas at the annual rates of 9.6, 9.5 and 4.7 per cent, respectively, between the years 1990 and 2014 [45]. The observed increase in the built-up class and the decline in vegetation and bare land classes, respectively, have several implications for sustainable urban planning of the area.

Firstly, the observed annual growth rate of 5.3 per cent for the built-up class in the study area surpassed the urban population growth rate estimated at 2.3 per cent between 1998 and 2018 [10,11]. This means that the urban growth in Blantyre City, with a land consumption ratio of 2.3, is becoming more expansive than compact [46]. This kind of growth creates profound repercussions for environmental sustainability in the city and also prevents the city from enjoying high social interaction due to close integration of communities, and easy access to social-economic facilities [47]. Further to that, the excessive increase in the built-up area indicates an increment of more impervious surfaces in the city. The increase of impervious surfaces causes a decrease in the groundwater recharge as well as high reflection of solar radiation back to the atmosphere, hence contributing to environmental problems such as urban flooding and urban heat islands [48]. This development puts the city at risk and calls for better management to guarantee sustainable urban development.

Secondly, the vegetation loss observed in this study signifies the loss of green spaces (such as forests and parks) [49]. The vegetation loss results in declining ecosystem services, such as air and water purification, flood mitigation services, noise reduction and climate regulation, including urban cooling [50]. It also causes soil degradation [51], which leads to the formation of gullies and derelict landscapes. In addition, such losses increase residents' vulnerability to environmental stress due to the loss of non-material benefits that people obtain from ecosystems through spiritual enrichment, cognitive development, reflection, recreation and aesthetic experience, as well as their role in supporting knowledge systems, social relations and aesthetic values [52].

Lastly, the decline in bare land exerts enormous pressure on land suitable for urban development. Recently, over 10,000 families were reported to have built their houses illegally in areas such as wetlands, steep slopes and river/stream buffer zones in the city [53]. Such unplanned settlements are prone to multiple hazards such as floods and landslides, which have increased in frequency and intensity in recent times due to climate change and climate variability [54]. Similar situations were also observed in Mzuzu City, northern Malawi, where people have encroached into areas prone to multiple hazards [55].

Furthermore, the analysis of the transitional matrix derived from the 1994 and 2018 classified maps, as shown in Table 5c, has revealed the two-way systematic pathway of changes as illustrated in Figure 5.

These systematic processes suggest that the gain in built-up targets bare land and avoids vegetation. This can be explained by the fact that most land under vegetated areas, such as Ndirande, Soche and Kanjedza forest reserves, as well as the Mudi catchment area, are protected leaving bare land vulnerable to conversion to built-up. Additionally, most vegetated areas have unsuitable land for development, such as wetlands, steep slopes and river/stream buffer zones.

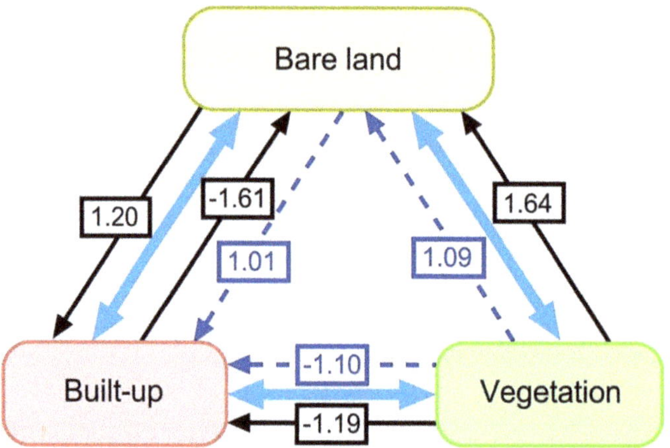

Figure 5. Systematic pathway of land use/land cover (LULC) changes from 1994 to 2018 in Blantyre city. Solid black arrows indicate expected gains, while the broken blue arrows indicate expected losses under the random processes of gains and losses.

Likewise, the gain in bare land systematically targets the vegetation class and avoids the built-up class. This systematic process has caused the loss of vegetative cover in the study area. This loss could be explained by anthropogenic activities, such as wood extraction for firewood, brick burning and lapses in management [56–58]. For example, the majority of Malawians have no access to a reliable energy source. The proportion of households with access to electricity in Malawi has increased from 8 per cent in 2010 to 11 per cent in 2017 [59]. Nevertheless, the most common source of cooking fuel in the country is firewood with 81 per cent, followed by charcoal (16 per cent), electricity (2 per cent) and crop residue at 1 per cent. Although 62.9 per cent of Blantyre city residences had access to electricity in 2017, only 15.5 per cent use it for cooking and heating [59]. While acknowledging that 90 per cent of the firewood supply in urban areas originates from rural areas, the remaining 10 per cent still comes from urban forests and other vegetation types [60] and may have contributed to the vegetation loss observed in this study. Secondly, brick burning also contributes to vegetation loss. Construction industries rely heavily on wood for burnt brick production, which remains the main walling construction material in Malawi. It is estimated that over 0.7 metric tons of firewood ae required to create 1000 bricks [61], leading to massive forest degradation [62]. Traditionally, bricks are usually made at the construction site because when made off-site, transportation increases their costs substantially [63]. As such, it is logical to attribute the production of burnt bricks as a possible cause of vegetation loss in Blantyre city during the study period. From a governance perspective, the authorities prior to the democratic era had protected forest reserves from poachers and squatters, however, this authoritative control was lost during the transitional period in the early 1990s. This was largely due to the austerity measures that were introduced under the structural adjustment program. These measures led to the reduced number of forest guards who were protecting forest reserves. This change, coupled with the severe economic downturn around the same time, induced people to overexploit forest resources to support their livelihoods [64]. By 1995, the Ndirande Mountain Forest Reserve was largely deforested [56], and other forest reserves in Blantyre, such as Kanjedza, Soche, Michiru and Bangwe, were not spared. Local residents also invaded the Mudi Dam catchment area owned by the Blantyre Water Board of the city, destroying its vegetation cover that resulted in the siltation of Mudi dam [65].

Limitations of the Study

This study used Landsat data with a spatial resolution of 30 m, meaning that the changes below this pixel size might have been missed during classification processes. Therefore, the use of high-resolution satellite data could have provided more detailed information about LULC changes with higher accuracy.

5. Conclusions

This study provides the details of urban LULC changes in Blantyre city from 1994 to 2018 characterised by an increase in built-up area and the decline of vegetation and bare land areas, respectively. The study further reveals a two-step temporal transition, firstly from the vegetation class to the bare land class, and from the bare land class to the built-up class. This clearly demonstrates the existence of inefficiencies in the management of urban growth in the city. Based on this information, urban development stakeholders can make policy and planning priorities to ensure sustainable urban development. In order to ensure sustainable urban development in Blantyre city, this study suggests that the authorities should expedite the allocation of all suitable land for development while safeguarding unauthorised development in risky areas. This study, further, calls for all planning authorities in Malawi at the city, district, regional and national levels to review physical plans more often and help to facilitate the timely supply of serviced land across the country.

Regarding the observed vegetation cover loss, some innovative policy measures, such as identifying crucial vegetation areas for protection from anthropogenic forces, could be implemented. There is also a need to enhance cooperation among urban stakeholders and local residents, including providing incentives that could encourage locals to conserve vegetation and allow natural regeneration on bare hills and other vegetation reserves within the city boundary.

Author Contributions: J.M., T.W. and R.A. designed the study, analysed the datasets, drafted the manuscript and addressed the reviewer's comments. All authors have read and agreed to the published version of the manuscript.

Funding: Japan International Cooperation Agency (JICA) provided financial help in ABE 4th Batch Scholarship for JM and in the filed survey for JM and TW. Part of the APC was covered by JICA.

Acknowledgments: This study was possible with the help from JICA and Japan International Cooperation Centre (JICE). The authors would also like to thank Atupyele Weston Komba, Ph.D. student, for the technical help during data analysis in ArcGIS Pro software.

Conflicts of Interest: The authors declare no conflict of interest.

References

1. Manandhar, R.; Odeh, I.O.A.; Pontius, R.G., Jr. Analysis of twenty years of categorical land transitions in the Lower Hunter of New South Wales, Australia. *Agric. Ecosyst. Environ.* **2010**, *135*, 336–346. [CrossRef]

2. Hegazy, I.R.; Kaloop, M.R. Monitoring urban growth and land use change detection with GIS and remote sensing techniques in Daqahlia governorate Egypt. *Int. J. Sustain. Built Environ.* **2015**, *4*, 117–124. [CrossRef]

3. United Nations. Department of Economic and Social Affairs (UNDESA)/Population Division. In *World Population Prospects: The 2018 Revision, Key Facts*; United Nations: New York, NY, USA, 2019.

4. United Nations. Department of Economic and Social Affairs (UNDESA)/Population Division. In *World Urbanization Prospects: The 2014 Revision*; United Nations: New York, NY, USA, 2015.

5. Akinyemi, F.O.; Pontius, R.G.; Braimoh, A.K. Land change dynamics: Insights from Intensity Analysis applied to an African emerging city. *J. Spat. Sci.* **2016**, *62*. [CrossRef]

6. Grimm, N.B.; Faeth, S.H.; Golubiewski, N.E.; Redman, C.L.; Wu, J.G. Global change and the ecology of cities. *Science* **2008**, *319*, 756–760. Available online: http://science.sciencemag.org/content/319/5864/756 (accessed on 15 November 2019). [CrossRef] [PubMed]

7. Hansen, A.J.; DeFries, R.; Turner, W. Land Use Change and Biodiversity: A Synthesis of Rates and Consequences during the Period of Satellite Imagery. In *Land Change Science: Observing, Monitoring, and Understanding Trajectories of Change on the Earth's Surface*; Gutman, G., Justice, C., Eds.; Springer: New York, NY, USA, 2004; pp. 277–299.

8. Masek, J.G.; Lindsay, F.E.; Goward, S.N. Dynamics of urban growth in the Washington DC metropolitan area, 1973–1996, from Landsat observations. *Int. J. Remote Sens.* **2000**, *21*, 3473–3486. [CrossRef]

9. Addae, B.; Oppelt, N. Land-Use/Land-Cover Change Analysis and Urban Growth Modeling in the Greater Accra Metropolitan Area (GAMA), Ghana. *Urban Sci.* **2019**, *3*, 26. [CrossRef]

10. National Statistics Office (NSO). *Malawi Population and Housing Census Main Report*; National Statistics Office: Zomba, Hungary, 2009.

11. National Statistical Office (NSO). *Population and Housing Census. Be Counted, Leave No One Behind*; Preliminary Report; National Statistics Office: Zomba, Hungary, 2018; Available online: http://www.nsomalawi.mw/index.php?option=com_content&view=article&id=226&Itemid=6 (accessed on 31 October 2019).

12. Manda, M. *Situation of Urbanization in Malawi (Consultancy Services to Prepare a National Urban Policy Framework)*; Ministry of Lands and Housing: Lilongwe, Malawi, 2013.

13. Munthali, M.G.; Davis, N.; Adeola, A.M.; Botai, J.O.; Kamwi, J.M.; Chisale, H.L.W.; Orimoogunje, O.O.I. Local Perception of Drivers of Land-Use and Land-Cover Change Dynamics across Dedza District, Central Malawi Region. *Sustainability* **2019**, *11*, 832. [CrossRef]

14. Jagger, P.; Perez-Heydrich, C. Land use and household energy dynamics in Malawi. *Environ. Res. Lett.* **2016**, *11*, 12. [CrossRef]

15. Pullanikkatil, D.; Palamuleni, L.G.; Ruhiiga, T.M. Land use/land cover change and implications for ecosystems services in the Likangala River Catchment, Malawi. *Phys. Chem. Earth* **2016**, *93*, 96–103. [CrossRef]

16. Haack, B.; Mahabir, R.; Kerkering, J. Remote sensing-derived national land cover land use maps: A comparison for Malawi. *Geocarto Int.* **2014**. [CrossRef]

17. Munthali, K.G.; Murayama, Y. Modeling Deforestation in Dzalanyama Forest Reserve, Lilongwe, Malawi: A Multi-Agent Simulation Approach. *GeoJournal* **2014**, *80*, 1–15. [CrossRef]

18. Palamuleni, L.G.; Annegarn, H.J.; Landmann, T. Land cover mapping in the Upper Shire River catchment in Malawi using Landsat satellite data. *Geocarto Int.* **2010**, *25*, 503–523. [CrossRef]

19. Yang, X.; Lo, C.P. Using a time series of satellite imagery to detect land use and land cover changes in the Atlanta, Georgia Metropolitan Area. *Int. J. Remote Sens.* **2002**, *23*, 1775–1798. [CrossRef]

20. Herold, M.; Couclelis, H.; Clarke, K.C. The role of spatial metrics in the analysis and modeling of urban land use change. *Comput. Environ. Urban Syst.* **2005**, *29*, 369–399. [CrossRef]

21. Bhatta, B. *Urban Growth Analysis and Remote Sensing: A Case Study of Kolkata, India 1980–2010*; Springer: Berlin/Heidelberg, Germany, 2012.

22. Fan, F.; Weng, Q.; Wang, Y. Land use land cover change in Guangzhou, China, from 1998 to 2003, based on Landsat TM/ETM+ imagery. *Sensors* **2007**, *7*, 1323–1342. [CrossRef]

23. Sundarakumar, K.; Harika, M.; Aspiya Begum, S.K.; Yamini, S.; Balakrishna, K. Land use and Land cover Change Detection and Urban Sprawl Analysis of Vijayawada City using Multi-temporal Landsat Data. *Int. J. Eng. Sci. Technol.* **2012**, *4*, 170–178.

24. Mwathunga, E.; Dolnaldson, R. Urban land contestation, challenges and planning strategies in Malawi's main urban centres. *Land Use Policy* **2018**, *77*, 1–8. [CrossRef]

25. Kalipeni, E. Contained urban growth in post-independence Malawi. *East Afr. Geogr. Rev.* **1997**, *19*, 49–66. [CrossRef]

26. Blantyre City Assembly (BCA). *Blantyre Urban Structure Plan Background Report 2000–2014*; Private Bag 67; Blantyre City Assembly: Blantyre, Malawi, 2000; Volume 1.

27. Willis, K.S. Remote sensing change detection for ecological monitoring in United States protected areas. *Biol. Conserv.* **2015**, *182*, 233–242. [CrossRef]

28. U.S. Geological Survey (USGS). *Landsat—Earth Observation Satellite (Version 1.1, August 2016)*; USGS: Reston, VA, USA, 2016; Volume 4.

29. Traore, A.; Mawenda, J.; Komba, A. Land-cover change analysis and simulation in Conakry (Guinea), using hybrid cellular-automata and markov model. *Urban Sci.* **2018**, *2*, 39. [CrossRef]

30. Anderson, J.; Hardy, E.; Roach, J.; Witmer, R. A land use and land cover classification system for use with remote sensor data. In *Geological Survey Professional Paper*; USGS: Reston, VA, USA, 1976; Volume 964.

31. Batar, A.K.; Watanabe, T.; Kumar, A. Assessment of land-use/land-cover change and forest fragmentation in the Garhwal Himalayan Region of India. *Environments* **2017**, *4*, 34. [CrossRef]

32. Foody, G.M. Thematic map comparison: Evaluating the statistical significance of differences in classification accuracy. *Photogramm. Eng. Remote Sens.* **2004**, *70*, 627–633. [CrossRef]

33. Congalton, R.; Green, K. *Assessing the Accuracy of Remotely Sensed Data: Principles and Practices*, 2nd ed.; Taylor and Francis Group: Abingdon, UK, 2009.

34. Puyravaud, J.P. Standardizing the Calculation of the Annual Rate of Deforestation. *For. Ecol. Manag.* **2003**, *177*, 593–596. [CrossRef]
35. Jensen, J.R. Digital Change Detection. In *Introductory Digital Image Processing—A Remote Sensing Perspective*, 3rd ed.; Keith, C., Ed.; Prentice Hall Series in Geographic Information Science; Prentice Hall: Upper Saddle River, NJ, USA, 2005; pp. 467–494.
36. Yuan, F. Land-cover change and environmental impact analysis in the Greater Mankato area of Minnesota using remote sensing and GIS modelling. *Int. J. Remote Sens.* **2008**, *29*, 1169–1184. [CrossRef]
37. Jiménez, A.; Vilchez, F.; González, O.; Flores, S. Analysis of the Land Use and Cover Changes in the Metropolitan Area of Tepic-Xalisco (1973–2015) through Landsat Images. *Sustainability* **2018**, *10*, 1860. [CrossRef]
38. Briones, P.S.; Sepúlveda-Varas, A. Systematic transitions in land use and land cover in a pre-Andean sub watershed with high human intervention in the Araucania Region, Chile. *Ciencia Investigación Agraria* **2016**, *43*, 396–407.
39. Ouedraogo, I.; Savadogo, P.; Tigabu, M.; Dayamba, S.D.; Odén, P.C. Systematic and random transitions of land-cover types in Burkina Faso, West Africa. *Int. J. Remote Sens.* **2011**, *32*, 5229–5245. [CrossRef]
40. Shoyama, K.; Braimoh, A.K. Analyzing about sixty years of land-cover change and associated landscape fragmentation in Shiretoko Peninsula, Northern Japan. *Landsc. Urban Plan.* **2011**, *101*, 22–29. [CrossRef]
41. Alo, C.A.; Pontius, R.G., Jr. Identifying systematic land cover transitions using remote sensing and GIS: The fate of forests inside and outside protected areas of Southwestern Ghana. *Environ. Plan.* **2008**, *35*, 280–295. [CrossRef]
42. Pontius, R.G., Jr.; Shusas, E.; McEachern, M. Detecting important categorical land changes while accounting for persistence. *Agric. Ecosyst. Environ.* **2004**, *101*, 251–268. [CrossRef]
43. Teixeira, Z.; Teixeira, H.; Marques, J.C. Systematic processes of land use/land cover change to identify relevant driving forces: Implications on water quality. *Sci. Total Environ.* **2014**, *470*, 1320–1335. [CrossRef] [PubMed]
44. Braimoh, A. Random and systematic land-cover transitions in northern Ghana. *Agric. Ecosyst. Environ.* **2006**, *113*, 254–263. [CrossRef]
45. Estoque, R.; Murayama, Y. Trends and spatial patterns of urbanization in Asia and Africa: A comparative analysis. In *Urban Development in Asia and Africa*; Springer: Berlin/Heidelberg, Germany, 2017; pp. 393–414.
46. Melchiorri, M.; Pesaresi, M.; Florczyk, A.J.; Corbane, C.; Kemper, T. Principles and Applications of the Global Human Settlement Layer as Baseline for the Land Use Efficiency Indicator–SDG 11.3. 1. *ISPRS Int. J. Geo-Inf.* **2019**, *8*, 96. [CrossRef]
47. UN-Habitat. *World Cities Report: Urbanisation and Development Emerging Futures*; UN-Habitat: Nairobi, Kenya, 2016.
48. Vargo, J.; Habeeb, D.; Stone, B., Jr. The importance of land cover change across urban–rural typologies for climate modelling. *J. Environ. Manag.* **2013**, *114*, 243–252. [CrossRef] [PubMed]
49. Mensah, C.A.; Eshun, J.K.; Asamoah, Y.; Ofori, E. Changing land use/cover of Ghana's oil city (Sekondi-Takoradi Metropolis): Implications for sustainable urban development. *Int. J. Urban Sustain. Dev.* **2019**, *11*, 223–233. [CrossRef]
50. Estoque, R.C.; Murayama, Y. Landscape pattern and ecosystem service value changes: Implications for environmental sustainability planning for the rapidly urbanizing summer capital of the Philippines. *Landsc. Urban Plan.* **2013**, *116*, 60–72. [CrossRef]
51. Konrad, C.P. *Effects of Urban Development on Floods*; USGS Fact Sheet FS-076-03; USGS: Reston, VA, USA, 2003.
52. Hirons, M.; Comberti, C.; Dunford, R. Valuing Cultural Ecosystem Services. *Annu. Rev. Environ. Resour.* **2016**, *41*. [CrossRef]
53. Chinele, J. Blantyre City Council U-Turns on Houses Demolition, The Times Group. 10 March 2018. Available online: https://www.times.mw/blantyre-city-council-u-turns-on-houses-demolition/ (accessed on 8 April 2019).
54. Government of Malawi. Malawi Growth and Development Strategy III, 2017–2022. In *Building a Productive, Competitive and Resilient Nation, November, 2017*; Government of Malawi: Lilongwe, Malawi, 2017.
55. Kita, S.M. Urban Vulnerability, Disaster Risk Reduction and Resettlement in Mzuzu City, Malawi. *Int. J. Disaster Risk Reduct.* **2017**, *22*, 158–166. [CrossRef]

56. Bone, R.A.; Parks, K.E.; Hudson, M.D.; Tsirinzeni, M.; Willcock, S. Deforestation since independence: A quantitative assessment of four decades of land-cover change in Malawi, Southern Forests. *J. For. Sci.* **2017**, *79*, 269–275. [CrossRef]

57. Mauambeta, D.D.C.; Chitedze, D.; Mumba, R.; Gama, S. *Status of Forests and Tree Management in Malawi*; A Position Paper Prepared for the Coordination Union for Rehabilitation of the Environment (CURE); CURE: Blantyre, Malawi, 2010.

58. Ngwira, S.; Watanabe, T. An Analysis of the Causes of Deforestation in Malawi: A Case of Mwazisi. *Land* **2019**, *8*, 48. [CrossRef]

59. National Statistical Office. *Fourth Integrated Household Survey, 2016–2017, Household Socio-Economic Characteristics Report*; National Statistical Office: Zomba, Hungary, 2017. Available online: http://www.nsomalawi.mw/index.php?option=com_content&view=article&id=225&Itemid=112 (accessed on 18 August 2018).

60. Government of Malawi. *Biomass Energy Strategy*; Government of Malawi: Lilongwe, Malawi, 2009. Available online: http://www.euei-pdf.org/en/seads/policy-strategy-and-regulation/biomass-energy-strategy-best-malawi (accessed on 28 October 2018).

61. Zingano, B.W. *The Problem of Fuel Wood Energy Demand in Malawi with Reference to the Construction Industry*; Zingano and Associates: Lilongwe, Malawi, 2005; Available online: https://www.joyhecht.net/mulanje/refs/Zingano-fuelwood-bricks-2005.pdf (accessed on 17 January 2019).

62. Wiyo, K.A.; Fiwa, L.; Mwase, W. Solving deforestation, protecting and managing key water catchments in Malawi using smart public and private partnerships. *J. Sustain. Dev.* **2015**, *8*, 251–261. [CrossRef]

63. UN Habitat. Malawi Urban Housing Sector Profile, Nairobi. 2010. Available online: https://unhabitat.org/books/malawi-urban-housing-sector-profile/ (accessed on 26 February 2019).

64. Riley, L. The political economy of urban food security in Blantyre, Malawi. In *African Institute Occasional Paper Series*; Africa Institute: London, UK, 2012; Volume 1, Available online: https://www.researchgate.net/profile/Liam_Riley/publication/265110167_The_political_economy_of_urban_food_security_in_Blantyre_Malawi/links/559eb7b708aeffab5687c833/The-political-economy-of-urban-food-security-in-Blantyre-Malawi.pdf (accessed on 7 April 2019).

65. Mangazi, C. Blantyre Water Board wins the battle against encroachers. *BWB Rev.* **2012**, *1*, 20.

 *sustainability* 

Article

Spatiotemporal Patterns and Driving Forces of Urban Expansion in Coastal Areas: A Study on Urban Agglomeration in the Pearl River Delta, China

Yichen Yan [1], Hongrun Ju [1,*], Shengrui Zhang [2,*] and Wei Jiang [3]

[1]   School of Tourism and Geography Science, Qingdao University, Qingdao 266071, China; yyc525geo@126.com
[2]   Management College, Ocean University of China, Qingdao 266100, China
[3]   China Institute of Water Resources and Hydropower Research, Beijing 100038, China; jiangwei@iwhr.com
*   Correspondence: juhr@qdu.edu.cn (H.J.); zhangshengrui@ouc.edu.cn (S.Z.)

Received: 24 November 2019; Accepted: 20 December 2019; Published: 25 December 2019

Abstract: Since the beginning of the 21st century, the spatial pattern of urban expansion and the mechanism of urbanization in coastal areas have undergone significant changes. This study aims to reveal the spatiotemporal patterns of urban land expansion and analyze the dynamic driving forces of urban agglomeration in the Pearl River Delta of China from 2000 to 2015. The urban-land-expansion intensity index, expansion difference index, and fractal dimension were used to study how the urban land in this area was developed, and the geographical detector was applied to explore the relative importance, expansion intensity, and interactions of physical and socioeconomic factors. The results revealed that the urban-land-expansion intensity of the Pearl-River-Delta urban agglomerations exhibit a downward trend, while cities exhibited a trend of developing more coordinately from 2000 to 2015. Physical factors determined the direction and scale of urban development, and the urban land expansion in the Pearl-River-Delta urban agglomeration is mainly distributed in plain areas that have an elevation below 120 m and a slope less than 5°. Socioeconomic factors have a greater influence on the expansion of urban land, and their effects have changed over time. Population growth and economic development has played a significant role in the expansion of urban land before 2005. Subsequently, the factor of GDP and distance to the core cities of Guangzhou and Shenzhen controlled the expansion to the greatest extent. The impacts of various factors tended to become balanced during 2010–2015. The majority of the factors enhanced each other via their interactions, and the distance to the rivers always exhibited a greater enhancement when there was interaction with other factors. The spatial and temporal analysis of the urban expansion and the mechanism of the Pearl River Delta urban agglomeration could provide useful information for coastal urban planning. This study also offers new knowledge regarding the interactions between different drivers of urban land expansion.

Keywords: urban land expansion; spatial pattern; driving forces; Pearl River Delta; urban agglomeration

---

1. Introduction

Coastal areas are commonly defined as the interface or transition areas between land and sea, and they comprise diverse functions and forms with no strict spatial boundaries [1]. Coastal areas are usually characterized by flat terrain, a moderate climate, booming economy, rich resources, and convenient accessibility to marine trade and transport. The combination of these features drives coastal migration and stimulates urban expansion [2]. The majority of the world's metropolises and urban agglomerations are situated in coastal areas [3], and many of these are in large deltas. The rates of urban expansion in coastal areas were significantly higher than that in the hinterland from 1970 to 2000 [4]. Unprecedented urbanization has resulted in profound changes in landscape [5], biodiversity [6], biogeochemical cycles [7], and energy flow [8] at multiple spatiotemporal scales. In China, urbanization in the coastal areas also grew and expanded faster than that in the western non-coastal areas [4]. The growth rate of the coastal urban land is three times higher than the national rate and has been driven by continued economic growth and specific policies that encourage coastal city development [9]. At the same time, rapid urbanization, increasing land utilization, and pollution have also put increasingly high pressure on coastal ecosystems and natural resources [10]. Therefore, understanding the spatiotemporal patterns of the rapidly expanding coastal cities in China is important for formulating sustainable land-use and urban-planning policies.

At present, geographic information systems (GIS) combined with remote sensing (RS) and landscape indexes have been widely applied for determining the spatiotemporal dynamics of urban growth patterns [11]. RS offers consistent and frequent data from study areas with positional spatial detail [12], which can be analyzed, displayed, and described using GIS [13]. Moreover, the development of the landscape ecology can be of immense help in quantitatively describing the urban expansion patterns, and the relationship between urbanized area structure and the urbanization process using landscape metrics [14]. With the rapid development of GIS, the valuable multispectral RS data set, and various indexes in landscape ecology, the specific spatial and temporal characteristics and driving forces of urban expansion can be better understood. The combination of these three methods is effective for updating the spatial data and obtaining accurate and timely geospatial information for illustrating the change patterns of urban land expansion [15].

The Pearl River Delta is one of three "National Optimized Development Zones" in China; however, the inclusion of Hong Kong and Macao in the Greater Pearl River Delta makes it the most globally integrated of the three. Since the start of the 21st century, the spatial structure and urbanization dynamic mechanism of the Pearl River Delta urban agglomeration (PRDUA) have undergone tremendous changes. The functions of cities have become diversified, and a multi-center urban system was formed with large urban and rural population immigration [16]. As China's urbanization and industrialization process continues to accelerate, the new impetus comes mainly from the development of the urban economy, strengthening of international and national connectivity, and growth of private capital. At the same time, as a new regional unit of the country's participation in global competition and the international division of labor [17], the urban agglomeration will determine the new pattern of the global political economy in the 21st century. However, there is a lack of studies comprising the use of spatially consistent data sets with a high temporal frequency of urban conditions after the year 2000 in the Pearl River Delta. Accordingly, it is necessary to form a clearer understanding of the urban land expansion in the PRDUA.

In coastal urban ecosystems, the interactions among geographical, ecological, economic and social factors exist widely at different spatio-temporal scales [18]. The functions of one factor can be influenced by the conditions of other factors in the same urban system. It has been proven that interactions between factors can improve the accuracy of the spatial prediction of urban growth [19]. The cities in an urban agglomeration usually have frequent interactions with each other in the form of economic, social, and ecological activities. Understanding the interactions among the aforementioned factors is vital to help simulate and predict urban growth patterns more precisely, and with greater detail [20]. However, the interactions between factors are rarely studied owing to the intricate functions of the urban system. The interactions of factors that influence the development of coastal urban agglomerations require further research.

This study is aimed at exploring the spatiotemporal patterns and driving forces of coastal urban expansion based on land-use data derived from RS images. The urban agglomeration of the Pearl River Delta is taken as an example area of this study. A geographical detector is used to identify the interactions of various driving forces and their changes. In this study, we first introduce the study area and the corresponding data. The details of the method used to describe the spatiotemporal pattern and the approach used with respect to geographical detector are presented later. We then present the results obtained in terms of the change pattern and the driving forces, especially the interactions, of the urban-land expansion in the PRDUA over the period of 2000–2015. Finally, we discuss the mechanism by which the policies act as factors affecting the urban-land use as implicated by the results.

2. Study Area and Data

2.1. Study Area

The PRDUA is one of the world's largest and fastest growing urban regions. This massive urban agglomeration is located in Guangdong in the South China region, and consists of 15 highly interconnected cities, namely, Guangzhou, Shenzhen, Foshan, Dongguan, Zhaoqing, Shaoguan, Qingyuan, Yunfu, Huizhou, Shanwei, Heyuan, Zhuhai, Zhongshan, Jiangmen, and Yangjiang (Figure 1). Among these cities, Guangzhou is the capital of the Guangdong province and is the cultural and political center of the Pearl River Delta. Shenzhen, as a special economic zone in China, is one of the fastest-growing cities in the world. Owing to its obvious geographical advantages, the PRDUA has always been the gateway to China. The flat lands of the delta are crisscrossed by a network of tributaries and distributaries of the Pearl River. This area has a humid subtropical climate characterized by hot and humid summers and cold-to-mild winters. Around the year 2000, with the handover of Hong Kong and Macau, and with China being a member of the World Trade Organization, the PRDUA increasingly attracted the nation's capital, technology, and talents. In the 21st century, the PRDUA ushered in new opportunities for economic integration and industrial restructuring. According to the statistical data obtained from the government, the permanent population of the area has increased by 21.04% from 43.92 million in 2000 to 53.16 million in 2015, and the gross domestic product (GDP) has increased dramatically from 845.41 billion RMB in 2000 to 6823.09 billion RMB in 2015 [21,22]. The market of the PRDUA is international and active because of its proximity to Hong Kong. Its significant economic development has attracted numerous global investors. Furthermore, the industrial cities in the Pearl River Delta have been called the Factory of the World owing to the presence of industrial parks populated with factories built using foreign investments. Rapid urbanization and industrialization in the PRDUA has made it the most populated area with a high proportion of immigration. However, rapid economic and population growth have also resulted in problems, such as environmental pollution, water shortages, cropland loss, and high costs of living and housing. It is necessary to study the dynamic spatial pattern and mechanism of urban expansion in the PRDUA to obtain useful information for effective urban planning.

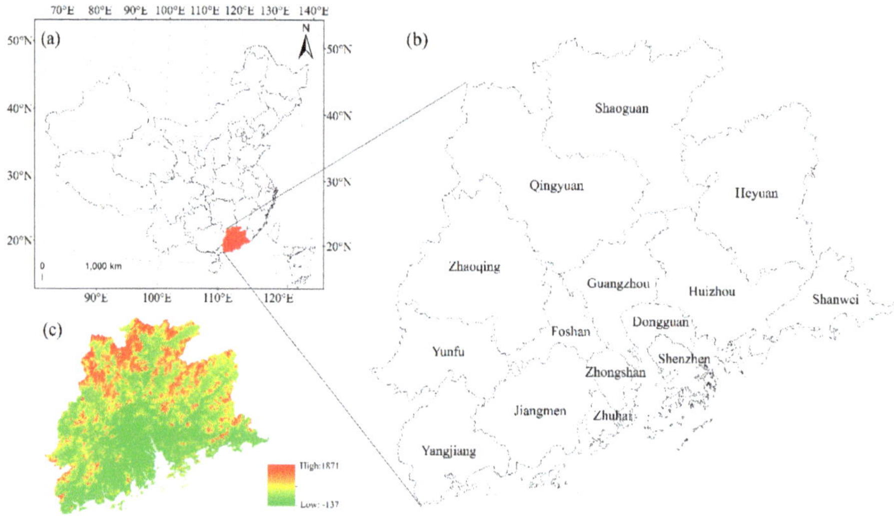

Figure 1. Location and administrative divisions of the Pearl River Delta urban agglomeration: (**a**) Location of the study area; (**b**) Administrative divisions of the study area; and (**c**) Topography of the study area.

2.2. Data Source

The data used in this study include the following: (i) land use raster data of the PRDUA in 2000, 2005, 2010, and 2015 having a spatial resolution of 100 m (Figure 2). The data are obtained from National Land Use/Cover Database of China. The land use types were visually interpreted from medium-resolution satellite images (Landsat MSS/TM/ETM, the China-Brazil Earth Resources Satellite and HJ-1A) and field surveys were conducted to verify the classification results [23]. The land use types are divided into six types of first-class: cropland, woodland, grassland, water bodies, built-up land, and unused land; and the built-up land contains three second-level types of urban land, rural settlement, and industry-traffic land. Urban land referred to land used for urban settlement, with a largely continuous area covered by urban construction and city facilities. (ii) The socioeconomic data of the PRDUA includes the population, GDP, secondary industry GDP, tertiary industry GDP, and total investment in fixed assets [21,22,24,25]. (iii) The vector data of the main roads, rivers, and coastlines are derived from the National Fundamental Geographical Information System of China. The elevation and slope data are derived from the 90 m elevation data of the Shuttle Radar Topography Mission.

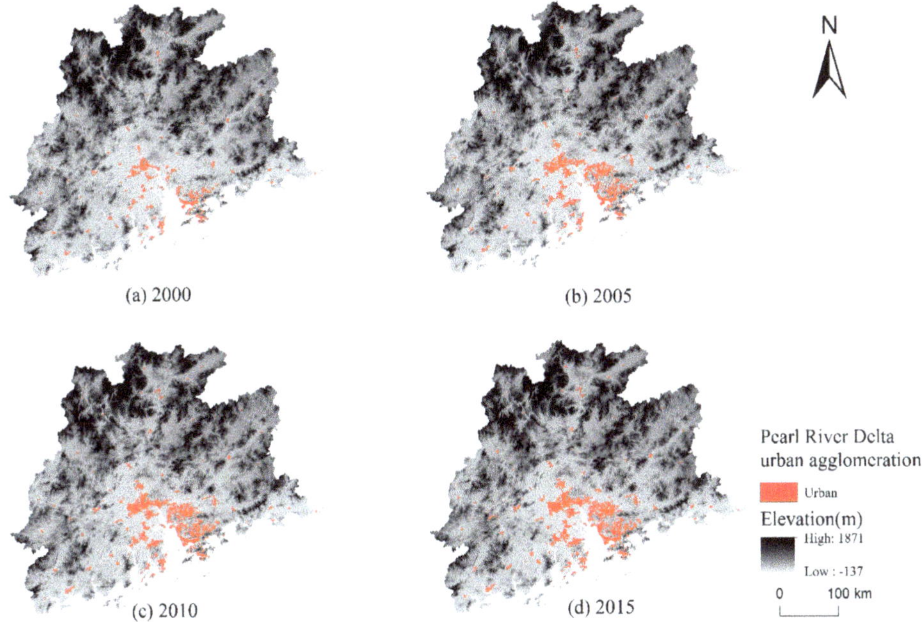

(a) 2000   (b) 2005

(c) 2010   (d) 2015

Figure 2. Spatial distribution of the urban land in the Pearl River Delta urban agglomeration from 2000 to 2015.

3. Methods

3.1. Spatial Pattern Indicators

3.1.1. Urban-Land Expansion Intensity Index

The urban-land expansion intensity index refers to the rate of expansion of the area of urban land within a certain period. After standardization, the speeds of urban-land expansion during different periods and in different spatial units are comparable. This index is defined as follows:

$$UEI_i = \frac{UA_i^{t+m} - UA_i^{t}}{UA_i^{t} \times m} \tag{1}$$

where UEI_i represents the urban land expansion intensity, UA_i^{t+m} and UA_i^{t} represent the urban land area in the spatial unit i of t and $t + m$ years, respectively, and m is the number of years in the study period.

3.1.2. Urban-Land Expansion Difference Index

The urban-land expansion difference index represents the ratio of the urban expansion rate of the cities in the urban agglomeration to the urban expansion rate of the entire urban agglomeration over a certain period, and it makes the urban-land expansion speed of different cities comparable. The index can also indicate the coordinate development among the cities. This index is defined as follows:

$$UEDI_i = \frac{\left|UA_i^{t+m} - UA_i^{t}\right| \times UA^{t}}{\left|UA^{t+m} - UA^{t}\right| \times UA_i^{t}} \tag{2}$$

where $UEDI_i$ represents the urban land expansion difference, and UA^{t+m} and UA^t represent the urban land area of the PRDUA during different time periods. Here, UA_i^{t+m}, UA_i^t and m are the same as those in Equation (1).

3.1.3. Fractal Dimension

The fractal dimension is the measurement of the form of urban land. Fractal dimension changes can reflect the spatial concentration and diffusion of the urban land [26]. The more complex the shape and structure of the urban land, the greater the fractal dimension. This index is defined as follows:

$$D_{it} = \frac{2\ln(0.25P_{it})}{\ln(A_{it})} \tag{3}$$

where D_{it}, P_{it} and A_{it} represent the fractal dimension, perimeter of the urban patch, and area of the urban patch, respectively, of the ith city in year t. The fractal dimension has a value range of 1–2. The complexity of the urban form is positively correlated with the value of the fractal dimension. When the value of the fractal dimension is greater than 1.5, the form of the urban land is complicated. In contrast, and the urban form is relatively simple [27].

3.2. Driving Force Analysis—Geographical Detector

A geographical detector comprises a set of statistical methods that are used to identify the spatial differentiation and reveal the driving mechanism behind geographical phenomena [28]. This method includes four detectors, namely: factor, risk, ecological, and interaction detectors. In this study, mainly the factor, risk, and interaction detectors are used to analyze the relative influence of the various factors in the urban expansion of the PRDUA, how each factor influences the expansion, and the interactions between the different factors.

(1) Factor Detector

The factor detector can quantitatively represent the relative importance of each possible factor. The power determinant (p) is defined as the difference of one and the ratio of the accumulated variance of the urban land growth area in the sub-regions of a factor to that over the entire study area [29]:

$$p = 1 - \frac{\sum\limits_{i=1}^{N} n_i \sigma_i^2}{n\sigma^2} \tag{4}$$

where N is the number of strata of the potential factor, and n_i and n are the number of grid elements of the strata i and the whole region, respectively. σ_i^2 and σ^2 are the variances of the urban-land area in the strata i and the whole region, respectively. The value range of p is [0, 1], and the greater the value of p, the greater the influence of the factor is.

(2) Risk Detector

In the case of the risk detector, the t-test is used to perform a comparison to determine whether the difference between the sub-regions divided by the potential factor is significant. In this study, the average expanded urban area of the grid cells in a sub-region Di is calculated as follows [20]:

$$U_d = \frac{1}{n_{Di}} \sum_{1}^{n_{Di}} y_{Di} \tag{5}$$

where y_{Di} denotes the urban expanded area of a grid in a strata Di, and n_{Di} denotes the number of grids in the strata. Using the U_d values, a comparison of the effects of different levels of a factor can be conveniently performed. The greater the value of U_d, the more rapid the urban-land expansion is.

(3) Interaction Detector

The interaction detector identifies how two different factors interact to exert an impact on the spatial expansion of the urban agglomeration. The interaction detector quantifies the interaction of the two possible influence factors as follows:

$$
\begin{aligned}
&\text{Nonlinear-weaken: } p(M{\cap}N) < \text{Min}(p(M), p(N)) \\
&\text{Uni-enhance/weaken: } \text{Min}(p(M), p(N)) < p(M{\cap}N) < \text{Max}(p(M), p(N)) \\
&\text{Bi-enhance: } \text{Max}(p(M), p(N)) < p(M{\cap}N) < (p(M) + p(N)) \\
&\text{Independent: } p(M{\cap}N) = p(M) + p(N) \\
&\text{Nonlinear-enhance: } p(M{\cap}N) > (p(M) + p(N))
\end{aligned}
\tag{6}
$$

where the symbol "$\cap$" denotes the interaction between factors M and N. In the Esri ArcGIS platform, the overlay of two influence factor (such as M and N) layers can be realized. When layer O is formed, the p values of M, N, and O are respectively input into the above formula to determine whether the two factors have interactions, and if the interactions are enhanced or weakened.

3.3. Driving Factors Selection

The factors affecting the expansion of urban land are diverse and the mechanism of land use change is complex. Based on a review of the extant research, the driving forces for urban-land expansion comprise three categories, namely physical, socioeconomic, and policy factors [20]. These factors interact and drive the expansion of urban land in urban agglomerations. The physical factors determine the background conditions for regional urban-land expansion [30,31]. The mountains and hills are sparsely distributed in the northern, eastern, and western parts of the PRDUA, which may restrict the expansion of the urban land (Figure 1c). As a coastal urban agglomeration, the sea could serve as a natural transportation hub to the outside areas, which could influence urban development. The Pearl River Delta is formed by three major rivers, including the Xi Jiang (West River), Bei Jiang (North River), and Dong Jiang (East River). Therefore, the terrain, distance to the sea, and river distribution may determine the potential, intensity, direction, and scale of the urban-land expansion. Socioeconomic factors are also vital factors affecting urbanization [32–34]. The rapidly growing population, increasing GDP, and investment in infrastructure all directly stimulate urban expansion. At the same time, as an effective means for the government to regulate urban development, policy factors have played an important role in urban planning. However, it is often difficult to quantify policy factors. In this study, the impact of the policy factors is revealed using physical and socioeconomic factors as geographic proxies. The so-called geographic proxies refer to factors that can be spatially represented in the real world owing to the determining factors behind the geographical phenomena [35]. The policy factors always interact with physical and socioeconomic factors, and thus the impact of the policy factors is reflected in the spatial pattern of land-use change and changes in the physical and socioeconomic factors. For example, the formulation of a policy may result in population gathering and economic growth, which would stimulate the expansion of urban land. Therefore, the effects of the policy are reflected by socioeconomic indicators such as population growth and GDP change. Based on the above considerations, in this study, 10 physical and socioeconomic factors are selected as the driving factors for urban-land expansion in the PRDUA (Table 1). Furthermore, these factors are introduced as geographical proxy factors for the policy factors in later analysis.

Table 1. Driving factors of urban land expansion in the Pearl River Delta urban agglomeration.

Category	Variable	Abbreviation	Unit
Physical factors	Elevation	ELE	m
	slope	SLP	°
	Distance to rivers	D_RV	km
	Distance to coastline	D_CL	km
Socioeconomic factors	Permanent population	P_POP	thousand persons
	Gross domestic product	GDP	billion RMB
	Proportion of secondary and tertiary industries in GDP	ST_GDP	%
	Total investment in fixed assets	T_FAI	billion RMB
	Distance to main roads	D_RD	km
	Distance to Guangzhou and Shenzhen	D_GS	km

4. Results

4.1. Spatial Pattern of Urban Land Expansion

The urban land expansion intensity indexes for the three time periods, 2000–2005, 2005–2010, and 2010–2015, and the entire time period are presented in Table 2. The results indicate that although the area of the urban land in the PRDUA continued to expand, its expansion intensity demonstrated a downward trend. The intensity of the urban-land expansion peaked in 2000–2005, and those of the majority of the cities were consistent with that of the whole area. The decreasing trend of the intensity of urban-land expansion may have been influenced by the "National Principal Function Zoning Plan" in 2010, which emphasized the optimization of construction-land expansion mode based on different function divisions. In 2005–2010 and 2010–2015, the difference in the intensity of expansion among the cities significantly reduced, thereby reflecting the coordinate development within the urban agglomerations. This change may demonstrate that the spatial structure of the PRDUA changed from two single-center structures in Guangzhou and Shenzhen to a multi-centered structure (Figure 2).

Table 2. Urban-land expansion intensity in Pearl River Delta urban agglomeration.

City	2000–2005	2005–2010	2010–2015	2000–2015
Guangzhou	0.141	0.036	0.019	0.080
Shenzhen	0.040	0.026	0.015	0.030
Foshan	0.057	0.015	0.001	0.026
Dongguan	0.037	0.015	0.011	0.023
Zhuhai	0.187	0.046	0.014	0.103
Zhongshan	0.031	0.048	0.028	0.042
Zhaoqing	0.023	0.022	0.045	0.034
Yunfu	0.086	0.075	0.023	0.079
Yangjiang	0.039	0.010	0.021	0.026
Shaoguan	0.065	0.018	0.016	0.037
Shanwei	0.068	0.022	0.075	0.069
Qingyuan	0.155	0.011	0.038	0.081
Huizhou	1.304	0.044	0.008	0.571
Heyuan	0.415	0.025	0.026	0.193
Jiangmen	0.087	0.027	0.142	0.119
PRDUA	0.153	0.032	0.016	0.081

Figure 3 shows the spatial pattern of the urban-land expansion difference of each city, and the five categories were defined as slow development, relatively slow development, medium-speed development, relatively rapid development, and rapid development according to Jenk's natural breaks. The speed of urban-land expansion varied between cities over different time periods. From 2000 to 2005, the cities of Dongguan, Zhongshan, Guangzhou, Foshan, and Qingyuan were developing rapidly

(Figure 3a). From 2005 to 2010, Guangzhou, Foshan, and Dongguan had a high expansion speed, which stimulated Huizhou and Jiangmen to become the new rapid-expansion zones (Figure 3b). From 2010 to 2015, the expansion speed of the southwestern part of the PRDUA was significantly higher than that of the central and eastern regions (Figure 3c). This phenomenon may have been influenced by the construction of Hong Kong-Zhuhai-Macao Bridge at the end of 2009, which strengthened the effect of Zhuhai causing an improvement in the development of the west of the Pearl River Estuary during this period. Overall, the expansion of urban land in the PRDUA was mainly concentrated in the central and southern regions, and the index showed a decreasing tendency as the distance to the Pearl River Estuary increased (Figure 3d).

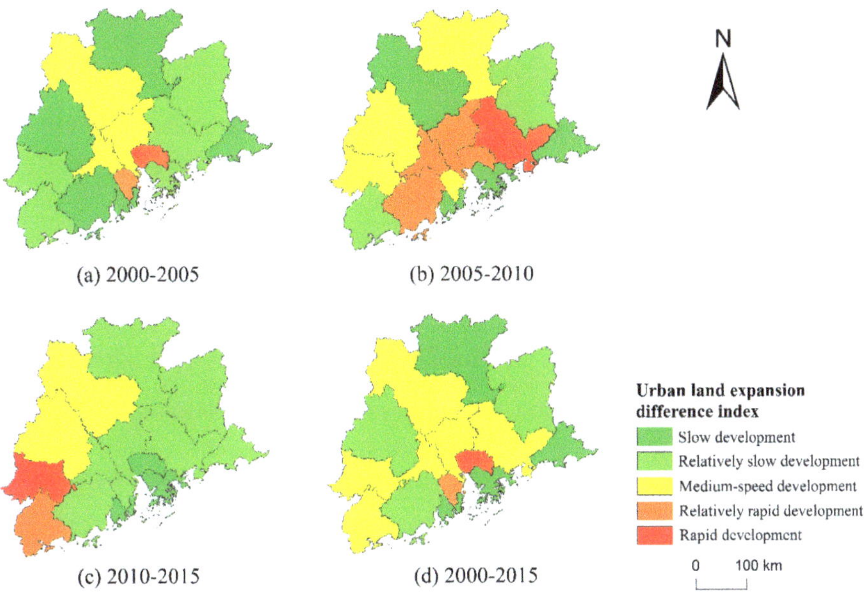

Figure 3. Spatial characteristics of the urban-land expansion difference index in the Pearl River Delta urban agglomeration.

During 2000–2015, the fractal dimensions of Shenzhen, Dongguan, and Foshan were significantly greater than those of other cities, which indicated that the urban form was more complicated in these cities. The maximum fractal dimension of all the cities during this period did not exceed 1.3, thus revealing that the urban form within the PRDUA was relatively simple and regular. During the study period, the fractal dimension of Guangzhou, Shenzhen, Zhaoqing, Shaoguan, Qingyuan, and Heyuan increased continuously, thus demonstrating that the shape of the cities became increasingly complicated, which was due to the increase in human interference activities. In contrast, the fractal dimension of the other cities first increased and then decreased, which may indicate that the urban development experienced a process of disorderly development followed by planned development during this period. From 2000 to 2015, the fractal dimensions of Shenzhen and Zhongshan changed the most, while the fractal dimension of Zhuhai changed the least. This phenomenon indicated that the shape of the Shenzhen and Zhongshan cities changed drastically in those 15 years, while the urban form of Zhuhai remained relatively stable (Table 3).

Table 3. Fractal dimension index of cities in the Pearl River Delta urban agglomeration.

City	2000	2005	2010	2015	Changes
Guangzhou	1.1687	1.2002	1.2057	1.2070	0.0383
Shenzhen	1.1926	1.2282	1.2328	1.2464	0.0538
Foshan	1.2005	1.2068	1.1837	1.1862	−0.0143
Dongguan	1.2131	1.2444	1.2392	1.2417	0.0286
Zhuhai	1.1643	1.1672	1.1701	1.1660	0.0017
Zhongshan	1.1518	1.2209	1.2098	1.2056	0.0538
Zhaoqing	1.1436	1.1485	1.1525	1.1595	0.0159
Yunfu	1.1491	1.1561	1.1539	1.1762	0.0271
Yangjiang	1.1879	1.2042	1.2062	1.2060	0.0181
Shaoguan	1.1204	1.1366	1.1471	1.1480	0.0276
Shanwei	1.1029	1.1330	1.1392	1.1387	0.0358
Qingyuan	1.1118	1.1404	1.1421	1.1433	0.0315
Huizhou	1.1369	1.1890	1.1757	1.1830	0.0461
Heyuan	1.1766	1.2118	1.2133	1.2159	0.0393
Jiangmen	1.1878	1.1980	1.2058	1.2042	0.0164

4.2. Results of Driving-Forces Analysis

4.2.1. Factor and Risk Detector

The factor detector is a measure of the relative importance of various factors to the expansion of the urban land. Between 2000 and 2005, socioeconomic factors had a significant effect on urban-land expansion. The increase in population had the greatest influence, followed by those of fixed asset investments and GDP. Between 2005 and 2010, the GDP still played as an important factor influencing the urban-land expansion. The distance to Guangzhou and Shenzhen became the second important driving force, thus indicating that the two cities played a significant role in the urban agglomeration during this period. However, the influence of distance to Guangzhou and Shenzhen decreased significantly between 2010 and 2015. In contrast, physical factors such as elevation, slope, and distance to the rivers had a relatively low effect on the urban-land expansion over the whole period. The distance to the coastline was the most influential physical factor during the first period and its influence decreased after 2005. Overall, the socioeconomic indicators, such as the influence of population, investment in fixed assets, GDP, and the distance to the core cities of the urban agglomeration had a higher impact at first, but the impact of the socioeconomic and physical factors gradually became even in 2010–2015 (Table 4).

Table 4. The p values of the driving factors from 2000 to 2015 (Acronyms are defined in Table 1).

Driving Factor	2000–2005	2005–2010	2010–2015
P_POP	44.83%	10.58%	2.39%
T_FAI	32.55%	8.51%	2.63%
GDP	29.94%	27.42%	5.40%
D_GS	22.14%	25.18%	6.93%
D_CL	9.04%	3.57%	3.31%
ST_GDP	6.16%	19.52%	5.67%
ELE	5.21%	15.28%	5.26%
SLP	4.83%	16.95%	4.75%
D_RV	3.96%	7.47%	3.95%
D_RD	2.96%	4.98%	2.25%

The risk detector presented a comparison of the mean values of the urban-land expansion in different sub-regions to determine where the urban-land expansion was severe. The expanded area of the urban land was basically positive in relation to the population growth during 2000–2015 (Figure 4a–c). The urban-land expansion rate was significantly greater than the population growth rate. During the study period, the area of the urban land increased by 121.6%, and the population increased by 21.0%. The growth rate of the urban-land area was 5.79 times that of the population, thus further demonstrating the uncoordinated development between the urban land and population in coastal areas [4,36].

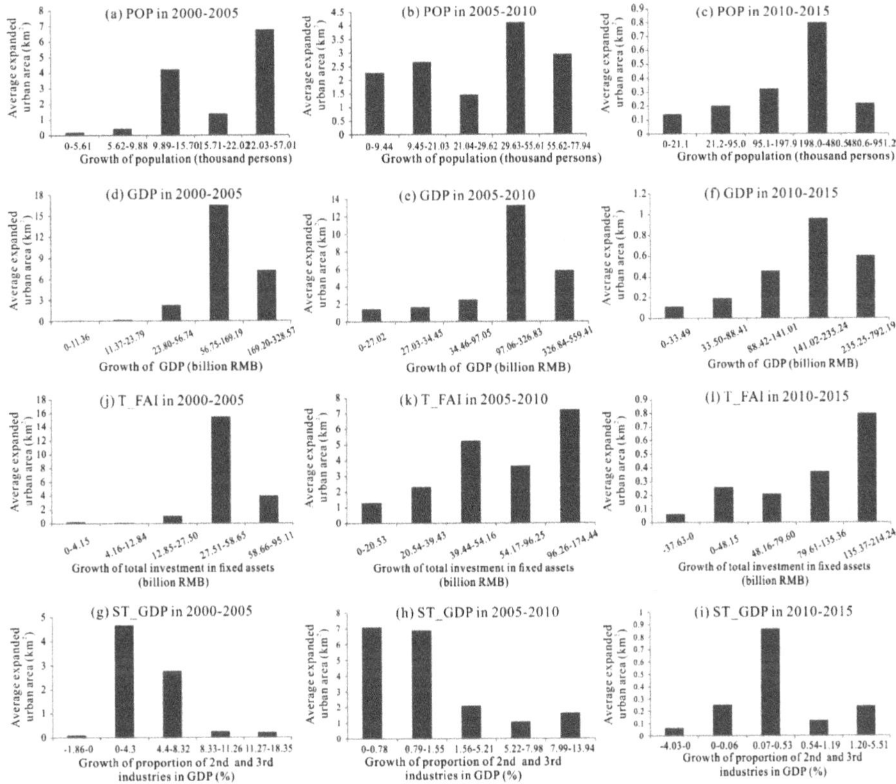

Figure 4. Average expanded urban area in different sub-regions from 2000 to 2015 (Factors of (**a**–**c**) growth of population, (**d**–**f**) growth of GDP, (**g**–**i**) growth of investment in fixed assets, and (**j**–**l**) growth of proportion of 2nd and 3rd industries in GDP. Acronyms are defined in Table 1).

Similarly, the greater the increase in GDP and total investment in fixed assets, the more intensive the urban-land expansion was (Figure 4d–i). During the study period, the GDP growth in the fourth sub-region was the most dramatic. The urban statistics showed that Dongguan and Foshan were both in this sub-region in three periods, and they were the two cities with the highest urbanization rate.

During the two initial periods, urban-land expansion was dynamic where the increase of the proportion of secondary and tertiary industries in GDP was small. This may have been because land resources were more important for the initial development of the secondary and tertiary industries. The demand for land in the stage during which the secondary and tertiary industries increased at a high speed was reduced because the emphasis was focused more on the quality development. During 2010–2015, the proportion of secondary and tertiary industries in GDP increased at a slower rate than in the former two periods. The urban area expanded the most where the increased value of the secondary and tertiary industries in GDP was 0.07%–0.53%, thus indicating that the optimization of the industrial structure promoted the expansion of urban land (Figure 4j–l).

The factors of regional slope and elevation had similar effects on urbanization. In areas with elevations below 120 m and a slope less than 5°, the average expanded area of each grid reached a maximum value. This is because urban expansion tended to occur at the least cost in areas where the terrain was low and flat. Furthermore, the population also tended to gather in flat and low coastal areas, which may have interactively enhanced the urban-land expansion (Figure 5a,b).

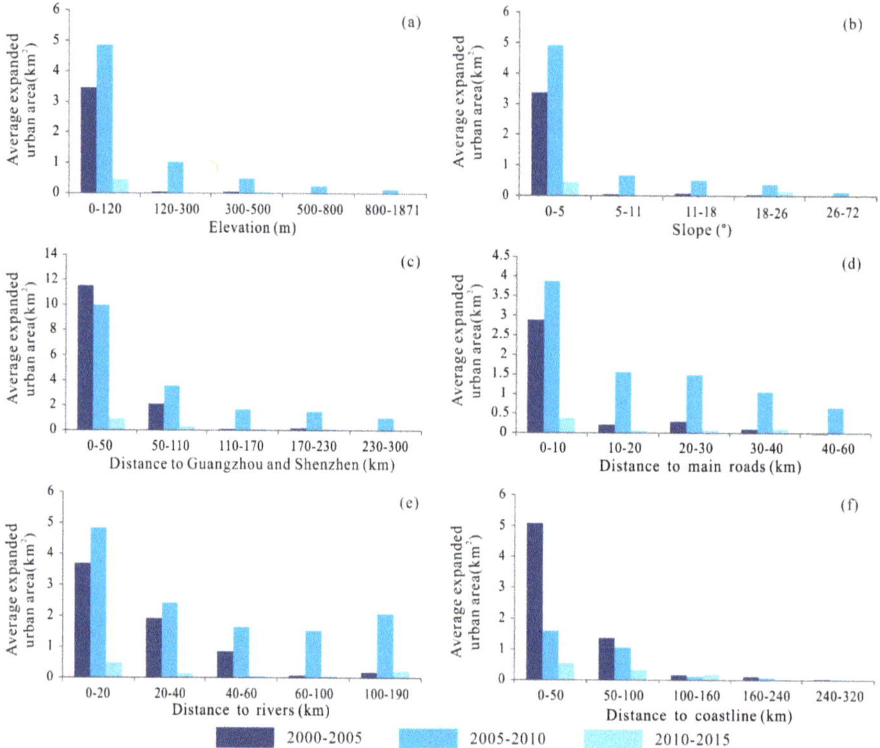

Figure 5. Average expanded urban area in different sub-regions from 2000 to 2015 (Factors of (**a**) Elevation, (**b**) Slope, (**c**) Distance to Guangzhou and Shenzhen, (**d**) Distance to main roads, (**e**) Distance to rivers and (**f**) Distance to coastline).

The distance to the two major cities of Shenzhen and Guangzhou had an apparent effect on the urban-land expansion. As the distance increased, the expansion intensity of the urban land decreased significantly during the first two periods. This non-linear negative correlation was mitigated in 2010–2015. This phenomenon reflected the core driving effects of the major cities on urbanization in the urban agglomeration during the initial expansion of the city. However, this effect tended to decrease when the urban agglomeration developed into a new stage with greater emphasis on quality and structure (Figure 5c).

Similarly, the impact of roads, rivers, and the coastline also showed a negative relationship with the urban-land expansion. The closer to the main roads, rivers, and coastline, the more intense the urban-land expansion was. The urban-land expansion was severe in regions within 10 km from main roads, 20 km from rivers, and 50 km from the coastline in the PRDUA (Figure 5d–f). This was because the construction of roads could improve the accessibility of the local areas, and rivers could serve as tourist attractions and provide convenient water and land connections. The distance to the coastline has a close relationship with the terrain and port transportation, and thus it showed a negative relationship with the urban land expansion.

4.2.2. Interaction Detector

There are 55 pairs of interactions between 10 influencing factors. These interactions were analyzed using three categories: pairs of socioeconomic factors, pairs of physical factors, and pairs of socioeconomic and physical factors. The results showed that if any two factors were superimposed, the explanatory power of urban land expansion was enhanced in the PRDUA.

We found that, between 2000 and 2005, the average interaction between pairs of socioeconomic factors was the strongest, whereas that between pairs of physical factors was the weakest (Table 5). For example, the top four interactions between pairs of socioeconomic factors included the population interacting with the following factors: distance to Guangzhou and Shenzhen (0.5201), total investment in fixed assets (0.5022), GDP (0.4988), and the proportion of secondary and tertiary industries in GDP (0.4983). At the same time, the greatest interaction between pairs of socioeconomic and physical factors was the pair of the population and the distance to coastline interacting (0.4970). In addition, the interaction between the distance to the coastline and the distance to rivers was the strongest in comparison the other pairs of physical factors in the periods.

Table 5. Interactions between the driving factors in 2000–2005 (Acronyms are defined in Table 1).

Factors	P_POP	GDP	ST_GDP	T_FAI	D_GS	D_RD	ELE	SLP	D_RV	D_CL
P_POP	0.4483									
GDP	0.4988	0.2994								
ST_GDP	0.4983	0.4910 *	0.0616							
T_FAI	0.5022	0.3332	0.5076 *	0.3255						
D_GS	0.5201	0.3348	0.2869 *	0.3424	0.2214					
D_RD	0.4551	0.3054	0.0952 *	0.3401	0.2338	0.0296				
ELE	0.4660	0.3169	0.1091	0.3489	0.2351	0.0823	0.0521			
SLP	0.4707	0.3229	0.0990	0.3733	0.2489	0.0754	0.0672	0.0483		
D_RV	0.4646	0.3144	0.1219 *	0.3830 *	0.2540	0.0699	0.0968 *	0.0974 *	0.0396	
D_CL	0.4970	0.4508	0.1243	0.4023	0.2842	0.1689 *	0.1139	0.1230	0.2308 *	0.0904

The symbol "*" denotes the nonlinear enhancement of two factors. The average interaction of each category: pairs of socioeconomic factors: 0.3830, pairs of physical factors: 0.0960, and pairs of socioeconomic and physical factors: 0.2785.

In 2005–2010, the interactions of the pairs of socioeconomic factors and pairs of physical factors still occupied the strongest and weakest positions, respectively, but the average interaction between the pairs of physical factors was significantly enhanced as compared to that of the previous period (Table 6). The top three interactions between pairs of socioeconomic factors comprised that of the distance to Guangzhou and Shenzhen with the population (0.4042), proportion of secondary and tertiary industries in GDP (0.3468), and GDP (0.3326). In addition, three of the four greatest interactions between the socioeconomic factors and physical factors were observed between the distance to rivers and the following factors: GDP (0.3684), distance to Guangzhou and Shenzhen (0.3602), and secondary and tertiary industries in GDP (0.3549).

Table 6. Interactions between the driving factors in 2005–2010 (Acronyms are defined in Table 1).

Factors	P_POP	GDP	ST_GDP	T_FAI	D_GS	D_RD	ELE	SLP	D_RV	D_CL
P_POP	0.1058									
GDP	0.3150	0.2742								
ST_GDP	0.3270 *	0.3104	0.1952							
T_FAI	0.1469	0.2965	0.3244 *	0.0851						
D_GS	0.4042 *	0.3326	0.3468	0.2645	0.2518					
D_RD	0.1475	0.2873	0.2262	0.1273	0.2755	0.0498				
ELE	0.2163	0.3438	0.2535	0.2156	0.3086	0.1829	0.1528			
SLP	0.2321	0.3699	0.2880	0.2327	0.3522	0.1929	0.2020	0.1695		
D_RV	0.1841 *	0.3684 *	0.3549 *	0.2068 *	0.3602 *	0.1267 *	0.2138	0.2369	0.0747	
D_CL	0.1216	0.2912	0.2123	0.1531 *	0.2777	0.1065 *	0.1703	0.1962	0.1984 *	0.0358

The symbol "*" denotes the nonlinear enhancement of two factors. The average interaction of each category: pairs of socioeconomic factors: 0.2755, pairs of physical factors: 0.1650, and pairs of socioeconomic and physical factors: 0.2480.

In 2010–2015, the average interaction between socioeconomic factors and physical factors was stronger than that between the other two types of interactions (Table 7). The strongest three interactions between pairs of socioeconomic and physical factors were observed between the distance to Guangzhou and Shenzhen and the distance to rivers (0.1125), GDP and the distance to rivers (0.1066), and the distance to Guangzhou and Shenzhen and the slope (0.0996). The strongest interaction between the pairs of socioeconomic factors was between the distance to Guangzhou and Shenzhen and the total investment in fixed assets (0.1095).

Table 7. Interactions between the driving factors in 2010–2015 (Acronyms are defined in Table 1).

Factor	P_POP	GDP	ST_GDP	T_FAI	D_GS	D_RD	ELE	SLP	D_RV	D_CL
P_POP	0.0239									
GDP	0.0887 *	0.0540								
ST_GDP	0.0753	0.1004	0.0567							
T_FAI	0.0415	0.0854 *	0.0680	0.0263						
D_GS	0.1078 *	0.0838	0.0988	0.1095 *	0.0693					
D_RD	0.0438	0.0710	0.0746	0.0495	0.0859	0.0225				
ELE	0.0686	0.0815	0.0870	0.0746	0.0922	0.0722	0.0526			
SLP	0.0770 *	0.0839	0.0946	0.0675	0.0996	0.0645	0.0688	0.0475		
D_RV	0.0776 *	0.1066 *	0.0993 *	0.0797 *	0.1125 *	0.0639 *	0.0894	0.0894 *	0.0395	
D_CL	0.0702 *	0.0777	0.0833	0.0669 *	0.0954	0.0760 *	0.0615	0.0676	0.0967 *	0.0331

The symbol "*" denotes the nonlinear enhancement of two factors. The average interaction of each category: pairs of socioeconomic factors: 0.0789, pairs of physical factors: 0.0646, and pairs of socioeconomic and physical factors: 0.1085.

The nonlinear enhancement, as the strongest type of enhancement, should be further analyzed. During 2000–2005, nonlinear enhancement was observed in the interactions related to the factors of the proportion of secondary and tertiary industries in GDP, distance to rivers, and distance to coastline, thus demonstrating that the interactions between these factors significantly promoted the expansion of urban land. From 2005 to 2015, the distance to river and the majority of the socioeconomic factors interacted nonlinearly, which indicated that the existence of rivers could act as a catalyst for urban expansion. It also showed that the number of pairs demonstrating nonlinear enhancement increased significantly during the study period, which indicated closer and more complicated interactions within the urban agglomeration (Tables 5–7).

5. Discussion: Implications of Political Effects

Policies and urban planning play a vital role in the expansion of urban land in the PRDUA, and their impact is reflected in the spatial pattern of urban land and changes in the driving factors. Based on the above results, several policies and major events were found to exert a significant impact on the expansion of urban land in the PRDUA.

The relatively high impact of the distance to Guangzhou and Shenzhen before 2010 proved that the spatial distribution of urban-land expansion was highly consistent with the distance to the core cities. This indicated that Guangzhou and Shenzhen are the economic cores of the area that stimulate regional development. However, the influence of the distance to Guangzhou and Shenzhen decreased during 2010–2015. This phenomenon indicated the effects of policy have promoted the spatial pattern of the PRDUA from the dual-center pattern to a relative integrated pattern. During the study period, the local government promoted three regional plans to manage and guide the urban development of the Pearl River Delta: Pearl River Delta Urban Cluster Coordinated Development Plan (2004–2020), The Outline of the Plan for the Reform and Development of the Pearl River Delta (2008–2020), and The Pearl River Delta Urban–Rural Integration Planning (2009–2020). In these three plans, the core ideas were to form a multi-center pattern of the region aimed at balancing the development of the eastern, western, and central areas, and forming a structure of large, middle, and small cites in the urban agglomeration. In these plans, the "Guangzhou-Foshan-Zhaoqing" was proposed to be considered as the central metropolitan area, "Shenzhen-Dongguan-Huizhou" as the eastern metropolitan area, and "Zhuhai-Zhongshan-Jiangmen" as the western metropolitan area. Guangzhou, Shenzhen, and Zhuhai were considered as the core cities of each metropolitan area. However, according to the urban–land expansion process observed in this study, the eastern, western, and central metropolitan areas in the PRDUA still developed in an unbalanced manner (Figure 6). In the central metropolitan area, Guangzhou and Foshan were relatively integrated, but Zhaoqing was lagging. In the eastern metropolitan area, Shenzhen and Dongguan were more developed with the development in Huizhou being marginal. In comparison, the western metropolitan area was least developed, and Zhuhai was less developed as compared with Guangzhou and Shenzhen. Overall, the decreasing influence of the distance to the Guangzhou and Shenzhen reflected the effect of the policies aimed at coordinating the development of the region, although the effects of the policies tended to exhibit a lag of approximately 5–10 years. In terms of urban-land development, more time may be required to realize multi-center integrated development in the PRDUA.

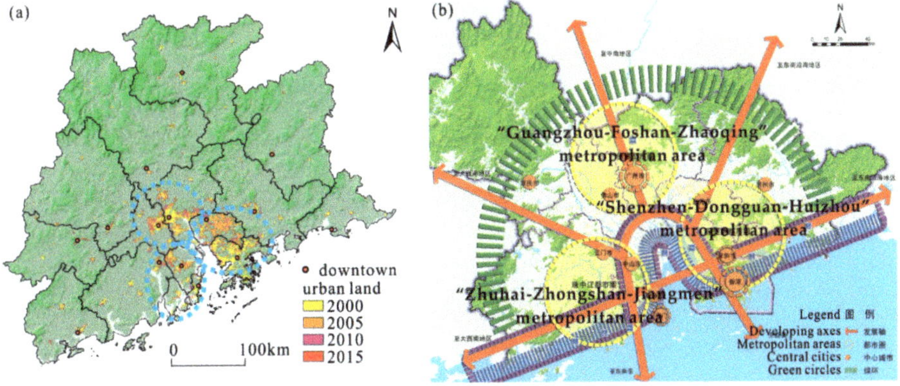

Figure 6. Comparison of (**a**) urban land expansion in 2000–2015 in the Pearl River Delta urban agglomeration with (**b**) the plan of The Pearl River Delta Urban–Rural Integration Planning (2009–2020).

The environmental protection policy also played an important role in the development of urban land in the PRDUA. The topographic factors were influential, and the expansion intensity was relatively low in higher and steeper districts in the north of the PRDUA (Table 4 and Figure 5). According to the division of "Guangdong Principal Functional Zoning Plan (2012–2020)," Shaoguan, Heyuan, and Qingyuan in the northern mountainous areas of the Pearl River Delta were the key ecological functional areas. The continuous mountainous forests acted as an ecological shelter and are important for maintaining a stable regional climate and environment. Therefore, the government restricted large-scale and high-intensity urbanization in these areas.

The total investment in fixed assets was also an influential socioeconomic factor during the study period (Table 4 and Figure 4). As the financial support for the urban construction, the fixed asset investment was largely determined by the government in China [20]. In 2004, Guangzhou was elected as the host city of the 2010 Asian Games, and Foshan, Dongguan, and Shanwei were the co-host cities. According to statistics, the total investment in fixed assets was mainly concentrated in Guangzhou, Foshan and Dongguan during this period [22,24,25]. It was found that the urban-land expansion in these cities was rapid between 2005 and 2010. The preparation and construction of the relevant venues attracted a large amount of investment, promoted infrastructure construction in these cities, boosted their economic growth, and accelerated the expansion of urban land.

6. Conclusions

Based the land use data derived from RS imagery in 2000, 2005, 2010, and 2015, the spatial and temporal patterns of urban land expansion in the coastal urban agglomeration of the Pearl River Delta were analyzed in this study. The geographical detector was used to analyze the driving mechanism of the urban-land expansion. New knowledge related to the urban-land expansion intensity, expansion difference, and fractal dimension of the PRDUA was obtained. Moreover, the interactions between a complex set of factors of urban-land expansion were analyzed quantitatively, and the influence of policy factors was discussed based on the physical and socioeconomic factors.

The results indicated that the area of the urban land in the PRDUA showed a continuous increase, but the speed and intensity of urban expansion slowed down as compared that in the initial period. The cities with faster development were mainly concentrated in the south-central part of the PRDUA. The difference in urban-land expansion gradually diminished. Moreover, the fractal dimension of the different cities was consistent, thus demonstrating the coordinated development of the cities within the urban agglomeration. The spatial structure of the PRDUA has gradually changed from having a dual core to a multi-center feature.

The driving force analysis showed that the terrain conditions such as elevation and slope affect the potential, intensity, direction, and scale of urban-land expansion. However, in the short-term study, the influence of geographical factors on the urban spatial expansion did not change significantly. The social and economic factors are the core driving forces of the expansion of urban land in the PRDUA. Between 2000 and 2005, factors such as population growth and economic development played the largest role in the expansion of the urban land in the PRDUA. Before 2010, the distance to the core cities had a relatively high explanatory power for the expansion of urban land. However, the influence of the core cities on urban-land expansion decreased in the last period. In addition, the interactions between the distance to rivers and other factors always demonstrated a non-linear enhancement in the urban-land expansion. During 2000–2010, the average interaction between pairs of socioeconomic factors was the strongest, whereas that between pairs of physical factors was the weakest. However, the average interaction between socioeconomic factors and physical factors was stronger than that between the other two types of interaction during 2010–2015. Moreover, the number of pairs demonstrating nonlinear enhancement increased significantly during the study period, which indicated closer and more complicated interactions within the urban agglomeration. Policies such as the regional urban plans, environmental protection policies, and major events exhibited a considerable impact on the expansion of the PRDUA's urban land, but the effects of the macro and long-term plans usually took 5–10 years to become effective, and even more time to realize the final prospects.

Based on these findings, we opine that the next challenge is to continuously propose specific alternative strategies and future plans for optimizing the spatial pattern of urban expansion in coastal areas. The influence of individual factors and their interactions may be applied to forecast the future urban development in the study area and elsewhere in coastal urban areas.

Author Contributions: Conceptualization, H.J. and S.Z.; Formal analysis, Y.Y. and H.J.; Methodology, H.J. and S.Z.; Supervision, H.J.; Validation, H.J., S.Z. and W.J.; Visualization, Y.Y.; Writing—original draft, Y.Y.; Writing—review & editing, Y.Y., H.J., S.Z. and W.J. All authors have read and agreed to the published version of the manuscript.

Funding: This research was funded by Fundamental Research Funds for the Central Universities: 2020.

Conflicts of Interest: The authors declare no conflict of interest.

References

1. Scialabba, N. *Integrated Coastal Area Management and Agriculture, Forestry and Fisheries*; Food and Agriculture Org.: Rome, Italy, 1998.
2. Seto, K.C. Exploring the dynamics of migration to mega-delta cities in Asia and Africa: Contemporary drivers and future scenarios. *Glob. Environ. Chang.* **2011**, *21*, S94–S107. [CrossRef]
3. Brown, S.; Nicholls, R.J.; Woodroffe, C.D.; Hanson, S.; Hinkel, J.; Kebede, A.S.; Neumann, B.; Vafeidis, A.T. *Sea-Level Rise Impacts and Responses: A Global Perspective*; Finkl, C.W., Ed.; Coastal Hazards; Springer: Dordrecht, The Netherlands, 2013; pp. 117–149.
4. Seto, K.C.; Fragkias, M.; Güneralp, B.; Reilly, M.K. A Meta-Analysis of Global Urban Land Expansion. *PLoS ONE* **2011**, *6*, e23777. [CrossRef] [PubMed]
5. Li, H.; Peng, J.; Liu, Y.; Hu, Y. Urbanization impact on landscape patterns in Beijing City, China: A spatial heterogeneity perspective. *Ecol. Indic.* **2017**, *82*, 50–60. [CrossRef]
6. Guetté, A.; Gaüzère, P.; Devictor, V.; Jiguet, F.; Godet, L. Measuring the synanthropy of species and communities to monitor the effects of urbanization on biodiversity. *Ecol. Indic.* **2017**, *79*, 139–154. [CrossRef]
7. Kaushal, S.S.; McDowell, W.H.; Wollheim, W.M. Tracking evolution of urban biogeochemical cycles: Past, present, and future. *Biogeochemistry* **2014**, *121*, 1–21. [CrossRef]
8. Kennedy, C.A.; Stewart, I.; Facchini, A.; Cersosimo, I.; Mele, R.; Chen, B.; Uda, M.; Kansal, A.; Chiu, A.; Kim, K.-G.; et al. Energy and material flows of megacities. *Proc. Natl. Acad. Sci. USA* **2015**, *112*, 5985–5990. [CrossRef]
9. McGranahan, G.; Balk, D.; Anderson, B. The rising tide: Assessing the risks of climate change and human settlements in low elevation coastal zones. *Environ. Urban* **2007**, *19*, 17–37. [CrossRef]
10. He, C.; Liu, Z.; Tian, J.; Ma, Q. Urban expansion dynamics and natural habitat loss in China: A multiscale landscape perspective. *Glob. Chang. Biol.* **2014**, *20*, 2886–2902. [CrossRef]
11. Wu, W.; Zhao, S.; Zhu, C.; Jiang, J. A comparative study of urban expansion in Beijing, Tianjin and Shijiazhuang over the past three decades. *Landsc. Urban Plan.* **2015**, *134*, 93–106. [CrossRef]
12. Blaschke, T.; Hay, G.J.; Weng, Q.; Resch, B. Collective Sensing: Integrating Geospatial Technologies to Understand Urban Systems—An Overview. *Remote Sens.* **2011**, *3*, 1743–1776. [CrossRef]
13. He, C.; Tian, J.; Shi, P.; Hu, D. Simulation of the spatial stress due to urban expansion on the wetlands in Beijing, China using a GIS-based assessment model. *Landsc. Urban Plan.* **2011**, *101*, 269–277. [CrossRef]
14. Herold, M.; Scepan, J.; Clarke, K.C. The Use of Remote Sensing and Landscape Metrics to Describe Structures and Changes in Urban Land Uses. *Environ. Plan. A Econ. Space* **2002**, *34*, 1443–1458. [CrossRef]
15. Haas, J.; Ban, Y. Urban growth and environmental impacts in Jing-Jin-Ji, the Yangtze, River Delta and the Pearl River Delta. *Int. J. Appl. Earth Obs. Geoinf.* **2014**, *30*, 42–55. [CrossRef]
16. Ye, L. Urban transformation and institutional policies: Case study of mega-region development in China's Pearl River Delta. *J. Urban Plan. Dev.* **2013**, *139*, 292–300. [CrossRef]
17. Fang, C. Important progress and future direction of studies on China's urban agglomerations. *J. Geogr. Sci.* **2015**, *25*, 1003–1024. [CrossRef]
18. Shao, J.-A.; Wei, C.-F.; Xie, D.-T. An insight on drivers of land use change at regional scale. *Chin. Geogr. Sci.* **2006**, *16*, 176–182. [CrossRef]
19. Fang, S.; Gertner, G.Z.; Sun, Z.; Anderson, A.A. The impact of interactions in spatial simulation of the dynamics of urban sprawl. *Landsc. Urban Plan.* **2005**, *73*, 294–306. [CrossRef]
20. Ju, H.; Zhang, Z.; Zuo, L.; Wang, J.; Zhang, S.; Wang, X.; Zhao, X. Driving forces and their interactions of built-up land expansion based on the geographical detector—A case study of Beijing, China. *Int. J. Geogr. Inf. Sci.* **2016**, *30*, 1–20. [CrossRef]
21. National Bureau of Statistics of the People's Republic of China. *China City Statistical Yearbook*; China Statistics Press: Beijing, China, 2001. (In Chinese)
22. National Bureau of Statistics of the People's Republic of China. *China City Statistical Yearbook*; China Statistics Press: Beijing, China, 2016. (In Chinese)

23. Zhang, Z.; Wang, X.; Zhao, X.; Liu, B.; Yi, L.; Zuo, L.; Wen, Q.; Liu, F.; Xu, J.; Hu, S. A 2010 update of National Land Use/Cover Database of China at 1:100,000 scale using medium spatial resolution satellite images. *Remote Sens. Environ.* **2014**, *149*, 142–154. [CrossRef]

24. National Bureau of Statistics of the People's Republic of China. *China City Statistical Yearbook*; China Statistics Press: Beijing, China, 2006. (In Chinese)

25. National Bureau of Statistics of the People's Republic of China. *China City Statistical Yearbook*; China Statistics Press: Beijing, China, 2011. (In Chinese)

26. Chen, Y.; Wang, J.; Feng, J. Understanding the Fractal Dimensions of Urban Forms through Spatial Entropy. *Entropy* **2017**, *19*, 600. [CrossRef]

27. Wang, H.; Zhang, B.; Liu, Y.L.; Liu, Y.F.; Xu, S.; Deng, Y.; Zhao, Y.; Chen, Y.; Hong, S. Multi-dimensional analysis of urban expansion patterns and their driving forces based on the center of gravity-GTWR model: A case study of the Beijing-Tianjin-Hebei urban agglomeration. *Acta Geogr. Sin.* **2018**, *73*, 1076–1092. (In Chinese)

28. Wang, J.; Xu, C. Geodetector: Principle and prospective. *Acta Geogr. Sin.* **2017**, *72*, 116–134.

29. Wang, J.; Li, X.; Christakos, G.; Liao, Y.; Zhang, T.; Gu, X.; Zheng, X. Geographical Detectors-Based Health Risk Assessment and its Application in the Neural Tube Defects Study of the Heshun Region, China. *Int. J. Geogr. Inf. Sci.* **2010**, *24*, 107–127. [CrossRef]

30. Shu, B.; Zhang, H.; Li, Y.; Qu, Y.; Chen, L. Spatiotemporal variation analysis of driving forces of urban land spatial expansion using logistic regression: A case study of port towns in Taicang City, China. *Habitat Int.* **2014**, *43*, 181–190. [CrossRef]

31. Braimoh, A.K.; Onishi, T. Spatial determinants of urban land use change in Lagos, Nigeria. *Land Use Policy* **2007**, *24*, 502–515. [CrossRef]

32. Ma, Y.; Xu, R. Remote sensing monitoring and driving force analysis of urban expansion in Guangzhou City, China. *Habitat Int.* **2010**, *34*, 228–235. [CrossRef]

33. Li, X.; Zhou, W.; Ouyang, Z. Forty years of urban expansion in Beijing: What is the relative importance of physical, socioeconomic, and neighborhood factors? *Appl. Geogr.* **2013**, *38*, 1–10. [CrossRef]

34. Wu, K.Y.; Zhang, H. Land use dynamics, built-up land expansion patterns, and driving forces analysis of the fast-growing Hangzhou metropolitan area, eastern China (1978–2008). *Appl. Geogr.* **2012**, *34*, 137–145. [CrossRef]

35. Wang, J.F.; Liu, X.; Christakos, G.; Liao, Y.L.; Gu, X.; Zheng, X.Y. Assessing local determinants of neural tube defects in the Heshun Region, Shanxi Province, China. *BMC Public Health* **2010**, *10*, 52. [CrossRef]

36. Liu, J.; Zhang, Z.; Xu, X.; Kuang, W.; Zhou, W.; Zhang, S.; Li, R.; Yan, C.; Yu, N.; Wu, S.; et al. Spatial patterns and driving forces of land use change in China during the early 21st century. *J. Geogr. Sci.* **2010**, *20*, 483–494. [CrossRef]

Article

Spatiotemporal Patterns and Drivers of the Surface Urban Heat Island in 36 Major Cities in China: A Comparison of Two Different Methods for Delineating Rural Areas

Lu Niu [1,2], **Ronglin Tang** [1,2,*], **Yazhen Jiang** [1,2] **and Xiaoming Zhou** [3]

[1] State Key Laboratory of Resources and Environment Information System, Institute of Geographic Sciences and Natural Resources Research, Chinese Academy of Sciences, Beijing 100101, China; niul.17s@igsnrr.ac.cn (L.N.); jiangyz@lreis.ac.cn (Y.J.)

[2] University of Chinese Academy of Sciences, Beijing 100049, China

[3] School of Civil Engineering, Lanzhou University of Technology, Lanzhou 730050, China; zhouxm0905@126.com

[*] Correspondence: trl_wd@163.com; Tel.: +86-106-488-8172

Received: 5 December 2019; Accepted: 5 January 2020; Published: 8 January 2020

Abstract: Urban heat islands (UHIs) are an important issue in urban sustainability, and the standardized calculation of surface urban heat island (SUHI) intensity has been a common concern of researchers in the past. In this study, we used the administrative borders (AB) method and an optimized simplified urban-extent (OSUE) algorithm to calculate the surface urban heat island intensity from 2001 to 2017 for 36 major cities in mainland China by using Moderate Resolution Imaging Spectroradiometer (MODIS) images. The spatiotemporal differences between these two methods were analyzed from the perspectives of the regional and national patterns and the daily, monthly, and annual trends. Regardless of the spatial or temporal scale, the calculation results of these two methods showed extremely similar patterns, especially for the daytime. However, when the calculated SUHI intensities were investigated through a regression analysis with multiple driving factors, we found that, although natural conditions were the main drivers for both methods, the anthropogenic factors obtained from statistical data (population and gross domestic product) were more correlated with the SUHI intensity from the AB method. This trend was probably caused by the spatial extent of the statistical data, which aligned more closely with the rural extent in the AB method. This study not only explores the standardization of the calculation of urban heat intensity but also provides insights into the relationship between urban development and the SUHI.

Keywords: surface urban heat island; MODIS; land cover; urban sustainability

---

1. Introduction

In 2014, over 54% of people in the world lived in densely populated urban areas, and it is expected that the global urban population will increase to 66% of the total population by 2050 [1]. Rapid urbanization is one of the most important human influences on the atmosphere and causes obvious disruptions of atmospheric energy [2,3]. The most widely known consequence of these influences is the generation of the urban heat island effect. The urban heat island concept was first proposed by Howard in 1833 [4], and it is used to describe the phenomenon where the temperature of an urban area is higher than that of the surrounding rural area. In the past several decades, a large number of scholars have studied the damage that urban heat islands bring to the natural environment and human life, including reducing biodiversity [5], destroying vegetation growth [6], reducing the quality of air and water [7], changing the climate [8,9], and increasing morbidity and mortality [10–12]. There is ample evidence that when these effects interact with global changes, the intensity of the impact is further increased [9].

Generally, previous studies have considered two types of urban heat islands: the atmospheric urban heat island (AUHI) and the surface urban heat island (SUHI) [13]. Data on the AUHI are generally obtained from the analysis of in situ air temperatures from weather stations [8]. However, the number of weather stations is low, which results in a low observation density, and they are unevenly distributed [11]. It is, therefore, difficult to obtain the spatial distribution of the AUHI for large areas. Conversely, remote sensing technology can periodically obtain comprehensive coverage of land surface temperatures (LSTs). Since the first analysis on the SUHI by Rao in 1972 [14], many scholars have conducted research on the SUHI, based on thermal infrared remote sensing data [15–22].

As the SUHI intensity is simple, efficient and universal, it has become the most common indicator for researching the SUHI [23,24]. The principle for estimating the SUHI intensity is to find an urban-rural pair and then calculate the temperature difference between the urban and rural areas. Therefore, the core problem for calculating SUHI intensity is delineating urban and rural areas [18]. Most studies have adopted a relatively consistent approach for identifying urban areas: the areas classified as built-up land in land cover maps are considered urban areas [13,25–27]. For the determination of rural areas, it is generally assumed that all areas within a certain range around the urban area are rural; thus, the determination of the rural area extent has been an important issue. In past studies, the most common methods were to set a buffer zone at a certain distance (CD), such as 1–50 km [13,19,28–31], or to set a buffer zone based on the size of the urban area (SUA), such as 0.5–3 times the area of the urban area [18,32–34]. However, regardless of which method is adopted, there is no widely recognized standard method.

The standardization of urban heat island intensity calculations is a scientific issue of great value, and considerable work has been performed on this topic. Schwarz et al. [35] first compared the temperature differences between core-rural areas and urban-other areas, which then became a milestone for remote sensing studies of the SUHI [36]. At the global scale, Clinton and Gong [19] found that the SUHI intensity became larger as the rural buffer became larger. Zhou et al. [32] estimated that the footprint of the UHI effect can reach 2.3 and 3.9 times more than the urban area size in, respectively, daytime and nighttime, with large spatiotemporal heterogeneities. Li et al. [13] characterized the significances and spatiotemporal dynamics of rural-extent-induced SUHI intensity variations in mainland China, and they found that differences in rural definitions may lead to significant uncertainties in the studies of SUHI intensity. Zhou et al. [36] proposed that large uncertainties associated with the definitions of urban areas and its references have been one of the most urgent issues in UHI studies.

According to previous studies, the current calculation method for SUHI intensity has the following problems. (1) It is not appropriate to simply adopt a buffer radius, especially in the case of the simultaneous investigation of multiple cities with different geographical locations and different levels of socioeconomic development [27]. (2) Choosing different rural extents will result in a larger or even opposite difference in the calculation results. For example, Peng et al. [18] defined the nearby suburbs as rural reference areas and found that the global average summer SUHI intensities for daytime and nighttime were 1.9 °C and 1.0 °C, respectively, while Zhang et al. [20] used the 20 km surrounding buffer zone as the extent of the rural area and obtained daytime and nighttime SUHI intensities of 2.6 °C and 1.6 °C, respectively. This limitation impedes the accurate characterization of the spatiotemporal patterns of the SUHI intensity and distorts the comparability between previous studies. (3) Although some studies have analyzed the relationship between the SUHI intensity obtained from a certain size buffer and the various statistical data obtained from administrative districts, this strategy is also inappropriate [37,38] due to the differing spatial extents of the two data representations.

Compared with the CD and the SUA, adopting an appropriate fixed boundary (FB) for the rural area can solve the above problems to some extent. Lai et al. [27] employed the groundbreaking administrative borders (AB) method to determine the extent of rural areas and explored the typical diurnal patterns of the surface urban heat islands in China. Chakraborty and Lee [25] proposed the simplified urban-extent (SUE) algorithm based on the urban area data provided by Natural Earth and compared the results with those of the previous multicity SUHI studies to verify the availability of the algorithm. This comparison was an important step in the process of the standardization of the calculation of SUHI intensity. However, no researchers have analyzed the differences between these two representative fixed boundaries in the existing UHI literature. There is an obvious need to better understand the UHI phenomenon by comparing the differences between these methods for SUHI intensity calculation to help city planners and other researchers make more sensible decisions about the future of urban sustainability.

The objectives of this research were (i) to compare the differences in spatiotemporal patterns of SUHI intensity between the AB and SUE methods in 36 Chinese major cities and (ii) to identify the factors associated with the SUHI intensity by considering a series of social, economic and natural factors with these two methods. Section 2 describes the study area, data and comparison method in this study. Section 3 shows the spatiotemporal patterns and drivers of the SUHI intensity in 36 Chinese cities. Section 4 discusses the results and highlights the notable problems.

2. Materials and Methods

2.1. Study Area

China is a large country located in East Asia that faces the western cost of the Pacific Ocean. Past research shows that there is a large precipitation gradient between Northwest and Southeast China [13], and different levels of the urban heat island (UHI) phenomenon have been observed among different areas in China. In this research, 36 major cities in China were selected as the research areas, and they contain 31 capital cities or direct-controlled municipalities and 5 important port cities (Shenzhen, Ningbo, Xiamen, Qingdao and Dalian). These major cities are the economic (such as Shanghai) or political (such as Beijing) centers of each region and have experienced significant urbanization in the past decade, which deserves further study to determine the associated changes of the UHI effect. The administrative areas of the above 36 cities were defined based on data from the National Geomatics Center of China (NGCC). In addition to studying the differences in SUHI intensity between cities in different regions according to natural conditions, climate conditions and social economic conditions, we divided China into six regions(www.resdc.cn). These 6 regions were the separate northern, northwestern, northeastern, eastern, central-southern and southwestern areas, as shown in Figure 1.

2.2. Data

This paper employed Terra and Aqua Moderate Resolution Imaging Spectroradiometer (MODIS) 8-day composite land surface temperature (LST) products (MOD11A2 and MYD11A2; collection 6 at 1 km × 1 km resolution) from 2001 (the Aqua data is from 2003–2017) to gain the LST. The LST product was retrieved from the clear-sky observations through the generalized split-window LST algorithm [39] at approximately 1:30 (nighttime, Aqua), 10:30 (daytime, Terra), 13:30 (daytime, Aqua) and 22:30 (nighttime, Terra) local solar time. The research of Wan [40] showed that the retrieval errors of MODIS LST were generally <1 k and the root mean square (RMS) was <0.5 k. Wan [41] then pointed out that the mean error of collection for 6 LST datasets was within 0.6 k in 10 validation data samples.

We use the land cover products that were obtained from the MODIS yearly land cover type L3 global 500 m products (MCD12Q1; collection 6 at 500 m × 500 m resolution) from 2001–2017 for the delineation of the urban and rural areas. Derived from MODIS Aqua and Terra, the product is a 17-category classification under the International Geosphere-Biosphere Program (IGBP) scheme. The Normalized Difference Vegetation Index (NDVI) product was obtained from the Aqua MODIS 16-day composite dataset (MYD13Q1, collection 6, at 250 m × 250 m resolution). To match the resolution of the LST, we aggregate the land cover products and NDVI products to 1 km. Similarly, we use the method to find the difference in the enhanced vegetation index (EVI) (ΔEVI), which is a proxy for green vegetation density differences between the urban and rural pixels. The data were obtained from the Aqua 16-day EVI dataset, which is available at the resolution of 250 m × 250 m (MYD13Q1) in the same time span.

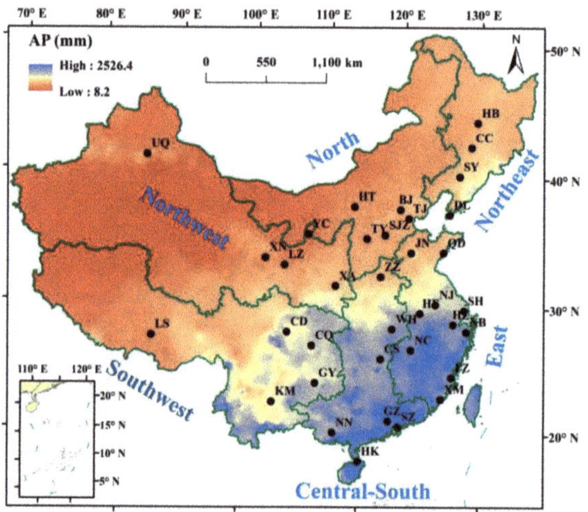

Figure 1. Geolocation of the selected 36 major cities and 6 regions across China. BJ (Beijing); CC (Changchun); CS (Changsha); CD (Chengdu); CQ (Chongqing); DL (Dalian); FZ (Fuzhou); GZ (Guangzhou); GY (Guiyang); HK (Haikou); HZ (Hangzhou); HB (Harbin); HF (Hefei); HT (Hohhot); JN (Jinan); KM (Kunming); LZ (Lanzhou); LS (Lhasa); NC (Nanchang); NJ (Nanjing); NN (Nanning); QD (Qingdao); SH (Shanghai); SY(Shenyang); SZ (Shenzhen); SJZ (Shijiazhuang); TY (Taiyuan); TJ (Tianjin); UQ (Urumqi); WH (Wuhan); XA (Xi'an); XN (Xining); YC (Yinchuan); and ZZ (Zhengzhou). The background information is annual precipitation (AP) across China.

To eliminate the influence of the elevation factor on the calculation for the SUHI intensity, we selected NASA Shuttle Radar Topography Mission (SRTM) data provided by the United States Geological Survey (USGS). Previous research has shown that these data have better vertical accuracy than the Global Multiresolution Terrain Elevation Data from 2010 (GMTED2010) and the advanced spaceborne thermal emission and reflection radiometer (ASTER) elevation data [42].

The urban area data (UAD), which were used in the SUE algorithm, were from Natural Earth (www.naturalearthdata.com). Natural Earth is a combination of the global urban land use database [43,44] and the Landscan population database of the Oak Ridge National Laboratory [45]. This dataset has already been validated using a Landsat-based map of 140 urban areas in different ecoregions and for different levels of population and economic development and has an overall accuracy of 93% [44]. More information about these data can be found from www.naturalearthdata.com.

The contributions of anthropogenic activity can be calculated by separately considering the major sources of economic development and human metabolism. This study used statistical datasets of the gross domestic products (GDPs) and populations of 36 major cities from 2001 to 2017, which were available from the National Bureau of Statistics (www.stats.gov.cn). We also collected a spatial interpolation dataset (at 1 km × 1 km resolution) of annual precipitation and annual moisture index from the Resource and Environment Data Cloud Platform (www.resdc.cn).

2.3. Method

This research work can be divided into three parts (Figure 2). First, the urban areas and rural areas were delineated with two different methods. Second, the SUE and AB were used to calculate the SUHI intensity of the long time-series data of 36 major cities in mainland China and then to analyze the spatiotemporal differences between the calculation results obtained by the two methods. Third, a correlation analysis was performed between the SUHI and its possible driving factors.

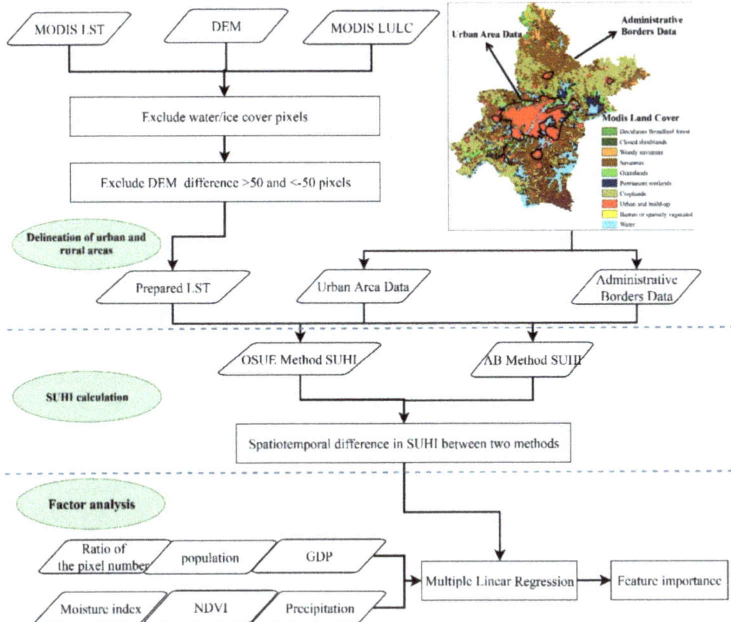

Figure 2. Flowchart of this article. Tianjin was used as an example to show the urban-rural definition in this research. LST: land surface temperature; DEM: digital elevation model; LULC: land use land cover; SUHI: surface urban heat island; AB: administrative borders; OSUE: optimized simplified urban-extent; GDP: gross domestic product; NDVI: normalized difference vegetation index.

2.3.1. Delineation of Urban and Rural Areas

In this study, we delineated urban and rural areas with the MODIS land cover product MCD12Q1. Within each city, we first removed certain types of pixels: the pixels classified as snow and ice and the pixels in an extremely high or low position (pixels with elevations higher or lower than 50 m than the average of built-up pixels). The removal of those pixels is necessary to eliminate the possible effects of temperature from water bodies and extreme positions. Excluding specific pixels is a common processing method in current SUHI calculations. Yao et al. [46] carefully analyzed the impact of removing these pixels in the SUHI intensity calculation and suggested that these pixels should be removed when calculating the multicity SUHI intensity. For each city with its rural extents, pixels classified as built-up among the remaining pixels were then given a flag as urban areas. Accordingly, we referred to the other pixels after that as rural areas. In this paper, the differences between the two methods (SUE and AB method) may occur based on only the rural area extent. The rural area extents in the SUE method were the largest UAD area within the boundaries of the administrative district of each city. The rural area extents of the AB method were the administrative boundaries of each city (Figure 2). It should be noted that different elevation data are input in the SUE algorithm implemented in this paper, unlike with the original version; therefore, we named it the optimized SUE to differentiate it from the previous method.

2.3.2. Calculation of the SUHI Intensity

After determining the rural area extent of each city with both methods, we first processed the *LST* image using the MODIS land cover products. We excluded the pixels classified as snow and ice at the beginning due to the possible miscalculation caused by extremely low temperatures. The pixels classified as water or permanent wetlands were also removed to avoid the influence from water temperature. In addition, based on the STRM data, we eliminated pixels in each city that were at elevations of more than ±50 m of that of the built-up pixels to suppress the impact of elevation on the *LST*. After the above processing, we used Equation (1) to calculate the SUHI intensity as follows:

$$SUHI = LST_{Urban} - LST_{Rural} \tag{1}$$

where *SUHI* represents the urban heat island intensity of the city, LST_{Urban} is the average land surface temperature in the urban area, and LST_{Rural} is the average temperature of the pixels in the rural area.

To compare the differences in the temporal trends between the AB method and the OSUE method SUHI, we combine the Theil–Sen median trend analysis and the Mann–Kendall test to analyze the calculated results. Theil–Sen median trend analysis is a robust statistical method for trend analysis [42], and it calculates the median slopes between all of the possible n(n − 1)/2 pairwise combinations throughout the time-series data. The method is based on nonparametric statistics and is particularly effective when the estimation of trends in small series is needed [43]. The Mann–Kendall test is applied to measure the significance of a given trend. It is a nonparametric statistical test with an advantage of being irrelevant to the distributions of its samples. The test is also free from the interference of outliers [44]. The effective combination of these two methods is an important method for judging the trends of long-term series data [45].

In this study, we measured the significance of SUHI intensity trends in the daytime and nighttime from 2003–2017 at a confidence level of 0.05. To facilitate the comparison of the differences between the different methods, we divided the trend from all cities into four categories according to the results obtained from the two aforementioned methods (Theil–Sen median trend analysis and the Mann–Kendall test): significant decrease, nonsignificant decrease, significant increase and nonsignificant increase.

2.3.3. Driving Factor Analysis

In past research, many scholars have studied the influencing factors of SUHI intensity and identified the main driving factors as natural factors, such as the NDVI, EVI, and precipitation, and social factors, such as the population and GDP. Among these factors, the data that characterize the natural conditions are usually raster data obtained by interpolation; and the statistical data that describe the socioeconomic development are mostly annual scale data obtained with the administrative area as the unit. We selected the NDVI, precipitation (Pr), and moisture index (M) to represent these factors specifically, and selected the two most important indicators of urban development level, GDP and population, and analyzed them with the SUHI intensity. In addition, one of the essential differences between the two calculation methods was the difference in the number of pixels between the urban and suburban areas involved in the calculation. To this end, we also defined a new indicator: the ratio of the number of urban pixels to the number of rural pixels involved in the calculation in each city (RN).

3. Results

3.1. Spatial Distribution

3.1.1. National Patterns

The annual SUHI, summer SUHI and winter SUHI have attracted the most attention from scholars in the past. Therefore, in this study, we also described the daytime and nighttime SUHI in 36 major cities in China (Figures 3 and 4), from 2001 to 2017, for these three conditions to understand the national distribution pattern of SUHI. (As Guangdong and Shenzhen have a high urbanization level and the distribution is centralized, they have been defined as one city in the OSUE method; there are 35 cities in Figure 4.) The daytime and nighttime intensities were consolidated values from the Terra and Aqua platforms.

Most cities had a positive annual daytime SUHI intensity. The AB and OSUE methods identified the proportions of SUHI intensity with positive values at 86% and 94%, respectively. The annual nighttime SUHI intensity values were all positive. The lower annual SUHI intensity values in the daytime were mainly in the northwestern, southwestern and northern areas, and the lower annual SUHI intensity values were distributed in the southeastern and central-southern areas; this distribution was opposite to the trend of the daytime values. The annual daytime SUHI in both methods showed negative values that were significantly lower than the national average level in Lanzhou. Lanzhou is a dry city, and according to the land cover maps, Lanzhou had a large number of pixels classified as the barren land cover type. The occurrence of this urban cold island phenomenon was consistent with the conclusions from the studies of Chakraborty and Lee [25], who indicated that the urban clusters with negative surface UHI were concentrated in the dry and desert areas.

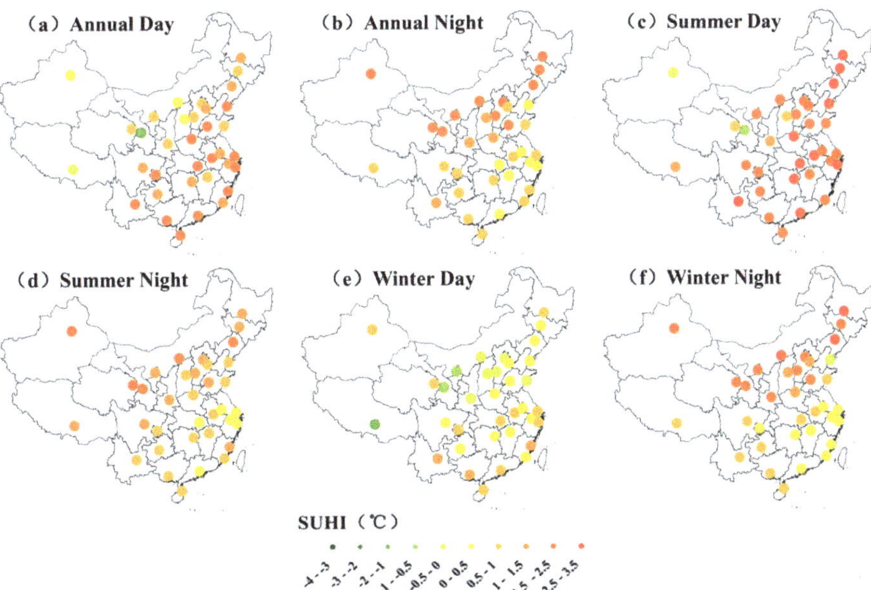

Figure 3. Spatial distribution of the daytime and nighttime SUHI from the optimized simplified urban-extent (OSUE) method. The daytime value is the mean of the UHI intensity at 1030 LT obtained from Terra (2001–2017) and the UHI intensity at 1330 LT obtained from Aqua (2003–2017). The nighttime value is the mean of the UHI intensity at 2230 LT obtained from Terra and the UHI intensity at 0130 LT obtained from Aqua.

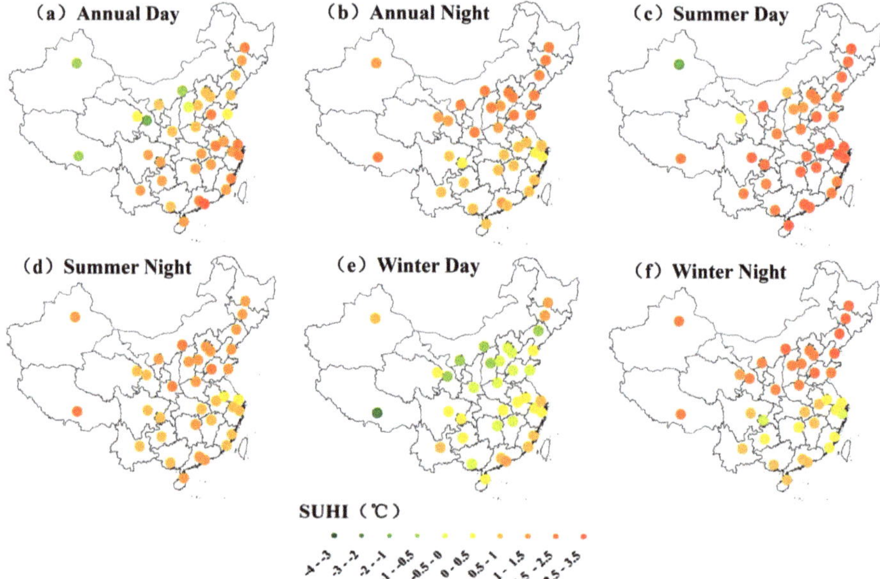

Figure 4. Spatial distribution of the daytime and nighttime SUHI from the administrative borders (AB) method. The daytime value is the mean of the SUHI intensity at 1030 LT obtained from Terra (2001–2017) and the SUHI intensity at 1330 LT obtained from Aqua (2003–2017). The nighttime value is the mean of the SUHI intensity at 2230 LT obtained from Terra and the SUHI intensity at 0130 LT obtained from Aqua.

The summer daytime SUHI in both methods was significantly higher than the winter daytime and annual daytime SUHI values. This outcome showed that the SUHI in most cities had an obvious seasonal effect. The winter daytime SUHI in both methods showed obvious differences between the northern and the southern areas, and the SUHI intensity of the northern cities was significantly higher than that of the southern cities. This outcome may be related to the central heating policy in northern China. Most cities in northern China adopt this approach, which significantly enhances the SUHI intensity during winter nights. The daytime winter SUHI intensity from the AB method showed that nearly half of the cities had negative values, while with the OSUE method, only one quarter of cities had negative values. This difference may be related to the fact that the surface temperature of vegetation during the day is higher than that of the buildings in the winter and the AB method usually includes more suburban areas.

3.1.2. Variations across Regions

To further observe the distribution of the annual SUHI intensity in different regions, we calculated the average SUHI intensity for all cities in each region from 2001–2017. It can be seen, from Figure S1, that although there are some differences in the numerical values of SUHI intensity calculated by the two methods, the relative differences between the different regions were basically the same. From the average of each region, the annual daytime SUHI intensity in the northern region was significantly lower than that in the southern region. In the nighttime, except for the northwestern region, the SUHI intensity value calculated by the AB method was higher than that of the OSUE method. The average value of the daytime SUHI intensity in the northwest was the lowest among all regions, although the standard deviation was too large (mainly due to the low SUHI intensity in Lanzhou City during the day). As a result, there was no clear point of reference for this average value. Significant differences within each region can lead to quite high standard error values in many cases. Increasing the number of cities in the study could improve the result to some extent, suggesting that the impact of climate on the SUHI intensity is limited.

3.1.3. Correlation Analysis of the Annual SUHI

We conducted a correlation analysis of the annual average SUHI intensities of 34 cities calculated by the AB and OSUE methods from 2001–2017 (the SUHI data obtained from the Terra platform was for 2003–2017) to further study the differences between the two methods. Figure 5 shows that the annual daytime SUHI intensity calculated by the two methods had a strong correlation with both the Terra platform and the Aqua platform, with the R^2 values reaching 0.662 and 0.643, respectively. When the SUHI intensity was low, the OSUE method had a certain degree of overestimation compared to the AB method.

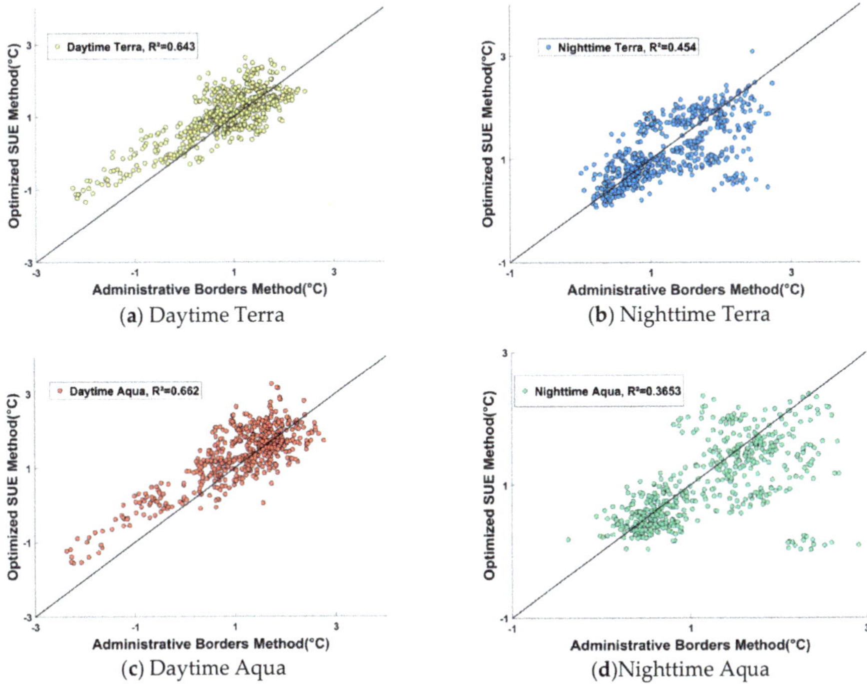

(**a**) Daytime Terra   (**b**) Nighttime Terra

(**c**) Daytime Aqua   (**d**) Nighttime Aqua

Figure 5. Differences in the spatial variations in the annual surface urban heat island intensity based on the administrative borders (AB) and optimized simplified urban-extent (OSUE) methods. (**a**) Daytime Terra; (**b**) Nighttime Terra; (**c**) Daytime Aqua; (**d**) Nighttime Aqua.

3.2. Temporal Patterns

3.2.1. Annual Trend

Figures 6 and 7 show that the calculation results from the two methods had similar variation tendencies in most areas. A completely opposite tendency was observed in the daytime SUHI intensities of Guiyang and Hefei and the nighttime SUHI intensity of Taiyuan with the application of the two methods. In the 34 cities, the proportion of cities with small differences in the daytime SUHI and nighttime SUHI calculated by the two methods was 77.8% and 88.9%, respectively. For these two methods, the cities with different daytime SUHI intensities and nighttime time-varying trends were mainly located in southeastern and northern China, which was primarily related to the variation of the SUHI drivers in northern and southern China.

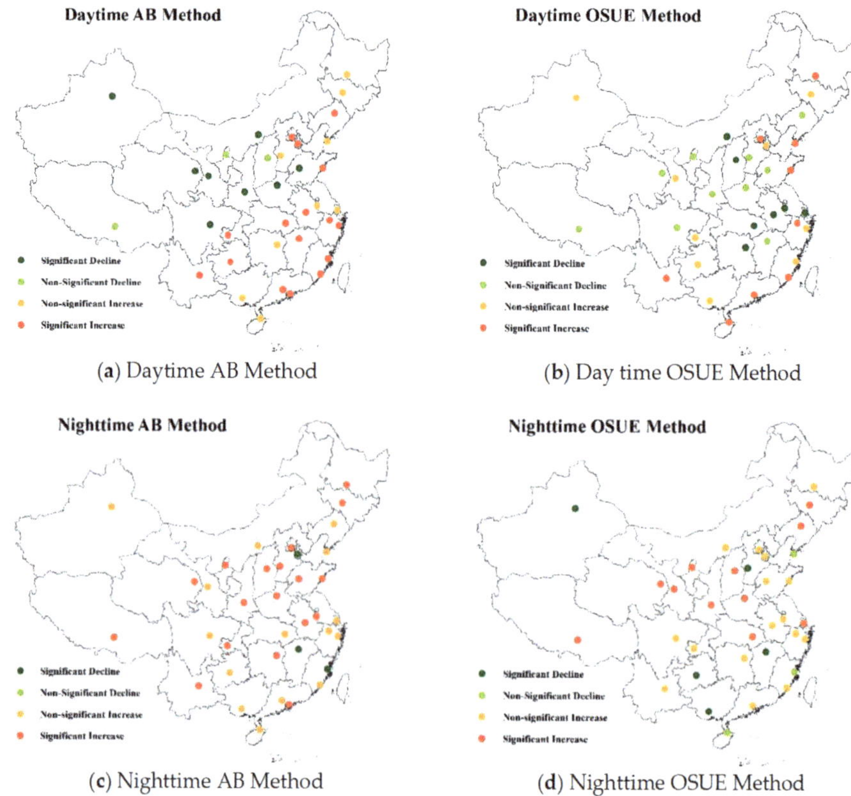

(**a**) Daytime AB Method

(**b**) Day time OSUE Method

(**c**) Nighttime AB Method

(**d**) Nighttime OSUE Method

Figure 6. Trends in the annual surface urban heat island (SUHI) in 36 major cities in China with the administrative borders (AB) and optimized simplified urban-extent (OSUE) methods based on Terra data. (**a**) Daytime AB Method; (**b**) Day time OSUE Method; (**c**) Nighttime AB Method; (**d**) Nighttime OSUE Method.

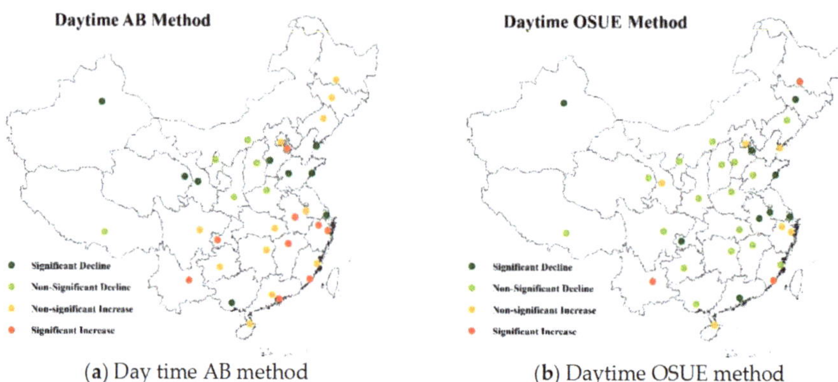

(**a**) Day time AB method

(**b**) Daytime OSUE method

Figure 7. *Cont.*

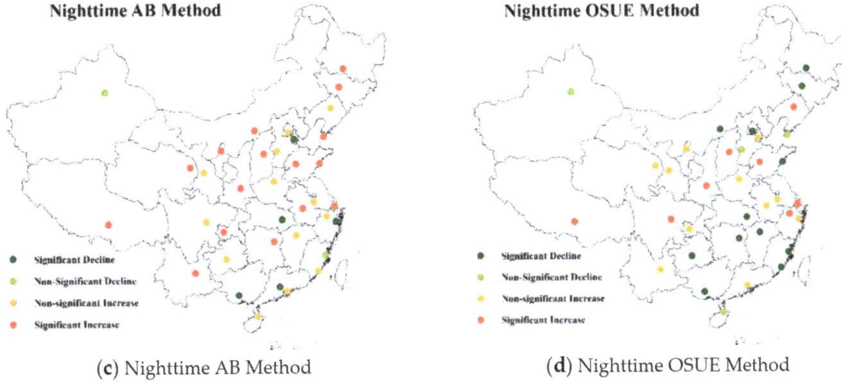

(**c**) Nighttime AB Method   (**d**) Nighttime OSUE Method

Figure 7. Trends of annual surface urban heat island (SUHI) in 36 major cities in China with the administrative borders (AB) and optimized simplified urban-extent (OSUE) methods based on Aqua data. (**a**) Day time AB method; (**b**) Daytime OSUE method; (**c**) Nighttime AB Method; (**d**) Nighttime OSUE Method.

3.2.2. Monthly Trends

The monthly average SUHI intensity for 12 months (one year) is calculated by the two methods and presented in Figure 8. The seasonal variation tendency of SUHI intensity by the two methods is very close. The daytime SUHI intensity is the highest in summer and the lowest in winter. The summer SUHI intensity calculated by the AB method is obviously higher than that calculated by the OSUE method, whereas the winter SUHI intensity calculated by the AB method is slightly lower than that calculated by the OSUE method. The seasonal variations of SUHI intensity are more obvious when applying the OSUE method.

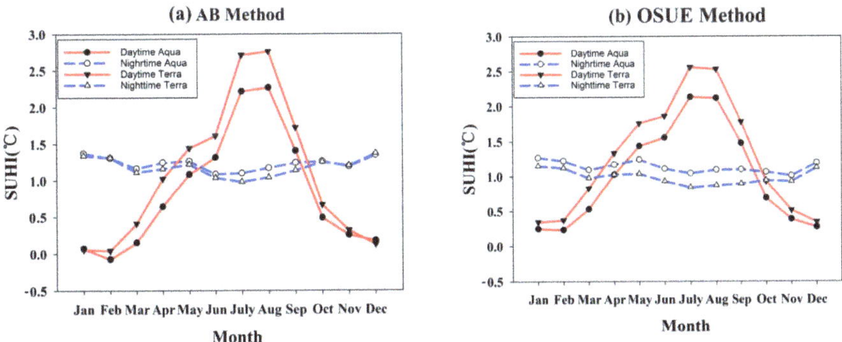

Figure 8. Trends of the monthly surface urban heat island (SUHI) intensity with the administrative borders (AB) and optimized simplified urban-extent (OSUE) methods in 36 major cities of China.

We conducted a correlation analysis to compare the SUHI intensity using the two methods with data obtained from the Aqua and Terra platforms (Figure 9). We found that the daytime monthly average SUHI intensity calculated by the two methods was very close regardless of the platform; the R^2 values were 0.979 and 0.965 for the daytime Terra and daytime Aqua data, respectively. However, the variation in the nighttime was obvious, and the R^2 values were 0.769 and 0.534 for the nighttime Terra and nighttime Aqua data, respectively.

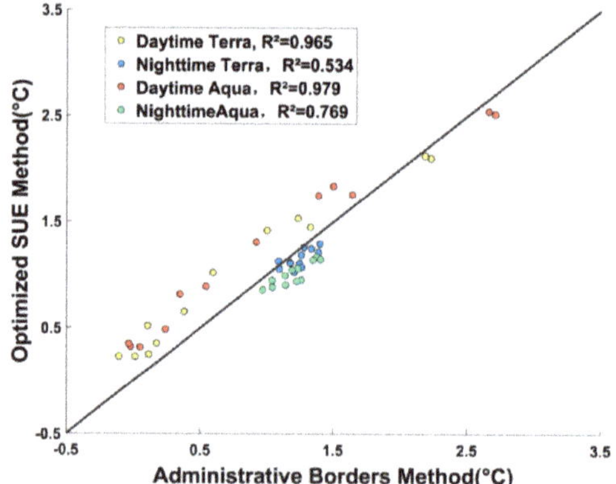

Figure 9. Differences between the two methods for the spatial variations in the monthly surface urban heat island intensity.

3.2.3. Daily Trends

We analyzed the average values for four different times of day in all Chinese cities based on the two calculation methods. We found that the SUHI intensity calculated by the AB method was slightly higher than that of the OSUE method at 0130 LT, whereas for all other times, the SUHI intensity calculated by the AB method was lower than that by the OSUE method (Figure 10). The strongest surface urban heat island time from the OSUE method within an entire day was similar to the time that had the highest temperature, namely, 1330 LT. However, the strongest urban heat island time was the time with the lowest temperature from the AB method, which was 1330 LT. In addition, in the AB method, the SUHI intensity reached its maximum value at approximately 01:30, and it was slightly higher (about 0.058 °C) than at 13:30. This outcome may have occurred because the rural area pixels occupied a high proportion in the AB method and the temperature of the barren land in rural areas decreases rapidly in nighttime, resulting in the high nighttime SUHI intensity from the AB method.

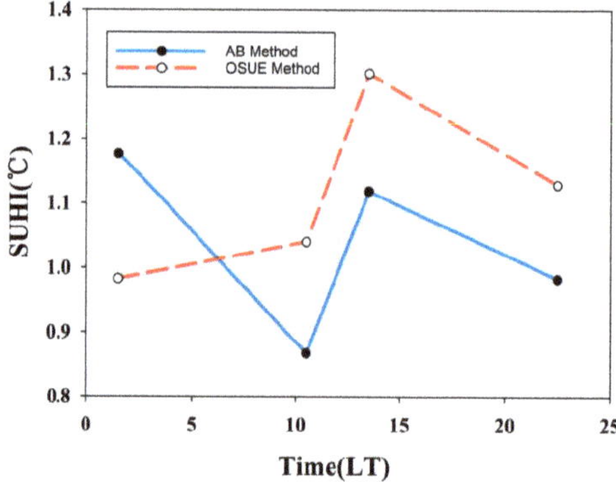

Figure 10. Trends of daily surface urban heat island (SUHI) in 36 major cities in China with the administrative borders (AB) and optimized simplified urban-extent (OSUE) methods.

In general, the driving factors of the SUHI intensity with these two methods were quite different, especially those related to social development, which suggested that the SUHI calculated by the AB method may be better able to help people study the social development factors that cause the UHI effect.

3.3. Predictive Models of SUHIs

We developed predictive models for the SUHI obtained from the AB and OSUE methods during the daytime and nighttime. The models based on the combination of the natural and social factors accounted for 31.8–74.3% of the SUHIs (Table S1). The natural condition factors included the Pr, M index, and NDVI. The social factors included the GDP, population and RN. The NDVI had negative effects on the daytime SUHI in both the AB and OSUE methods, while the RN had a positive influence. The NDVI and Pr had positive effects on the nighttime SUHI with both the AB and OSUE methods, while M had a negative influence. The remaining factors had opposite effects on the SUHI based on the two methods.

The relative contributions of the potential drivers can be quantified by the standard coefficients of the regression models (Figure 11). The explanation rates of natural conditions for the SUHIs with the AB method were 50.9% (daytime Terra), 54.4% (daytime Aqua), 42.1% (nighttime Terra), and 47% (nighttime Aqua). The explanation rates of social development for the SUHIs with the AB method were 22.1% (daytime Terra), 19.9% (daytime Aqua), 12.3% (nighttime Terra), and 11.6% (nighttime Aqua). The explanation rates of natural conditions for the SUHIs with the OSUE method were 60% (daytime Terra), 53.5% (daytime Aqua), 26.6% (nighttime Terra), and 34.7% (nighttime Aqua). The explanation rates of social development for the SUHIs with the OSUE method were 5.1% (daytime Terra), 8.9% (daytime Aqua), 5.2% (nighttime Terra), and 5.8% (nighttime Aqua). The results showed that vegetation had a significant impact as a temperature regulator of the daytime SUHI, whether with the AB method or the OSUE method. The daytime SUHI calculated with the AB method was significantly more affected by the social development factors than that calculated by the OSUE method. In the AB method, RN is an important factor, which means that the proportion of urban and rural pixels participating in the calculation of the SUHI would have a relatively large impact on the results.

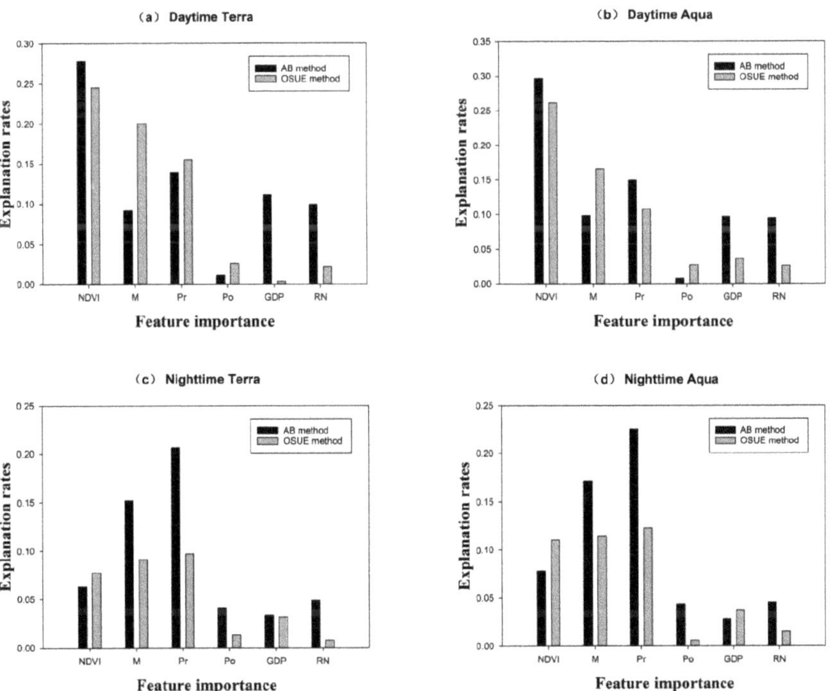

Figure 11. Contributions of the 6 factors to the variability of the urban heat islands in 36 major Chinese cities.

4. Discussion

Although the SUHI intensity is primarily controlled by diverse rural extents, the methods for selecting appropriate rural extents for future research remain largely unclear. Few previous studies on the uncertainty of SUHI intensity have focused on the differences in the driving factors of the calculations from different methods, with most of the comparison generally focusing on the spatiotemporal distribution [27,46]. However, the most important purpose of UHI research is to explore the drivers of the UHI phenomenon and to propose further mitigation suggestions. Therefore, in this study, we not only compared the differences between the AB and OSUE methods from the perspective of the spatiotemporal patterns but also analyzed the differences in the driving factors.

In Section 3, the differences between the two methods were fully demonstrated. We can clearly see that, in most cases, the two methods exhibited similar temporal and spatial patterns, especially on the national scale (Figures 3 and 4) and the monthly scale (Figures 8 and 9). Specifically, the daytime in the southeastern areas and the nighttime in the northern areas had higher SUHI intensities than the rest of China, which was consistent with previous conclusions on the SUHI in China in the literature [21]. From the perspective of the driving factors, both methods showed a very large daily difference due to the different driving factors of UHI during the daytime and nighttime, which were mentioned in several previous studies [21,32,47,48]. A comparison of the above two points, with the conclusions of previous studies, shows that the FB and CD or SUA showed differences in SUHI intensity, although the differences in the spatiotemporal modes were negligible. The two methods showed great differences in terms of anthropogenic factors. The most likely reason for these differences was that the spatial extents (administrative borders) of the data used in the AB method for the driving factor analysis were more closely aligned with the true borders.

On this basis, we conducted a thought experiment from the perspective of the standardization of the calculation of the SUHI, that of, is the AB method a valuable option for the calculation of SUHI intensity in multiple cities? Although some researchers argue that it is not appropriate [38], the AB method is simple and versatile. Moreover, two additional reasons support our use of the AB method in the SUHI intensity calculations: (i) for Chinese cities, the rural areas within the administrative borders typically correspond well to the associated size of the urban areas [27] (ii) and some studies have confirmed that the administrative division has a significant impact on the spatial form of an urban area [49]. The above two points support the rationality of using the AB method to calculate the SUHI intensity. According to the above points, we have sufficient evidence that the AB method has great application potential for future UHI studies, especially in regard to the analysis of the anthropogenic driving factors.

The strength of the UHI effect is generally measured based on the SUHI intensity, and this method can be used in most cities in China or other parts of the world. Furthermore, both the AB method and the OSUE method used in this paper have their own advantages; thus, these two methods can be applied for research around the world. We need to point out that each city has its own unique characteristics, including but not limited to the climate, development mode, urban form and urban planning. Our conclusions based on a spatial-temporal analysis may not be universally suitable to all regions or time spans. However, the conclusions obtained here by evaluating the potential of the AB method in the analysis of SUHI driving factors provide a valuable reference for future research.

It is important to note that several factors should be investigated in future studies. The first is the uncertainty created by the remote sensing data. Although the accuracy of the MODIS products has been demonstrated in many studies, obtaining more reliable temperature data is an important direction for urban heat island research. The second factor is the exploration of other driving factors. It has been reported that air pollution [48] and energy consumption [17,50] will lead to an increase in urban heat islands, but the factors that have been identified thus far cannot fully explain the occurrence of urban heat islands. Future research on the factors that influence the UHI and are closely related to urban development is of great significance. The third point is the exploration of the calculation methods for urban heat island intensity. There is still no unified calculation standard for the SUHI intensity, which is an important direction that can be explored in the future.

5. Conclusions

By comparing the differences between the two methods of AB and OSUE, we not only mapped the spatiotemporal patterns in 36 major cities in mainland China, but also explored whether administrative borders represent an appropriate standard range for the rural extent in SUHI intensity calculations. The SUHI intensity obtained from the two methods did not show significantly different spatial distributions or temporal trends, especially seasonally. The daytime in the southeastern areas and the nighttime in the northern areas had higher SUHI intensities than the rest of China and the daytime SUHI intensity is the highest in summer and the lowest in winter. However, in the regression analysis with the driving factors, the anthropogenic factors had a more significant influence on the SUHI intensity of the AB method. In the AB method, GDP and population were two of the most important drivers of SUHI intensity during the day and night, respectively. These conclusions stress the necessity of further research of the UHI drivers. We suggest that in future studies on the socioeconomic drivers of UHIs such as the analysis of negative spillover effect from UHIs by using a spatial econometrics model, more attention should be paid to the application of the AB method. However, uncertainties remain in the current study and the standardized calculation for SUHI intensity needs deeper research through attribution analysis or numeric modeling.

As administrative divisions are a universal phenomenon in human society, the conclusions of this study also have a certain reference value at the global scale and are worthy of further expansion such as adding more ponderable drivers in future research.

Supplementary Materials: The following are available online at http://www.mdpi.com/2071-1050/12/2/478/s1, Table S1. Model equations of the surface urban heat islands (SUHI) intensities. NDVI: differences in normalized difference vegetation index between urban and rural areas; R^2: coefficient of determination; M, differences in moisture index between urban and rural areas; Pr: differences in moisture index between urban and rural areas; Po: resident population; GDP: gross domestic product; RN: the ratio of the number of urban pixels to the number of rural pixels involved in the calculation in each city. *, ** or *** indicates significance at the 1%, 5% or 10% levels, respectively. Figure S1. Daytime and nighttime SUHI intensity (mean ± standard error) from the administrative borders (AB) and the optimized simplified urban-extent (OSUE) methods for each region.

Author Contributions: Conceptualization, L.N. and R.T.; methodology, L.N.; software, L.N.; data curation, L.N.; writing—original draft preparation, L.N.; writing—review and editing, X.Z., Y.J. and R.T.; supervision, R.T.; funding acquisition, R.T. All authors have read and agreed to the published version of the manuscript.

Funding: This research was funded by the Beijing Municipal Science and Technology Project under Grant Z181100005318003, the National Key R&D Program of China under Grant 2018YFA0605401, the Youth Innovation Promotion Association CAS under Grant 2015039, and the National Natural Science Foundation of China under Grant 41571351.

Conflicts of Interest: The authors declare no conflict of interest.

References

1. Nations, U. World urbanization prospects: The 2014 revision, highlights. In *Department of Economic and Social Affairs*; United Nations: New York, NY, USA, 2014.

2. DeFries, R. Terrestrial Vegetation in the Coupled Human-Earth System: Contributions of Remote Sensing. *Annu. Rev. Environ. Resour.* **2008**, *33*, 369–390. [CrossRef]

3. Grimm, N.B.; Faeth, S.H.; Golubiewski, N.E.; Redman, C.L.; Wu, J.G.; Bai, X.M.; Briggs, J.M. Global change and the ecology of cities. *Science* **2008**, *319*, 756–760. [CrossRef]

4. Howard, L. Climate of London deduced from meteorological observation. *Harvey Darton* **1833**, *1*, 1–24.

5. Reid, W.V. Biodiversity hotspots. *Trends Ecol. Evol.* **1998**, *13*, 275–280. [CrossRef]

6. Chen, X.L.; Zhao, H.M.; Li, P.X.; Yin, Z.Y. Remote sensing image-based analysis of the relationship between urban heat island and land use/cover changes. *Remote Sens. Environ.* **2006**, *104*, 133–146. [CrossRef]

7. Li, H.D.; Meier, F.; Lee, X.H.; Chakraborty, T.; Liu, J.F.; Schaap, M.; Sodoudi, S. Interaction between urban heat island and urban pollution island during summer in Berlin. *Sci. Total Environ.* **2018**, *636*, 818–828. [CrossRef] [PubMed]

8. Arnfield, A.J. Two decades of urban climate research: A review of turbulence, exchanges of energy and water, and the urban heat island. *Int. J. Climatol.* **2003**, *23*, 1–26. [CrossRef]

9. Bernstein, L.; Bosch, P.; Canziani, O.; Chen, Z.; Christ, R.; Riahi, K. *IPCC, 2007: Climate Change 2007: Synthesis Report*; IPCC: Geneva, Switzerland, 2008.

10. Patz, J.A.; Campbell-Lendrum, D.; Holloway, T.; Foley, J.A. Impact of regional climate change on human health. *Nature* **2005**, *438*, 310–317. [CrossRef]

11. Lafortezza, R.; Carrus, G.; Sanesi, G.; Davies, C. Benefits and well-being perceived by people visiting green spaces in periods of heat stress. *Urban For. Urban Green.* **2009**, *8*, 97–108. [CrossRef]

12. Gong, P.; Liang, S.; Carlton, E.J.; Jiang, Q.W.; Wu, J.Y.; Wang, L.; Remais, J.V. Urbanisation and health in China. *Lancet* **2012**, *379*, 843–852. [CrossRef]

13. Li, K.N.; Chen, Y.H.; Wang, M.J.; Gong, A. Spatial-temporal variations of surface urban heat island intensity induced by different definitions of rural extents in China. *Sci. Total Environ.* **2019**, *669*, 229–247. [CrossRef]

14. Rao, P.K. Remote Sensing of Urban Heat Islands from an Environmental Satellite. *Bull. Am. Meteorol. Soc.* **1972**, *53*, 647–648.

15. Roth, M.; Oke, T.; Emery, W. Satellite-derived urban heat islands from three coastal cities and the utilization of such data in urban climatology. *Int. J. Remote Sens.* **1989**, *10*, 1699–1720. [CrossRef]

16. Gallo, K.P.; Tarpley, J.D.; McNab, A.L.; Karl, T.R. Assessment of urban heat islands: A satellite perspective. *Atmos. Res.* **1995**, *37*, 37–43. [CrossRef]

17. Imhoff, M.L.; Zhang, P.; Wolfe, R.E.; Bounoua, L. Remote sensing of the urban heat island effect across biomes in the continental USA. *Remote Sens. Environ.* **2010**, *114*, 504–513. [CrossRef]

18. Peng, S.; Piao, S.; Ciais, P.; Friedlingstein, P.; Ottle, C.; Breon, F.M.; Nan, H.; Zhou, L.; Myneni, R.B. Surface urban heat island across 419 global big cities. *Environ. Sci. Technol.* **2012**, *46*, 696–703. [CrossRef]

19. Clinton, N.; Gong, P. MODIS detected surface urban heat islands and sinks: Global locations and controls. *Remote Sens. Environ.* **2013**, *134*, 294–304. [CrossRef]

20. Zhang, P.; Imhoff, M.L.; Wolfe, R.E.; Bounoua, L. Characterizing urban heat islands of global settlements using MODIS and nighttime lights products. *Can. J. Remote Sens.* **2010**, *36*, 185–196. [CrossRef]

21. Zhou, D.; Zhao, S.; Liu, S.; Zhang, L.; Zhu, C. Surface urban heat island in China's 32 major cities: Spatial patterns and drivers. *Remote Sens. Environ.* **2014**, *152*, 51–61. [CrossRef]

22. Zhou, D.; Bonafoni, S.; Zhang, L.; Wang, R. Remote sensing of the urban heat island effect in a highly populated urban agglomeration area in East China. *Sci. Total Environ.* **2018**, *628*, 415–429. [CrossRef]

23. Voogt, J.A.; Oke, T.R. Thermal remote sensing of urban climates. *Remote Sens. Environ.* **2003**, *86*, 370–384. [CrossRef]

24. Lai, J.; Zhan, W.; Huang, F.; Quan, J.; Hu, L.; Gao, L.; Ju, W. Does quality control matter? Surface urban heat island intensity variations estimated by satellite-derived land surface temperature products. *ISPRS J. Photogramm. Remote Sens.* **2018**, *139*, 212–227. [CrossRef]

25. Chakraborty, T.; Lee, X. A simplified urban-extent algorithm to characterize surface urban heat islands on a global scale and examine vegetation control on their spatiotemporal variability. *Int. J. Appl. Earth Obs. Geoinf.* **2019**, *74*, 269–280. [CrossRef]

26. Yao, R.; Wang, L.C.; Huang, X.; Chen, J.P.; Li, J.R.; Niu, Z.G. Less sensitive of urban surface to climate variability than rural in Northern China. *Sci. Total Environ.* **2018**, *628*, 650–660. [CrossRef]

27. Lai, J.; Zhan, W.; Huang, F.; Voogt, J.; Bechtel, B.; Allen, M.; Peng, S.; Hong, F.; Liu, Y.; Du, P. Identification of typical diurnal patterns for clear-sky climatology of surface urban heat islands. *Remote Sens. Environ.* **2018**, *217*, 203–220. [CrossRef]

28. Yao, R.; Wang, L.C.; Huang, X.; Niu, Z.G.; Liu, F.F.; Wang, Q. Temporal trends of surface urban heat islands and associated determinants in major Chinese cities. *Sci. Total Environ.* **2017**, *609*, 742–754. [CrossRef]

29. Zhang, X.Y.; Friedl, M.A.; Schaaf, C.B.; Strahler, A.H.; Schneider, A. The footprint of urban climates on vegetation phenology. *Geophys. Res. Lett.* **2004**, *31*. [CrossRef]

30. Li, X.M.; Zhou, Y.Y.; Asrar, G.R.; Imhoff, M.; Li, X.C. The surface urban heat island response to urban expansion: A panel analysis for the conterminous United States. *Sci. Total Environ.* **2017**, *605*, 426–435. [CrossRef]

31. Quan, J.L.; Zhan, W.F.; Chen, Y.H.; Wang, M.J.; Wang, J.F. Time series decomposition of remotely sensed land surface temperature and investigation of trends and seasonal variations in surface urban heat islands. *J. Geophys. Res. Atmos.* **2016**, *121*, 2638–2657. [CrossRef]

32. Zhou, D.C.; Zhao, S.Q.; Zhang, L.X.; Sun, G.; Liu, Y.Q. The footprint of urban heat island effect in China. *Sci. Rep.* **2015**, *5*, 11160. [CrossRef]

33. Meng, Q.Y.; Zhang, L.L.; Sun, Z.H.; Meng, F.; Wang, L.; Sun, Y.X. Characterizing spatial and temporal trends of surface urban heat island effect in an urban main built-up area: A 12-year case study in Beijing, China. *Remote Sens. Environ.* **2018**, *204*, 826–837. [CrossRef]

34. Zhou, D.; Zhao, S.; Zhang, L.; Liu, S. Remotely sensed assessment of urbanization effects on vegetation phenology in China's 32 major cities. *Remote Sens. Environ.* **2016**, *176*, 272–281. [CrossRef]

35. Schwarz, N.; Lautenbach, S.; Seppelt, R. Exploring indicators for quantifying surface urban heat islands of European cities with MODIS land surface temperatures. *Remote Sens. Environ.* **2011**, *115*, 3175–3186. [CrossRef]

36. Zhou, D.C.; Xiao, J.F.; Bonafoni, S.; Berger, C.; Deilami, K.; Zhou, Y.Y.; Frolking, S.; Yao, R.; Qiao, Z.; Sobrino, J.A. Satellite Remote Sensing of Surface Urban Heat Islands: Progress, Challenges, and Perspectives. *Remote Sens.* **2019**, *11*, 48. [CrossRef]

37. Cui, Y.; Xu, X.; Dong, J.; Qin, Y. Influence of Urbanization Factors on Surface Urban Heat Island Intensity: A Comparison of Countries at Different Developmental Phases. *Sustainability* **2016**, *8*, 706. [CrossRef]

38. Sun, R.H.; Lu, Y.H.; Yang, X.J.; Chen, L.D. Understanding the variability of urban heat islands from local background climate and urbanization. *J. Clean. Prod.* **2019**, *208*, 743–752. [CrossRef]

39. Wan, Z.M.; Dozier, J. A generalized split-window algorithm for retrieving land-surface temperature from space. *IEEE Trans. Geosci. Remote Sens.* **1996**, *34*, 892–905.

40. Wan, Z.M. New refinements and validation of the MODIS Land-Surface Temperature/Emissivity products. *Remote Sens. Environ.* **2008**, *112*, 59–74. [CrossRef]

41. Wan, Z.M. New refinements and validation of the collection-6 MODIS land-surface temperature/emissivity product. *Remote Sens. Environ.* **2014**, *140*, 36–45. [CrossRef]

42. Li, X.C.; Zhang, Y.J.; Jin, X.L.; He, Q.N.; Zhang, X.P. Comparison of digital elevation models and relevant derived attributes. *J. Appl. Remote Sens.* **2017**, *11*, 046027. [CrossRef]

43. Schneider, A.; Friedl, M.A.; Potere, D. A new map of global urban extent from MODIS satellite data. *Environ. Res. Lett.* **2009**, *4*, 044003. [CrossRef]

44. Schneider, A.; Friedl, M.A.; Potere, D. Mapping global urban areas using MODIS 500-m data: New methods and datasets based on 'urban ecoregions'. *Remote Sens. Environ.* **2010**, *114*, 1733–1746. [CrossRef]

45. Dobson, J.E.; Bright, E.A.; Coleman, P.R.; Durfee, R.C.; Worley, B.A. LandScan: A global population database for estimating populations at risk. *Photogramm. Eng. Remote Sens.* **2000**, *66*, 849–857.

46. Yao, R.; Wang, L.; Huang, X.; Niu, Y.; Chen, Y.; Niu, Z. The influence of different data and method on estimating the surface urban heat island intensity. *Ecol. Indic.* **2018**, *89*, 45–55. [CrossRef]

47. Yao, R.; Wang, L.; Huang, X.; Zhang, W.; Li, J.; Niu, Z. Interannual variations in surface urban heat island intensity and associated drivers in China. *J. Environ. Manag.* **2018**, *222*, 86–94. [CrossRef]

48. Cao, C.; Lee, X.; Liu, S.; Schultz, N.; Xiao, W.; Zhang, M.; Zhao, L. Urban heat islands in China enhanced by haze pollution. *Nat. Commun.* **2016**, *7*, 12509. [CrossRef]

49. Gao, X.; Long, C.X. Cultural border, administrative border, and regional economic development: Evidence from Chinese cities. *China Econ. Rev.* **2014**, *31*, 247–264. [CrossRef]

50. McCarthy, M.P.; Best, M.J.; Betts, R.A. Climate change in cities due to global warming and urban effects. *Geophys. Res. Lett.* **2010**, *37*. [CrossRef]

 *sustainability*

Article

The Impacts of the Expansion of Urban Impervious Surfaces on Urban Heat Islands in a Coastal City in China

Lizhong Hua [1], Xinxin Zhang [1], Qin Nie [1], Fengqin Sun [1] and Lina Tang [2,*]

1   College of Computer and Information Engineering, Xiamen University of Technology, 600 Ligong Road, Xiamen 361024, China; Lzhua@xmut.edu.cn (L.H.); 2013110709@xmut.edu.cn (X.Z.); 2014110705@xmut.edu.cn (Q.N.); fengqinsun@xmut.edu.cn (F.S.)
2   Key Lab of Urban Environment and Health, Institute of Urban Environment, Chinese Academy of Sciences, Xiamen 361021, China
*   Correspondence: lntang@iue.ac.cn; Tel.: +86-592-619-0681

Received: 15 November 2019; Accepted: 5 January 2020; Published: 8 January 2020

Abstract: The effect of the expansion of urban impervious surfaces on surface urban heat islands (UHIs) has attracted research attention due to its relevance for studies of local climatic change and habitat comfort. In this study, using five satellite images of Xiamen city, Southeast China (four images from the Landsat 5 Thematic Mapper (TM) and one from the Landsat 8 Operational Land Imager/Thermal Infrared Sensor (OLI/TIRS)) acquired in summer between 1989 and 2016, together with spatial statistical methods, the changes in impervious surface area (ISA) were investigated, the spatiotemporal variation of the intensity of urban heat islands (UHIs) was explored, and the relationships between land surface temperature (LST) and the percentage of impervious surface area (ISA%), the normalized difference vegetation index (NDVI), and fractional vegetation coverage (Fv) were investigated. The results showed the following: (1) According to the biophysical composition index (BCI) combined with an ISA post-processing method, Xiamen has witnessed a substantial increase in ISA, showing a 6.1-fold increase from 1989 to 2016. The direction of ISA expansion was consistent throughout the study period in each of the five districts of Xiamen; (2) a bay-like UHI form is observed in the study area, which is remarkably distinct from the central-radial UHI form observed in previous studies of other cities; (3) the extent of UHIs in Xiamen greatly increased between 1989 and 2016, experiencing a 4.7-fold increase in UHI areas during this time. However, during the same period, the urban heat island ratio index (URI)—that is, the ratio of UHI area to ISA—decreased slightly. The UHI area decreased in some urban parts of Xiamen due to a significant increase in vegetation coverage, urban village redevelopment, and the construction of new parks; (4) sea ports and heavy industrial zones are the greatest contributor to surface UHI, followed by urban villages; and (5) LST is strongly positively correlated with ISA%. Each 10% increase in ISA was associated with an increase in summer LST of 0.41 to 0.91 K, which compares well with the results of related studies. This study presents valuable information for the development of regional urban planning strategies to mitigate the effects of UHIs during rapid urbanization.

Keywords: urban heat island; impervious surface area; biophysical composition index; remote sensing; coastal city; Xiamen

1. Introduction

Over the past several decades, urbanization has progressed at an unprecedented rate throughout the world [1]. In 1950, only 29% of the global population lived in urban areas; however, by 2018, this proportion was 55%. Furthermore, the UN Population Division estimates that the global percentage of urban residents will increase to 68% by 2050 [2]. In 1978, China's urban population was 170 million, representing 17.4% of China's population at the time; however, by the end of 2018, more than 831 million Chinese lived in cities, accounting for 59.6% of the country's population. Coastal urban regions in China have undergone particularly rapid population growth and strong economic growth [3]. Approximately 11.0% of China's population lives in the Yangtze River Delta, which represents China's largest "super city" and contributes approximately 23.5% of the country's GDP (2018 data) [4]. Urbanization inevitably involves the transition from natural vegetation coverage to impervious surfaces (e.g., paved roads, building roofs, and parking lots) [5]. Thus, although it brings great social and economic benefits, urbanization leads to a number of environmental issues, such as urban heat islands (UHIs), the degradation of water quality, and loss of biodiversity [6]. The term "surface UHI" refers to the phenomenon whereby the land surface temperature (LST) in urban areas is higher compared to the surrounding rural areas [7]. The high temperature of UHIs influences not only the local and mesoscale climate but also human health, air quality, and ecosystem functions [8]. Therefore, UHIs have received large research attention around the world [9]. Remote sensing technology can be used to obtain detailed images of land cover with a high temporal resolution, allowing a synoptic and uniform means of mapping urban impervious areas and studying the effects of surface UHIs [10].

In recent years, impervious surface area (ISA) has increasingly been used as a key environmental indicator. Impervious surface areas can reduce water quality and lead to greater soil dryness and higher air temperatures [9]. Numerous methodologies have been constructed for the extraction of the ISA from remote sensing images with various spatial scales in order to map ISA and evaluate their dynamics. Three major classes of algorithms have been used for the extraction of ISA from satellite images, namely machine learning methods, spectral unmixing techniques, and spectral index (SI) methods. Machine learning techniques include artificial neural networks (ANNs) [11], support vector machines (SVMs) [12], decision tree classification (DTC), classification and regression tree (CART) analysis [6], and regression modeling [13]. Methods for the determination of empirical relationships between various spectral and spatial characteristics include linear spectral mixture analysis (LSMA) [14–16], and the multiple endmember spectral mixture analysis (MESMA) method [17]. A number of SIs have been constructed to quantify the biophysical characteristics of the earth's surface, including the impervious surface area index [18], the normalized difference impervious surface index (NDISI) [19], the biophysical composition index (BCI) [20,21], the modified NDISI [22], and the combinational built-up index (CBI) [23]. Spectral indices can be used to derive ISA, usually using a simple linear composition formula for the original image bands, reflected image bands, or other feature index bands obtained from imagery. This index-based method does not require much a priori knowledge of image pre-processing. Therefore, compared to machine learning methods and spectral unmixing techniques, SIs have the advantages of uncomplicated implementation and computational convenience in practical applications [20,21]. Furthermore, some SIs (e.g., the BCI) can be estimated using a wide range of satellite imagery with various spatial resolutions, ranging from coarse resolution (e.g., Moderate Resolution Imaging Spectroradiometer (MODIS)), to medium resolution (e.g., Landsat 5, Sentinel 2), to high resolution (e.g., IKONOS) satellite imagery [20].

Satellite-derived LST has been used to study UHIs, and has been used to inform various strategies for the mitigation of UHIs, climate modeling, and studies of local and global-scale climate change. In recent years, there has been increasing concern regarding the effect of ISA on LST [14,19,24,25]. However, there have been few studies of the effect of urban development on surface UHIs using satellite images taken at different times in the same season, especially in coastal cities.

Therefore, this study aimed to use remote sensing images and spatial statistical methods to assess the effects of rapid urbanization—and particularly the rapid expansion of ISA—on UHIs in the city of Xiamen, China, for a 27-year period from 1989 to 2016. This study focused on the changes in ISA, investigated the area and intensity of UHIs, and revealed the relationships between LST and (1) the percentage of impervious surface area (ISA%), (2) the normalized difference vegetation index (NDVI), and (3) the fractional vegetation cover (Fv). The results of this study could be helpful for the development of regional urban planning strategies to mitigate the effects of UHIs during rapid urbanization.

2. Study Area and Data

2.1. Study Area

Xiamen city is situated in the SE coast of Fujian Province, China (117°53′–118°26′ E and 24°23′–24°54′ N), and covers an area of 1699.39 km^2 (Figure 1). The city has undulating terrain, an elevation range of 0 to 1175 m, and has a subtropical monsoon climate and a mean annual temperature of 21 °C. The city consists of two main areas—Xiamen Island and four suburban districts, namely Jimei, Haicang, Tong'an, and Xiang'an.

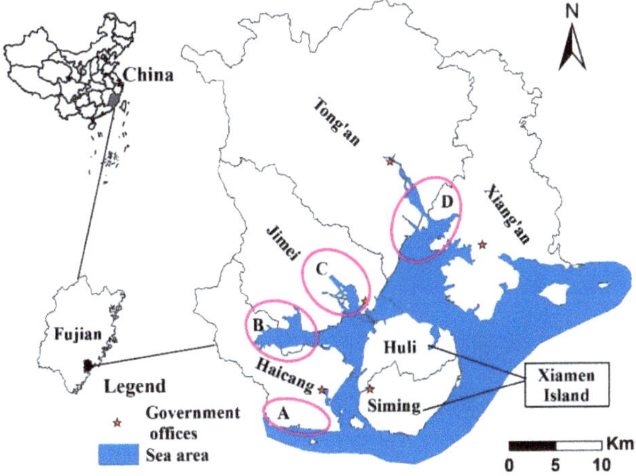

Figure 1. The location of Xiamen city. A, B, C, and D show Haicang Bay, Maluan Bay, Xinglin Bay, and Tong'an Bay, respectively.

Xiamen is one of the four oldest special economic zones in China. In recent decades, due to the high quality of its urban ecological environment and its diversified and growing economy, the city has continuously attracted a huge number of new residents and has experienced significant urban growth [26]. Xiamen has evolved from an "island-like" city to a "bay-like city" with an urban center on Xiamen Island and various surrounding towns and suburban agglomerations located in the bays and plains on the mainland [27]. National census data indicate that Xiamen's resident population grew from 1.18 million to 4.11 million between 1990 and 2018. The city's dramatic and rapid urbanization has led to environmental issues, such as the deterioration of water quality, loss of farmland, and the eutrophication of offshore waters [28,29]. This urbanization, combined with the city's diversity of land-use/land-cover, makes Xiamen an ideal area to investigate the effects of surface urban heat islands.

2.2. Data Collection and Processing

Four Landsat 5 Thematic Mapper (TM) images and one Landsat 8 image (Table 1) were acquired from the website of the United States Geological Survey (http://earthexplorer.usgs.gov/). Digital elevation models (DEMs) were acquired from the Geospatial Data Cloud website (http://www.gscloud.cn/). Landsat 8 is equipped with two instruments, the Operational Land Imager (OLI) and the Thermal Infrared Sensor (TIRS). The Landsat 8 OLI data consist of nine spectral bands. Bands 1–7 and band 9 have a spatial resolution of 30 m, and band 8 is a panchromatic band with a 15-m spatial resolution. The four Landsat 5 TM images have a cloud cover of less than 1.81% and covered remote mountainous areas with forests.

Table 1. Details of the Landsat images used in this work.

Image Type	Image Acquisition Date	Sun Elevation Angle (Degrees)	Path/Row	Cloud Cover (%)
Landsat 8 OLI/TIRS	27 July 2016	66.42	119/43	1.81
Landsat 5 TM	06 Jun 2009	66.34	119/43	0.31
Landsat 5 TM	23 May 2004	64.68	119/43	0.00
Landsat 5 TM	20 July 1996	56.60	119/43	0.16
Landsat 5 TM	15 Jun 1989	61.40	119/43	1.71

Note: OLI: Operational Land Imager; TIRS: Thermal Infrared Sensor; TM: Thematic Mapper.

Detailed descriptions of the images used in this study are shown in Table 1. The images were all acquired in the same season with adjacent sun elevation angles and clear sky conditions. All images were divided using the Universal Transverse Mercator projection (zone 50N). The images were georegistered with root-mean-square (RMS) errors of less than 0.5 pixels. The images were pre-processed via an atmospheric correction procedure that is based on the dark object subtraction (DOS) method [30,31].

3. Methods

3.1. Image-Based Atmospheric Correction Method

The DOS method for atmospheric correction [30,31] was applied to both the Landsat 5 TM and Landsat 8 OLI images. It can be expressed as:

$$\rho = \frac{(L_{sat} - L_{haze}) \cdot \pi \cdot d^2}{ESUN \cdot cos\theta}, \tag{1}$$

where $ESUN$ represents the exo-atmospheric solar spectral irradiance; ρ represents the surface reflectance; L_{sat} represents the spectral radiance at the aperture of the sensor (W/(m^2·sr·μm)); L_{haze} represents the path radiance, which is derived using continuous relative scatter [30] for Landsat 5 TM images and using the lowest valid value attribute table method [32] for Landsat 8 OLI images; θ represents the solar zenith angle (°); and D represents the normalized Earth–Sun distance.

L_{sat} can be written as:

$$Lsat = Gain \cdot DN + Bias, \tag{2}$$

where DN represents the remotely sensed digital number, and $Gain$ and $Bias$ represent the radiometric gain and bias gain for a specific band, respectively, and were obtained from the head files of images.

3.2. Retrieval of Land Cover

To determine the recent land surface cover in Xiamen, the main land cover was first retrieved, namely the impervious built land, vegetation, bare land, and water bodies (sea, ponds, beach, rivers, and lakes). Therefore, the land cover in the study was divided into two classes, namely impervious areas and non-impervious areas (i.e., vegetation, bare land, and water bodies).

3.2.1. Retrieval of Vegetation Index

The NDVI was used to extract vegetation [33]. This index is given by:

$$NDVI = (\rho_{NIR} - \rho_{Red}) / (\rho_{NIR} + \rho_{Red}), \tag{3}$$

where ρ_{NIR} and ρ_{Red} represent the spectral reflectance of the red and NIR bands, respectively, for the Landsat 5 TM and Landsat 8 OLI images.

3.2.2. Retrieval of Water Index

The modified normalized difference water index (MNDWI) of Xu [34,35] was applied to delineate open water features. The index is given by:

$$MNDWI = (\rho_{Green} - \rho_{SWIR1}) / (\rho_{Green} + \rho_{SWIR1}),$$
(4)

where ρ_{Green} and ρ_{SWIR1} are the spectral reflectance of the green and short-wave infrared 1 bands, respectively, for TM/OLI images.

3.2.3. Retrieval of Bare Land Index

The modified normalized difference bare index (MNDBaI) [36] was used to extract bare land. The index is expressed as follows:

$$MNDBaI = (\rho_{Red} - \rho_{Blue}) / (\rho_{Red} + \rho_{Blue}),$$
(5)

where ρ_{Red} and ρ_{Blue} represent the spectral reflectance of the red and blue bands, respectively, for TM/OLI images.

3.2.4. Retrieval of Impervious Surface Area

Deng et al. [20] proposed the BCI, the calculation of which involved a modification of the Tasseled Cap (TC) transformation, to effectively distinguish impervious surfaces or other impervious features from non-impervious features (e.g., soil and vegetation) based on a vegetation–impervious surface–soil (V-I-S) triangle model. In this study, the BCI was used to distinguish impervious surfaces in order to avoid the shortcomings of spectral mixture analysis (SMA) methods regarding the selection of endmembers and endmember signature specification [15]. The BCI can be calculated as follows:

$$BCI = \frac{(H+L)/2 - V}{(H+L)/2 + V},$$
(6)

where H represents the normalized first tasseled cap (TC1) (i.e., high albedo); V represents the normalized second tasseled cap (TC2) (i.e., vegetation); and L represents the normalized third tasseled cap (TC3) (i.e., low albedo). These three factors can be given by the following equations:

$$H = \frac{TC1 - TC1_{min}}{TC1_{max} - TC1_{min}},$$
(7)

$$V = \frac{TC2 - TC2_{min}}{TC2_{max} - TC2_{min}},$$
(8)

$$L = \frac{TC3 - TC3_{min}}{TC3_{max} - TC3_{min}},$$
(9)

where TCi (i = 1,2,3) represent the first three TC components; and $TCimin$ and $TCimax$ represent the minimum and maximum values of the ith TC component, respectively.

The TC transformation was performed by constructing linear combinations of the original or reflected image bands. The TC transformation reflectance coefficients for the TM and OLI images (Table 2) were obtained from Crist [37] and Li et al. [38], respectively:

Water bodies for five different periods from 1989 to 2016 were masked out before estimating ISA using the MNDWI. Then, BCI maps of the study area were produced for the five different periods using Equation (6). Subsequently, impervious surface features were extracted from the BCI maps using optimum threshold values, which were manually adjusted with the aid of Google Earth (GE) images. However, it was found that the Landsat 5 and Landsat 8 BCI images were not very effective for separating impervious surfaces, beach land, and bare soil due to the similarity of the spectral characteristics of these three surfaces in Landsat imagery. This problem was also reported by Deng et al. [20], who determined the BCI in Wisconsin, USA, using Landsat Enhanced Thematic Mapper Plus (ETM+) and IKONOS images. During rapid urbanization, a large amount of bare land is converted from vegetation and water bodies. Furthermore, beach land is one of the most important land cover types for a coastal city, such as Xiamen. To address the problem of their spectral confusion, we first separated ISA and bare soil using Equation (5) and then separated ISA and beach land using a combination of TC3 and the DEM [36].

Table 2. Reflectance coefficients the first three Tasseled Cap (TC) components for Landsat sensors.

Sensor	TCs	Coastal	Blue	Green	Red	NIR	SWIR1	SWIR2
Landsat 5 TM	TC1	-	0.2043	0.4158	0.5524	0.5741	0.3124	0.2303
	TC2	-	−0.1603	−0.2819	−0.4934	0.7940	−0.0002	−0.1446
	TC3	-	0.0315	0.2021	0.3102	0.1594	−0.6806	−0.6109
Landsat 8 OLI/TIRS	TC1	0.2540	0.3037	0.3608	0.3564	0.7084	0.2358	0.1691
	TC2	−0.2578	−0.3064	−0.3300	−0.4325	0.6860	−0.0383	−0.2674
	TC3	0.1877	0.2097	0.2038	0.1017	0.0685	−0.7460	−0.5548

Note: SWIR: short-wave infrared.

The overall accuracy (OA) of the three extracted ISA images was assessed. Three higher-resolution images for the corresponding years were used as reference data to assess the accuracy of the ISA classification, namely GE images with a spatial resolution of 1 m, which were acquired in December 2003, October 2009, and June 2016. Six hundred pixels were sampled in each image at random. The determined overall accuracies were 90.67% (2004), 90.50% (2009), and 92.17% (2016), with corresponding Kappa values of 0.813, 0.810, and 0.843. These results suggest that the ISA classification has a relatively high accuracy and is appropriate for the purposes of this study.

3.2.5. Impervious Surface Expansion Index (ISEI) and Fan Analysis

Equation (10) was used to calculate the impervious surface expansion index (ISEI) to determine the intensity of urban growth.

$$ISEI_j = ISEI_i + \Delta E \quad \Delta E = \frac{IS_j - IS_i}{IS_i(j-i)}, \tag{10}$$

where $ISEI_i$ and $ISEI_j$ are the ISEI in year i and j, respectively; ΔE is the inrease of $ISEI$ from year i to year j; and IS_i and IS_j represent the impervious areas in the study area in year i and j, respectively.

An equal fan analysis [39] was used to characterize the expansion direction of the urban ISA in Xiamen between 1989 and 2016. The government offices of Xiamen Island and the four districts of Xiamen city were chosen as the centers of ISA (Figure 1), respectively, and then five circles were drawn based on the appropriate radius for each area. Additionally, each circle was divided into 16 fans (equally spaced rays). Radar graphs showing the expansion of ISA were drawn by summarizing the ISEI in each fan for Xiamen Island and the four districts for each study year.

3.3. Retrieval of Land Surface Temperature (LST)

The LST is an important indicator for identifying surface UHIs. Therefore, the LST across Xiamen city in each study year was derived to study the effect of ISA expansion on surface UHIs. For Landsat 5 TM images, the thermal band 6 was used to retrieve the LST using the algorithm of Artis and Carnahan [40]. For Landsat 8 TIRS images, thermal band 10 was used in order to retrieve the LST using the modified single channel (MSC) algorithm of Jiménez-Muñoz et al. [41]. The two methods have the following three common processing steps: (1) The digital numbers (DNs) of the thermal band were first converted to a satellite spectral radiance (*L*) using Equation (11); (2) then, L was further converted to the at-satellite brightness temperature (T, in Kelvin) using Equation (12); and (3) T was converted to the LST using the land surface emissivity (ε), which was estimated from the NDVI [25,41,42]:

$$L = G \cdot Q_{cal} + B, \tag{11}$$

$$T = K_2 / \ln(K_1 / L + 1)), \tag{12}$$

where G and B represent, respectively, the rescaled gain factor and bias factor of the thermal band, in W/(m²·sr·m); and K_1 and K_2 represent the calibration constants. G, B, K_1, and K_2 were obtained from the head files of the Landsat 5 TM and Landsat 8 TIRS images. Q_{cal} represents the quantized calibrated pixel value (in DN) of the thermal band.

The Artis and Carnahan algorithm is given in Equation (13):

$$LST = T / [1 + (\lambda T / \rho) ln\varepsilon], \tag{13}$$

where λ represents the wavelength of the central band (λ = 11.5 μm); ρ = 1.438 × 10⁻² m·K; ε represents the land surface spectral emissivity of band 6 of Landsat 5 TM; and Ts is the land surface temperature (K).

The equation for the MSC algorithm is as follows:

$$LST = \gamma \left[\varepsilon^{-1} (\psi_1 L + \psi_2) + \psi_3 \right] + \delta, \tag{14}$$

where ψ_1, ψ_2, and ψ_3 represent the atmospheric functions; τ represents the atmospheric transmission; and γ and δ represent two parameters expressed by the following equations:

$$\gamma \approx T^2 / (b_\gamma L) \; \delta \approx T - T^2 / b_\gamma, \tag{15}$$

$$\psi_1 = 1 / \tau, \psi_2 = -L^\downarrow - \frac{L^\uparrow}{\tau}, \psi_3 = L^\downarrow, \tag{16}$$

where $b\gamma$ is 1324 for the TIRS band 10; τ represents the atmospheric transmissivity; and $L\uparrow$ and $L\downarrow$ represent the upwelling atmospheric radiance and the downwelling atmospheric radiance, respectively. The three parameters, τ, $L\uparrow$ and $L\downarrow$, were obtained using an atmospheric correction parameter calculator on the NASA (National Aeronautics and Space Administration) website (https://atmcorr.gsfc.nasa.gov/). This calculator requires the following inputs: (1) Geographical location; and (2) the date and time of the Landsat 8 satellite overpass.

3.4. LST Normalization and Determination of Urban Heat Island Ratio Index (URI)

The thermal environment of the urban surface was represented by the LST [43]. To facilitate the comparison of the LST over a period of time, a normalized LST equation was used, which reduces the temporal variability of the LST [25,44]:

$$U_i = \frac{T_i - T_{min}}{T_{max} - T_{min}}, \tag{17}$$

where U_i is the value of the i-th pixel of the normalized LST image; T_i is the initial value of pixel i in the LST image; and Tmin and Tmax are the minimum and maximum LSTs on the LST image.

The normalized LST image in urban areas characterized by impervious surface areas was further divided into six levels using a density slice technique based on the mean and standard deviation of the LST in the normalized LST image (Table 3) [45,46]. The combination of level 5 and level 6 was taken to be the UHI distribution zone. Thus, the spatial distribution and temporal variation of UHIs could be simply determined using a single graded urban LST map.

Table 3. Details of the density slice technique which was used to divide the normalized land surface temperature (LST) image.

Level	U_i Value	UHI Meaning	Description
1	$\leq T_m - 1.5\text{Std}$	Very low-temperature zone	
2	$>T_m - 1.5\text{Std}$ and $\leq T_m - 0.5\text{Std}$	Low-temperature zone	No UHI
3	$>T_m - 0.5\text{Std}$ and $\leq T_m$	Moderate temperature zone	distribution zone
4	$>T_m$ and $\leq T_m + 0.5\text{Std}$	Sub-high-temperature zone	
5	$>T_m + 0.5\text{Std}$ and $\leq T_m + 1.5\text{Std}$	High-temperature zone	UHI distribution
6	$>T_m + 1.5\text{Std}$	Very high-temperature zone	zone

Note: T_m and Std are the mean and standard deviation of the LST in the normalized LST image, respectively; and U_i is the value of pixel i of the normalized LST image. UHI: urban heat island.

We introduced the urban heat island ratio index (URI) proposed by Xu and Chen [44] in Equation (18), below. The URI is the ratio of the UHI area to the urban impervious area, which has been a key indicator of assessing an urban thermal environment in Technical Criterion for Eco-environmental Status Evaluation (trial) issued by the Ministry of Ecology and Environment of China. The URI can be used to investigate the variations in the UHI intensity using multi-date remote sensing data. Generally, the larger the URI, the stronger the UHI effect is:

$$URI = \frac{1}{m}\sum_{i=1}^{n} w_i P_i, \tag{18}$$

where m represents the number of levels in the normalized LST image ($m = 6$ in this study); i is the level of the UHI; m is the number of normalizations; n is the level of the UHI areas ($n = 2$ in this study); w is the weight using the value of corresponding level i ($w = 5$ or 6 in this study); and P_i represents the percentage of level i.

The mean intensity of UHIs is defined as the difference between the LST of the urban area and the background LST of the rural area:

$$I_m = T_{IS} - T_{\text{non-IS}}, \tag{19}$$

where I_m represents the mean intensity of the UHI effect (in K) in the study area, T_{IS} represents the mean LST (in K) of the urban ISA, and $T_{\text{non-IS}}$ represents the mean LST (in K) in the non-ISA except for the sea area.

3.5. Statistical Analysis

Regression analysis was employed to quantify the LST relationships to NDVI, ISA% (Equation (20)), and Fv (Equation (21)). A zonal analysis method [14] was used to determine the mean LST at each 0.01 increment of the NDVI from −1 to 1, each 0.01 increment of the Fv from 0 to 1, and each 1% increment of ISA% from 0% to 100%. At each increment of NDVI, Fv, and ISA%, a mean LST value was derived from all corresponding pixels.

The BCI was used to determine ISA%, which was expressed as follows:

$$ISA\% = (BCI_i - BCI_{min})/(BCI_{max} - BCI_{min}), \tag{20}$$

where BCI*nor* is the normalized BCI; BCI*i* is the original BCI value of pixel i; and BCI*min* and BCI*max* represent the minimum value and maximum value of the original BCI, respectively.

Fractional vegetation coverage was calculated after Gutman and Ignatov [47,48]:

$$f_v = (NDVI - NDVI_0)/(NDVIs - NDVI_0), \tag{21}$$

where NDVI is the NDVI value of a pixel, NDVI0 is the NDVI value for bare soil, and NDVIs is the NDVI value of a surface with an Fv of 100% (which in this study was obtained from extremely dense forest).

4. Results and Discussion

4.1. Changes in Urban Impervious Surface Area (ISA)

The results show that the ISA in Xiamen city increased dramatically over the study period, from 63.06 km^2 in 1989 to 447.04 km^2 in 2016, a net increase of 384.01 km^2 (Figure 2). This corresponds to an average annual increase rate of 14.22%. The majority of the increase in ISA occurred between 2004 and 2016, when the ISA grew by 238.11 km^2. The increase in ISA was mostly concentrated in the areas surrounding Xinglin Bay, Tong'an Bay, Malyuan Bay, and Xiamen Island (Figure 2). Between 1989 and 2016, Xiamen underwent a transformation from an "island-like" city to a "bay-like" city. There may be two main reasons for this transformation. Firstly, a development plan was launched in 2003 to combat overpopulation and unbalanced economic development between rural and urban areas in Xiamen, which prioritized the development of areas around the four aforementioned bays to form new urban centers [26]. Secondly, Xiamen's ISA growth is constrained by its proximity to the coast. The bay-like city form is distinctly different from inland cities, where the main pattern of ISA expansion approximately follows the form of concentric circles surrounding a central area (i.e., "urban sprawl") [27].

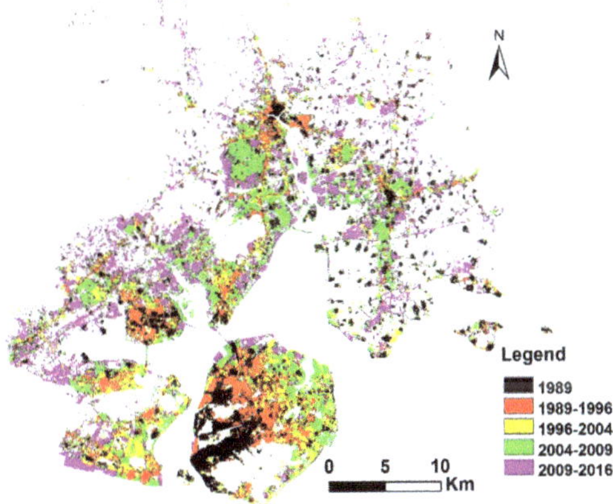

Figure 2. The calculated expansion of impervious surface area (ISA) in Xiamen city during 1989–2016.

As shown in Figure 3, the increase in ISA in Xiamen between 1989 and 2016 came at the cost of vegetation and water areas. A total of 82.53% of the added ISA, which was constructed between 1989 and 2016, was converted from vegetated areas, representing a loss of vegetated land at a rate of over 11.76 km^2 per year. Meanwhile, 17.26% of the added ISA was converted from water areas, mostly sea areas adjacent to land. Only 0.21% of the added ISA was converted from bare land and other land-cover categories. In many coastal cities, coastal reclamation is one of the most important approaches to supply land resources to support an expanding population. However, coastal reclamation can lead to marine environmental problems, such as a reduction in the quality of the marine environment, habitat degradation, and a reduction in coastal biodiversity. In 2016, the Xiamen local government designated 981 km^2 of ecologically sensitive areas (such as sea, reservoir water sources, ecological forests, and basic farmland, etc.) as an ecological protection "red line" in an attempt to strictly protect ecological resources and control urban expansion.

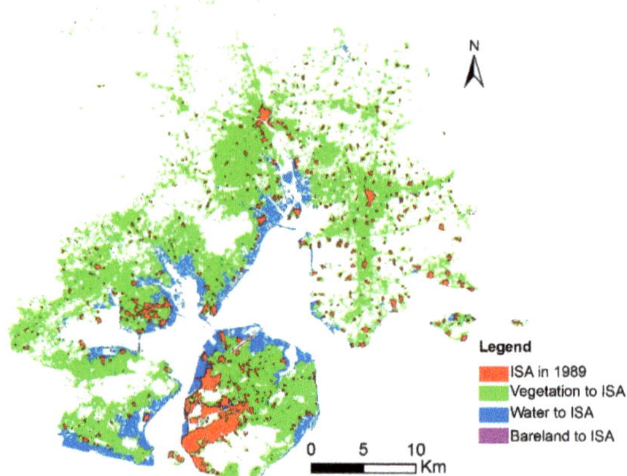

Figure 3. Map detailing the conversion of water, vegetation, and bare land to ISA in Xiamen between 1989 and 2016.

The spatial distribution of the expansion of ISA in Xiamen city during the study period is presented in Figure 4 and Table 4. In Xiamen Island, the ISA expanded multi-directionally before 2009, that is, to the northeast, east-northeast, and east. However, between 2009 and 2016, the expansion was more uni-directional, that is, towards the northeast, and lower ISEI values were observed. In the Jimei, Haicang, Xiang'an, and Ton'g'an districts, the ISA expansion was uni-directional before 2004; however, between 2009 and 2016, the expansion became multi-directional, and higher ISEI values were observed. In Jimei District, the ISA mainly expanded to the west between 1989 and 2004, and then expanded gradually to the west-northwest and northwest between 2004 and 2016. In Haicang District, the ISA expanded to the northwest between 1989 and 1996, to the west-southwest between 1996 and 2004, and to the west-southwest and northwest–west-northwest during 2004–2016. In Xiang'an District, the ISA mainly expanded to the north between 1989 and 2004, and expanded significantly to the north, north-northeast, and north-northwest between 2004 and 2015. In Tong'an District, the ISA mainly expanded to the south-southwest between 1989 and 1996, expanded to the south and south-southwest between 1996 and 2004, and expanded gradually to the south, south-southwest, and southwest between 2004 and 2016. In Xiamen Island, the ISA mainly expanded to the northeast, east-northeast (ENE), and east between 1989 and 2009. However, between 2009 and 2016, the ISA expansion sharply decreased in all directions except the northeast.

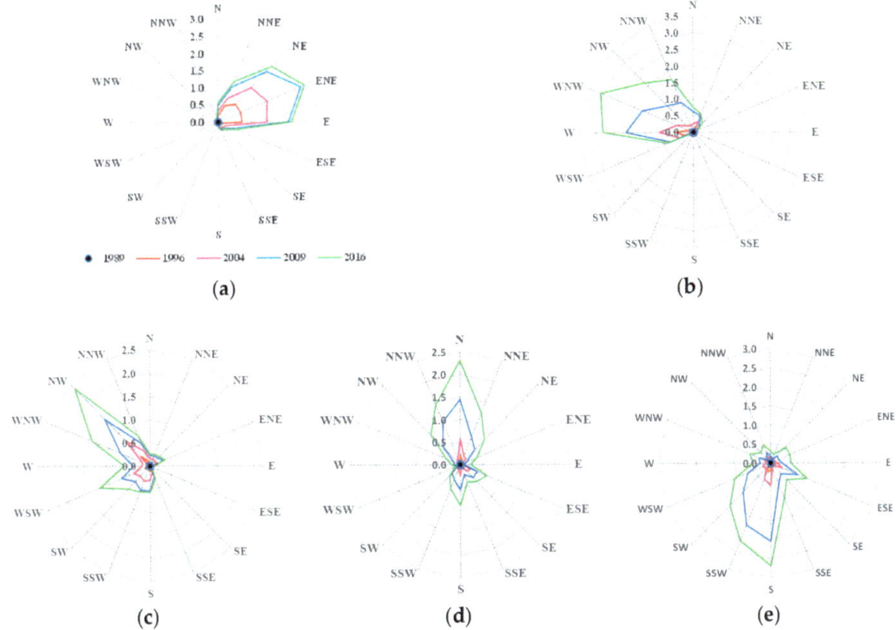

Figure 4. Spatial distribution of the expansion of ISA in various areas of Xiamen city during 1989–2016. (**a**) Xiamen Island. (**b**) Jimei District. (**c**) Haicang District. (**d**) Xiang'an District. (**e**) Tong'an District.

Table 4. Spatial characteristics of the expansion of impervious surface area (ISA) in Xiamen city during 1989–2016.

District	Period	Increase in ISA (km^2)	ΔISEI	Expansion Direction
Xiamen Island	1989–1996	22.94	0.205	NE-ENE-E
	1996–2004	25.81	0.202	NE-ENE-E
	2004–2009	16.67	0.208	NE-ENE-E
	2009–2016	5.89	0.053	NE
Jimei	1989–1996	10.40	0.093	W
	1996–2004	15.45	0.121	W
	2004–2009	24.65	0.308	W-WNW-NW
	2009–2016	29.90	0.267	W-WNW-NW
Haicang	1989–1996	8.41	0.075	NW
	1996–2004	19.81	0.155	NW, WSW
	2004–2009	15.44	0.193	NW-WNW, WSW
	2009–2016	21.26	0.190	NW-WNW, WSW
Tong'an	1989–1996	9.17	0.082	SSW
	1996–2004	15.22	0.119	S-SSW
	2004–2009	31.51	0.394	S-SSW-SW
	2009–2016	34.45	0.308	S-SSW-SW
Xiang'an	1989–1996	5.07	0.045	N
	1996–2004	13.52	0.106	N
	2004–2009	23.10	0.289	NNW-N-NNE
	2009–2016	30.78	0.275	NNW-N-NNE

Note: ISEI: Impervious Surface Expansion Index.

4.2. Changes in Urban Heat Islands

Figure 5a–e shows the distribution of UHIs in Xiamen city over the study period. As can be seen in the figure, similar to the changes in ISA shown in Figure 2, the distribution of UHIs varied significantly.

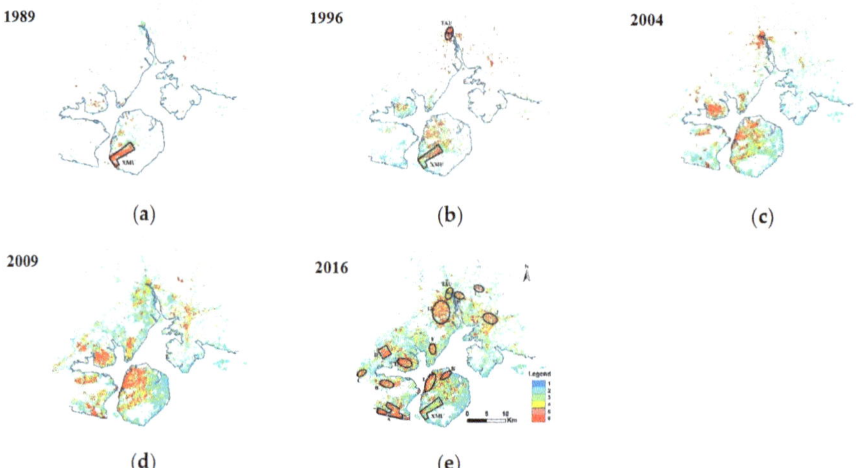

Figure 5. Graded urban land surface temperature (LST) maps indicating the distribution of urban heat islands (UHIs) in Xiamen city from 1989 to 2016. The different colors represent LST levels 1–6, representing increasing LST. XMU: Xiamen old downtown area; TAU: Tong'an old downtown area; A: Haicang bonded port zone; B: Xinyang industrial area (IA); C: Dongfu IA; D: Guankou IA; E: Xinglin IA; F: Northern industrial zone in Jimei District; G: Tong'an IA; H: Tong'an Chengdong IA; I: Xiangbei IA; J: Xiang'an IA; K: Airport industrial and logistics park; L: Xiangyu bonded zone with seaport and logistics park. The blue line indicates the coastline.

Table 5 shows statistics of the LST of impervious and non-impervious areas (NIAs) in Xiamen from 1989 to 2016. As shown in the table, in general, impervious areas exhibited a higher LST than non-impervious areas. The mean intensities of the UHIs (that is, the mean difference in LST between the urban ISA and the NIAs) were, on average, 3.56, 2.92, 3.01, 3.93, and 6.36 K in 1989, 1996, 2004, 2009, and 2016, respectively. The large standard deviations of LST for ISA indicate that these surfaces experience a large variation in LST due to the use of diverse construction materials. Furthermore, the higher standard deviation of the LST for NIAs in 2009 and 2016 shows that great temperature fluctuations occurred over the study area in these years (Table 5). These results indicate that urbanization increased the LST by an average of 3.56 and 6.36 K in 1989 and 2016, respectively, by replacing natural environments (vegetation and water) with non-transpiring, non-evaporating surfaces such as stone, asphalt, and concrete. This observation suggests that the construction of ISA could contribute to the development of UHIs much more than the construction of NIAs (such as forestland, water, and cultivated land), and that urban expansion raises the land surface temperature. Therefore, the effects of UHIs can be mitigated by optimizing land development, land-use planning, and constructing areas of natural vegetation.

Table 5. Statistics of the LST (in K) of impervious and non-impervious areas in Xiamen city from 1989 to 2016.

Year	Non-Impervious Areas				Impervious Areas			
	Tmin	Tmax	Tmean	SD	Tmin	Tmax	Tmean	STD
1989	297.10	307.91	301.06	1.56	298.05	307.91	304.62	1.56
1996	295.35	304.74	299.22	1.55	296.73	304.74	302.14	1.23
2004	295.44	307.56	299.54	1.31	295.44	307.56	302.55	1.68
2009	295.80	312.64	301.49	1.74	298.68	312.64	305.42	1.94
2016	296.45	323.11	307.20	3.07	301.55	323.11	313.56	2.39

Note: Tmin, Tmax, and Tmean refer to the maximum LST, minimum LST, and mean LST. SD represents one standard deviation of Tmean.

The distribution and intensities of UHIs in Xiamen city changed greatly from 1989 to 2016. In 1989, the UHI hotspots (i.e., clusters of neighboring thermal patches that have higher UHI grades) [49] mainly occurred in the old downtown area of the city (Figure 5a), which contains a number of urban villages and traditional heavy industries. As urbanization continued, new hotspots appeared in the districts of Jimei, Tong'an, Haicang, and Xiang'an. By 2016, the number of UHI hotspots in the urban area had increased to 12 (Figure 5e). The new hotspots were likely due to the rapid development of airports, seaports (such as the Xiamen Xiangyu Bonded Zone and the Haicang Bonded Port), and heavy industries, such as the food, automobile, petrochemical, and machine-building industries (Figure 5). For example, there are plans to construct an industrial area with a size of approximately 12 km^2 in the main industrial zone in Tong'an District, which would become the largest industrial area in Xiamen city. Almost all of the UHI hotspots in Xiamen exist along the coastline of Xiamen Island or in the four bay areas, except for two hotspots in Xiang'an District (Figure 5e). The spatial distribution of UHI hotspots results in a bay-like UHI form. This form is obviously distinct from the central-radial UHI pattern caused by ISA expansion in the form of concentric circles around a central area that was observed in other cities in previous studies. In the central-radial UHI pattern, the highest UHI grades appear in the central area and UHI grades gradually decrease with increasing distance from the center. The central-radial UHI pattern was observed by Mathew et al. [50] in Chandigarh City, India. The UHI form in Xiamen is similar to that reported for Hong Kong by Li and Zhang [51].

This work suggests that seaports and heavy industrial activities may increase the temperature of the urban environment. For example, seaports that are built on massive areas of reclaimed wetland have extremely high ISA, and could lead to the presence of large UHIs. As shown in Figure 6(b1),(c1), there is a great difference in the UHI level between zones of heavy industry (Guankou industrial area) and zones of high-technology industry (Xiamen software park). Zones of heavy industry have high impervious surface areas, large areas of endothermic roofing, and produce large amounts of waste heat, and could therefore also contribute to the creation of UHIs. This was also noted by Zhao et al. [49]. On the contrary, zones of high-technology industry form "cool" islands as they have low impervious surface areas, lower endothermic roofing area, no waste heat release, and large areas of vegetation. Additionally, from Figure 6(a1), a clear difference can be observed between "cool" islands in high-rise residential areas and heat islands in urban villages. Figure 6(a1) indicates that high-density low-rise urban villages with low vegetation cover are another important thermal source for Xiamen city. There are a number of urban villages in Xiamen. Therefore, re-planning of the old downtown area, from urban villages to high-rise residential areas, is an important approach to mitigate UHIs.

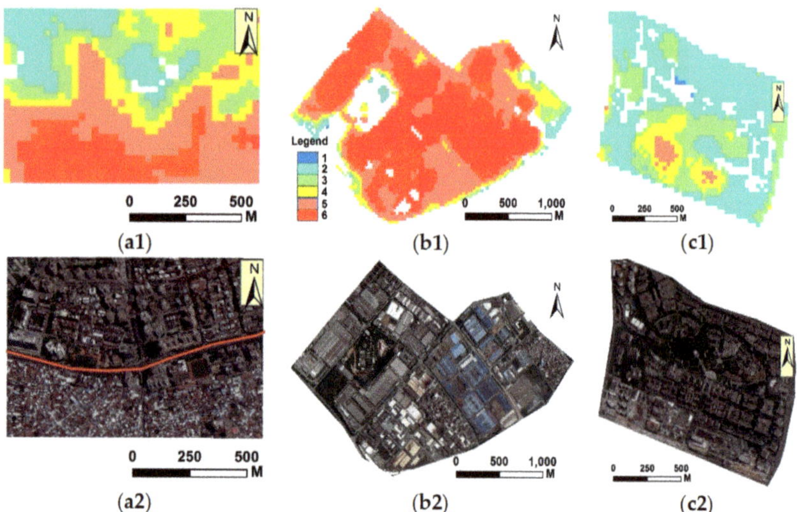

Figure 6. Graded urban land surface temperature (LST) maps for three impervious surface types ("1" suffix) for 2016, and their corresponding Google Earth images ("2" suffix). (**a**) Residential areas (urban villages and high-rise areas). (**b**) Guankou industrial area. (**c**) Xiamen software park. The red line in (**a2**) indicates the division between urban villages and high-rise residential areas.

In this study, the UHI distribution zone was defined as the superposition of levels 5 and 6 of the normalized LST data. The total areas of the UHI distribution zone were 20.07, 39.47, 56.73, 82.88, and 114.98 km^2 in 1989, 1996, 2004, 2009, and 2016, respectively, accounting for 1.18%, 2.32%, 3.34%, 4.88%, and 6.77% of the total area of the city in that year, respectively. As shown in Figure 7, before 2009, Xiamen Island had the largest UHI area among the five studied zones (Xiamen Island and Jimei, Haicang, Tong'an, and Xiang'an districts). In 2016, the area of UHIs in Tong'an District exceeded that in Xiamen Island, and the area of level-6 UHIs in Jimei District was greater than that in Xiamen Island. This finding suggests that the greatest intensity of UHIs may not always be located in Xiamen Island, and that UHI centers had spread from Xiamen Island to outside of the island by 2016. In 2016, Xiamen Island contained 51.6% of the population of Xiamen city and had a high population density of 12,823 people per square kilometer. However, in the same year, the UHI area in Xiamen Island only accounted for 22.04% of that of Xiamen city. This indicates that population density is not a key factor related to surface UHI in Xiamen city.

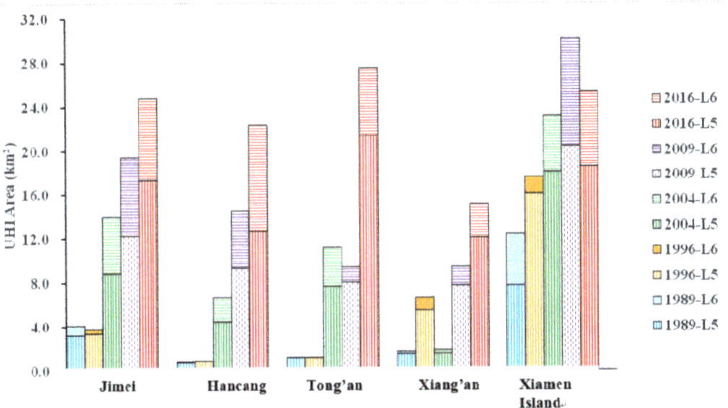

Figure 7. Variations in the area of UHIs in Xiamen city with an LST of level 5 (L5) and level 6 (L6) during 1989–2016.

The URI values were 0.281, 0.285, 0.239, 0.231, and 0.229 in 1989, 1996, 2004, 2009, and 2016, respectively; that is, while the area of UHIs increased over the study period, the URI decreased. In 1989, 9.23% of the urban area of Xiamen city corresponded to the highest LST level (level 6) while 22.6% corresponded to level 5; by 2016, the same values had reduced to 7.48% and 18.57%, respectively. This can be clearly observed in the old downtown area by comparing Figure 5a,b,e. This indicates that the relative area of UHIs slightly reduced in the old downtown area during the study period. As shown in Figure 5, in 1989, there was a large concentrated area with a level-6 LST in this area; however, the LST of this area had largely reduced to level 4 or 5 by 2004 and had reduced even further by 2016. This change can largely be attributed to the fact that the local government and the public took measures to improve the living environment of the old downtown area. First, in the early 1990s, Xiamen's local government proposed a new policy for the development of industry, which encouraged the development of the tertiary industry in Xiamen Island and the removal of primary and secondary industries to areas outside of Xiamen Island. Therefore, a large number of heavy industry factories were moved out of Xiamen Island. Second, since 1989, Xiamen's ecological green space has been greatly expanded. For example, in 2001, Xiamen had 31 parks with a combined area of 4.51 km² while in 2016, it had 120 parks with a combined area of 26.04 km². Moreover, by the end of 2020, the number of parks in Xiamen city (including integrated parks, mountain parks, special parks, and community parks) is expected to reach 342 and cover a total area of 124.20 km², including 40.52 km² of green buffers and 83.68 km² of park green. Additionally, in the 1990s, plans for the redevelopment of urban villages were proposed by Xiamen's local government. Some urban villages in Xiamen city have been transformed from high-density low-rise residential areas with scarce vegetation cover to low-density high-rise residential areas with high vegetation cover. All of the above efforts have contributed significantly to mitigating the effects of UHIs in Xiamen city.

4.3. Relationships between LST and ISA%, Vegetation Fraction, and NDVI

Impervious surfaces and water and vegetation areas are all critical biophysical components of urban ecosystems, and the interactions between these three components can significantly affect the urban temperature.

As shown in Figure 8(a1–a5), a very strong linear relationship ($R^2 \geq 0.93$) was observed between the mean LST and the ISA% throughout the study period, with the R^2 value exceeding 0.96 between 1989 and 2009. The steepest slope of the linear correlation between LST and ISA% was observed for the Landat-8 image acquired in summer 2016, indicating that the magnitude of UHIs (i.e., the difference in LST between urban and rural areas) was largest in summer 2016. The lowest difference between the LST of rural and urban areas was found for the image acquired in summer 1996, which indicates that the lowest UHI magnitude occurred at this time. Each 10% increase in ISA was associated with an increase in summer LST of 0.41 to 0.91 K, which compares well with the results of related studies. Yuan and Bauer [14] reported a 0.95 K increase in summer LST corresponding to a 10% increase in ISA in the Twin Cities Metropolitan Area, Minnesota, USA while Li et al. [15] observed a 0.52 K increase in summer LST corresponding to a 10% increase in ISA in Shanghai, China.

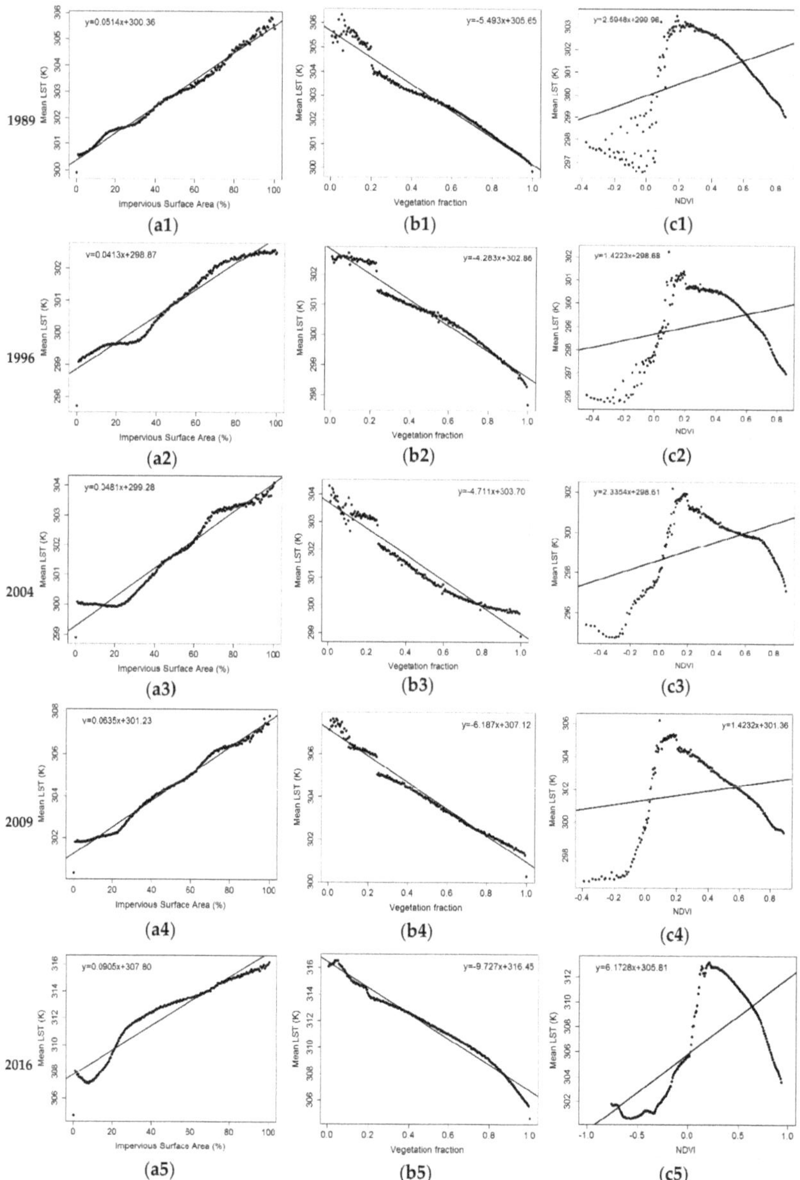

Figure 8. Relationships between mean LST and (**a**) ISA%, (**b**) vegetation fraction, and (**c**) the normalized difference vegetation index (NDVI) during 1989–2016.

The relationships between the mean LST and the NDVI are presented in Figure 8(c1–c5). Overall, a negative correlation is found between LST and the NDVI from 1989 to 2016. However, the trends are not stable, with a significant change in behavior being observed at an NDVI of around 0.2 for all the studied years. This indicates heterogeneous effects from various non-vegetated areas, such as wetland, bare soil, and high-density urban surfaces. These observations suggest that using the NDVI to investigate UHIs is complicated since the NDVI is sensitive to land use changes caused by rapid urbanization. Furthermore, previous research found that the relationship between LST and the NDVI experiences strong seasonal changes [14,15].

The Fv has been considered to be a very accurate representation of vegetation abundance [16]. The results of the zonal analysis (Figure 8(b1–b5)) indicate that a strong negative correlation exists between LST and Fv, with R^2 values exceeding 0.94 for all the studied years. Each 0.1 increase in Fv was associated with a decrease in summer LST of 0.43 to 0.97 K. This result also compares well with the results of previous studies. Li et al. [15] reported a 0.61 K decrease in summer LST and a 0.31 K decrease in spring LST corresponding to a 0.1 increase while Ma et al. [16] reported a 0.80 K decrease in summer LST corresponding to a 0.1 increase.

Yuan and Bauer [14] and Li et al. [15] observed a strong linear relationship between LST and ISA% using zonal analysis and the LSMA. Conversely, another study [22] found an exponential relationship between these two parameters based on a per-pixel analysis and the NDISI. The results of the present study suggest that a linear relationship exists between LST and ISA% using zonal analysis and the BCI, similar to the findings of Yuan and Bauer [14] and Li et al. [15].

5. Summary and Conclusions

Landsat 5 TM and Landsat 8 TIRS/OLI thermal infrared images are of great use in the analysis of UHIs. Using quantitative remote sensing images and statistical methods, this study investigated the impacts of rapid urban expansion on the UHIs of a coastal city (Xiamen city, China) between 1989 and 2016.

The ISA can be a key environmental indicator. In this study, the ISA was estimated from the remote sensing images using the BCI method. It is essential to distinguish impervious surfaces from bare soil and beach land when using the BCI method to estimate urban ISA in places that have experienced rapid urbanization, especially in coastal cities; however, this is complicated due to the similar spectral properties of these three surface types. The use of the BCI combined with an ISA post-processing method produced satisfactory results in terms of ISA identification accuracy (overall accuracies ≥90.50% and Kappa values ≥0.810).

The results show that the ISA of Xiamen city increased 6.1-fold between 1989 and 2016. This rapid expansion of ISA was mainly attributable to the reclamation of coastal, arable, and forest land. The ISA expansion was gradual, and was mostly focused along Haicang Bay, Xinglin Bay, Tong'an Bay, Malyuan Bay, and the coastline of Xiamen Island. The directions of ISA extension were generally consistent throughout the study period in all of the five districts of Xiamen city.

The results of this study indicate that the intensity, size, and distribution of UHIs in Xiamen city changed during the study period due to urban expansion. The expansion of ISA is a major contributor to the size and intensity of UHIs, and therefore to urban climatic warming. Moreover, the results of this study show that the spatiotemporal variation of UHIs is very likely to be due to surface changes caused by the rapid expansion of ISA, the reduction of vegetated areas, and the development of industries that produce large amounts of waste heat. Due to the fact that ISA expansion is constrained by the coastline, combined with urban development planning, Xiamen has a bay-like UHI form. This UHI form is markedly distinct from the central-radial UHI form observed in other cities in previous studies.

In Xiamen, sea ports and heavy industries are the most important contributors to surface UHIs while another important contributor is urban villages. In contrast, high-rise residential areas and new-technology industry produce relatively "cool" temperature islands. This result suggests that certain measures (e.g., the appropriate distribution of industry, increasing the coverage of forest and park lands, the use of less endothermic roofing material, and the re-planning of urban villages) could effectively alleviate the effect of UHIs, and could thus increase the healthiness of the urban environment and promote sustainable development. Between 1989 and 2016, the total area of UHIs increased 4.7-fold (i.e., the area of UHIs increased at a lower rate than the ISA did). However, during the same period, the ratio of the UHI area to the total ISA decreased slightly. Furthermore, the size and intensity of UHIs reduced in some parts of Xiamen city during the study period due to a significant increase in vegetation coverage, the re-planning of the old downtown area, and the construction of new parks.

The results of the zonal analysis suggest that a strong positive correlation exists between LST and ISA%, consistent with the findings of recent research. Each 10% increase in ISA was associated with an increase in summer LST of 0.41 to 0.91 K, which compares well with the results of related studies. The relationship between LST and the NDVI is more complicated, since the NDVI is sensitive to variations in land use cover and seasonal variations; however, at higher NDVI values, a negative linear correlation is observed between LST and the NDVI. This study indicates that the analysis of the ISA is useful for investigating changes in surface UHIs over time.

This study presents useful information for the sustainable planning of urban areas and for the mitigation of the effects of UHIs. Urban ISA growth patterns may affect the distribution patterns of UHIs. Bay-like cities may be less compact than central-radial cities. Determining which ISA growth pattern(s) are helpful to mitigate UHIs strongly merits future research. Furthermore, it is recommended that more remote sensing images with a higher spatial resolution (e.g., from the Sentinel-2 or Gaofen-1 satellites) and covering more metropolitan areas in different climatic zones should be extensively investigated to improve the quantitative remote sensing of land surface temperature in order to better determine the effects of urban development on the thermal environment of urban areas.

Author Contributions: L.H. made essential contributions to the research design and the image data processing and analysis. L.T. provided important concepts for the study. X.Z. modified the paper critically. Q.N. and F.S. provided original Landsat images, offered analysis tools, and carried out data cleaning. All authors have read and agreed to the published version of the manuscript.

Funding: This research was funded by the National Natural Science Foundation of China (Nos. 41471366, 41501447), the Educational Commission of Fujian Province of China (JAT170420) and the Fujian Natural Science Foundation, China (2017J01469, 2017J01666, 2017J05069).

Conflicts of Interest: The authors declare no conflict of interest.

References

1. Vaz, E.; Aversa, J. A Graph theory approach for geovisualization of land use change: An application to Lisbon. *J. Spat. Organ. Dyn.* **2013**, *1*, 254–264.
2. 2018 Revision of World Urbanization Prospects. Available online: https://www.un.org/development/desa/publications/2018-revision-of-world-urbanization-prospects.html (accessed on 16 May 2018).
3. Residency Shift Part of Plan to Boost Urban Base. Available online: http://en.ce.cn/main/latest/201904/09/t20190409_31819121.shtml (accessed on 4 April 2019).
4. National Bureau of Statistics of China. *China Statistical Yearbook 2019*; China Statistics Press: Beijing, China, 2019.
5. Xu, H.; Lin, D.; Tang, F. The impact of impervious surface development on land surface temperature in a subtropical city: Xiamen, China. *Int. J. Climatol.* **2013**, *33*, 1873–1883. [CrossRef]
6. Xian, G.; Crane, M. An analysis of urban thermal characteristics and associated land cover in tampa bay and Las Vegas using landsat satellite data. *Remote Sens. Environ.* **2006**, *104*, 147–156. [CrossRef]
7. Voogt, J.A.; Oke, T.R. Thermal remote sensing of urban climates. *Remote Sens. Environ.* **2003**, *86*, 370–384. [CrossRef]
8. Grimm, N.B.; Faeth, S.H.; Golubiewski, N.E.; Wu, J.; Bai, X.M.; Briggs, J.M. Global change and the ecology of cities. *Science* **2008**, *319*, 756–760. [CrossRef] [PubMed]
9. Wu, H.; Ye, L.P.; Shi, W.Z.; Clarke, K.C. Assessing the effects of land use spatial structure on urban heat islands using HJ-1B remote sensing imagery in Wuhan, China. *Int. J. Appl. Earth Obs. Geoinf.* **2014**, *32*, 67–78. [CrossRef]

10. Mountrakis, G.; Luo, L. Enhancing and replacing spectral information with intermediate structural inputs: A case study on impervious surface detection. *Remote Sens. Environ.* **2011**, *115*, 1162–1170. [CrossRef]

11. Voorde, T.V.D.; Roeck, T.D.; Canters, F. A comparison of two spectral mixture modelling approaches for impervious surface mapping in urban areas. *Int. J. Remote Sens.* **2009**, *30*, 4785–4806. [CrossRef]

12. Okujeni, A.; Linden, S.V.D.; Hostert, P. Extending the vegetation–impervious–soil model using simulated EnMAP data and machine learning. *Remote Sens. Environ.* **2015**, *158*, 69–80. [CrossRef]

13. Mohapatra, R.P.; Wu, C. High resolution impervious surface estimation. *Photogramm. Eng. Remote Sens.* **2010**, *76*, 1329–1341. [CrossRef]

14. Yuan, F.; Bauer, M.E. Comparison of impervious surface area and normalized difference vegetation index as indicators of surface urban heat island effects in Landsat imagery. *Remote Sens. Environ.* **2007**, *106*, 375–386. [CrossRef]

15. Li, J.; Song, C.; Cao, L.; Zhu, F.G.; Meng, X.L.; Wu, J.G. Impacts of landscape structure on surface urban heat islands: A case study of Shanghai, China. *Remote Sens. Environ.* **2011**, *115*, 3249–3263. [CrossRef]

16. Ma, Y.; Kuang, Y.; Huang, N. Coupling urbanization analyses for studying urban thermal environment and its interplay with biophysical parameters based on TM/ETM+ imagery. *Int. J. Appl. Earth Obs.* **2010**, *12*, 110–118. [CrossRef]

17. Powell, R.L.; Roberts, D.A.; Dennison, P.E.; Hess, L.L. Sub-pixel mapping of urban land cover using multiple endmember spectral mixture analysis: Manaus, Brazil. *Remote Sens. Environ.* **2007**, *106*, 253–267. [CrossRef]

18. Carlson, T.N.; Arthur, S.T. The impact of land use-land cover changes due to urbanization on surface microclimate and hydrology: A satellite perspective. *Glob. Planet. Chang.* **2000**, *25*, 49–65. [CrossRef]

19. Xu, H.Q. Analysis of Impervious Surface and its impact on urban heat environment using the normalized difference impervious surface index (NDISI). *Photogramm. Eng. Remote Sens.* **2010**, *76*, 557–565. [CrossRef]

20. Deng, C.; Wu, C. BCI: A biophysical composition index for remote sensing of urban environments. *Remote Sens. Environ.* **2012**, *127*, 247–259. [CrossRef]

21. Meng, Q.; Zhang, L.; Sun, Z.; Meng, F.; Wang, L.; Sun, Y.X. Characterizing spatial and temporal trends of surface urban heat island effect in an urban main built-up area: A 12-year case study in Beijing, China. *Remote Sens. Environ.* **2018**, *204*, 826–837. [CrossRef]

22. Liu, C.; Shao, Z.F.; Chen, M.; Luo, H. MNDISI: A multi-source composition index for impervious surface area estimation at the individual city scale. *Remote Sens. Lett.* **2013**, *4*, 803–812. [CrossRef]

23. Sun, G.Y.; Chen, X.L.; Jia, X.P.; Yao, Y.J.; Wang, Z.J. Combinational Build-Up Index (CBI) for effective impervious surface mapping in urban areas. *IEEE J. Sel. Top. Appl. Earth Observ. Remote Sens.* **2016**, *9*, 2081–2092. [CrossRef]

24. Onishi, A.; Cao, X.; Ito, T.; Shi, F.; Imur, H. Evaluating the potential for urban heat-island mitigation by greening parking lots. *Urban For. Urban Green.* **2010**, *9*, 323–332. [CrossRef]

25. Xiong, Y.Z.; Huang, S.P.; Chen, F.; Ye, H.; Wang, C.P.; Zhu, C.B. The Impacts of rapid urbanization on the thermal environment: A remote sensing study of Guangzhou, south China. *Remote Sens.* **2012**, *4*, 2033–2056. [CrossRef]

26. Hua, L.Z.; Liao, J.F.; Chen, H.X.; Chen, D.K.; Shao, G.F. Assessment of ecological risks induced by land use and land cover changes in Xiamen city, China. *Int. J. Sustain. Dev. World Ecol.* **2018**, *25*, 439–447. [CrossRef]

27. Hua, L.Z.; Tang, L.N.; Cui, S.H.; Yin, K. Simulating urban growth using the SLEUTH model in a coastal peri-urban district in China. *Sustainability* **2014**, *6*, 3899–3914. [CrossRef]

28. Tang, L.N.; Zhao, Y.; Yin, K.; Zhao, J. City profile: Xiamen. *Cities* **2013**, *31*, 615–624. [CrossRef]

29. Hua, L.Z.; Li, X.Q.; Tang, L.N.; Yin, K.; Zhao, Y. Spatio-temporal dynamic analysis of island-city landscape: A case study of Xiamen Island, China. *Int. J. Sustain. Dev. World Ecol.* **2010**, *17*, 273–278. [CrossRef]

30. Chavez, P.S. Image-based atmospheric corrections—Revisited and revised. *Photogramm. Eng. Rem. Sens.* **1996**, *62*, 1025–1036.

31. Lu, D.; Mausel, P. Assessment of atmospheric correction methods applicable to Amazon basin LBA research. *Int. J. Remote Sens.* **2002**, *23*, 2651–2671. [CrossRef]

32. Ke, Y.; Im, J.; Lee, J.; Lee, J.; Gong, H.; Ryu, Y. Characteristics of Landsat 8 OLI-derived NDVI by comparison with multiple satellite sensors and in-situ observations. *Remote Sens. Environ.* **2015**, *164*, 298–313. [CrossRef]

33. Huete, A.R. A Soil-adjusted vegetation index (SAVI). *Remote Sens. Environ.* **1988**, *25*, 295–309. [CrossRef]

34. Xu, H.Q. Modification of normalised difference water index (NDWI) to enhance open water features in remotely sensed imagery. *Int. J. Remote Sens.* **2006**, *27*, 3025–3033. [CrossRef]

35. Wang, M.Y.; Xu, H.Q. Temporal and spatial changes of urban impervious surface and its influence on urban ecolo-gical quality: A comparison between Shanghai and New York. *J. Appl. Ecol.* **2018**, *29*, 236–247.
36. Hua, L.Z.; Zhang, X.X.; Chen, X.; Kai, K.; Tang, L.N. A feature-based approach of decision tree classification to map time series urban land use and land cover with Landsat 5 TM and Landsat 8 OLI in a coastal city, China. *ISPRS Int. J. Geo. Inf.* **2017**, *6*, 331. [CrossRef]
37. Crist, E.P. A TM tasseled cap equivalent transformation for reflectance factor data. *Remote Sens. Environ.* **1985**, *17*, 301–306. [CrossRef]
38. Li, B.; Ti, C.; Zhao, Y.; Yan, X. Estimating soil moisture with Landsat data and its application in extracting the spatial distribution of winter flooded paddies. *Remote Sens.* **2016**, *8*, 38. [CrossRef]
39. Man, W.; Nie, Q.; Hua, L.Z.; Wu, X.W.; Li, H. Spatio-temporal variations in impervious surface patterns during urban expansion in a coastal city: Xiamen, China. *Sustainability* **2019**, *11*, 2404. [CrossRef]
40. Artis, D.A.; Carnahan, W.H. Survey of emissivity variability in thermography of urban areas. *Remote Sens. Environ.* **1982**, *12*, 313–329. [CrossRef]
41. Jimenez-Munoz, J.C.; Cristobal, J.; Sobrino, J.A.; Soria, G.; Ninyerola, M.; Pons, X. Revision of the Single-Channel Algorithm for Land Surface Temperature Retrieval from Landsat Thermal-Infrared Data. *IEEE Trans. Geosci. Remote Sens.* **2009**, *47*, 339–349. [CrossRef]
42. Sobrino, J.A.; Jimenez-Muoz, J.C.; Soria, G.; Romaguera, M.; Guanter, L.; Moreno, J.; Plaza, A.; Martinez, P. Land surface emissivity retrieval from different VNIR and TIR sensors. *IEEE Trans. Geosci. Remote Sens.* **2008**, *46*, 316–327. [CrossRef]
43. Cai, Y.; Zhang, H.; Zheng, P.; Pan, W. Quantifying the impact of land use/land cover changes on the urban heat island: A case study of the natural wetlands distribution area of Fuzhou City, China. *Wetlands* **2016**, *36*, 285–298. [CrossRef]
44. Xu, H.Q.; Chen, B.Q. Remote sensing of the urban heat island and its changes in Xiamen city of SE China. *Int. J. Remote Sens.* **2004**, *16*, 276–281.
45. Liu, Y.X.; Peng, J.; Wang, Y.L. Diversification of land surface temperature change under urban landscape renewal: A case study in the main city of Shenzhen, China. *Remote Sens.* **2017**, *9*, 919. [CrossRef]
46. Cheng, S.L.; Wang, T.X. Comparison analyses of equal interval method and mean-standard deviation method used to delimitate urban heat island. *Geo Inf. Sci.* **2009**, *11*, 145–150.
47. Gutman, G.; Ignatov, A. The derivation of the green vegetation fraction from NOAA/AVHRR data for use in numerical weather prediction models. *Int. J. Remote Sens.* **1998**, *19*, 1533–1543. [CrossRef]
48. Xu, H.Q.; Hu, X.; Guan, H.; Guan, H.D.; Zhang, B.B.; Wang, M.Y.; Chen, S.M.; Chen, M.H. A remote sensing based method to detect soil erosion in forests. *Remote Sens.* **2019**, *11*, 513. [CrossRef]
49. Zhao, X.F.; Huang, J.; Ye, H.; Wang, K.; Qiu, Q.Y. Spatiotemporal changes of the urban heat island of a coastal city in the context of urbanisation. *Int. J. Sustain. Dev. World Ecol.* **2010**, *17*, 311–316. [CrossRef]
50. Mathew, A.; Khandelwal, S.; Kaul, N. Spatial and temporal variations of urban heat island effect and the effect of percentage impervious surface area and elevation on land surface temperature: Study of Chandigarh City, India. *Sustain. Cities Soc.* **2016**, *26*, 264–277. [CrossRef]
51. Liu, L.; Zhang, Y. Urban heat island analysis using the Landsat TM data and ASTER data: A case study in Hong Kong. *Remote Sens.* **2011**, *3*, 1535–1552. [CrossRef]

 *sustainability* 

Article

The Impacts of Landscape Changes on Annual Mean Land Surface Temperature in the Tropical Mountain City of Sri Lanka: A Case Study of Nuwara Eliya (1996–2017)

Manjula Ranagalage [1,2,*], Yuji Murayama [1], DMSLB Dissanayake [2,3,*] and Matamyo Simwanda [4]

[1] Faculty of Life and Environmental Sciences, University of Tsukuba, 1-1-1, Tennodai, Tsukuba, Ibaraki 305-8572, Japan; mura@geoenv.tsukuba.ac.jp

[2] Department of Environmental Management, Faculty of Social Sciences and Humanities, Rajarata University of Sri Lanka, Mihintale 50300, Sri Lanka

[3] Graduate School of Life and Environmental Sciences, University of Tsukuba, 1-1-1, Tennodai, Tsukuba, Ibaraki 305-8572, Japan

[4] Department of Plant and Environmental Sciences, School of Natural Resources, Copperbelt University, P.O. Box 21692, Kitwe 10101, Zambia; matamyo@gmail.com

* Correspondence: manjularanagalage@ssh.rjt.ac.lk or manjularanagalage@gmail.com (M.R.); dissanayakedmslb@gmail.com (D.D.)

Received: 23 August 2019; Accepted: 1 October 2019; Published: 6 October 2019

Abstract: Although urbanization has contributed to improving living conditions, it has had negative impacts on the natural environment in urbanized areas. Urbanization has changed the urban landscape and resulted in increasing land surface temperature (LST). Thus, studies related to LST in various urban environments have become popular. However, there are few LST studies focusing on mountain landscapes (i.e., hill stations). Therefore, this study investigated the changes in the landscape and their impacts on LST intensity (LSTI) in the tropical mountain city of Nuwara Eliya, Sri Lanka. The study utilized annual median temperatures extracted from Landsat data collected from 1996 to 2017 based on the Google Earth Engine (GEE) interface. The fractions of built-up (BL), forested (FL) and agricultural (AL) land, were calculated using land use and cover maps based on urban–rural zone (URZ) analysis. The urban–rural margin was demarcated based on the fractions of BL (<10%), and LSTI that were measured using the mean LST difference in the urban–rural zone. Besides, the mixture of land-use types was calculated using the AL/FL and BL/FL fraction ratios, and grid-based density analysis. The results revealed that the BL in all URZs rapidly developed, while AL decreased during the period 1996 to 2017. There was a minimal change in the forest area of the Nuwara Eliya owing to the government's forest preservation policies. The mean temperature of the study area increased by 2.1 °C from 1996 to 2017. The magnitude of mean LST between urban–rural zones also increased from 1.0 °C (1996) to 3.5 °C (2017). The results also showed that mean LST was positively correlated with the increase and decrease of the BL/FL and AL/FL fraction ratios, respectively. The grid-based analysis showed an increasing, positive relationship between mean LST and density of BL. This indicated that BL density had been a crucial element in increasing LST in the study area. The results of this study will be a useful indicator to introduce improved landscape and urban planning in the future to minimize the negative impact of LST on urban sustainability.

Keywords: land use and cover; land surface temperature; built-up land; agricultural land; gradient analysis; Nuwara Eliya; Sri Lanka

1. Introduction

In recent decades, population growth and economic development have directly affected landscapes' transformations in developing countries [1]. Rapid changes in the landscape have resulted in the conversion of natural vegetation and agricultural land into built-up (impervious) land, such as buildings, parking lots, roads, and other constructions [1–4]. This has caused several environmental problems at local, regional, and global scales [5], such as decreases in agricultural land [6]; habitat destruction [5,7,8]; air, soil, and water contamination [9–11]; increases in vector-borne diseases, such as malaria and dengue [12]; decreases in green space [13,14]; and increases in land surface temperature (LST) [15–17].

Increasing temperatures in urban environments are largely an outcome of rapid urbanization and anthropogenic activities [18]. Built-up areas have been formed using several materials, such as concrete, flooring, pebbles, stone, and gravel, which decrease evapotranspiration, increase the sensitivity of the city and notably affect its local climate [16,19,20]. However, studying temperature changes based on the air temperature is challenging, as there is a lack of meteorological stations, especially in developing countries [21]. Thus, satellite remote-sensing data provides vital information for observing the temperature patterns in urban areas [21]. Several studies have been conducted by using vast range of remote sensing data, such as Modis data [22–25], Synthetic Aperture Radar (SAR) data [26], Nightlights data [27,28], Landsat data [17,29,30], Land Scan data [31–33], and fossil fuel CO_2 emission data [33,34], to understand LST patterns. Many urban landscapes from small to large scales have been studied worldwide [16,35,36], including coastal cities [16,33,37,38], desert cities [20,39], and mountain cities [13,14]. Mountain cities are attractive for rich people, as they have a cold climate and comfortable living conditions, which have resulted in rapid urban development [14]. Thus, studies focusing on mountain landscapes are vital for understanding the changing pattern of LST to introduce mitigation measures for comfortable living conditions.

A large number of studies have used two or more satellite images from different time points to analyze an LST pattern due to the unavailability of cloud-free images. However, the difference in the acquisition times might influence any resulting LST pattern due to varying environmental factors (wind speed, the Sun's radiation, surface moisture, and humidity) [4,32]. The use of more satellite images captured in multiple time points can potentially provide more specific information to understand the changing pattern of LST [4]. Still, it is not easy to analyze extensive earth observation data sets, because of issues of spatial and temporal resolution [40,41]. Thus, big data analysis platforms can be used as an alternative for conducting accurate results [40]. The Google Earth Engine (GEE) provides the potential to process a large number of satellite images and researchers can easily access free public data archives for more than thirty years of historical data [40]. Hence, in this study, we used GEE to extract the annual median LST for three-time points such as 1996, 2006, and 2017 based on several images captured during the selected years. We hypothesized that the use of many images captured in multiple time points would provide a more precise pattern of LST in the study area.

The spatial distribution of the LST intensity (LSTI) provides essential environmental information for understanding a temperature pattern in detail. The literature provides two methods that can be used to study LSTI. The first one is the categorization of land use and cover as the local climate zone and then following the cross cover comparison method to calculate the LSTI [13,14,42–44]. The second method involves determining the difference in the mean LST between urban and rural zones based on the urban–rural gradient analysis [13,14,45]. In this study, we used the second method. Other studies utilizing the urban–rural gradient analysis have shown that the urban–rural demarcation is essential for determining the temperature differences [13,14].

Mountain cities in Asia have been developing since the colonization period of the 19th and 20th centuries [46,47]. The cool climate and natural landscapes became the most prominent factors driving the development of mountain cities. During the colonial period, cool climates were preferred, as they allowed the colonials to maintain their "western lifestyle." Mountain cities were also selected to avoid wasting illnesses, episodic pestilence, sunstroke, and depression [46,48]. As a result, most mountain cities have been experiencing rapid urbanization since the end of World War II until now. Notwithstanding the vast literature on the impacts of urbanization on landscape changes and other associated consequences, like increasing LST, mountain cities remain one of the landscapes that have not been extensively studied.

Nuwara Eliya is a historical city that developed during the British colonial period, as reflected in the city's architecture [49]. The British people preferred a cold climate and Nuwara Eliya was referred to as "Little England" [49]. They used Nuwara Eliya as a meeting place for wealthy British families [50]. Currently, several recreational sites, such as Lake Gregory, golf links, racecourses, and a large number of clubs, remain in Nuwara Eliya [50]. After gaining independence in 1948, Nuwara Eliya became renowned worldwide as a tourist destination in Sri Lanka, and thousands of tourists from Sri Lanka and overseas visited this area to enjoy the cold climate and natural beauty [49]. This has resulted in several environmental problems associated with urbanization in Nuwara Eliya. One of the critical environmental impacts of urbanization that has likely affected the mountain city Nuwara Eliya is the Urban Heat Island (UHI) resulting from variations in LST [13,14]. Still, previous studies related to LST have been focused on the Colombo City [3,4,37] and Kandy [13,51] in Sri Lanka.

Thus, in this study, we hypothesized that the rapid changes in the urban landscape of Nuwara Eliya influenced the patterns of LST. The study examined the changes in the urban landscape and their impacts on the spatiotemporal changes of LST, in order to contribute to efforts aimed at introducing proper landscape and urban planning in Nuwara Eliya. The study has three objectives to: (1) monitor urban landscape changes and their impacts on the spatial and temporal variations of LST intensity from 1996–2017 based on the urban–rural gradient analysis; (2) identify relationships between mean LST and the agricultural land (AL)/forested land (FL) and built-up land (BL)/FL fraction ratios; and (3) study the relationship between mean LST and densities of BL, FL, and AL based on the grid-based analysis method. The results of this study could be useful for enhancing capacity and knowledge to minimize the possible negative impacts of rapid urbanization in the study area.

2. Materials and Methods

2.1. Study Area: Nuwara Eliya, Sri Lanka

Nuwara Eliya is located in the central province of Sri Lanka and resides at latitudes of 6°54′21.94″ to 7°2′28.53″ N and longitudes of 80°50′5.89″ to 80°41′57.23″ E (Figure 1). The study area includes the landscape within a 7.5 km^2 radius of the center of Nuwara Eliya. Nuwara Eliya is surrounded by one of the tallest mountains (~2524 m high) in the study area, known as Pidurutalagala. The average elevation of Nuwara Eliya is about 1800 m. Monthly rainfall in Nuwara Eliya ranges from approximately 71.5 to 226.8 mm, with an annual average of 1905 mm. The lowest average monthly rainfall is recorded in January, February, and March, resulting in a relatively dry period. According to the Meteorology Department of Sri Lanka, the daily mean temperature is about 15.9 °C, with average minimum and maximum temperatures of 11.6 °C and 20.2 °C, respectively. Nuwara Eliya is the most popular site in Sri Lanka among local and foreign tourists due to the cold climate that persists throughout the year. In terms of the urban development pattern, Nuwara Eliya exhibits a single locus.

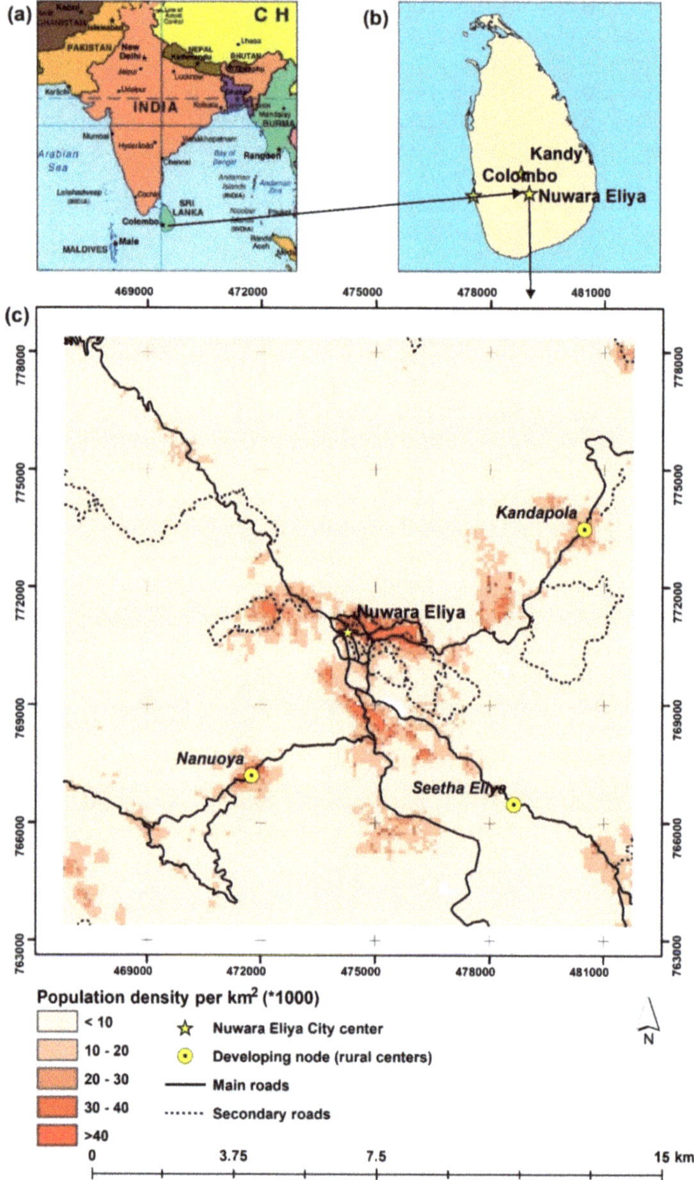

Figure 1. Study area: (**a**) map from part of South Asia (http://www.maps-world.net); (**b**) location of Nuwara Eliya; and (**c**) population density of Nuwara Eliya (100 × 100 m).

2.2. Overall Workflow

Figure 2 shows the overall workflow of the study to achieve the objectives described above. The study workflow included five major steps: (i) extraction of median upper-atmosphere brightness temperature from the thermal bands of Landsat images (Figure 2a); (ii) retrieval of LST (Figure 2b); (iii) land use/land cover (LULC) classification into five classes, forest land (FL), built-up (BL), agriculture land (AL), other land, and water using machine learning algorithms, based on supervised classifications

(Figure 2c); (iv) urban–rural gradient, statistical, and intensity analyses, based on the BL, FL, and AL fractions, AL/FL and BL/FL fraction ratios, and mean LST; and finally (v) grid analysis based on the mean LSTs and the densities of BL, FL, and AL.

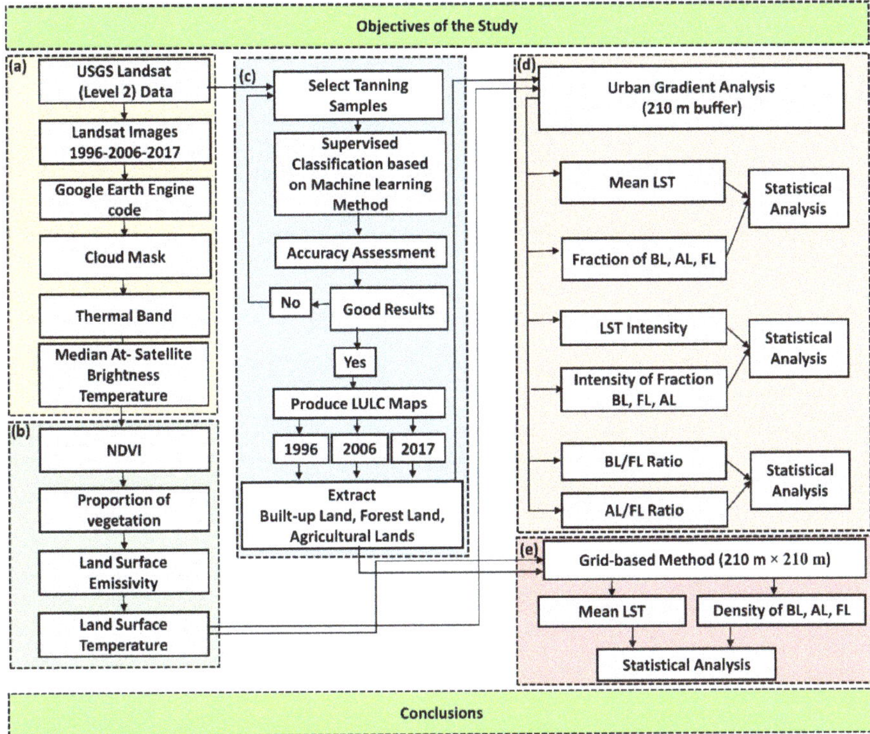

Figure 2. The workflow of the study. Note BL, FL, and AL refer to built-up land, forested land, and agricultural land, respectively.

2.3. *Calculating Annual Median At-Satellite Brightness Temperature Using the Google Earth Engine (GEE)*

The study employed the GEE to calculate annual median at-satellite brightness temperatures using atmospherically corrected pre-processed Landsat datasets (Level 2) [52]. We hypothesized that the use of more images would produce clearer output images to understand the LST patterns in the study area. Past studies have shown the aptness of this method to generate median LSTs in different areas [40,41]. In this process, several steps were performed as follows.

(i) The study area was defined and imported as an "Asset" in GEE, and subsequently used as the primary geometry source throughout the process. Masking was then conducted due to cloud disturbance in the available Landsat imageries. The cloud disturbance could be attributed to Nuwara Eliya being located in a tropical area.

(ii) The Image Collection tool in GEE was used to prepare the imagery for the study, including 15 images from 1996 (Landsat 5), 17 images from 2006 (Landsat 5), and 20 images from 2017 (Landsat 8) (Table A1).

(iii) Afterwards, median temperatures of each pixel were extracted for 1996, 2006, and 2017 by using "ee.reducer" methods available in GEE [53]. Extraction of the pixel based median temperatures was done to take care of inaccurate extraction caused by disturbance from clouds in the study area. Figure 3 shows the graphical illustration of creating image collection and the code used to generate the annual, median, at-satellite brightness temperature for each pixel of each year (provided in Annex Code A1 (Landsat 5TM) and A2 (Landsat 8)).

(iv) The extracted median temperatures based on the upper-atmosphere brightness temperatures from the thermal bands (in Kelvin) (band 6 for Landsat 5 and band 10 for Landsat 8) were used to calculate the LSTs, as described in Section 2.4.

(v) All extracted images were georectified using the WGS84/UTM 44N projection system before further processing.

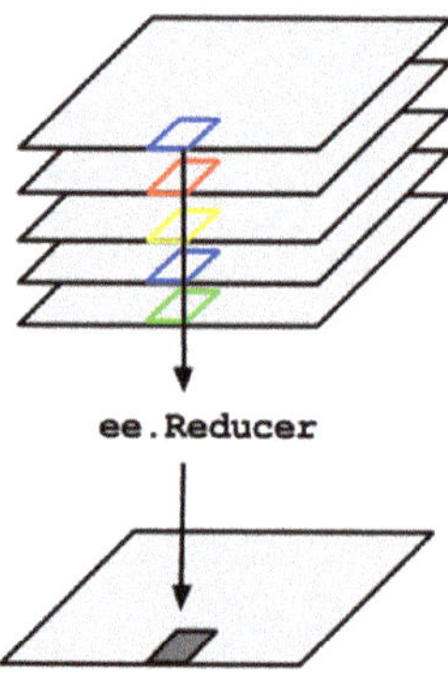

Figure 3. Graphical illustration of image collection [53].

2.4. LST Calculations

The extracted annual, median, at-satellite brightness temperatures, as mentioned in Section 2.3, were scaled using the land surface emissivity derived from Equation (1) [54].

$$\varepsilon = \{mPV + n\} \tag{1}$$

where m = $(\varepsilon - \varepsilon) - (1 - \varepsilon\sigma)$; F$\varepsilon$v and n = εs + $(1 - \varepsilon$s) Fεv. εs and εv are the soil emissivity and vegetation emissivity, respectively. In this study, we used the result of [54] for m = 0.004 and n = 0.986. The proportion of vegetation (P$_v$) was calculated using the normalized difference vegetation index (NDVI) based on Equation (2) [55].

$$NDVI = \frac{\rho_{NIR} - \rho_{Red}}{\rho_{NIR} + \rho_{Red}} \tag{2}$$

where ρ_{NIR} refers to the surface reflectance values of bands 4 (Landsat-5) and 5 (Landsat-8); and ρ_{Red} refers to the surface reflectance values of bands 3 (Landsat-5), and 4 (Landsat-8 OLI).

P$_v$ was extracted using Equation (3).

$$P_v = ((NDVI - NDVI_{min})/(NDVI_{max} - NDVI_{min}))^2 \tag{3}$$

where, P$_v$ is the proportion of vegetation, NDVI is original NDVI value calculated using Equation (2), and NDVI$_{min}$ and NDVI$_{max}$ are the minimum and maximum values of the NDVI dataset, respectively.

The emissivity corrected images were used to extract LSTs using Equation (4) [3,30].

$$LST = T_b/1 + (\lambda \times T_b/\rho)\ln \varepsilon \tag{4}$$

where T_b is the at-satellite brightness temperature in degrees Kelvin; λ is the central band wavelength of emitted radiance (11.5 μm for band 6 [30] and 10.8 μm for band 10 [13]; ρ is h × c/σ (1.438 × 10^{-2} m K), with σ as the Boltzmann constant (1.38 × 10^{-23} J/K), h as Planck's constant (6.626 × 10^{-34} J *s), and c as the velocity of light (2.998×10^8 m/s); and ε is the land-surface emissivity estimated using the NDVI method [54]. The calculated LST value (Kelvin) is then converted to °C.

2.5. Land Use/Land Cover (LULC) Mapping

Machine learning methods, such as support vector machines (SVM), K-nearest neighbor, random forest, and neural networks, have been widely used to classify LULC [13,14]. Among these methods, the SVM has provided higher overall accuracy [56–59]. Thus, in this study, we used the SVM algorithm to conduct LULC mapping. The classification scheme used in the study included five LULC categories; namely, built-up land (BL), forested land (FL), agricultural land (AL), other lands, and water. The other lands category was comprised of a combination of grasslands and bare lands.

We produced three LULC maps containing the five LULC categories (BL, FL, AL, other land and water) for the years 1996, 2006, and 2017 with overall accuracies of 85%, 93%, and 92%, respectively (see Tables A2–A4). The accuracy was assessed by using 500 reference points generated by a stratified random sampling method [14] for all the LULC categories. Google Earth imageries were used to assess the accuracy of the classified LULC maps for 2006 and 2017. The accuracy assessment for the 1996 LULC map was conducted with the aid of available topographic maps and different band combinations of Landsat imageries.

2.6. LST Intensity (LSTI) Measurement

The LSTI was calculated based on the urban–rural gradient analysis approach, which involves the creation of concentric rings or buffer zones around the city center with standard distance intervals extending to the rural areas [13,14,42]. Urban–rural gradient analysis has been conducted to identify the spatial and temporal variations in environmental variables in many previous environmental studies [13,14,16,20,60,61]. In this study, thirty-five 210 m buffer zones (hereafter referred to as urban–rural zones (URZs)) were created for the study area. The mean LST of each URZ was then extracted using zonal statistics. The fractions of BL, FL, and AL were determined by calculating their respective proportions in each URZ. Of note is that previous studies combined AL and FL into one LULC category defined as green space [13,14,16]. However, AL has been one of the key drivers of LULC changes in the study area. Therefore, AL in Nuwara Eliya and its surroundings was considered separately.

The magnitude of the LSTI along the urban–rural gradient (LSTI $_{U-R}$) was determined based on the Δ mean LST, Δ fraction of BL, Δ fraction of FL, and Δ fraction of AL, following methodology proposed by Estoque and Murayama [14]. To calculate LSTI$_{U-R}$, we first determined the Δ mean LST by finding the difference between the mean LST in the URZ with the highest fractions of BL, FL and AL (defined as URZ$_1$) and other URZ$_s$ (i.e., URZ$_1$ − URZ$_{2\ldots35}$). We then applied the same method to determine the Δ fraction of BL, Δ fraction of FL, and Δ fraction of AL along the urban–rural gradient. We used the same threshold for delineating the urban and rural zones as the authors of [13,14]; i.e., from the city center, all URZs with an >10% fraction of BL were considered as urban and those beyond the first URZ with <10% fraction of BL were considered as rural.

2.7. AL/FL and BL/FL Fraction Ratios and Their Intensities

The AL/FL and BL/FL fraction ratios were calculated using the URZs created in Section 2.6. Ranagalage et al., 2018 [13] proposed the green space (GS)/impervious surface (IS) fraction ratio and its intensity, and we followed their methodology to extract AL/FL and BL/FL in each URZ using Equations (5) and (6), respectively.

$$\text{AL/FL fraction ratio} = \frac{\text{ALz}}{\text{FLz}} \tag{5}$$

$$\text{BL/FL fraction ratio} = \frac{\text{BLz}}{\text{FLz}} \tag{6}$$

2.8. Grid-Based Analysis

The grid-based method was used to analyze the spatial distributions of the densities of BL, FL, AL, and mean LSTs in 1996, 2006, and 2017. In this analysis, we used 210m × 210 m grids (7 × 7 pixels) to demarcate the relationships between BL, FL, and AL density with mean LST. The 210 × 210 m grid size was used in previous studies and showed a high correlation with mean LST [16,39,44,62]. After the creation of the set of grids, the mean LST and the densities of BL, FL, and AL in each grid were calculated. Then, scatter plots were created and linear regression analysis was performed to identify the relationships between mean LST and densities of BL, FL, and AL.

3. Results

3.1. Landscape Changes and LST Distribution of Nuwara Eliya

The results revealed that Nuwara Eliya experienced rapid urbanization in the last 21 years. The built-up area increased from 289.9 ha to 2,080.4 ha from 1996 to 2017 with an annual growth rate of 85.3 ha per year (Figure 4 and Tables 1 and 2). The forest area did not change significantly due to the implementation of forest reserves by the government [63]. It was observed that the rapid changes in the built-up land negatively affected the agricultural sector. The area of agricultural land decreased from 8503.2 ha to 6583.9 ha from 1996 to 2017 (Table 1).

Table 1. Details of the land use/land cover (LULC) changes in Nuwara Eliya (1996, 2006, and 2017).

Land Use/Cover	1996		2006		2017	
	Area (ha)	%	Area (ha)	%	Area (ha)	%
Built-up	289.9	1.3	785.5	3.5	2080.4	9.3
Forest	13,076.7	58.2	13,502.1	60.1	13,234.3	58.9
Agricultural Land	8503.2	37.9	8085.5	36	6583.9	29.3
Other Land	511.8	2.3	6	0	481.8	2.1
Water	73.4	0.3	75.9	0.3	74.6	0.3
Total	22,455	100	22,455	100	22,455	100

Table 2. LULC changes during 1996–2006, 2006–2017, and 1996–2017.

Land Use/Cover	1996–2006		2006–2017		1996–2017	
	Land Use/Cover Changes (ha)	Annual Growth Rate (ha per year)	Land Use/Cover Changes (ha)	Annual Growth Rate (ha per year)	Land Use/Cover Changes (ha)	Annual Growth Rate (ha per year)
Built-up	495.6	49.6	1294.9	117.7	1790.6	85.3
Forest	425.3	42.5	−267.8	−24.3	157.6	7.5
Agricultural Land	−417.7	−41.8	−1501.7	−136.5	−1919.3	−91.4
Other Land	−505.8	−50.6	475.7	43.2	−30.1	−1.4
Water	2.5	0.3	−1.3	−0.1	1.3	0.1

Figure 4d–f shows the spatial distribution of LST in Nuwara Eliya for the years 1996, 2006, and 2017, respectively. In 1996, LST ranged between 9.1 and 29.2 °C with a mean value of 18.9 °C. In 2006, LST ranged between 5.9 and 30.2 °C with a mean value of 17.9 °C. In 2017, LST ranged between 14.3 and 31.0 °C with a mean value of 21.0 °C. Therefore, the mean LST exhibited an increasing trend from 1996 to 2017. The mean LST increased by 2.1 °C during the past 21 years. Figure 4d–f shows increasing temperatures in the city center and in the southwestern and northeastern areas. The observed pattern of increasing temperatures was similar to the increase in BL.

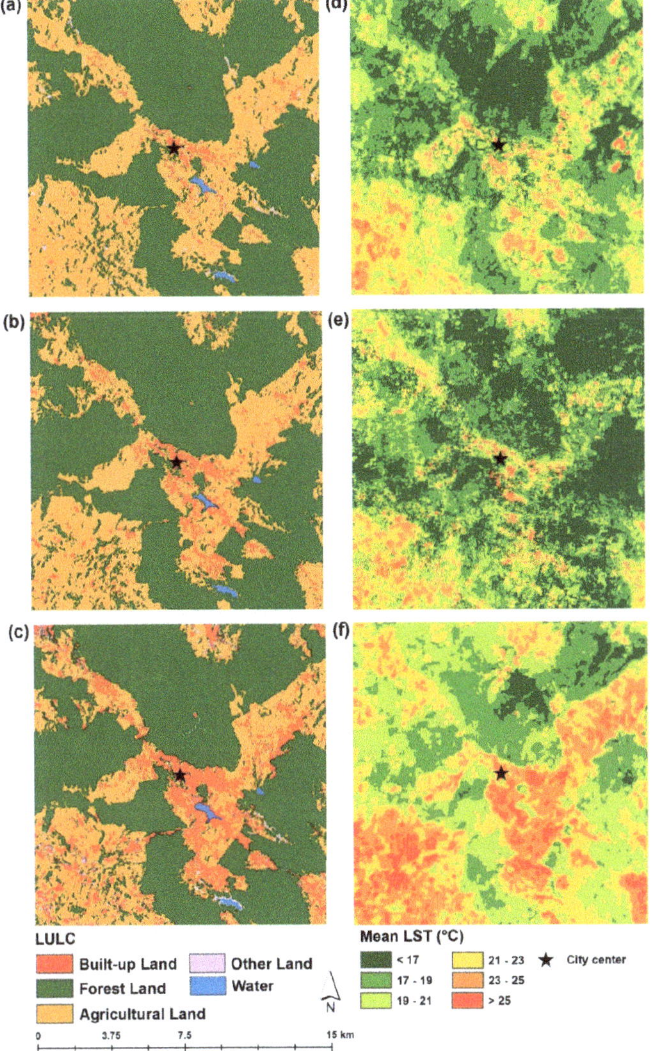

Figure 4. Land use/cover maps and land surface temperature (LST) values in Nuwara Eliya: land use/cover in (**a**) 1996, (**b**) 2006, and (**c**) 2017; and LST in (**d**) 1996, (**e**) 2006, and (**f**) 2017.

3.2. *Magnitude and Trend of LSTI*

3.2.1. $LSTI_{U-R}$ along the Urban–Rural Gradient

Figure 5a–c shows the spatial pattern of mean LSTs and the fractions of BL, FL, and AL along the urban–rural gradient. The highest mean LST value was recorded in the URZ_1 near the city center. In 1996, the mean LST of URZ_1 was 19.8 °C which decreased to 18.9 °C in 2006 and increased to 24.4 °C in 2017. Conversely, the mean LST of all other URZs increased from 1996 to 2017. The results revealed that the mean LST of all other URZs ranged from 18.9 °C to 21.2 °C between 1996 and 2017. The results further revealed that the lowest temperatures were recorded in URZ_{12} in 1996, URZ_{10} in 2006, and URZ_{19} in 2017 (Figure 5).

The fraction of BL increased rapidly between 1996 and 2017. In URZ$_1$, the fraction of BL increased by 32.4%, 48.7%, and 58.7%, in 1996, 2006, and 2017, respectively. The fraction of FL did not significantly change during the last 21 years. However, rapid changes in AL occurred in all URZs throughout the study's temporal extent. The lowest fraction of AL was continually recorded in URZ$_1$ and exhibited a decreasing pattern. The AL fraction in URZ$_1$ was 43.6% in 1996, 29.6% in 2006, and 15.8% in 2017.

Figure 5b shows the results of the statistical analysis of mean LST with the fractions of BL, FL, and AL. The mean LST had a strongly significant, positive correlation ($p < 0.001$) with the fraction of BL at all the three time-points. The fraction of FL exhibited a strongly significant, negative correlation ($p < 0.001$) with mean LST. On the other hand, mean LSTs exhibited dissimilar relationships with the fractions of AL in 1996, 2006, and 2017. In 1996, the mean LST had a significant, positive correlation ($p < 0.001$) with the fraction of AL. In 2006, the mean LST and the fraction of AL had a weak relationship ($p < 0.5$), and in 2017, the mean LST had a significant, negative correlation with the fraction of AL ($p < 0.1$).

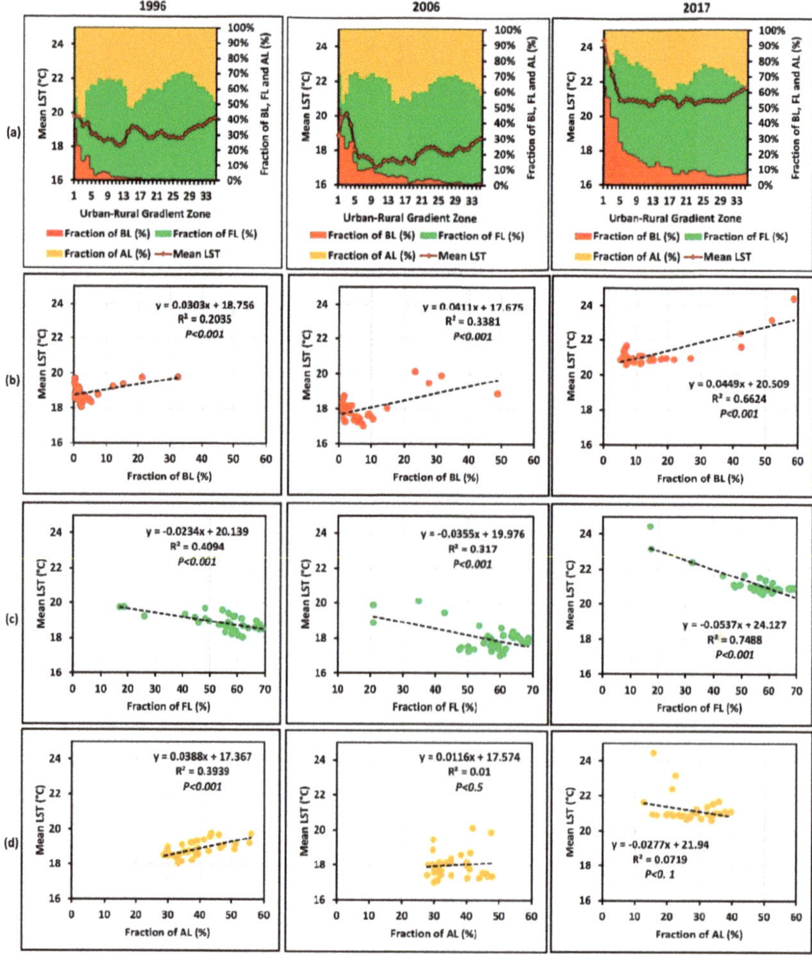

Figure 5. (**a**) Spatial distribution of mean LST, and the fractions of BL, FL, and AL along the urban–rural gradient; and scatter plots of the mean LST and fractions of (**b**) BL, (**c**) FL, and (**d**) AL. Note: changes in BL, FL, and AL types of land are not independent of each other.

3.2.2. The Magnitude of LSTI$_{U-R}$ along the Urban–Rural Gradient

The magnitudes of the mean LST and fractions of BL, FL, and AL were calculated based on URZ$_1$ as described in Section 2.6 above. The Δ mean LST in URZ$_{13}$ showed a considerable decrease in 1996 and 2006. However, the decrease shifted to URZ$_{20}$ in 2017 (Figure 6a). The Δ fractions of BL, FL, and AL also showed a similar pattern. The Δ mean LST had a significant relationship (positive) with the Δ fraction of BL, and a strong relationship (negative) with the Δ fraction of FL. The Δ mean LST exhibited a significant, positive relationship in 1996 ($p < 0.001$), and in 2006 ($p < 0.5$), with the Δ fraction of AL. However, the relationship changed in 2017, showing a weak, negative correlation ($p < 0.5$) with the Δ fraction of AL.

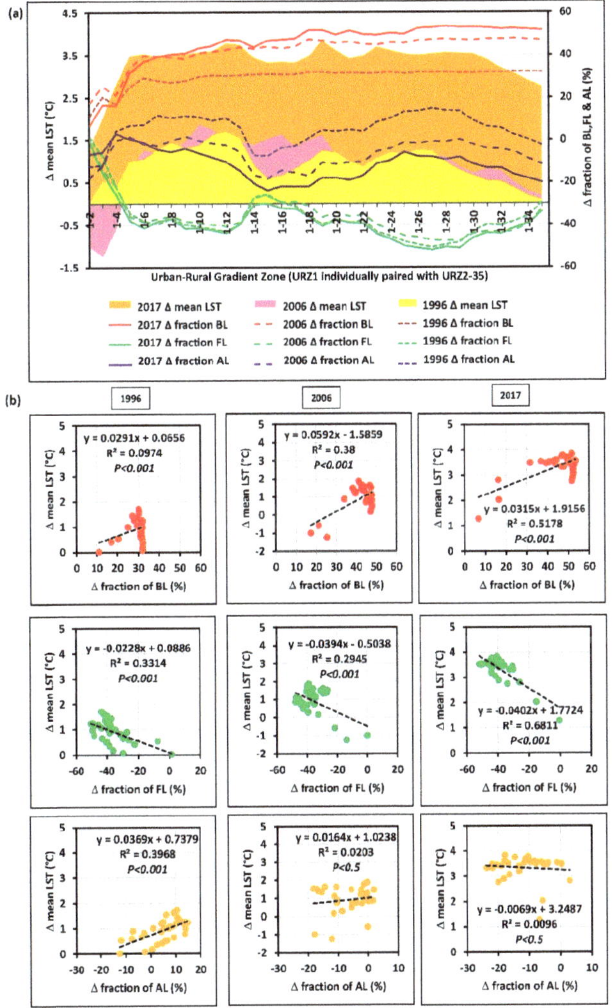

Figure 6. (**a**) Magnitudes of the trends of LST intensity (LSTI)$_{U-R}$ (°C), and fractions of BL, FL, and AL along the URZ; and (**b**) statistical relationships between the magnitudes of mean LSTs and fractions of BL, FL, and AL in Nuwara Eliya during 1996, 2006, and 2017. Note: changes in BL, FL, and AL types of land are not independent of each other.

3.3. AL/FL and BL/FL Ratios

An annual decrease in AL of 91.6 ha was observed from 1996 to 2017 (Table 2). FL was stable due to the protection measures implemented by the Forest Department of Sri Lanka (Table 1). A rapid decreasing trend in the AL/FL fraction ratio was observed from 1996 to 2017. The AL/FL fraction ratio decreased in all zones from URZ_1 to URZ_{35} with average values of 0.84, 0.7, and 0.53 in 1996, 2006, and 2017, respectively (Figure 7a). The mean LST values had a significant ($p < 0.001$), positive relationship with the AL/FL fraction ratio. Figure 7c shows the statistical relationship between the change in mean LST and Δ(AL/FL) fraction ratio along the urban–rural gradient, and the result shows a significant positive relationship in all the three time-points in this study.

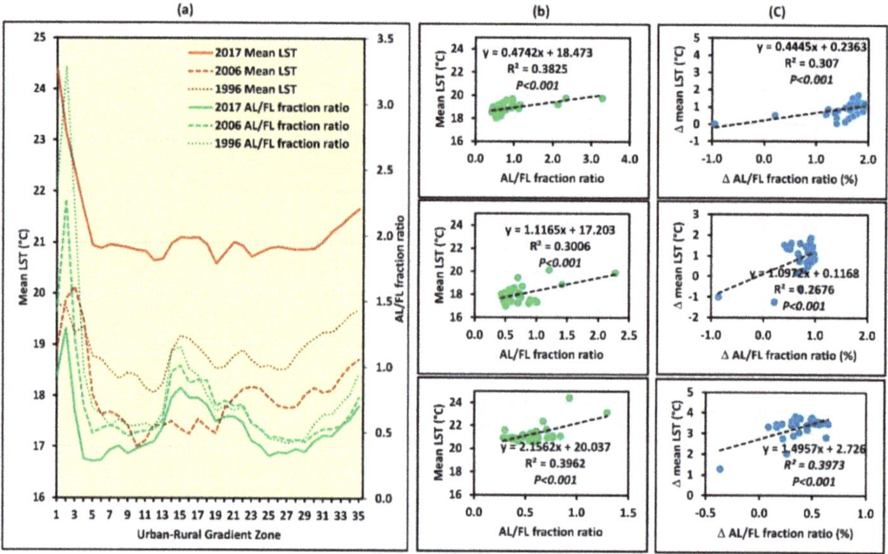

Figure 7. (**a**) Spatial distribution of the mean LST, AL/FL fraction ratio; (**b**) scatter plots of the mean LST and AL/FL fraction ratio; and (**c**) the Δ(mean LST) (°C) and Δ(AL/FL) fraction ratio in Nuwara Eliya in 1996, 2006, and 2017.

The BL/FL ratio increased from 1996 to 2017, with average values of 0.13, 0.22, and 0.42 in 1996, 2006, and 2017, respectively (Figure 8a). The mean LST exhibited a significant ($p < 0.001$), positive relationship with the BL/FL fraction ratio (Figure 8b). The relationship between the change in mean LST and Δ(BL/FL) fraction ratio is presented in Figure 8c. The results revealed that there was a strongly significant, positive relationship between the change in mean LST and Δ(BL/FL) fraction ratio for all the three time-points.

3.4. The Density of BL, FL, and AL Compared to Mean LST

Figure 9a shows the spatial distribution of mean LST and the density of BL, FL, and AL in 1996, 2006, and 2017. High temperatures were observed around the city center and the northwestern and southwestern parts of the study area. The BL density was also high around the city center area in 1996 and it spread in the southern, southwestern and southeastern directions within the Nuwara Eliya area in 2006 and 2017. The FL density had an opposite pattern, showing low FL densities around all areas of high BL densities in all the three-time points. However, AL density showed a decreasing trend, especially in the city center area and southwestern direction of the study area.

Figure 9b shows the relationship between the densities of BL, FL, AL, and mean LST based on the grid-based analysis. The mean LST was statistically significant with the densities of BL, FL, and AL in all the three-time points ($p < 0.001$). There was an increasing trend of a positive correlation between BL density and mean LST from 1996 to 2017. The correlation between mean LST and the density of FL showed a significant negative relationship throughout the study period. In addition to that, there was a decreasing trend of a positive correlation between AL density and mean LST from 1996 to 2017.

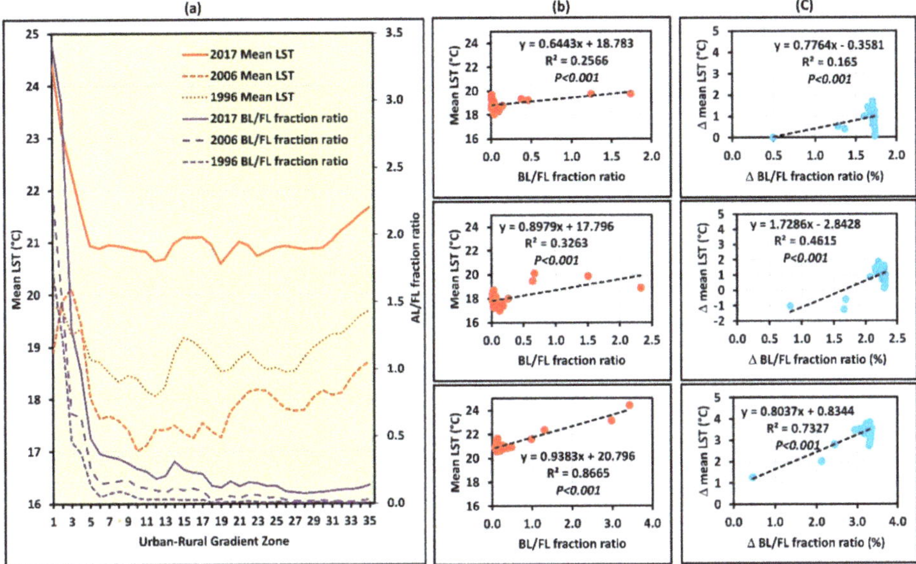

Figure 8. (**a**) Spatial distribution of the mean LST and BL/FL fraction ratio; (**b**) scatter plots of the mean LST and BL/FL fraction ratio; and (**c**) Δ(mean LST) (°C) and Δ(BL/FL) fraction ratio in Nuwara Eliya in 1996, 2006, and 2017.

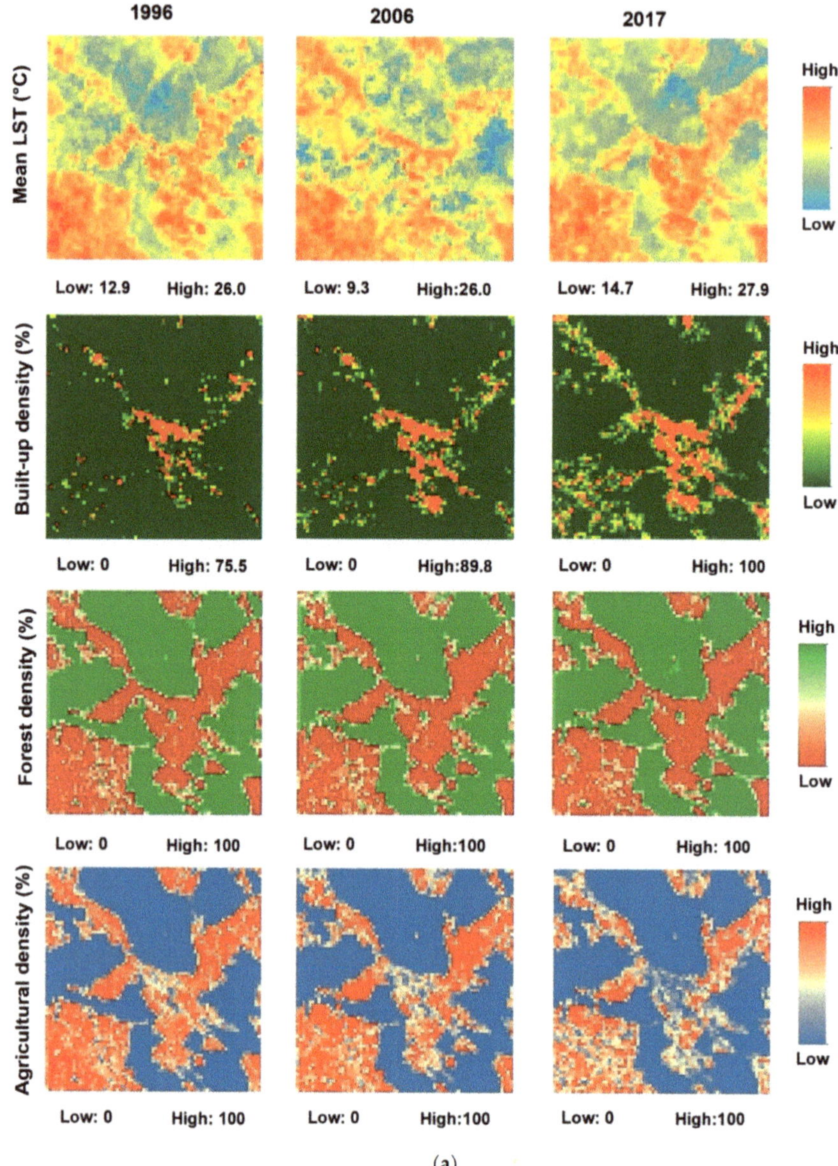

(a)

Figure 9. *Cont.*

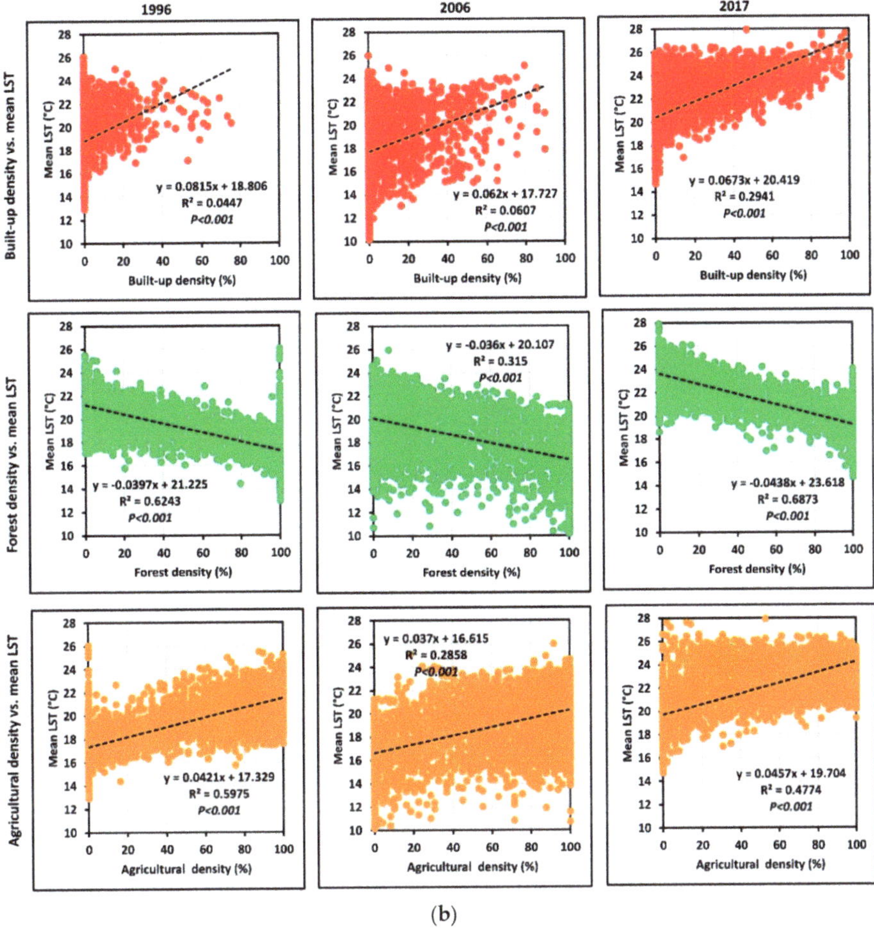

(b)

Figure 9. (**a**) Graphical view of the 210 × 210 m polygon grid showing maps of mean LST, built-up density, forest density, and agricultural density in 1996, 2006, and 2017; (**b**) Scatterplots showing the relationship between the densities of BL, FL, AL, and mean LST in 1996, 2006, and 2017.

4. Discussion

In this study, we examined the landscape changes and their impacts on LST in Nuwara Eliya, Sri Lanka over a 21-year period (1996 to 2017). The results revealed that the urbanization pattern of Nuwara Eliya was concentrated around the city center areas in 1996 which then expanded towards the southern, southwestern, northwestern, and eastern parts by 2017. The development pattern of Nuwara Eliya is similar to that of other mountain cities (e.g., Kandy) in Sri Lanka [13]. The built-up land has expanded rapidly in the past 21 years (Tables 1 and 2, and Figure 4) with an overall increase of 1790.6 ha and an annual increase of 85.3 ha. The tourism industry has been reported to be the main driving force of the observed rapid development in the study area, as thousands of local and foreign tourists have been visiting Nuwara Eliya [49]. Previous studies, such as Baguio in the Philippines [14,46], Kandy in Sri Lanka [13,51], Bengaluru in India [64], Bogor in Indonesia [46], and Dalat in Vietnam [46], revealed that green spaces, such as forests, declined due to the urban development pressure in their respective study areas. However, our study revealed that the forest areas were intact throughout the study period, owing to strong government policies on forest preservation. Most of the forest areas located in Nuwara

Eliya are protected. That has resulted in other LULC types other than forests being consumed. Our results demonstrate that BL expanded mainly at the expense of agricultural areas (Figure 5a).

The results revealed that the fraction of BL greatly influenced the LSTI in Nuwara Eliya. During the past 21 years, annual mean LST increased by 2.1 °C (18.9 °C in 1996 and 21.0 °C in 2017). Additionally, the mean annual temperature difference between the urban (URZ_1) and rural (<10% BL) zones showed an increasing pattern, recording values of 1.0 °C (19.8 °C in URZ_1 and 18.8 °C in URZ_5) in 1996, 1.3 °C in 2006 (18.9 °C in URZ_1 and 17.6 °C in URZ_6), and 3.5 °C in 2017 (24.4 °C in URZ_1 and 21.0 °C in URZ_{18}) (Figure 5a). These results show an interesting pattern of LST in the study area. Urbanization has considerably influenced the increasing annual mean LST of the study area. The highest mean LST values were recorded around the city center (URZ_1) in all the three-time points. The fractions of BL increased by 32.4%, 48.7%, and 58.7%, while the fraction of AL decreased by 43.6%, 29.6%, and 15.8% in 1996, 2006, and 2017, respectively. The mean LST had a significant, positive relationship with the fraction of BL and significant negative relationship with the fraction of FL ($p < 0.001$) in all the three time-periods (Figure 5b). These results are similar to previous research findings in other mountain cities in Asia, such as Kandy in Sri Lanka [13], and Baguio in the Philippines [14]. The mean LST of all URZs (i.e., urban and rural combined) also showed an increasing pattern throughout the temporal extent of the study (1996–2017). The patterns observed in Nuwara Eliya not only showed changes in the landscape, but also agreed with other researchers that have reported an increasing trend of global temperatures at high-elevation regions [65–67]. It is thus plausible to speculate that these global temperatures in high-elevation regions could have also contributed to the observed LST increases in the study area and other urban environments worldwide.

The statistical relationship between mean LST and fraction of AL was different in all the three time-points. The mean LST showed a strongly significant, positive relationship with the fraction of AL in 1996, while in 2006, the relationship was significant, but weakly positive. However, in 2017, the relationship became weakly negative. We observed that poorly managed agricultural lands were converted into built-up lands during the 21 years of this study. The average fraction of AL exhibited a decreasing pattern, with values of 38.7%, 35.8%, and 27.2% in 1996, 2006, and 2017, respectively. This result is similar to the observations in other mountain cities in Sri Lanka [13]. The observed pattern is critical evidence for policymakers to introduce proper mechanisms for minimizing the adverse impacts associated with the rapid changes in the urban landscape of the study area. The negative impacts of urban development on agricultural land could cause several environmental issues [59], such as increased soil erosion [68], and landslides [69–71]. The rapid development of built-up land, resulting in increased levels of LST, could also negatively affect the tourism industry. Policies should be implemented to maintain the natural beauty of Nuwara Eliya for the sustainable tourism industry. Most tourists wish to enjoy the landscape and comfortable climate of the area [49,59].

The change in the urban–rural pattern provides a clear picture of the urbanization pattern in the study area. The urban–rural zones were demarcated using the fraction of BL (<10%). The rural zone exhibited a shifting pattern from 1996 to 2017. The rural zone located in URZ_5 (1 km from the city center) in 1996 and the rural zone located in URZ_6 (1.2 km from the city center) in 2006, had shifted to URZ_{18} by 2017 (3.8 km from the city center) (Figure 6a). Rapid changes to built-up land resulted in a shift in the rural zone away from the city center. Previous researchers have shown a similar trend in the shifting rural zone in mountain cities [13,14]. Results showed that the temperature of URZ_1 was higher than that of the rural zones. The change in mean LST had a strong, positive relationship with the Δ fraction of BL and a strong negative relationship with the Δ fraction FL. However, the relationship between the Δ mean LST and Δ fraction of AL shifted from positive (1996 and 2006) to negative (2017). Policymakers should pay much attention to the shifting of the urban zone.

The mixture of land-use types, which is important for controlling LST [72], was investigated based on two indices: the AL/FL and BL/FL fraction ratios. The results showed that the AL/FL fraction ratio decreased from 1996 to 2017 (Figure 7a). The maximum AL/FL fraction ratio declined with values of 3.3, 2.3, and 1.3 in 1996, 2006, and 2017, respectively. The mean LST exhibited a significantly strong positive relationship with the AL/FL fraction ratio in 1996 and 2006. However, the degree of the relationship became weaker in 2017 due to the changes in the AL of the study area. The change in mean LST exhibited a positive relationship with the Δ(AL/FL) fraction ratio over the three time-points (Figure 7c). On the other hand, the result of the BL/FL fraction ratio provided an indicator of the LSTI pattern in the study area. Changes in built-up land have been proven to be vital in understanding the LST changing pattern in urban areas [20,33,73]. The maximum BL/FL fraction ratio increased with values of 1.7, 2.3, and 3.4 (URZ_1) in 1996, 2006, and 2017, respectively. This result could be the reason for the observed increase in the mean LST over the 21-year study period (Figure 8a). The increasing BL/FL fraction ratio directly affected the LSTI in the study area, as the mean LST had a strongly significant relationship with the BL/FL fraction ratio over the 21-year study period.

The spatial distribution pattern of the BL, FL, and AL densities provided vital information for understanding the distribution pattern of the study area. The linear regression analysis results also revealed the vital formation of capturing the land use mixture of the study area. The average density of BL increased with values of 1.3%, 3.5%, and 9.2% while the density of AL decreased with values 37.6%, 35.9%, and 29.2% in 1996, 2006, and 2017, respectively (Figure 9a). This pattern indicates that the BL density has been increasing in the last 21 years. The linear regression analysis results showed that the coefficient of determination (R^2) between mean LST and the density of BL increased from 0.04 to 0.29 from 1996 to 2017. This shows that the density of BL had a significant influence on the increase of LST in the study area. The R^2 between the mean LST and the density of AL declined from 0.59 to 0. 47 from 1996 to 2017. This also shows that AL density had a positive influence on the increase of LST. The density of FL, on the other hand, influenced the control of LST in the study area, showing a negative, strongly significant relationship between the two (Figure 9b). Most of the previous studies have combined AL as green spaces, but AL of the Nuwara Eliya area has been positively influencing to the increase of LST in the three-time points. The policymakers and planners need to consider maintaining a good mixture of land use to control the LST pattern in the future.

On the whole, the rapid increase of built-up areas has been the main force driving the high temperatures recorded in 2017 compared to 2006 and 1996. The Δ mean LST exhibited a positive relationship with the Δ(BL/FL) fraction ratio over the three time-points (Figure 8c). Previous research has demonstrated that having a mixture of land-use types can reduce temperatures in urban areas [72]. Based on the results from this study, maintenance of forest coverage will continue at a similar level due to governmental preservation policies. However, rapidly developing urban areas require proper urban planning to maintain sustainability. We have noticed that the present urban development pattern is unplanned. The walls and roofs of most of the buildings are also green in color; something which could reduce the indoor temperatures [74,75]. To achieve sustainable development, we recommend that urban planners and policymakers should consider constructing green belts along the main road, schools, hospitals, and other government buildings. The development pattern observed was mainly horizontal rather than vertical. Vertical development could reduce pressure on land and leave a lot of free land for improving greenspace coverage. Vertical development can also be used as a proxy indicator to understand urban-development intensity [76,77]. In addition, greenspace cover plays a vital role in reducing the impacts of LST in urban areas [78]. We believe that the current urban planning must focus on protecting this popular tourist destination in Sri Lanka's sustainably. The findings of this study can be used as an indicator for reorienting current urban planning policies to improve the natural and social environment in Nuwara Eliya.

Overall, this study successfully used annual mean LST to quantify landscape changes and their impacts on LSTI from 1996 to 2017. However, there are two main limitations of the study that should be considered. Firstly, comparing the mean LST change magnitudes with changes in observational air temperature measured from metrological stations could have helped in verifying our findings. However, that was challenging due to the unavailability of air temperature observational data from meteorological stations. Nonetheless, we used previous similar studies to validate our results, and we recommend that more studies should be conducted to confirm the increasing temperature in mountain cities of Sri Lanka. The lack of observational air temperature data in most developing countries has been the basis for the proliferation of UHI studies based on LST retrieved from remote sensing satellite imagery, compared to atmospheric UHI studies based on measured air temperature. Secondly, we believe that horizontal and vertical development of buildings both play a vital role in controlling the annual mean LST of the study area. However, vertical data on various types of buildings and built-structures were not used in this study due to unavailability in Sri Lanka. Thus, the results should be interpreted in the context of those limitations.

5. Conclusions

The spatial distribution of LST provides vital information for understanding the local climatic variation of cities and can be used as a proxy indicator to introduce sufficient measures to minimize the negative impacts associated with high LST. This study revealed that Nuwara Eliya had undergone rapid urban development over the 21 years studied. The study used multiple images to extract the median annual LSTs for 1996, 2006, and 2017. The results revealed that the annual mean LST increased by 2.1 °C over the past 21 years. The annual mean LST difference between the urban and rural zones increased from 1.0 °C to 3.5 °C from 1996 to 2017. The study shows that rapid development was spreading towards rural zones, and the fraction of BL influenced the increase in annual mean LST. The government policies have been protecting the FL, resulting in minimal changes which have been the major contributor to the control of the LST pattern. This study recommends that having a mixture of land-use types would considerably control the increasing LST in the study area. Thus, policymakers need to consider the importance of the land use mixture to reduce the impacts of a high LST in the study area. The future urban policy must focus on maintaining the natural splendor of the Nuwara Eliya to promote tourism in the study area and Sri Lanka. Nuwara Eliya is renowned as a tourist destination in Sri Lanka. Thus, the findings of this study can be used as an indicator to introduce sustainable future landscape and urban planning to protect the world-renowned tourist hub of Sri Lanka.

Author Contributions: The author to receive correspondence, M.R., proposed the topic and spearheaded the data processing and analysis, as well as the writing of the manuscript. Y.M., D.D., and M.S. helped in the design, research implementation, analysis, and the writing of the manuscript.

Funding: This study was supported by the Japan Society for the Promotion of Science (JSPS) through Grant-in-Aid for Scientific Research (B) 18H00763 (2018-20).

Acknowledgments: The authors are grateful to the anonymous reviewers for their helpful comments and suggestions to improve the quality of this paper.

Conflicts of Interest: The authors declare no conflict of interest.

Appendix A

Table A1. Properties of the Landsat images (Level 2) used in this study.

Sensor	Scene ID	Acquisition Date	Time (GMT)	Cloud Cover (%) in Landsat Title
	LT05_L1TP_141055_19960221_20170106_01_T1	1996-02-21	03:59:45	19
	LT05_L1TP_141055_19960308_20170106_01_T1	1996-03-08	04:00:49	6
	LT05_L1TP_141055_19960324_20170105_01_T1	1996-03-24	04:01:51	11
	LT05_L1TP_141055_19960409_20170105_01_T1	1996-04-09	04:02:51	46
	LT05_L1TP_141055_19960425_20170104_01_T1	1996-04-25	04:03:49	9
	LT05_L1TP_141055_19960511_20170104_01_T1	1996-05-11	04:04:46	9
	LT05_L1TP_141055_19960527_20170104_01_T1	1996-05-27	04:05:41	29
Landsat 5 TM	LT05_L1TP_141055_19960730_20170103_01_T1	1996-07-30	04:09:06	78
	LT05_L1TP_141055_19960815_20170103_01_T1	1996-08-15	04:09:56	66
	LT05_L1TP_141055_19960831_20170103_01_T1	1996-08-31	04:10:48	41
	LT05_L1TP_141055_19960916_20170102_01_T1	1996-09-16	04:11:41	70
	LT05_L1TP_141055_19961002_20170102_01_T1	1996-10-02	04:12:33	48
	LT05_L1TP_141055_19961103_20170102_01_T1	1996-11-03	04:14:09	38

Table A1. *Cont.*

Sensor	Scene ID	Acquisition Date	Time (GMT)	Cloud Cover (%) in Landsat Title
	LT05_L1TP_141055_19961119_20170101_01_T1	1996-11-19	04:14:53	19
	LT05_L1TP_141055_19961205_20170101_01_T1	1996-12-05	04:15:40	84
	LT05_L1TP_141055_20060131_20161123_01_T1	2006-01-31	04:44:06	46
	LT05_L1TP_141055_20060216_20161123_01_T1	2006-02-16	04:44:28	28
	LT05_L1TP_141055_20060304_20161122_01_T1	2006-03-04	04:44:49	41
	LT05_L1TP_141055_20060405_20161123_01_T1	2006-04-05	04:45:26	10
	LT05_L1TP_141055_20060421_20161122_01_T1	2006-04-21	04:45:41	68
	LT05_L1TP_141055_20060507_20161122_01_T1	2006-05-07	04:45:55	33
	LT05_L1TP_141055_20060523_20161121_01_T1	2006-05-23	04:46:07	31
	LT05_L1TP_141055_20060608_20161121_01_T1	2006-06-08	04:46:23	27
Landsat 5 TM	LT05_L1TP_141055_20060624_20161121_01_T1	2006-06-24	04:46:39	72
	LT05_L1TP_141055_20060710_20161120_01_T1	2006-07-10	04:46:53	30
	LT05_L1TP_141055_20060811_20161119_01_T1	2006-08-11	04:47:17	79
	LT05_L1TP_141055_20060827_20161119_01_T1	2006-08-27	04:47:29	45
	LT05_L1TP_141055_20060912_20161119_01_T1	2006-09-12	04:47:41	57
	LT05_L1TP_141055_20060928_20161119_01_T1	2006-09-28	04:47:52	55
	LT05_L1TP_141055_20061014_20161118_01_T1	2006-10-14	04:48:03	22
	LT05_L1TP_141055_20061030_20161118_01_T1	2006-10-30	04:48:13	30
	LT05_L1TP_141055_20061115_20161118_01_T1	2006-11-15	04:48:21	60
	LC08_L1TP_141055_20170113_20170311_01_T1	2017-01-13	04:54:05	3
	LC08_L1TP_141055_20170129_20170214_01_T1	2017-01-29	04:53:59	40
	LC08_L1TP_141055_20170214_20170228_01_T1	2017-02-14	04:53:52	68
	LC08_L1TP_141055_20170302_20170316_01_T1	2017-03-02	04:53:46	64
	LC08_L1TP_141055_20170318_20170328_01_T1	2017-03-18	04:53:36	13
	LC08_L1TP_141055_20170403_20170414_01_T1	2017-04-03	04:53:29	16
	LC08_L1TP_141055_20170419_20170501_01_T1	2017-04-19	04:53:20	15
	LC08_L1TP_141055_20170505_20170515_01_T1	2017-05-05	04:53:13	32
	LC08_L1TP_141055_20170606_20170616_01_T1	2017-06-06	04:53:34	54
Landsat 8	LC08_L1TP_141055_20170622_20170630_01_T1	2017-06-22	04:53:40	28
OLI/TIRS	LC08_L1TP_141055_20170708_20170716_01_T1	2017-07-08	04:53:43	44
	LC08_L1TP_141055_20170724_20170809_01_T1	2017-07-24	04:53:49	38
	LC08_L1TP_141055_20170809_20170824_01_T1	2017-08-09	04:53:56	47
	LC08_L1TP_141055_20170825_20170913_01_T1	2017-08-25	04:54:00	38
	LC08_L1TP_141055_20170910_20170927_01_T1	2017-09-10	04:54:02	61
	LC08_L1TP_141055_20170926_20171013_01_T1	2017-09-26	04:54:07	82
	LC08_L1TP_141055_20171012_20171024_01_T1	2017-10-12	04:54:12	76
	LC08_L1TP_141055_20171028_20171108_01_T1	2017-10-28	04:54:13	19
	LC08_L1TP_141055_20171113_20171121_01_T1	2017-11-13	04:54:10	38
	LC08_L1TP_141055_20171215_20171223_01_T1	2017-12-15	04:54:05	32

Code A1. The code used to generate median temperature of Landsat 5 TM images for 1996 and 2006.

```
//Export Landsat 5 SR data
/**
* Function to mask clouds based on the pixel_qa band of Landsat SR data.
* @param {ee.Image} image Input Landsat SR image
* @return {ee.Image} Cloudmasked Landsat image
*/
var cloudMaskL457 = function(image) {
var qa = image.select('pixel_qa');
//If the cloud bit (5) is set and the cloud confidence (7) is high
//or the cloud shadow bit is set (3), then it's a bad pixel.
var cloud = qa.bitwiseAnd(1 << 5)
.and(qa.bitwiseAnd(1 << 7))
.or(qa.bitwiseAnd(1 << 3));
//Remove edge pixels that don't occur in all bands
var mask2 = image.mask().reduce(ee.Reducer.min());
return image.updateMask(cloud.not()).updateMask(mask2);
```

```
};
//Load the Landsat 5 data collection - example for B6
{
var col = ee.ImageCollection('LANDSAT/LT05/C01/T1_SR')
.map(cloudMaskL457)
.filterDate('1996-01-01','1996-12-30')
.select(['B6'])
.filterBounds(geometry);
}
//Compute the median of the collection -example for B6
{
var image = col.median().clip(geometry);
print(image, 'Selected band');
Map.addLayer(image);
}
//Export the B6, specifying scale and region.
Export.image.toDrive({
image: image,
description: 'B6',
scale: 30,
region: geometry,
maxPixels:1e13,
folder: 'Landsat 5 data collection',
skipEmptyTiles: true
});
//Export the B6 information as table.
Export.table.toDrive({
collection: col,
description: 'B6_information',
fileFormat: 'CSV',
folder:'Landsat 5 data collection'
});
//Get the number of collections.
var count = col.size();
print('Count: ', count);
//Get the date range of images in the collection.
var range = col.reduceColumns(ee.Reducer.minMax(), ["system:time_start"])
print('Date range: ', ee.Date(range.get('min')), ee.Date(range.get('max')))
```

Code A2. The code used to generate median temperature of Landsat 8 images for 2017.

```
//Export Landsat 8 SR data
/**
* Function to mask clouds based on the pixel_qa band of Landsat 8 SR data.
* @param {ee.Image} image input Landsat 8 SR image
* @return {ee.Image} cloudmasked Landsat 8 image
*/
function maskL8sr(image) {
//Bits 3 and 5 are cloud shadow and cloud, respectively.
var cloudShadowBitMask = (1 << 3);
var cloudsBitMask = (1 << 5);
```

```javascript
//Get the pixel QA band.
var qa = image.select('pixel_qa');
//Both flags should be set to zero, indicating clear conditions.
var mask = qa.bitwiseAnd(cloudShadowBitMask).eq(0)
.and(qa.bitwiseAnd(cloudsBitMask).eq(0));
return image.updateMask(mask);
}
//Load the Landsat 8 data collection - example for B10
{
var col = ee.ImageCollection('LANDSAT/LC08/C01/T1_SR')
.map(maskL8sr)
.filterDate('2018-01-01','2018-12-30')
.select(['B10'])
.filterBounds(geometry);
}
//Compute the median of the collection - example for B10
{
var image = col.median().clip(geometry);
print(image, 'Selected band');
Map.addLayer(image);
}
//Export the B10, specifying scale and region.
Export.image.toDrive({
image: image,
description: 'B10',
scale: 30,
region: geometry,
maxPixels:1e13,
folder: 'Landsat 8 data collection',
skipEmptyTiles: true
});
//Export the B10 information as table.
Export.table.toDrive({
collection: col,
description: 'B10_information',
fileFormat: 'CSV',
folder:'Landsat 8 data collection'
});
//Get the number of collections.
var count = col.size();
print('Count: ', count);
//Get the date range of images in the collection.
var range = col.reduceColumns(ee.Reducer.minMax(), ["system:time_start"])
print('Date range: ', ee.Date(range.get('min')), ee.Date(range.get('max')))
```

Table A2. Error matrix for the classified 1996 land use/cover map, classified.

Classified Data	Reference Data					Total	User's Accuracy (%)
	Built-up	Forest	Agricultural Land	Other	Water		
Built-up	72	8	6	3	0	89	80.9
Forest	2	165	10	3	3	183	90.2
Agricultural Land	5	17	140	2	2	166	84.3
Other Land	2	2	3	34	1	42	81.0
Water	0	2	2	0	16	20	80.0
Total	81	194	161	42	22	500	
Producer's Accuracy (%)	88.9	85.1	87.0	81.0	72.7		

Overall accuracy (%) = 85%.

Table A3. Error matrix for the classified 2006 land use/cover map, classified.

Classified Data	Reference Data					Total	User's Accuracy (%)
	Built-up	Forest	Agricultural Land	Other	Water		
Built-up	98	2	5	0	0	105	93.3
Forest	3	182	8	0	0	193	94.3
Agricultural Land	4	5	155	1	1	166	93.4
Other Land	0	0	1	5	1	7	71.4
Water	0	1	1	1	26	29	89.7
Total	105	190	170	7	28	500	
Producer's Accuracy (%)	93.3	95.8	91.2	71.4	92.9		

Overall accuracy (%) = 93%.

Table A4. Error matrix for the classified 1996 land use/cover map, classified.

Classified Data	Reference Data					Total	User's Accuracy (%)
	Built-up	Forest	Agricultural Land	Other	Water		
Built-up	110	3	3	1	0	117	94.0
Forest	5	160	7	1	0	173	92.5
Agricultural Land	3	2	135	2	1	143	94.4
Other Land	2	3	2	33	1	41	80.5
Water	0	1	1	0	24	26	92.3
Total	120	169	148	37	26	500	
Producer's Accuracy (%)	91.7	94.7	91.2	89.2	92.3		

Overall accuracy (%) = 92%.

References

1. Chen, X.; Zhang, Y. Impacts of urban surface characteristics on spatiotemporal pattern of land surface temperature in Kunming of China. *Sustain. Cities Soc.* **2017**, *32*, 87–99. [CrossRef]
2. Thi Van, T.; Bao, H.D.X. Study of the impact of urban development on surface temperature using remote sensing in Ho Chi Minh City, Northern Vietnam. *Geogr. Res.* **2010**, *48*, 86–96. [CrossRef]
3. Ranagalage, M.; Estoque, R.C.; Murayama, Y. An urban heat island study of the Colombo Metropolitan Area, Sri Lanka, based on Landsat data (1997–2017). *ISPRS Int. J. Geo-Inf.* **2017**, *6*, 189. [CrossRef]
4. Ranagalage, M.; Estoque, R.C.; Zhang, X.; Murayama, Y. Spatial changes of urban heat island formation in the Colombo District, Sri Lanka: Implications for sustainability planning. *Sustainability* **2018**, *10*, 1367. [CrossRef]
5. Son, N.-T.; Chen, C.-F.; Chen, C.-R.; Thanh, B.-X.; Vuong, T.-H. Assessment of urbanization and urban heat islands in Ho Chi Minh City, Vietnam using Landsat data. *Sustain. Cities Soc.* **2017**, *30*, 150–161. [CrossRef]
6. Hou, H.; Wang, R.; Murayama, Y. Scenario-based modelling for urban sustainability focusing on changes in cropland under rapid urbanization: A case study of Hangzhou from 1990 to 2035. *Sci. Total Environ.* **2019**, *661*, 422–431. [CrossRef] [PubMed]
7. Alphan, H. Land-use change and urbanization of Adana, Turkey. *Land Degrad. Dev.* **2003**, *14*, 575–586. [CrossRef]
8. Shalaby, A.; Ghar, M.A.; Tateishi, R. Desertification impact assessment in Egypt using low resolution satellite data and GIS. *Int. J. Environ. Stud.* **2004**, *61*, 375–383. [CrossRef]

9. Simwanda, M.; Murayama, Y. Spatiotemporal patterns of urban land use change in the rapidly growing city of Lusaka, Zambia: Implications for sustainable urban development. *Sustain. Cities Soc.* **2018**, *39*, 262–274. [CrossRef]

10. El Araby, M. Urban growth and environmental degradation. *Cities* **2002**, *19*, 389–400. [CrossRef]

11. Debela, T.H.; Beyene, A.; Tesfahun, E.; Getaneh, A.; Gize, A.; Mekonnen, Z. Fecal contamination of soil and water in sub-Saharan Africa cities: The case of Addis Ababa, Ethiopia. *Ecohydrol. Hydrobiol.* **2018**, *18*, 225–230. [CrossRef]

12. Knudsen, A.B.; Slooff, R. Vector-borne disease problems in rapid urbanization: New approaches to vector control. *Bull. World Health Organ.* **1992**, *70*, 1–6. [PubMed]

13. Ranagalage, M.; Dissanayake, D.; Murayama, Y.; Zhang, X.; Estoque, R.C.; Perera, E.; Morimoto, T. Quantifying surface urban heat island formation in the World Heritage Tropical Mountain City of Sri Lanka. *ISPRS Int. J. Geo-Inf.* **2018**, *7*, 341. [CrossRef]

14. Estoque, R.C.; Murayama, Y. Monitoring surface urban heat island formation in a tropical mountain city using Landsat data (1987–2015). *ISPRS J. Photogramm. Remote Sens.* **2017**, *133*, 18–29. [CrossRef]

15. Voogt, J.A.; Oke, T.R. Thermal remote sensing of urban climates. *Remote Sens. Environ.* **2003**, *86*, 370–384. [CrossRef]

16. Estoque, R.C.; Murayama, Y.; Myint, S.W. Effects of landscape composition and pattern on land surface temperature: An urban heat island study in the megacities of Southeast Asia. *Sci. Total Environ.* **2017**, *577*, 349–359. [CrossRef] [PubMed]

17. Weng, Q. A remote sensing—GIS evaluation of urban expansion and its impact on surface temperature in the Zhujiang Delta, China. *Int. J. Remote Sens.* **2014**, *22*, 1999–2014.

18. Singh, P.; Kikon, N.; Verma, P. Impact of land use change and urbanization on urban heat island in Lucknow city, Central India. A remote sensing based estimate. *Sustain. Cities Soc.* **2017**, *32*, 100–114. [CrossRef]

19. Mirzaei, P.A. Recent challenges in modeling of urban heat island. *Sustain. Cities Soc.* **2015**, *19*, 200–206. [CrossRef]

20. Rousta, I.; Sarif, M.O.; Gupta, R.D.; Olafsson, H.; Ranagalage, M.; Murayama, Y.; Zhang, H.; Mushore, T.D. Spatiotemporal analysis of land use/land cover and its effects on surface urban heat island using Landsat data: A case study of Metropolitan City Tehran (1988–2018). *Sustainability* **2018**, *10*, 4433. [CrossRef]

21. Teferi, E.; Abraha, H. Urban heat island effect of Addis Ababa City: Implications of urban green spaces for climate change adaptation. In *Climate Change Adaptation in Africa*; Springer: Berlin, Germany, 2017.

22. Tran, H.; Uchihama, D.; Ochi, S.; Yasuoka, Y. Assessment with satellite data of the urban heat island effects in Asian mega cities. *Int. J. Appl. Earth Obs. Geoinf.* **2006**, *8*, 34–48. [CrossRef]

23. Van Nguyen, O.; Kawamura, K.; Trong, D.P.; Gong, Z.; Suwandana, E. Temporal change and its spatial variety on land surface temperature and land use changes in the Red River Delta, Vietnam, using MODIS time-series imagery. *Environ. Monit. Assess.* **2015**, *187*, 464. [CrossRef] [PubMed]

24. Ayanlade, A. Seasonality in the daytime and night-time intensity of land surface temperature in a tropical city area. *Sci. Total Environ.* **2016**, *557–558*, 415–424. [CrossRef] [PubMed]

25. Zhou, D.; Zhang, L.; Hao, L.; Sun, G.; Liu, Y.; Zhu, C. Spatiotemporal trends of urban heat island effect along the urban development intensity gradient in China. *Sci. Total Environ.* **2016**, *544*, 617–626. [CrossRef] [PubMed]

26. Marconcini, M.; Metz, A.; Esch, T.; Zeidler, J. Global urban growth monitoring by means of SAR data. In Proceedings of the 2014 IEEE Geoscience and Remote Sensing Symposium, Quebec, QC, Canada, 13–18 July 2014; pp. 1477–1480.

27. Paranunzio, R.; Ceola, S.; Laio, F.; Montanari, A. Evaluating the effects of urbanization evolution on air temperature trends using nightlight satellite data. *Atmosphere* **2019**, *10*, 117. [CrossRef]

28. Zhou, Y.; Smith, S.J.; Zhao, K.; Imhoff, M.; Thomson, A.; Bond-Lamberty, B.; Asrar, G.R.; Zhang, X.; He, C.; Elvidge, C.D. A global map of urban extent from nightlights. *Environ. Res. Lett.* **2015**, *10*, 2000–2010. [CrossRef]

29. Weng, Q. Thermal infrared remote sensing for urban climate and environmental studies: Methods, applications, and trends. *ISPRS J. Photogramm. Remote Sens.* **2009**, *64*, 335–344. [CrossRef]

30. Weng, Q.; Lu, D.; Schubring, J. Estimation of land surface temperature-vegetation abundance relationship for urban heat island studies. *Remote Sens. Environ.* **2004**, *89*, 467–483. [CrossRef]

31. Bhaduri, B.; Bright, E.; Coleman, P.; Urban, M.L. LandScan USA: A high-resolution geospatial and temporal modeling approach for population distribution and dynamics. *GeoJournal* **2007**, *69*, 103–117. [CrossRef]
32. Zhang, X.; Estoque, R.C.; Murayama, Y. An urban heat island study in Nanchang City, China based on land surface temperature and social-ecological variables. *Sustain. Cities Soc.* **2017**, *32*, 557–568. [CrossRef]
33. Dissanayake, D.; Morimoto, T.; Ranagalage, M.; Murayama, Y. Impact of urban surface characteristics and socio-economic variables on the spatial variation of land surface temperature in Lagos City, Nigeria. *Sustainability* **2019**, *11*, 25. [CrossRef]
34. Bhargava, A.; Lakmini, S.; Bhargava, S. Urban Heat Island Effect: It's Relevance in Urban Planning. *J. Biodivers. Endanger. Species* **2017**, *5*, 5–187.
35. Xu, S. An approach to analyzing the intensity of the daytime surface urban heat island effect at a local scale. *Environ. Monit. Assess.* **2009**, *151*, 289–300. [CrossRef] [PubMed]
36. Li, Y.Y.; Zhang, H.; Kainz, W. Monitoring patterns of urban heat islands of the fast-growing Shanghai metropolis, China: Using time-series of Landsat TM/ETM+ data. *Int. J. Appl. Earth Obs. Geoinf.* **2012**, *19*, 127–138. [CrossRef]
37. Senanayake, I.P.; Welivitiya, W.D.D.P.; Nadeeka, P.M. Remote sensing based analysis of urban heat islands with vegetation cover in Colombo city, Sri Lanka using Landsat-7 ETM+ data. *Urban Clim.* **2013**, *5*, 19–35. [CrossRef]
38. Sakakibara, Y.; Owa, K. Urban–rural temperature differences in coastal cities: Influence of rural sites. *Int. J. Climatol.* **2005**, *25*, 811–820. [CrossRef]
39. Myint, S.W.; Brazel, A.; Okin, G.; Buyantuyev, A. Combined effects of impervious surface and vegetation cover on air temperature variations in a rapidly expanding desert city. *GIScience Remote Sens.* **2010**, *47*, 301–320. [CrossRef]
40. Ravanelli, R.; Nascetti, A.; Cirigliano, R.V.; Di Rico, C.; Leuzzi, G.; Monti, P.; Crespi, M. Monitoring the impact of land cover change on surface urban heat island through Google Earth Engine: Proposal of a global methodology, first applications and problems. *Remote Sens.* **2018**, *10*, 1488. [CrossRef]
41. Parastatidis, D.; Mitraka, Z.; Chrysoulakis, N.; Abrams, M. Online global land surface temperature estimation from landsat. *Remote Sens.* **2017**, *9*, 1208. [CrossRef]
42. Stewart, I.; Oke, T. Classifying urban climate field sites by "local climate zones": The case of nagano, japan. In Proceedings of the Seventh International Conference on Urban Climate, Yokohama, Japan, 29–30 June 2009.
43. Oke, T.R. *Initial Guidance to Obtain Representative Meteorological Observations at Urban Sites*; World Meteorological Organization: Geneva, Switzerland, 2004.
44. Ranagalage, M.; Estoque, R.C.; Handayani, H.H.; Zhang, X.; Morimoto, T.; Tadono, T.; Murayama, Y. Relation between urban volume and land surface temperature: A comparative study of planned and traditional cities in Japan. *Sustainability* **2018**, *10*, 2366. [CrossRef]
45. Priyankara, P.; Ranagalage, M.; Dissanayake, D.; Morimoto, T.; Murayama, Y. Spatial process of surface urban heat island in rapidly growing Seoul metropolitan area for sustainable urban planning using Landsat Data. *Climate* **2019**, *7*, 110. [CrossRef]
46. Estoque, R.C.; Murayama, Y. Quantifying landscape pattern and ecosystem service value changes in four rapidly urbanizing hill stations of Southeast Asia. *Landsc. Ecol.* **2016**, *31*, 1481–1507. [CrossRef]
47. Estoque, R.C.; Murayama, Y. City Profile: Baguio. *Cities* **2013**, *30*, 240–251. [CrossRef]
48. Crossette, B. *The Great Hill Stations of Asia*; Basic Books: New York, NY, USA, 1999; Volume 73.
49. Jayasinghe, M.K.D.; Gnanapala, W.K.A.; Sandaruwani, J.A.R. Factors affecting tourists' perception and satisfaction in Nuwara Eliya, Sri Lanka. *Ilorin J. Econ. Policy* **2015**, *2*, 1–15.
50. Weerasinghe, W.W.K. Transformation of the landscape of Nuwara-Eliya. Ph.D. Thesis, University of Moratuwa, Colombo, Sri Lanka, 2003.
51. Dissanayake, D.; Morimoto, T.; Ranagalage, M.; Murayama, Y. Land-use/land-cover changes and their impact on surface urban heat islands: Case study of Kandy City, Sri Lanka. *Climate* **2019**, *7*, 99. [CrossRef]
52. Google Erath Engine, Landsat Collection Structure. Available online: https://developers.google.com/earth-engine/landsat (accessed on 25 May 2019).
53. Google Erath Engine, Image Collection Reductions. Available online: https://developers.google.com/earth-engine/reducers_image_collection (accessed on 25 May 2019).

54. Sobrino, J.A.; Jiménez-Muñoz, J.C.; Paolini, L. Land surface temperature retrieval from LANDSAT TM 5. *Remote Sens. Environ.* **2004**, *90*, 434–440. [CrossRef]

55. Zhang, Y.; Odeh, I.O.A.; Han, C. Bi-temporal characterization of land surface temperature in relation to impervious surface area, NDVI and NDBI, using a sub-pixel image analysis. *Int. J. Appl. Earth Obs. Geoinf.* **2009**, *11*, 256–264. [CrossRef]

56. Huang, C.; Davis, L.S.; Townshend, J.R.G. An assessment of support vector machines for land cover classification. *Int. J. Remote Sens.* **2002**, *23*, 725–749. [CrossRef]

57. Yang, X. Parameterizing Support Vector Machines for land cover classification. *Photogramm. Eng. Remote Sens.* **2011**, *77*, 27–37. [CrossRef]

58. Shi, D.; Yang, X. Support vector machines for land cover mapping from remote sensor imagery. In *Monitoring and Modeling of Global Changes: A Geomatics Perspective*; Springer: Dordrecht, The Netherlands, 2015; pp. 265–279.

59. Ranagalage, M.; Wang, R.; Gunarathna, M.H.J.P.; Dissanayake, D.; Murayama, Y.; Simwanda, M. Spatial forecasting of the landscape in rapidly urbanizing hill stations of South Asia: A case study of Nuwara Eliya, Sri Lanka (1996–2037). *Remote Sens.* **2019**, *11*, 1743. [CrossRef]

60. Gunaalan, K.; Ranagalage, M.; Gunarathna, M.H.J.P.; Kumari, M.K.N.; Vithanage, M.; Srivaratharasan, T.; Saravanan, S.; Warnasuriya, T.W.S. Application of geospatial techniques for groundwater quality and availability assessment: A case study in Jaffna Peninsula, Sri Lanka. *ISPRS Int. J. Geo-Inf.* **2018**, *7*, 20. [CrossRef]

61. Simwanda, M.; Ranagalage, M.; Estoque, R.C.; Murayama, Y. Spatial analysis of surface urban heat islands in four rapidly growing African Cities. *Remote Sens.* **2019**, *11*, 1645. [CrossRef]

62. Dissanayake, D.; Morimoto, T.; Murayama, Y.; Ranagalage, M. Impact of landscape structure on the variation of land surface temperature in Sub-Saharan Region: A case study of Addis Ababa using Landsat data (1986–2016). *Sustainability* **2019**, *11*, 2257. [CrossRef]

63. Department of Forest. *Forest Ordinance*; Department of Forest: Sri Jayawardenepura Kotte, Sri Lanka, 1908.

64. Sultana, S.; Satyanarayana, A.N.V. Urban heat island intensity during winter over metropolitan cities of India using remote-sensing techniques: Impact of urbanization. *Int. J. Remote Sens.* **2018**, *39*, 6692–6730. [CrossRef]

65. Pepin, N.; Bradley, R.S.; Diaz, H.F.; Baraer, M.; Caceres, E.B.; Forsythe, N.; Fowler, H.; Greenwood, G.; Hashmi, M.Z.; Liu, X.D.; et al. Elevation-dependent warming in mountain regions of the world. *Nat. Clim. Chang.* **2015**, *5*, 424–430.

66. Diaz, H.F.; Bradley, R.S.; Ning, L. Climatic changes in mountain regions of the American Cordillera and the tropics: Historical changes and future outlook. *Arct. Antarct Alp. Res.* **2014**, *46*, 735–743. [CrossRef]

67. Bradley, R.S.; Keimig, F.T.; Diaz, H.F.; Hardy, D.R. Recent changes in freezing level heights in the Tropics with implications for the deglacierization of high mountain regions. *Geophys. Res. Lett.* **2009**, *36*, 2–5. [CrossRef]

68. Dissanayake, D.; Morimoto, T.; Ranagalage, M. Accessing the soil erosion rate based on RUSLE model for sustainable land use management: A case study of the Kotmale watershed, Sri Lanka. *Model. Earth Syst. Environ.* **2018**, *4*, 291–306. [CrossRef]

69. Ranagalage, M. Landslide hazards assessment in Nuwara Eliya District in Sri Lanka. In Proceedings of the Japanese Geographical Meeting, Tsukuba, Japan, 28–30 March 2017; p. 100336.

70. Perera, E.N.C.; Jayawardana, D.T.; Jayasinghe, P.; Ranagalage, M. Landslide vulnerability assessment based on entropy method: A case study from Kegalle district, Sri Lanka. *Model. Earth Syst. Environ.* **2019**, 1–15. [CrossRef]

71. Perera, E.N.C.; Jayawardana, D.T.; Ranagalage, M.; Jayasinghe, P. Spatial multi criteria evaluation (SMCE) model for landslide hazard zonation in tropical hilly environment: A case study from Kegalle. *Geoinform. Geostat. Overv.* **2018**, *S3*. [CrossRef]

72. Thanh Hoan, N.; Liou, Y.-A.; Nguyen, K.-A.; Sharma, R.; Tran, D.-P.; Liou, C.-L.; Cham, D. Assessing the effects of land-use types in surface urban heat islands for developing comfortable living in Hanoi City. *Remote Sens.* **2018**, *10*, 1965. [CrossRef]

73. Yuan, F.; Bauer, M.E. Comparison of impervious surface area and normalized difference vegetation index as indicators of surface urban heat island effects in Landsat imagery. *Remote Sens. Environ.* **2007**, *106*, 375–386. [CrossRef]

74. Galagoda, R.U.; Jayasinghe, G.Y.; Halwatura, R.U.; Rupasinghe, H.T. The impact of urban green infrastructure as a sustainable approach towards tropical micro-climatic changes and human thermal comfort. *Urban For. Urban Green.* **2018**, *34*, 1–9. [CrossRef]
75. Manawadu, L.; Ranagalage, M. Urban heat islands and vegetation cover as a controlling factor. In Proceedings of the International Forestry and Environment Symposium 2013 of the Department of Forestry and Environmental Science, University of Sri Jayewardenepura, Thulhiriya, Sri Lanka, 10–11 January 2014; p. 125.
76. Estoque, R.C.; Murayama, Y.; Ranagalage, M.; Hou, H.; Subasinghe, S. Validating ALOS PRISM DSM-derived surface feature height: Implications for urban volume estimation. *Tsukuba Geoenviron. Sci.* **2017**, *13*, 13–22.
77. Ranagalage, M.; Murayama, Y. Measurement of urban built-up volume using remote sensing data and geospatial techniques. *Tsukuba Geoenviron. Sci.* **2018**, *14*, 19–29.
78. Handayani, H.H.; Murayama, Y.; Ranagalage, M.; Liu, F.; Dissanayake, D. Geospatial analysis of horizontal and vertical urban expansion using multi-spatial resolution data: A case study of Surabaya, Indonesia. *Remote Sens.* **2018**, *10*, 1599. [CrossRef]

 *sustainability*

Article

Impact of Landscape Structure on the Variation of Land Surface Temperature in Sub-Saharan Region: A Case Study of Addis Ababa using Landsat Data (1986–2016)

DMSLB Dissanayake [1,2,*], Takehiro Morimoto [3], Yuji Murayama [3] and Manjula Ranagalage [1,2,*]

1   Graduate School of Life and Environmental Sciences, University of Tsukuba, 1-1-1, Tennodai, Tsukuba, Ibaraki 305-8572, Japan

2   Department of Environmental Management, Faculty of Social Sciences and Humanities, Rajarata University of Sri Lanka, Mihintale 50300, Sri Lanka

3   Faculty of Life and Environmental Sciences, University of Tsukuba, 1-1-1, Tennodai, Tsukuba, Ibaraki 305-8572, Japan; tmrmt@geoenv.tsukuba.ac.jp (T.M.); mura@geoenv.tsukuba.ac.jp (Y.M.)

*   Correspondence: dissanayakedmslb@gmail.com or s1730220@u.tsukuba.ac.jp (D.D.); manjularanagalage@gmail.com or manjularanagalage@ssh.rjt.ac.lk (M.R.); Tel.: +81-029-853-4211 (D.D. & M.R.)

Received: 13 March 2019; Accepted: 10 April 2019; Published: 15 April 2019

Abstract: Urbanization has bloomed across Asia and Africa of late, while two centuries ago, it was confined to developed regions in the largest urban agglomerations. The changing urban landscape can cause irretrievable changes to the biophysical environment, including changes in the spatiotemporal pattern of the land surface temperature (LST). Understanding these variations in the LST will help us introduce appropriate mitigation techniques to overcome negative impacts. The research objective was to assess the impact of landscape structure on the variation in LST in the African region as a geospatial approach in Addis Ababa, Ethiopia from 1986–2016 with fifteen-year intervals. Land use and land cover (LULC) mapping and LST were derived by using pre-processed Landsat data (Level 2). Gradient analysis was computed for the pattern of the LST from the city center to the rural area, while intensity calculation was facilitated to analyze the magnitude of LST. Directional variation of the LST was not covered by the gradient analysis. Hence, multidirectional and multitemporal LST profiles were employed over the orthogonal and diagonal directions. The result illustrated that Addis Ababa had undergone rapid expansion. In 2016, the impervious surface (IS) had dominated 33.8% of the total lands. The IS fraction ratio of the first zone (URZ_1) has improved to 66.2%, 83.7%, and 87.5%, and the mean LST of URZ_1 has improved to 25.2 °C, 26.6 °C, and 29.6 °C in 1986, 2001, and 2016, respectively. The IS fraction has gradually been declining from the city center to the rural area. The behavior of the LST is not continually aligning with a pattern of IS similar to other cities along the URZs. After the specific URZs (zone 17, 37, and 41 in 1986, 2001, and 2016, respectively), the mean LST shows an increasing trend because of a fraction of bare land. This trend is different from those of other cities even in the tropical regions. The findings of this study are useful for decision makers to introduce sustainable landscape and urban planning to create livable urban environments in Addis Ababa, Ethiopia.

Keywords: LST; urban-rural gradient; sub-Saharan region; Addis Ababa; Ethiopia

1. Introduction

At the turn of the 20th century, there were 371 cities with one million or more inhabitants around the world and this increased to 548 cities in 2018 [1]. It is projected that in 2030, there will be 706 such cities. The number of global urban residents surpassed the global rural population in 2007. Moreover, about 60% and 66% of the global population is expected to be lodged in urban areas in 2030 and 2050 respectively [2]. From the urbanization viewpoint, Africa and Asia have been notable regions since 1950, and Ethiopia has been identified as a country with relatively rapid urbanization similar to that of developing counties [3]. The population of Addis Ababa, the capital city of Ethiopia, is more than 3 million and is predicted to be 12 million in 2024 [4]. Changing landscape and biophysical attributes of the urban environment are some of the offshoots of overpopulation in the urban area. They resulted in rapid changes in the urban landscape by converting natural vegetation into the impervious surface (IS) [5].

Landscape structure consists of anthropogenic and natural components and their spatial pattern [6,7]. Building materials and non-vegetative surface (bare soil) can trap solar radiation [8] in the daytime and then re-radiate during the nighttime due to the decline in albedo [9]. They are one of the reasons for the increasing land surface temperature (LST), which can also be the cause of the fluctuation of surface energy balance [10]. Generally, cities produce urban heat island due to the high LST regardless of size and location. However, the magnitude of the LST can depend on the size of the city and it often decreases as city size decreases [11]. Hence, LST is the most crucial parameter for various kinds of applications at the local and global level, including the physical, chemical, and biological [12,13]. Changing landscape structure affects LST and has negative consequences for the socioeconomic and environmental attributes in an urban area. Elevated temperature or heatwaves in an urban area can affect the natural environmental and human health. The air quality decreases as the surface air gets warmer [14]. Heat waves can affect the wellbeing of urban dwellers [13,15,16]. Hence, LST and its causal factors have been becoming a significant research topic among the scientific community [17,18]. To the best of our knowledge, studies related to the LST combining with urban landscape structure and geospatial application are scarce. Hence, focusing research on addressing this limitation is an important, timely task.

Previous studies have shown various approaches for assessing landscape structure and its impact on the variation of the LST over the urban area [19–22]. A few popular approaches include (i) observing the pattern of the LST along the urban-rural zone (URZ) and (ii) studying the variation of the LST concerning land use and land cover [23–27]. However, these approaches necessitate the classification of Land use and land cover(LULC) and the computation of LST (LULC as a local climate zone). Geospatial and remote sensing (RS) techniques are facilitated for the classification and investigation of the changes in LULC in both spatial and temporal viewpoints, as conventional approaches such as ground-based methods are limited as they are time- and capital-consuming [28–30]. Similarly, calculation of the LST from thermal infrared RS data is a large-scale time-consuming approach and a solution for the absence of ground-based temperature data [13,31]. Assessing the composition of the LULC provides necessary information to understand the spatial and temporal distribution pattern of the landscape. Variation in the magnitude of both LST and LULC can be observed through intensity calculation. The influence of LST on the dynamic urban landscape structure is difficult to study from a single perspective, and a multidirectional and multitemporal approach can overcome this obstacle by generating a comprehensive LST profile.

In this context, Addis Ababa, the capital of Ethiopia and a diplomatic landmark of Africa, is one of the fastest growing cities in the continent [4,16]. The city has an abundance of sunshine as it lies in the tropical region, close to the equator, which is an essential attribute for LST studies [13]. The minimum and maximum mean annual temperature are about 12 °C and 24 °C, respectively [32]. Urban expansion and overpopulation are some of the anthropogenic activities that resulted in changes in the landscape structure, while physical attributes such as the dry season and the desert [1] (location of the sub-Saharan region) could affect LST. Keeping the objective of this research was to assess the impact of the landscape structure for the variation of LST in the African region as a geospatial approach. The results of this study will be useful as a proxy indicator for sustainable urban planning, and the methodology could be applied to other similar cities, especially in the African region.

2. Materials and Methods

2.1. Study Area

Addis Ababa is the capital and also the largest city in Ethiopia that lies in the central highland of the Ethiopian federal government and on the western edge of the Rift valley in the Eastern Africa region. The climate is generally sub-tropical with an average temperature of 10–15 °C in the night and 20–24 °C in the daytime in the dry period [33]; rainfall peaks during the summer from July to August and is minimum during the winter (December–February) [34]. The mean center of the Central Business District (CBD) of Addis Ababa is defined as the central point (9.02130° N, 38.75163° E) of the study area. We have selected a 30 km × 30 km geographical area as the study area with a 15 km radius from the city center covering 900 km^2 (Figure 1), bound by latitude 9.141622° N to 8.811627° N and longitude 38.605921° E to 38.906661° E.

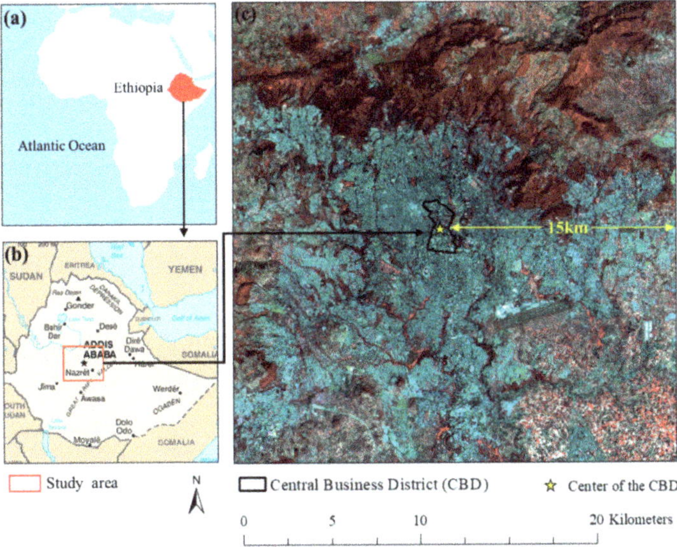

Figure 1. Study area. (**a**) Map of the African continent showing Ethiopia; (**b**) map of Ethiopia with Addis Ababa; and (**c**) Landsat-8 image in a false color composite (band 5, 4, 3) (9 December 2016) with central business district (CBD).

2.2. Datasets and Data Pre-Processing

Radiometric-calibrated and atmospheric-corrected Landsat Level 2 (On-Demand) data were obtained from the United States Geological Survey (USGS) official website, including Normalized Difference Vegetation Index (NDVI). Landsat-8 TIRS (Band 10) and Landsat-5 TM (Band 6) were provided as the atmospheric brightness temperature in Kelvin (K), which is generated from Landsat top of atmosphere [35,36]. All multispectral bands (Landsat-8 OLI and Landsat-5 TM) were provided as surface reflectance [36]. In this process, the authors were responsible for maintaining the dry season daytime data and cloud-free images or minimum cloud cover data (<10%). Furthermore, the temporal resolution of the data was maintained as much as possible by considering the availability of free data, but it is important to note that the same temporal resolution data were not available. Hence, the same month of the dry season [37] (December) was selected to maintain the temporal uniformity over the study period. However, this issue did not significantly affect the final result because (i) the research emphasized the LST pattern along the URZs rather than the absolute value and (ii) necessary atmospheric and radiometric calibrations were conducted by USGS. Table 1 shows the comprehensive summary of the raw data.

Table 1. Experimental data collection.

Sensor	Landsat-5 TM	Landsat-5 TM	Landsat-8 OLI/TIRS
Landsat Sensor ID	LT51680541986357XXX10	LT51680542001350RSA00	LC81680542016344LGN01
Temporal resolution	23 December 1986	16 December 2001	9 December 2016
Spatial resolution	30 × 30 m * (except for pan, TIR, and TIRS)		
Row/Path	54/168		
Time (GMT) **	06:59:25	07:19:50	07:40:41

* The spatial resolution of the pan (15 m) on Landsat-8 OLI, TIR (120 m) of the Landsat-5 TM, and TIRS (100 m) of the Landsat-8 TIRS. ** GMT is known as Greenwich Mean Time. GMT is 3 h behind Addis Ababa, Ethiopia time.

2.3. Land Use/Cover Mapping

An attempt is made to conduct four types of classifications which are facilitated by R software (support vector machine, K-nearest neighbor, random forest, and neural networks); both overall accuracy and kappa statistics were the highest for the neural networks method [38]. Hence, this method was selected. The three images used were classified into six categories: Impervious surface (IS), forest, grassland, bareland, cropland, and water. The authors used (i) IS, (ii) green space 1 (GS1) as forest, (iii) green space 2 (GS2) as grassland and cropland, and (iv) bareland (BL) as the same bareland for the current study.

2.4. Computation of LST

First, the proportion of vegetation was calculated (Equation (1)) by using the NDVI data downloaded from the USGS as explained in Section 2.2.

$$P_v = ((\text{NDVI} - \text{NDVI}_{min})/(\text{NDVI}_{max} - \text{NDVI}_{min}))^2 \tag{1}$$

where P_v represents the amount of vegetation, NDVI_{min} represents the minimum values of the NDVI, and NDVI_{max} represents the maximum value of the NDVI.

Second, land surface emissivity (ε) was computed by using Equation (2).

$$\varepsilon = m\,P_v + n \tag{2}$$

where ε represents land surface emissivity; m represents $(\varepsilon_v - \varepsilon_s) - (1 - \varepsilon_s)\,F\varepsilon_v$; P_v represents the amount of vegetation; n represents $\varepsilon_s + (1 - \varepsilon_s)\,F\varepsilon_v$; ε_s is the soil emissivity; ε_v is the vegetation emissivity; and F is a shape factor whose mean value, assuming different geometrical distributions, is 0.55 [39]. In this study, we adopted m as 0.004 and n as 0.986 based on previous results [33].

Finally, emissivity-corrected LST was calculated by using Equation (3) as follows:

$$\text{LST} = T_b/1 + \left(\lambda \times \frac{T_b}{\rho}\right)\ln \varepsilon \tag{3}$$

where T_b is the at-satellite brightness temperature in degrees Kelvin; λ is the central band wavelength of emitted radiance (11.5 μm for Band 6 and 10.8 μm for Band 10 [40]); ρ is $h \times c/\sigma$ (1.438 × 10^{-2} m K) with σ being the Boltzmann constant (1.38 × 10^{-23} J/K), h is Planck's constant (6.626 × 10^{-34} J·s), and c is the velocity of light (2.998 × 10^8 m/s) [6]; and ε is the land surface emissivity estimated using Equation (3). Then, the calculated LST values (Kelvin) were converted to degrees Celsius (°C).

2.5. Urban-Rural Gradient Analysis

Several steps have been taken for the formation of urban-rural zones (URZ) from the city center to the rural area [41]. A set of polygon grids with the same snapped with LST map were created, and the dimensions of each grid were 210 m × 210 m [41,42]. The mean center of the CBD conceded as the center grid, and its value was 0; 70 zones were designed from the centroid grid to cover the whole study area.

2.5.1. The Fraction of LULC and Urban-Rural Zone

The fraction ratio of LULC in each zone was computed to assess the effect of the composition of LULC on LST variation. In this process, fraction ratios of IS, GS1, GS2, and BL were computed. The urban and rural boundary was demarcated based on the fraction ratio of IS from the city center if the composition of IS fraction is less than 10 (<10%) when compared to the zone designated as rural area [41]. This method was introduced by Estoque and Murayama in 2017 [41]. Various studies have shown that this method generated reliable results [43].

2.5.2. LST Intensity

LST variation among the gradient zone is determined as the LST intensity (LSTI), or it can be expressed as the inter-zone temperature difference [41,43,44]. In this study, the mean LST of each LULC type in each gradient zone was determined by following the methodology explained in Section 2.5. Mean LST of the IS, GS1, GS2, and BL in each gradient zone was calculated, except for the water class. The reason for this exclusion is that it does not represent the study area well (Table 1); hence, there will not be a significant influence on the result due to this exclusion.

2.5.3. The Magnitude of the LSTI

Relative mean LST and fraction ratios of IS, GS1, GS2, and BL were subtracted with URZ_1, and the method was applied for all zones (URZ_1–URZ_2, URZ_1–URZ_3, and URZ_1–$URZ \ldots _{n(70)}$). This methodology was proposed by Estoque and Murayama in 2017 [41]; Ranagalage et al. obtained reliable results with this method in 2018 [43]. The method illustrates changes in the magnitude of mean LST and magnitude of the fraction ratios of IS, GS1, GS2, and BL when compared with the URZ_1 along the urban-rural gradient.

2.6. LST Profile

Multidirectional and multitemporal LST profiles were computed by following the orthogonal and diagonal directions. First, LST values were extracted to a set of grids, with the dimensions of 210 m × 210 m, by following the orthogonal direction to cover the north-south and east-west direction. Second, LST values of the northeast-southwest and northwest-southeast direction were also extracted along the diagonal by following the same grid [41]. Finally, four profiles were made in eight directions.

All the derived values were used to prepare various kinds of graphs and statistical analyses to visualize the results and assess the statistical significance.

2.7. Population Data

This study used 100 m × 100 m continuous spatial population data (2006) provided by the Geo-data Institute of the University of Southampton, UK [45]. However, this source could not provide continuous data to cover the study period. Hence, Ethiopia's total population and urban population data were used as a proxy indicator to visualize the population growth pattern (world population prospects 2017 by UN) [46]. Some studies have adopted this population density data approach effectively [47–49]. The population or population density does not directly affect the LST variation or its intensity, but overpopulation can enhance some attributes of LST including the expansion of built-up area [6]. In additional to that, some spaces are required for settlements, industrial zones, and infrastructure to cater to the urban community. These improvements are directly influenced by the changes in landscape structure. By considering these facts, population data were incorporated.

3. Results

3.1. The Spatiotemporal Distribution Pattern of the LST

The spatial distribution pattern of the LST of Addis Ababa is shown in Figure 2. On 23 December 1986, the LST ranged from 10.8–35.4 °C, with an average of 24.8 °C. It ranged from 11.2–37.4 °C, and the average was 26.2 °C on 16 December 2001. On 9 December 2016, it ranged from 14.6–38.8 °C, and the average was 27.9 °C. A significant LST distribution pattern can be observed both in the southern and eastern parts of the study area. However, an improvement can be seen in the high LST area in the northern part in 2006 (Figure 2b) compared to 1986, and it matured in 2016, as shown by Figure 2c.

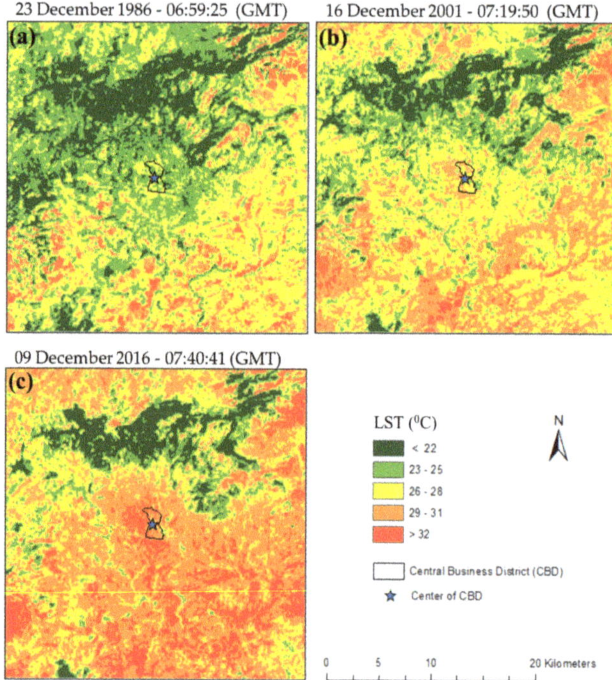

Figure 2. Land surface temperature (LST) map of Addis Ababa; (**a**) LST in 1986; (**b**) LST in 2001; and (**c**) LST in 2016.

3.2. Landscape Pattern and Its Changes

The classification was performed by following the methodology explained in Section 2.3. To determine the accuracy of the LULC classification, 970 training samples (40% of the total) were selected to cover the six LULC types, and Google Earth image was used as reference data for accuracy assessment. The overall accuracy of the classified LULC was 91%, 89%, and 90% in 1986, 2001, and 2016, respectively. Further, the kappa statistic was 0.88 in 1986, 0.87 in 2001, and 0.88 in 2016. This method is commonly used in similar studies and details of which can be found elsewhere [43,50,51].

The results of the LULC classification (Figure 3) shows that rapid urbanization has taken place over the past 30 years. It can be observed that IS has dramatically increased from 6262 ha to 30,700 ha, contributing to 34.1% of the total area in 2016 (Table 2). The annual growth rate of IS was 11 ha^{-1} year^{-1} (1986–2016). Croplands and grasslands declined and registered a total net loss of 5982 ha and 13,493 ha, over the study period. The results of the LULC classification show that forest area remains only over the mountain region (Entoto mountain). Significant efforts have been made in the past decade through forest plantation [52] to make the green environment in the central mountain region.

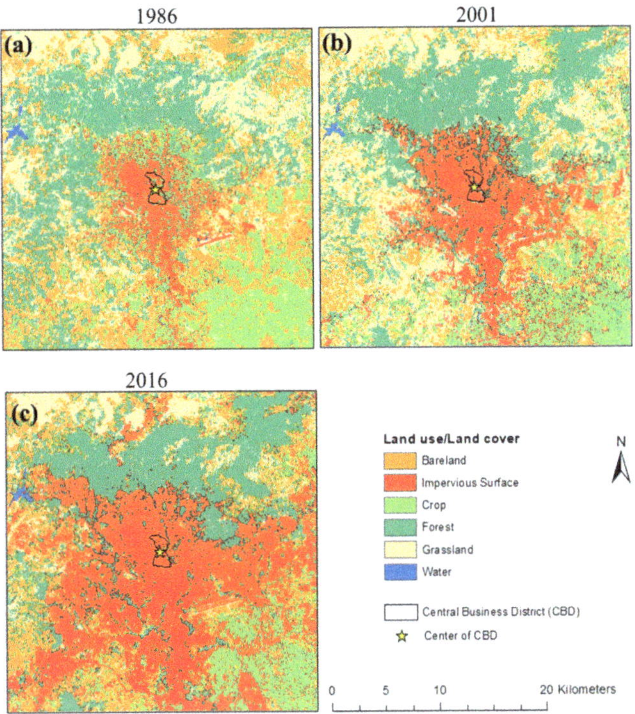

Figure 3. Land use and land cover (LULC) map of Addis Ababa; (**a**) LULC in 1986; (**b**) LULC in 2001; and (**c**) LULC in 2016.

Table 2. Details of LULC changes in Addis Ababa in 1986, 2001, and 2016.

LULC	1986		2001		2016	
	Area (ha.)	%	Area (ha.)	%	Area (ha.)	%
Bare land	24,372.0	27.1	12,723.0	14.1	19,200.0	21.3
Built-up	6262.3	7.0	14,033.0	15.6	30,700.0	34.1
Crop	17,002.2	18.9	12,544.0	13.9	11,020.0	12.2
Forest	18,406.0	20.5	20,957.0	23.3	18,713.0	20.8
Grassland	23,721.3	26.4	29,611.0	32.9	10,228.0	11.4
Water	236.2	0.3	132.0	0.1	139.0	0.2
Total	90,000.0	100	90,000.0	100	90,000.0	100

3.3. The Composition of the LULC and Expansion of IS

Spatial and temporal variations of the composition of IS, GS1, GS2, BL and variation of the LST distribution pattern along the URZs are shown in Figure 4. The figure clearly illustrates the pattern of the expansion of IS and urban-rural demarcation boundary. As explained in Section 2.5.1, the urban area was demarcated based on the fraction ratio of the IS. URZ_{26} in 1986, URZ_{43} in 2001, and URZ_{69} in 2016 are the first zones with IS < 10% that are observed as rural areas, as shown in Figure 4.

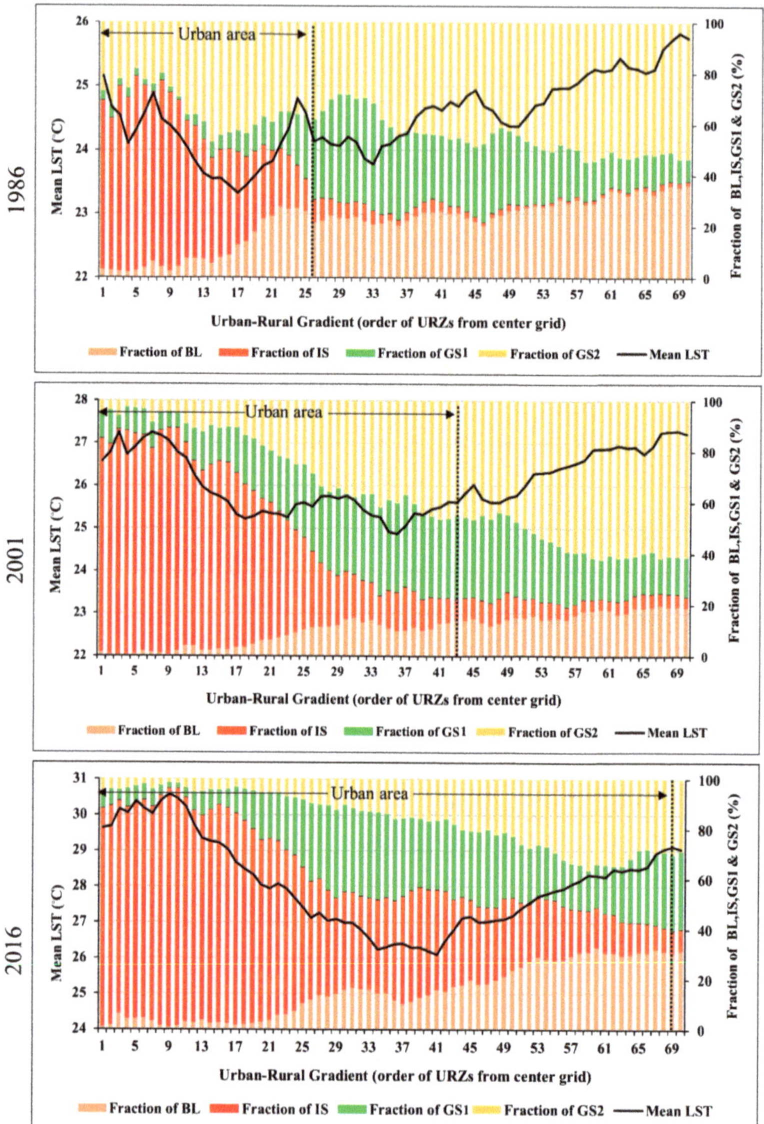

Figure 4. Distribution of the LULC composition and LST along the urban-rural zones (URZs) and expansion of the urban area in Addis Abba from 1986–2016.

3.4. LST Behavior Pattern

The fraction ratio of IS has gradually declined when moving away from the city center to the rural area along the URZs, while contrasting the fraction ratio of BL. GS2 fraction has gradually increased along the URZs except in 1986, but the irregular pattern could be observed for the fraction ratio of GS1 throughout the study period.

In general, the behavior of the LST in an urban area declines to move away from the city center to the rural area [41,43], but this pattern cannot be observed in Addis Ababa. It is declining from the city center to some extent and shows an increasing trend moving to the suburban and rural area. This is a significant observation in understanding the power that landscape structure affects the LST variation. By observing this pattern, mean LST turning point has been determined, as shown in Figure 5a. To this end, the lowest value of LST in each year was considered: 23.3 °C, 24.9 °C, and 26.5 °C in 1986, 2001, and 2016, respectively. The zone that ranges from the center grid to the LST turning point is designated as "Region 1" (R1), and the area spread away from the LST turning poin toward the rural area is designated as "Region 2" (R2), as shown in Figure 5a. In 1986, the R1 extended from URZ_1 to URZ_{17}, and R2 ranged from URZ_{18} to URZ_{70}. In 2001, the first 36 zones (URZ_{36}) belonged to R1, and the rest of the zones (URZ_{37-70}) belonged to R2. The R1 in 2016 extended from URZ_1 to URZ_{41} and R2 included RUZ_{42-70}. The linear regression analysis plotted based on all 70 URZs in the three-time points shows a significant ($p < 0.001$) positive relationship between mean LST and fraction ratios of IS, BL, and GS2. The statistical data on the composition of the LULC is provided in Table 3. It is worthy to note that the major fraction of the R1 is IS over the three-time points and that R2 has largely been dominated by GS2, except in 2016.

Table 3. The distribution pattern of the LULC in R1 and R2 in 1986, 2001, and 2016.

| LULC | Fraction Ratio | | | | | |
| | 1986 | | 2001 | | 2016 | |
	R1 (URZ_{1-17})	R2 (URZ_{18-70})	R1 (URZ_{1-37})	R2 (URZ_{37-70})	R1 (URZ_{1-41})	R2 (URZ_{42-70})
BL	5.6	26.8	6.3	15.4	6.6	24.0
IS	59.0	6.1	54.9	7.5	70.8	25.5
GS1	4.0	23.4	20.0	24.2	16.4	25.4
GS2	31.4	43.7	18.8	52.9	6.2	25.1
Mean LST (°C)	24.2	24.6	25.9	26.3	28.3	27.8

The magnitude of the mean LST with BL, IS, GS1, and GS2 along the URZs is shown in Figure 6. Any positive values of the mean LST or fraction ratio of BL, IS, GS1, and GS2 in any zone mean that the magnitude of the corresponding zone is less than that of the previous one ($Z_1-Z_2 > 0$). On the other hand, it shows a declining trend along the URZ. Any negative values of the mean LST or fraction ratio of IS, BL, GS1, and GS2 represent the opposite scenario ($Z_1-Z_2 < 0$). It is observed that the magnitude of the mean LST in a rural area is higher than in the urban area. The magnitudes of the BL (Figure 6a) and GS2 (Figure 6d) gradually incline from the city center to the rural area, while the magnitude of IS declines (as shown as Figure 6b); moreover, the magnitude of the GS1 shows a pattern opposing that of the LST: It is low where the magnitude of IS is high, as shown in Figure 6c.

Multidirectional and multitemporal profiles of the LST in Figure 7 show the cross-sectional outline of surface temperature described in Section 2.6. The reason could be that the fraction ratio of IS was weak in 1986 because of a lower degree of urbanization (Table 2), but it matured in 2016, as shown by the increase in IS fraction. Mean LST has declined in the area where natural forest is located (the North and Northwest direction), especially in the central mountain area. Figure 5b shows the types of LULC and their respective geographical locations overlaid onto the multidirectional and multi-temporal LST profile following chronological order.

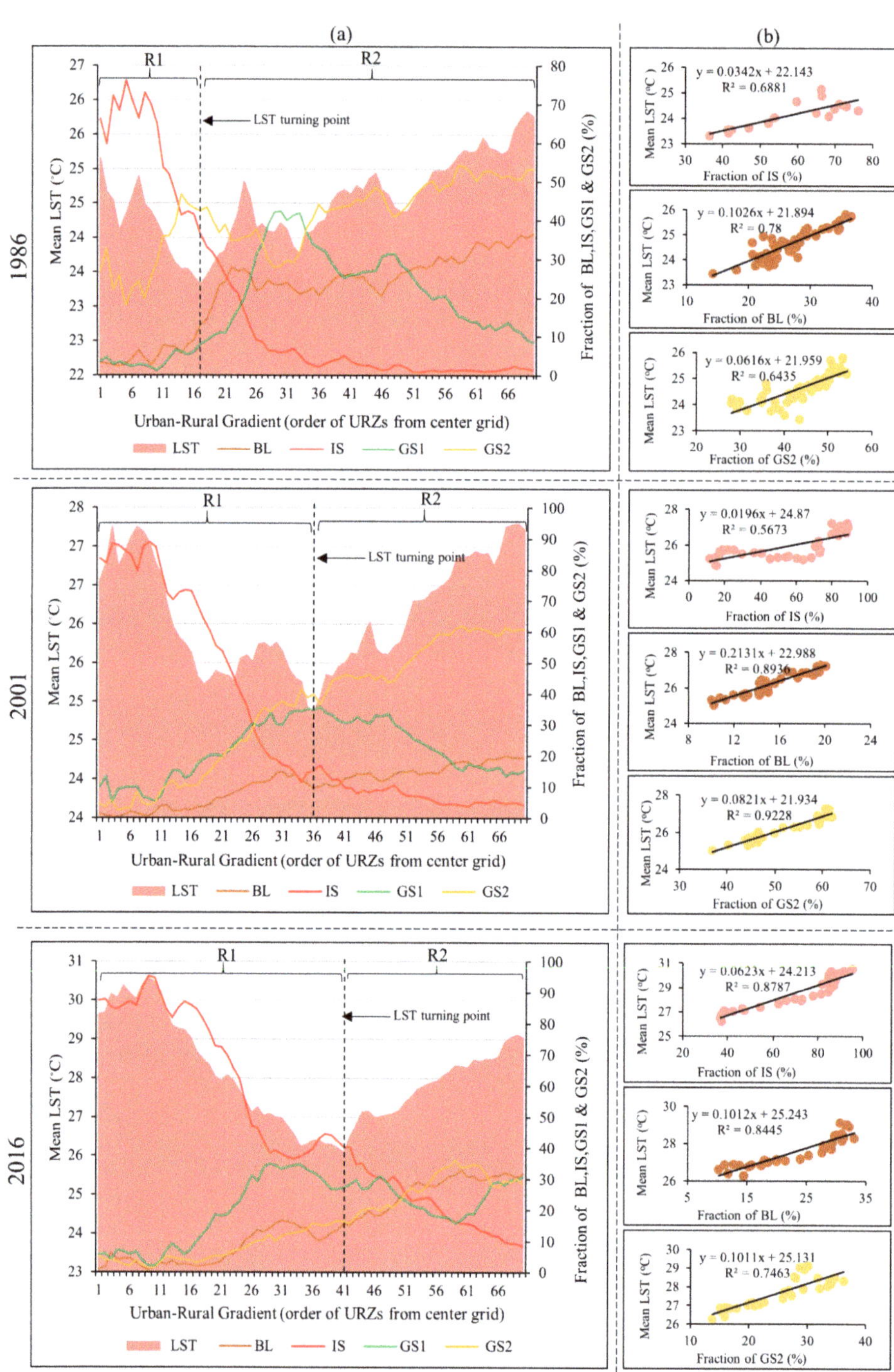

Figure 5. (**a**) Relationship between LST and LULC composition in 1986, 2001, and 2016; and (**b**) scatter plot diagram between LST and LULC in 1986, 2001, and 2016. Scatter plot diagrams between mean LST and IS fraction only belongs to R1, and the remaining two belong to R2.

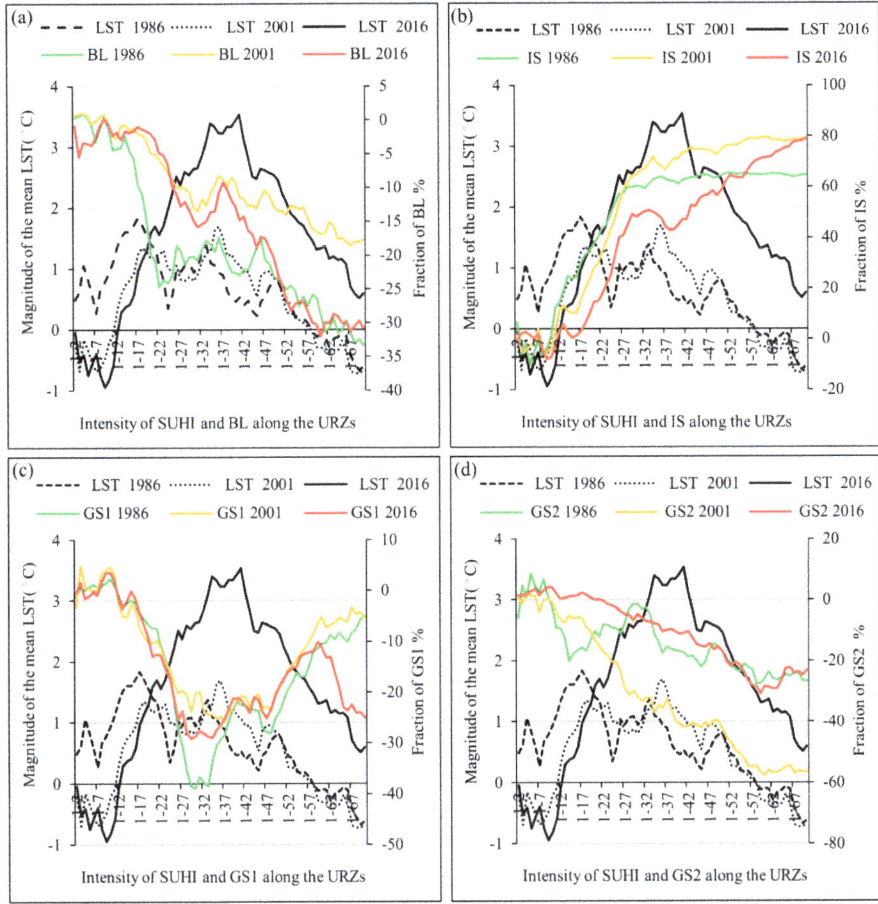

Figure 6. The magnitude of LST along the URZs with a fraction ratio of (**a**) BL, (**b**) IS, (**c**) GS1, and (**d**) GS2 in Addis Ababa.

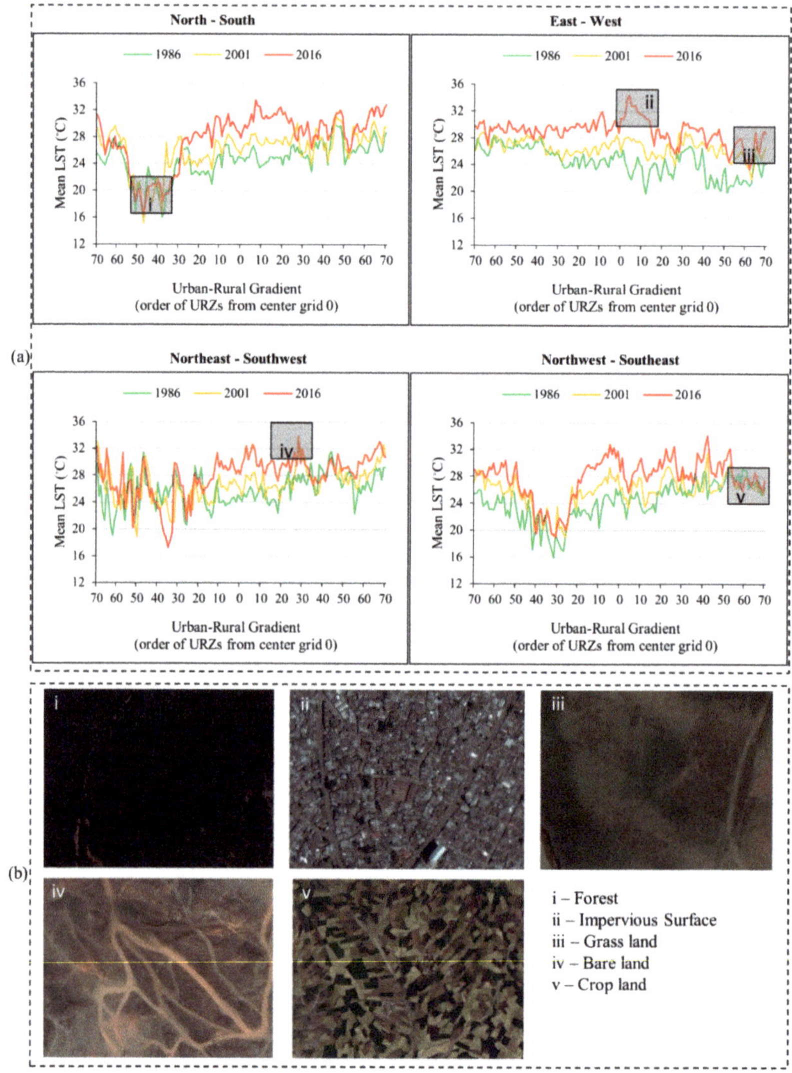

Figure 7. (**a**) Multidirectional and multitemporal land surface temperature intensity (LSTI) profile of Addis Ababa; (**b**) Different kinds of LULC illustrations along the orthogonal and diagonal direction in 2016. (Image Source: Google Earth, accessed on 7 February 2019).

4. Discussion

4.1. Urbanization and Changes in the LULC Structure

Addis Ababa is the capital city of the Federal Democratic Republic of Ethiopia, an essential administrative and economic hub not only for Ethiopia but also for the whole of Africa [53]. It has made impressive socioeconomic progress in the past three decades. According to the official statistics, the gross domestic product (GDP) was 10.9 while per capita income was USD 691 for the past one-and-a-half decades [53]. Employment opportunities and infrastructure development have encouraged people to live in the urban area or close to the urban area, as result of which the population of Addis Ababa is rapidly increasing, and it will continue in the future following the same increasing pattern (Figure 8). Rural–urban migration has also been a significant influence on the growth of the urban population [32]. Both total population and urban population show an increasing trend from 1980–2024, as depicted in Figure 8a. Proportion to the total population, the urban population, was 11.6% in 1986, 14.92% in 2001, and 19.86% in 2016. However, Ethiopia's Data Ecosystem annual report published by UNDP shows that 25% of the Ethiopians live in Addis Ababa [54]. Hence, it can be realized that the proxy data have mirrored the pattern of the urban population in Addis Ababa.

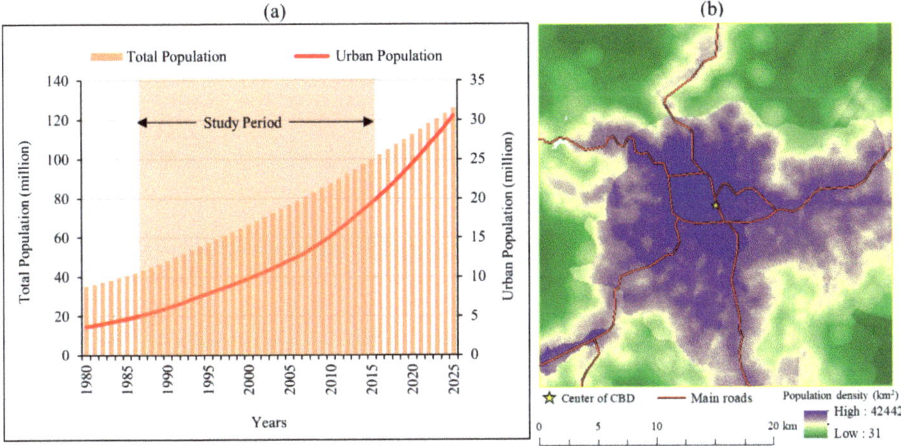

Figure 8. (**a**) Total and urban population in Ethiopia from 1980–2025 including the forecasted population (2019–2025) [46]; and (**b**) spatial distribution of the 2006 population with the main transportation line [45].

In Addis Ababa, the population has nearly doubled every decade; it was 1.4 million in 1984 and 2.6 million in 1994. It will be 12 million in 2024 as predicted by UN-Habitat [4]. However, population or its density is not directly influenced by the LST variation or its intensity, but overpopulation can enhance some attributes of LST, such as the expansion of IS [6]. Additional space is required for settlements, industrial zones, and infrastructure to cater to the urban community. Consequences of this scenario are the expansion of the urban area by acquiring space from other LULC. Our analysis shows that the IS area has expanded. In 2016, it comprised 34.1% (30,700 ha) of the total land (Table 2), but in 1986, it was only 6262.3 ha, which is 7% of the total land. The growth rate of the IS fraction was 11 ha^{-1} year^{-1} from 1986 to 2016. The urban area with an IS fraction greater than 10% (IS > 10%) was expanded to 5.46 km (URZ$_{26}$), 9.03 km (URZ$_{43}$), and 14.49 km (URZ$_{69}$) in 1986, 2001, and 2016, respectively. In other words, this indicates a diminishing rural area. The expanded IS fraction has directly resulted in a decrease in cropland and grassland (GS2) as illustrated by Table 2: A decline from 31.4% to 18.8% and 6.2% in 1986, 2006 and 2016, respectively at the rate of −4.2 ha^{-1} year^{-1}. These results demonstrate that urban expansion has an influence not only on the urban landscape but also on the suburban landscape.

4.2. The Relationship between LST and LULC Composition

Behavior pattern of the LST in Addis Ababa over the study period is in contrast to those of other cities as explained in Section 3.4. Two separate regions (R1 and R2) were identified (Figure 5a). It can be observed that R1 has shifted toward the periphery or rural area, with a distance of 3.57 km (URZ$_{17}$), 7.56 km (URZ$_{37}$), and 8.61 km (URZ$_{41}$) in 1986, 2001, and 2016, respectively. The relative mean LST of these three zones were reported to be 23.3 °C in 1986, 25.1 °C in 2001, and 26.9 °C in 2016. The changes could be the result of urbanization as depicted in Figure 4. The IS fraction in R1 has dominated more than half of the land in all three-time points as illustrated in Table 3, and the IS fraction positively correlated with the mean LST with high R^2 values (0.68 in 1986, 0.56 in 2001, and 0.87 in 2016), as shown in Figure 5a. However, the fraction of IS in R2 is less than that in R1, but it can also be observed that mean LST has increased in R2 (Figure 5a). If so, the mean LST behavior pattern of R2 is debatable and causal factors should be explored further. Our analysis shows that the fraction of BL (bare land) and GS2 (cropland and grassland) together dominate nearly 50% of the total land. In 1986, it was 70.5% (26.8% from BL and 43.7% from GS2), in 2001 it was 68.3% (15.4% from BL and 52.9% from GS2), and in 2016 it was 49.1% (24% from BL and 25.1% from GS2). Scatter plots confirm their significant ($p < 0.001$) positive correlation with LST, as illustrated by Figure 5b. Hence, it can be realized that BL and GS2 are the factors that influence LST in suburban and rural areas.

Addis Ababa as a sub-Saharan city shows the signs of desert attributes including bare land/bare soil. Economic and Social Council of the United Nations stated in 2007 that *"Seventy percent of Ethiopia is reported to be prone to desertification"* [55]. Non-vegetated land can be directly exposed to the sun, and the results could be more solar radiation back to the atmosphere, called "heat back to the atmosphere" [43]. Nevertheless, GS2 represents grassland and cropland, but we observed that it does not look like as green as that of healthy vegetation. We have selected dry season Landsat images for the LULC classification, as explained in Section 2.2 and Table 1. In the dry season, grassland and cropland can be seen as off-green or open land because of the low chlorophyll content, as shown by Figure 7b. The LST profile can be further visualized by displaying the cross-section characteristics of the landscape structure in the rural area. Figure 7b, Grassland (Figure 7b(iii)), bare land (Figure 7b(iv)) and cropland (Figure 7b(v)) where located rural areas show high LST value compared to the forest (Figure 7b(i)). The LST profiles depict the power of BL and GS2 in influencing the enhancement of magnitude of LST in rural areas.

4.3. The Implication of the Results for Urban Sustainability

The negative effect of the LSTI resulted not only in the expansion of IS but also in losing green cover. The vegetated urban fraction can regulate the effect of LSTI by providing shade to the land and accelerate higher albedo through evapotranspiration to generate a cool island against the heat island [41,56], as shown by Figure 7b. This study also identified lower LST areas that have covers with natural forest: As an example, the Entoto mountain area and the green parks located in the heart of the city center close to the National Palace. Furthermore, the rapid decrease in the LST profile along the north, northeast, and northwest (Figure 7a) provides evidence for the ability of vegetated land to maintain a lower level of LST. The results are consistent with past research findings [13]. Uncomforted air which resulted in LST will be a cause for the increase of energy demand for making comfort indoor air using an air conditioner or related equipment both day and night [6,43]. Heat-related illnesses can affect the city dwellers [6,14]. Losing the green fraction may result in food security and ecosystem services of the urban area.

To overcome these obstacles, appropriate measures need to be taken as indicated by our results, which are not only consistent with past research also but are also sustainable oriented [41,57]. In the viewpoint of sustainability of Addis Ababa, an environmentally friendly, socially acceptable, and economically accountable land administration process is compulsory. Past research has shown that green walls can mitigate indoor temperatures in tropical countries by 2.4 °C [58]. Addis Ababa—as a city in a tropical country—will be able to apply these concepts to generate low indoor temperatures and increase the green zone in the urban area. Additionally, green belts along the transportation line and green parks in residential areas will result in a cool island and encourage the natural air regulation process to make a comfortable living environment. From this perspective, the plan should be green-oriented aligning with "goal 11" of sustainable development [59]. Vertical, rather than horizontal, urban development is a space-preserving technique [60] that can be applied to Addis Ababa also.

5. Conclusions

This study has examined the influence of landscape structure on the spatiotemporal variation in the LST along the urban-rural gradient as a geospatial application by using the Landsat data from 1986–2016 with fifteen-years intervals in Addis Ababa, Ethiopia. We have adopted the relative variation in LST rather than the absolute value to understand the influence of landscape structure for the variation of mean LST. It is noted that the urban area has expanded and LST has grown more intense with a substantial loss of green fraction. When IS fraction improved to 27% with an annual growth rate of 11.1 ha^{-1} $year^{-1}$, the relative mean LST of URZ_1 increased by 4.5 °C over the study period. The behavior pattern of the LST is specific to the city as we discussed by applying R1 and R2. Previous studies have shown that the phenomenon does not appear in the sub-Saharan region. It is a new trend of LST behavior in the urban area, especially in the sub-Saharan region. However, shifting the R1 region toward the peripheral area in parallel to urban shifting is designated that LST becomes stronger in the future.

Hence, future city planning should consider this scenario to overcome the effect of LST and its intensity. Our results prove that improving the green fraction as much as possible is one of the mitigation measures. Urban expansion should not violate the balance of the landscape structure. As shown by the statistical plots, bare land has made a significant influence on the improvement of LST in rural areas than other land use types. However, more studies are required to examine the mechanism for managing geographical and climatological attributes that occur in the sub-Saharan region. Indeed, the outcomes of this research indicate the dissimilarity in the LST results of the structure of the landscape. We conclude that the overall findings of this study can be used as a proxy indicator for the sustainability of Addis Ababa.

Author Contributions: The corresponding author, D.D., has proposed the topic and spearheaded the data processing and analysis, as well as the writing of the manuscript. T.M., Y.M., and M.R. helped in the design, research implementation and analysis, and writing of the manuscript.

Funding: This study was supported by the Japan Society for the Promotion of Science (JSPS) through Grant-in-Aid for Challenging Exploratory Research 16K12816 and Scientific Research (B) 18H00763.

Acknowledgments: The authors express their gratefulness to the anonymous reviewers for their valuable comments and suggestions.

Conflicts of Interest: The authors declare no conflict of interest.

References

1. United Nations. *The World's Cities in 2018*; United Nations, Department of Economic and Social Affairs, Population Division: New York, NY, USA, 2018.
2. United Nations. *World Urbanization Prospects: The 2014 Revision, Highlights*; United Nations, Department of Economic and Social Affairs, Population Division: New York, NY, USA, 2015.
3. United Nations. *The Speed of Urbanization around the World*; United Nations: New York, NY, USA, 2018; pp. 1–2.
4. United Nations. *Addis Ababa—Urban Profile*; United Nations, Regional and Technical Cooperation Division, United Nations Human Settlements Programme (UN-HABITAT): New York, NY, USA, 2008.
5. Guo, L.; Liu, R.; Men, C.; Wang, Q.; Miao, Y.; Zhang, Y. Quantifying and simulating landscape composition and pattern impacts on land surface temperature: A decadal study of the rapidly urbanizing city of Beijing, China. *Sci. Total Environ.* **2019**, *654*, 430–440. [CrossRef] [PubMed]
6. Dissanayake, D.; Morimoto, T.; Murayama, Y.; Ranagalage, M.; Handayani, H.H. Impact of urban surface characteristics and socio-economic eariables on the spatial variation of land surface temperature in Lagos City, Nigeria. *Sustainability* **2019**, *11*, 25. [CrossRef]
7. Feyissa, G.; Gebremariam, E. Mapping of landscape structure and forest cover change detection in the mountain chains around Addis Ababa: The case of Wechecha Mountain, Ethiopia. *Remote Sens. Appl. Soc. Environ.* **2018**, *11*, 254–264. [CrossRef]

8. Staniec, M.; Nowak, H. The application of energy balance at the bare soil surface to predict annual soil temperature distribution. *Energy Build.* **2016**, *127*, 56–65. [CrossRef]
9. Arsiso, B.K.; Mengistu Tsidu, G.; Stoffberg, G.H.; Tadesse, T. Influence of urbanization-driven land use/cover change on climate: The case of Addis Ababa, Ethiopia. *Phys. Chem. Earth.* **2018**, *105*, 212–223. [CrossRef]
10. Munn, R.E. *Descriptive Micrometeorology*; Atmospheric Environment Service; Academic Press: Toronto, ON, Canada, 1966; p. 198.
11. Oke, T.R. *Boundary Layer Climates*; Methuen and Co., Ltd.: London, UK, 1978; p. 372.
12. Li, Z.-L.; Duan, S.-B. Land Surface Temperature. *Compr. Remote Sens.* **2018**, *5*, 264–283.
13. Geremeskel, T.; Abera, M. The need for transformation: Local perception of climate change, vulnerability and adaptation versus 'Humanitarian' response in Afar Region, Ethiopia. In *Climate Change Adaptation in Africa. Climate Change Management*; Leal Filho, W., Belay, S., Kalangu, J., Menas, W., Munishi, P., Musiyiwa, K., Eds.; Springer: Cham, UK, 2017.
14. IPCC. *Climate Change 2013: The Physical Science Basis. Contribution of Working Group I (WGI) to the Fifth Assessment Report (AR5) of the Intergovernmental Panel on Climate Change (IPCC)*; Cambridge University Press: Cambridge, UK, 2013.
15. World Health Organization. *WHO Guidance to Protect Health from Climate Change through Health Adaptation Planning*; WHO, Department of Public Health Environment and Social Determinants of Health: Geneva, Switzerland, 2014.
16. Feyisa, G.L.; Dons, K.; Meilby, H. Efficiency of parks in mitigating urban heat island effect: An example from Addis Ababa. *Landsc. Urban Plan.* **2014**, *123*, 87–95. [CrossRef]
17. Ranagalage, M.; Estoque, R.C.; Murayama, Y. An urban heat island study of the Colombo Metropolitan Area, Sri Lanka, Based on Landsat Data (1997–2017). *Isprs Int. J. Geo-Inf.* **2017**, *6*, 189. [CrossRef]
18. Ranagalage, M.; Estoque, R.C.; Handayani, H.H.; Zhang, X.; Morimoto, T.; Tadono, T.; Murayama, Y. Relation between urban volume and land surface temperature: A comparative study of planned and traditional cities in Japan. *Sustainability* **2018**, *10*, 2366. [CrossRef]
19. Zhou, X.; Wang, Y.C. Dynamics of land surface temperature in response to land-use/cover change. *Geogr. Res.* **2011**, *49*, 23–36. [CrossRef]
20. Handayani, H.H.; Murayama, Y.; Ranagalage, M.; Liu, F.; Dissanayake, D. Geospatial analysis of horizontal and vertical urban expansion using multi-spatial resolution data: A case study of Surabaya, Indonesia. *Remote Sens.* **2018**, *10*, 25. [CrossRef]
21. Zhao, H.; Zhang, H.; Miao, C.; Ye, X.; Min, M. Linking heat source–sink landscape patterns with analysis of urban heat islands: Study on the fast-growing Zhengzhou City in Central China. *Remote Sens.* **2018**, *10*, 1268. [CrossRef]
22. Li, J.; Song, C.; Cao, L.; Zhu, F.; Meng, X.; Wu, J. Impacts of landscape structure on surface urban heat islands: A case study of Shanghai, China. *Remote Sens. Environ.* **2011**, *115*, 3249–3263. [CrossRef]
23. Pal, S.; Ziaul, S. Detection of land use and land cover change and land surface temperature in English Bazar urban centre. *Egypt. J. Remote Sens. Sp. Sci.* **2017**, *20*, 125–145. [CrossRef]
24. Tarawally, M.; Xu, W.; Hou, W.; Mushore, T.D. Comparative analysis of responses of land surface temperature to long-term land use/cover changes between a coastal and Inland City: A case of Freetown and Bo Town in Sierra Leone. *Remote Sens.* **2018**, *10*, 112. [CrossRef]
25. Fitts, Y.; Comby, J.; Renard, F.; Hadjiosif, A.; Alonso, L. Evaluation of the effect of urban redevelopment on surface urban heat islands. *Remote Sens.* **2019**, *11*, 299.
26. Gunaalan, K.; Ranagalage, M.; Gunarathna, M.; Kumari, M.; Vithanage, M.; Srivaratharasan, T.; Saravanan, S.; Warnasuriya, T.W.S. Application of geospatial techniques for groundwater quality and availability assessment: A case study in Jaffna Peninsula, Sri Lanka. *Isprs Int. J. Geo-Inf.* **2018**, *7*, 20. [CrossRef]
27. Wijesundara, N.C.; Abeysingha, N.S.; Dissanayake, D.M. GIS-based soil loss estimation using RUSLE model: A case of Kirindi Oya river basin, Sri Lanka. *Model. Earth Syst. Environ.* **2018**, *4*, 251–262. [CrossRef]
28. Demissie, F.; Yeshitila, K.; Kindu, M.; Schneider, T. Land use/land cover changes and their causes in Libokemkem District of South Gonder, Ethiopia. *Remote Sens. Appl. Soc. Environ.* **2017**, *8*, 224–230. [CrossRef]
29. Kindu, M.; Schneider, T.; Teketay, D.; Knoke, T. Land use/land cover change analysis using object-based classification approach in Munessa-Shashemene landscape of the ethiopian highlands. *Remote Sens.* **2013**, *5*, 2411–2435. [CrossRef]

30. Dissanayake, D.; Morimoto, T.; Ranagalage, M. Accessing the soil erosion rate based on RUSLE model for sustainable land use management: A case study of the Kotmale watershed, Sri Lanka. *Model. Earth Syst. Environ.* **2018**, *5*, 291–306. [CrossRef]
31. Liu, Y.; Peng, J.; Wang, Y. Efficiency of landscape metrics characterizing urban land surface temperature. *Landsc. Urban Plan.* **2018**, *180*, 36–53. [CrossRef]
32. Kifle Arsiso, B.; Mengistu Tsidu, G.; Stoffberg, G.H.; Tadesse, T. Climate change and population growth impacts on surface water supply and demand of Addis Ababa, Ethiopia. *Clim. Risk Manag.* **2017**, *18*, 21–33. [CrossRef]
33. Wubneh, M. Addis Ababa, Ethiopia—Africa's diplomatic capital. *Cities* **2013**, *35*, 255–269. [CrossRef]
34. Abo-El-Wafa, H.; Yeshitela, K.; Pauleit, S. The use of urban spatial scenario design model as a strategic planning tool for Addis Ababa. *Landsc. Urban Plan.* **2018**, *180*, 308–318. [CrossRef]
35. United States Geological Survey (USGS) Earth Explorer Home Page. Available online: https://earthexplorer.usgs.gov (accessed on 7 February 2019).
36. United States Geological Survey (USGS) Landsat Missions: Landsat Science Products. Available online: https://www.usgs.gov/land-resources/nli/landsat/landsat-science-products (accessed on 7 February 2019).
37. Ebabu, K.; Tsunekawa, A.; Haregeweyn, N.; Adgo, E.; Meshesha, D.T.; Aklog, D.; Masunaga, T.; Tsubo, M.; Sultan, D.; Fenta, A.A.; et al. Effects of land use and sustainable land management practices on runoff and soil loss in the Upper Blue Nile basin, Ethiopia. *Sci. Total Environ.* **2019**, *648*, 1462–1475. [CrossRef]
38. Jin, B.; Ye, P.; Zhang, X.; Song, W.; Li, S. Object-Oriented method combined with deep convolutional neural networks for land-use-type classification of remote sensing images. *J. Indian Soc. Remote Sens.* **2019**, *3*, 1–15. [CrossRef]
39. Sobrino, J.A.; Jiménez-Muñoz, J.C.; Paolini, L. Land surface temperature retrieval from LANDSAT TM 5. *Remote Sens. Environ.* **2004**, *90*, 434–440. [CrossRef]
40. Weng, Q.; Lu, D.; Schubring, J. Estimation of land surface temperature-vegetation abundance relationship for urban heat island studies. *Remote Sens. Environ.* **2004**, *89*, 467–483. [CrossRef]
41. Estoque, R.C.; Murayama, Y. Monitoring surface urban heat island formation in a tropical mountain city using Landsat data (1987–2015). *Isprs J. Photogramm. Remote Sens.* **2017**, *133*, 18–29. [CrossRef]
42. Myint, S.W.; Brazel, A.; Okin, G.; Buyantuyev, A. Combined effects of impervious surface and vegetation cover on air temperature variations in a rapidly expanding Desert City. *GIScience Remote Sens.* **2010**, *47*, 301–320. [CrossRef]
43. Ranagalage, M.; Dissanayake, D.; Murayama, Y.; Zhang, X.; Estoque, R.C.; Perera, E.; Morimoto, T. Quantifying surface urban heat island formation in the World Heritage Tropical Mountain City of Sri Lanka. *Isprs Int. J. Geo-Inf.* **2018**, *7*, 341. [CrossRef]
44. Stehman, S.V. Sampling designs for accuracy assessment of land cover. *Int. J. Remote Sens.* **2009**, *30*, 5243–5272. [CrossRef]
45. World Population Data Website. Available online: https://www.worldpop.org/geodata/summary?id=2880 (accessed on 7 February 2019).
46. United Nations—Population Division. World Population Prospects 2017. Available online: https://population.un.org/wpp/ (accessed on 17 February 2019).
47. Tatem, A.J. WorldPop, open data for spatial demography. *Sci. Data* **2017**, *4*, 2–5. [CrossRef] [PubMed]
48. Wei, C.; Taubenböck, H.; Blaschke, T. Measuring urban agglomeration using a city-scale dasymetric population map: A study in the Pearl River Delta, China. *Habitat Int.* **2017**, *59*, 32–43. [CrossRef]
49. Ye, T.; Zhao, N.; Yang, X.; Ouyang, Z.; Liu, X.; Chen, Q.; Hu, K.; Yue, W.; Qi, J.; Li, Z.; et al. Improved population mapping for China using remotely sensed and points-of-interest data within a random forests model. *Sci. Total Environ.* **2019**, *658*, 936–946. [CrossRef] [PubMed]
50. Estoque, R.C.; Murayama, Y.; Akiyama, C.M. Pixel-based and object-based classifications using high- and medium-spatial-resolution imageries in the urban and suburban landscapes. *Geocarto Int.* **2015**, *30*, 1113–1129. [CrossRef]
51. Ye, S.; Pontius, R.G.; Rakshit, R. A review of accuracy assessment for object-based image analysis: From per-pixel to per-polygon approaches. *Isprs J. Photogramm. Remote Sens.* **2018**, *141*, 137–147. [CrossRef]
52. Bewket, W.; Abebe, S. Land-use and land-cover change and its environmental implications in a tropical highland watershed, Ethiopia. *Int. J. Environ. Stud.* **2013**, *70*, 126–139. [CrossRef]
53. Melka, S.G. *Ethiopia's Data Ecosystem*; United Nations, African Data Report: New York, NY, USA, 2016.

54. World Bank Group. *Addis Ababa, Ethiopia Enhancing Urban Resilience*; World Bank Group: Washington, DC, USA, 2015.

55. United Nation. *Africa Review on Drought and Desertification*; Economic and Social Council: New York, NY, USA, 2007.

56. Manawadu, L.; Ranagalage, M. Urban Heat Islands and Vegetation Cover as a Controlling Factor. In Proceedings of the International Forestry and Environment Symposium 2013 of the Department of Forestry and Environmental Science, University of Sri Jayewardenepura, Thulhiriya, Sri Lanka, 10–11 January 2014; p. 125.

57. Ranagalage, M.; Estoque, R.C.; Zhang, X.; Murayama, Y. Spatial changes of urban heat island formation in the Colombo District, Sri Lanka: Implications for sustainability planning. *Sustainability* **2018**, *10*, 1367. [CrossRef]

58. Galagoda, R.U.; Jayasinghe, G.Y.; Halwatura, R.U.; Rupasinghe, H.T. The impact of urban green infrastructure as a sustainable approach towards tropical micro-climatic changes and human thermal comfort. *Urban For. Urban Green.* **2018**, *34*, 1–9. [CrossRef]

59. United Nations. *Transforming Our World: The 2030 Agenda for Sustainable Development*; United Nations General Assembly: New York, NY, USA, 2015; Volume 16301, pp. 1–35.

60. Lin, J.; Huang, B.; Chen, M.; Huang, Z. Modeling urban vertical growth using cellular automata-Guangzhou as a case study. *Appl. Geogr.* **2014**, *53*, 172–186. [CrossRef]

 *sustainability*

Article

Analysis of Life Quality in a Tropical Mountain City Using a Multi-Criteria Geospatial Technique: A Case Study of Kandy City, Sri Lanka

DMSLB Dissanayake [1,*], Takehiro Morimoto [2], Yuji Murayama [2], Manjula Ranagalage [1,2,*] and ENC Perera [3]

1   Department of Environmental Management, Faculty of Social Sciences and Humanities, Rajarata University of Sri Lanka, Mihintale 50300, Sri Lanka

2   Faculty of Life and Environmental Sciences, University of Tsukuba, 1-1-1, Tennodai, Tsukuba, Ibaraki 305-8572, Japan; tmrmt@geoenv.tsukuba.ac.jp (T.M.); mura@geoenv.tsukuba.ac.jp (Y.M.)

3   Department of Regional Science and Planning, SANASA Campus, Kegalle 71000, Sri Lanka; chinssu@gmail.com

*   Correspondence: dissanayakedmslb@gmail.com or dissanayakedmslb@ssh.rjt.ac.lk (D.D.); manjularanagalage@gmail.com or manjularanagalage@ssh.rjt.ac.lk (M.R.)

Received: 28 February 2020; Accepted: 2 April 2020; Published: 6 April 2020

Abstract: The blooming of urban expansion has led to the improvement of urban life, but some of the negative externalities have affected the life quality of urban dwellers, both directly and indirectly. As a result of this, research related to the quality of life has gained much attention among multidisciplinary researchers around the world. A number of attempts have been made by previous researchers to identify, assess, quantify, and map quality of life or well-being under various kinds of perspectives. The objectives of this research were to create a life quality index (LQI) and identify the spatial distribution pattern of LQI in Kandy City, Sri Lanka. Multiple factors were decomposed, a hierarchy was constructed by the multi-criteria decision making (MCDM) method, and 13 factors were selected under two main criteria—environmental and socioeconomic. Pairwise comparison matrices were created, and the weight of each factor was determined by the analytic hierarchy process (AHP). Finally, gradient analysis was employed to examine the spatial distribution pattern of LQI from the city center to the periphery. The results show that socioeconomic factors affect the quality of life more strongly than environmental factors, and the most significant factor is transportation. The highest life quality zones (26% of the total area) were distributed around the city center, while the lowest zones represented only 9% of the whole area. As shown in the gradient analysis, more than 50% of the land in the first five kilometers from the city center comes under the highest life quality zone. This research will provide guidance for the residents and respective administrative bodies to make Kandy City a livable city. It the constructed model can be applied to any geographical area by conducting necessary data calibration.

Keywords: life quality index (LQI); Kandy city; AHP; MCDM; gradient analysis; Sri Lanka

---

1. Introduction

In mid-2019, the world's population reached 7.7 billion, having added approximately 1 billion people in the last two decades [1]. Twenty-three percent of the world's population were living in cities in 2018, and this percentage is predicted to reach 66% by 2050 [2]. This means that 43% of the global population will be living in urban regions within roughly 40 years, and it is estimated that the world's fastest-growing cities will be in Asia and Africa [3]. Cities provide environmentally sound and socially accepted living environments in various ways, but they generate urbanization issues that severely impact the well-being of city dwellers [2,4]. Although developed countries have managed those issues to some extent with their capital power, the situation in most developing countries, especially in Asia, is dilapidated. The lack of socioeconomic and environmental resources in developing countries presents considerable challenges in an urban area. Thus, conducting research related to the well-being of inhabitants in complex urban areas is an important, timely task.

To ensure the well-being of urban inhabitants, various kinds of approaches have been taken following various viewpoints, the details of which can be explored elsewhere [5–8]. The United Nation's Habitat Conference presented the concept of city livability (CL) in 1996 and stated that every city should be habitable [9]. Moreover, the following index and approaches provide some indicators to measure the level of livability in a city, such as the gross domestic product (GDP) and the human development index (HDI) [9]. The term "life quality" (LQ) is also used to describe the general well-being of societies and people [10,11]. As LQ is a theoretical concept that remains a significant subjective element, it is difficult to use it to compute measurable dimensions. As a result, there is no accepted definition or adequate measurements of LQ [4]. However, it is widely applied in various subjects using different types of elements and dimensions. Recently, research related to the LQ in urban areas has been given a benchmark by using various kinds of approaches that numerous studies have demonstrated [4,9,12].

Assessing the quality of life or constructing its index is quite a challenging task because it has no pre-defined factors or attributes [4]. On the other hand, measuring LQ is not a one-size-fits-all approach because life quality is the result of a combination of multifactorial attributes that can consist of social, economic, and environmental factors. These factors can depend on each other, or they may be connected directly or indirectly. Moreover, a multifactorial approach can be used in a complex decision-making environment, because each criterion does not contribute to results equally; rather, different criteria make different contributions [9]. Hence, there should be a scientifically sound and mathematically trustworthy approach to evaluate the contribution criteria based on their importance or weight. Nevertheless, humans have some difficulty with simultaneously addressing decisions using many criteria or alternatives because criteria have several uses [9]. The available literature shows that there are various methods for decision making in a multi-criteria environment, including the analytic hierarchy process (AHP) [12,13].

The AHP is a technique used to derive ratio gages from paired comparisons, which were initially developed by Professor Thomas L. Saaty in the 1970s [14]. It allows the criteria weights to be measured through pairwise comparison, which depends on the decisions of experts to regulate the priority ranges [15,16]. In the AHP technique, a hierarchical form, including goals, criteria, sub-criteria, and factors are used to solve every problem [15]. One of the notable leading factors is that the AHP approach allows users to define the weights of variables in the construction of the solution of a multi-criteria problem as a multi-criteria decision making (MCDM) method. Other than this, conventional research based on field surveys is not cost-effective and is a time-consuming task [17]. Because of its applicability, correctness, theoretical suitability, and capacity for addressing any criteria (intangible and tangible), the integration of AHP with MCDM in LQ analysis is still interesting for multi-disciplinary researchers [15,16]. However, such research is lacking in Sri Lanka.

In view of the above aspect, we hypothesized that socioeconomic and environmental factors are mainly influenced by the determination of LQ in urban dwellers in Kandy City, Sri Lanka. Over the past two decades, urbanization has led to rapid development, and several environmental problems, such as urban heat islands [18], increased energy consumption (R1), and reduced air quality [19], have occurred. Thus, LQ requires further study to enhance future urban planning activities. The objectives of this research are as follows: (i) to create an LQ index (LQI) for Kandy City and (ii) to identify the spatial distribution pattern of LQ and its composition along the gradient zones (GZs). We believe that our point of novelty is the incorporation of socioeconomic data to build a socially accepted LQI in Kandy City.

2. Materials and Methods

2.1. Study Area

Kandy is the capital and administrative city of the central province. It is the second-largest city in Sri Lanka, located 116 km away from Colombo City [5]. Because it is the home of the temple of the Tooth Relic (Sri Dalada Maligawa) and is a tropical mountain green city, Kandy is identified as an important tourist destination and was declared to be a world heritage site by the United Nations Educational, Scientific and Cultural Organization (UNESCO) [5]. Available literature shows that the estimated population is approximately 1.7 million, and the daily transient population is approximately one million in Kandy City [5]. As a result of the expansion of impervious surfaces and population size, Kandy City has undergone high urbanization growth in the last two decades. However, some of the topographical obstacles (mountain ranges and slopes) and malfunctioning of socioeconomic attributes (infrastructure and services) have been caused by the scarcity of livable resident land. The mean daytime temperature ranges from 28–32 °C, while the average rainfall is approximately 2085 mm (long-term) [20] and 52–398 mm (monthly) [18]. Except for a short, dry period, Kandy City is like other green cities and has a tropical equatorial climate and a monsoon rainfall pattern.

In this study, a geographical area of 400 km^2 covering a 20 × 20 km grid with a 10 km radius from the city center was selected as the study area, and it is bound by latitudes from 7.225195° N to 7.360744° N and longitudes from 80.566965° E to 80.702955° E, as illustrated in Figure 1.

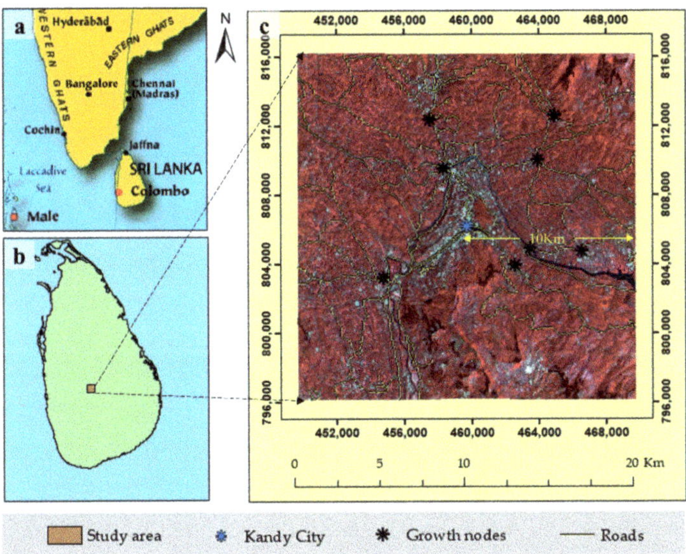

Figure 1. Study area. (**a**) South Asia; (**b**) the Democratic Socialist Republic of Sri Lanka, including the study area; and (**c**) the study area represented by a Landsat-8 image with a false-color composite (bands 5, 4, 3).

2.2. Types of Data and Sources

The research framework was constructed using two strategies; first, we observed regional requirements by consulting with a group of experts and administrative bodies in Kandy City, and second, the residents' opinions regarding well-being were also observed by conducting informal interviews in both urban and peripheral areas. Then, we realized that the identification of the spatial distribution patterns of LQ would help both administrative bodies and inhabitants to achieve the goal of sustainable urban development.

Regarding this context, 13 factors extracted from socioeconomic and environmental (physical) criteria were evaluated using MCDM model coupling with AHP on the GIS platform. Both primary and secondary data were collected to assess the LQ of residents in Kandy City. In the process, the following steps were completed. To gather primary data, (i) formal interviews were conducted with a group of experts and respective administration bodies to determine the factors related to the AHP and MCDM (Figure 2), (ii) expert's judgments were collected to complete the AHP questionnaire, and (iii) field observations were conducted to identify the current situation of the study area while interviewing some residents. As an approach for collecting secondary data, (i) literary information was collected, focusing on similar studies, and (ii) geospatial data were collected from respective sources, as explained in Table 1.

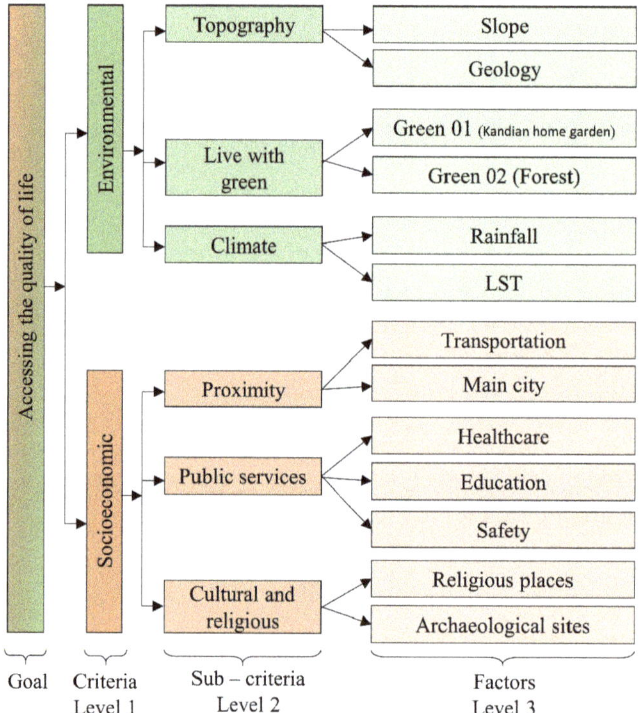

Figure 2. Criteria and factors related to the analytic hierarchy process (AHP).

2.3. Data Pre-Processing

Both primary and secondary data were carefully checked, and fault-free data sets were rectified. Subsequently, raw data were pre-preceded in order to build factors.

2.3.1. Landsat-8 OLI/TIRS Data

We hypothesized that the use of more Landsat data in order to cover all months in the year would produce a reliable data set. There might be slight seasonal changes in the study area that could affect land use land cover (LULC) and land surface temperature (LST). Hence, obtaining annual average basic data sets is primarily important to generalize the model's results for the whole year. In this perspective, 2018 was selected as the investigated year, and the following simple steps were accomplished to prepare basic remote sensing (RS) data.

i.   The Google Earth Engine (GEE) was selected because of its free tools and facilitates for radiometric-calibrated and atmospheric-corrected Landsat-8 Level 2 data sets powered by the United States Geological Survey (USGS) [8]. Additionally, it has many functional facilities for handling remote sensing data.

ii. The research area was imported and checked for areas with cloud disturbance in the available Landsat imagery. Because the study area is located in a tropical region, there was cloud disturbance in the Landsat data [8]. Hence, the masking method was applied to remove the cloud.

iii. The annual median at-satellite brightness temperature (in Kelvin) and multispectral bands (in radiance values) were computed by using image collection and ee.reducer functions in the GEE [21].

iv. Prepared data sets were downloaded. Then, (a) calculation of the land surface temperature (LST) and (b) classification of the LULC were carried out.

2.3.2. Land Use Land Cover Classification

i.   A pixel-based supervised classification method was chosen [22]. Because medium- resolution Landsat data were selected, the best approach was to select a pixel-oriented classification method. Other than this, in order to identify precious LULC information as much as possible, the level was very important for the decision-making process.

ii. Four classification techniques were employed with R software (open source) [23]: (i) support vector machine, (ii) K-nearest neighbor, (iii) random forest, and (iv) neural networks. As a result of the classification, four LULC maps were produced.

iii. The resultant LULC maps were sorted [5]. Due to the highest values (overall accuracy and kappa coefficient), the map generated by the random forest method was chosen.

iv. The issues of misclassification error or salt-and-pepper noise made by spectral confusion were resolved by using majority filters and hybrid classification methods [24]. As evidenced by the literature [25–27], past researchers have also adopted this method and have gained reliable results. Finally, an accuracy assessment was conducted to test whether the classification results could be trusted.

2.3.3. Retrieval of Land Surface Temperature

The median temperatures (Section 2.3.1) extracted from thermal band 10 (in Kelvin) were used to calculate the LST, as follows:

i.   The land surface emissivity (ε) was calculated using Equation (1):

$$\varepsilon = \{mPV + n\} \tag{1}$$

where m = $(\varepsilon - \varepsilon) - (1 - \varepsilon\sigma)$; F$\varepsilon$v and n = εs + $(1 - \varepsilon$s$)$ Fεv. εs and εv are the soil emissivity and vegetation emissivity, respectively. In this study, we used the results of [28] for m = 0.004 and n = 0.986. The proportion of vegetation (Pv) was calculated using Equation (2).

ii. Using Equation (2), the proportion of vegetation (Pv) was computed.

$$P_v = ((NDVI - NDVI_{min})/(NDVI_{max} - NDVI_{min}))^2 \tag{2}$$

where P_v denotes the proportion of vegetation, NDVI refers to the Normalized Difference Vegetation Index, which is calculated using the original NDVI values calculated in Equation (3), and $NDVI_{min}$ and $NDVI_{max}$ are the minimum and maximum values of the NDVI dataset, respectively.

iii. The NDVI was calculated from Equation (3):

$$NDVI = \frac{\rho_{NIR} - \rho_{Red}}{\rho_{NIR} + \rho_{Red}} \tag{3}$$

where ρNIR refers to the surface reflectance values of band 5, and ρRed refers to the surface reflectance values of band 4 from the Landsat-8 OLI data.

iv.   The emissivity corrected LST was calculated using Equation (4):

$$LST = T_b / 1 + (\lambda \times T_b / \rho) \ln \varepsilon. \tag{4}$$

where T_b is the at-satellite brightness temperature in Kelvin; λ is the central-band wavelength of the emitted radiance (11.5 µm for band 6 and 10.8 µm for band 10); ρ is h $\times$ c/σ (1.438 $\times$ 10^{-2} m K) with σ being the Boltzmann constant (1.38 $\times$ 10^{-23} J/K), h being Planck's constant (6.626 $\times$ 10^{-34} J·s), and c being the speed of light (2.998 $\times$ 10^8 m/s) [29]; ε is the land-surface emissivity estimated using Equation (3). Then, the calculated LST values (Kelvin) were converted to degrees Celsius (°C).

2.4. Preparation of Criteria and Factors in the AHP

Criteria and factors were determined based on the results of the field observations, past research, and information gained from experts [9]. To get wider image and deeper understanding, a discussion with experts was conducted to define the criteria, sub-criteria, and factors related to the AHP. The experts, including professional researchers and officers who engage with public services and the welfare of both the government and private sectors were interviewed. Gathering information from experts is a common approach in the AHP and references can be found elsewhere [15,16,30]. In addition to that, we also met with the respective administrative bodies of settlement constructors and land selling companies and carried out discussions.

Subsequently, the problem was decomposed by constructing a hierarchy of inter-related decision sections. From the highest to the lowest level, the hierarchical structure was made to interrelate and chain all decision sections, and the goal of the study was placed at the upper part of the hierarchical structure. The bottom level of the hierarchical structure was used to present more detailed factors inter-related with the criteria in the next upper level. The hierarchical structure is presented in Figure 2. In the structure, physical factors governed by nature are represented by environmental criteria, while man-made and modified natural environment factors are classes as socioeconomic factors. A detailed description of each factor and its relevance for the study is presented in Table 1. All factors were constructed using secondary data, and the data pre-procedure was dependent on the types of raw data, as shown in Table 1.

Table 1. Factors and their relevance including the factor derivation method.

Factors	Purpose and Relevance	Raw Data	Data Sources	Data Processing/Method
Slope	The high slope area is not convenient for the settlement because of the risk of landslides [31] and the high construction cost. Considering that Kandy is a hill country city, slope is a relevant factor for the study.	DEM	SDOSL	Construction of slope from digital elevation model (DEM) and reclassification.
Geology	A geological structural fault causes the vulnerability of the settlement and leads to damage to infrastructure [32,33]. Settlements located on poor geological structures can be vulnerable to landslides [34], and this may cause many problems, notably land instability and road damage [29]. This information emphasizes that geological structural faults create an unpleasant environment for lodging and directly affect the LQ in Kandy city.	Geology map	SDOSL	The suitability of construction was reclassified into a geology map.
Green 01 (Kandian Home garden [35,36] which is known as secondary forest in Sri Lanka)	Kandyan Home Gardens (KHG), which is known as Kandyan Forest Gardens (KFG) in Sri Lanka denotes a traditional system of wealth cropping that has been in practice for several centuries. It refers to mixed settlements with a variety of crops from economically valuable groups (spices, fruits, medicinal plants, and timber species), where the climate and edaphic environment support luxurious lodges [35]. All types of forest (restricted and reserved) are excluded from the Kandian Home garden.	LULC Map (Section 2.3.2)	USGS	Proximity analysis was used by applying the buffer distance to the Kandian home gardens.
Green 02 (Forest)	A high-density green environment is important not only for scenic beauty, but also, as a comfortable living environment [6]. Further, rich green environments are reliable in terms of scenic aspects, aesthetic beauty, water recharging, and boosting the natural air regulation procedure to generate a comfortable breathing environment [23,37].			Proximity analysis was conducted by applying the buffer distance to the forest area.
Rainfall	Vegetated land, streams, and channels are sustained by rainfall. However, heavy rainfall can cause several hazards, for example, flooding, damage to buildings and infrastructure, loss of crops and livestock, landslides [31,38], and risk to human life that can threaten the quality of life. Thus, the consideration of rainfall as a factor is important in LQ studies.	Annual average rainfall over the last twenty years (1998–2018)	MDOSL	Rain gauging station (21 stations) data were transformed into spatial data by using an inverse distance weighted (IDW) interpolation technique [17]. Then, this was reclassified into five classes.
Land Surface Temperature (LST)	Kandy, as a tropical city, has short seasonal changes, where the dry period is from January to April. The average daytime ambient temperature ranges from 28 to 32 °C, and the relative daytime humidity is in the range of 63%–83% [5]. High temperatures can negatively affect the ability to lead a healthy life. They may lead to a low LQ [39,40]. Due to the absence of continuous air temperature data measured at ground level, remote sensing based LST was derived [5,8]	Landsat-8 OLI/TIRS data (Section 2.3.3)	USGS	Reclassification by LST based on suitability class

Table 1. *Cont.*

Factors	Purpose and Relevance	Raw Data	Data Sources	Data Processing/Method
Transportation	Transportation facilities are crucial not only for mobility but also for access to resources. From the perspective of quality of life, the level of service is based on the available transport facilities [41]. Additionally, the price of residential land and social status are influenced by transportation facilities. Thus, considering transportation as a one-factor in LQ research is important.	Roads and railway	SDOSL	Proximity analysis was used by applying buffer distances, as noted in Table 2
Main city	When it comes to choosing a place to lodge, everyone has their priorities and subjective tastes. Still, there are specific attributes, among them, the distance to the main city. This is the principal factor because the main city provides many services including those related to livelihood and infrastructure [42]. Additionally, living in the city is prestigious to residents and is an indicator of social class. Hence, the distance from the main city is an important factor for measuring LQ.	Kandy City		
Healthcare	Healthcare is an essential factor in the quality of life, and the availability of hospital facilities can reduce life risk by accidental damage [9]. Both private and public hospitals were considered.	Teaching and regional hospitals		
Education	Schools are education hubs that provide not only subject knowledge but also improvement of attitude [9].	Main schools		
Safety	The safety factor was selected by considering police stations because they provide national security at the regional and local levels [9].	All police stations		
Religious places *	Engaging with religious activities brings spiritual power, which is vital for mental health to build a perfect life [43]. As noted in Section 2.1, Kandy is a famous cultural city because it is home to the temple of the Tooth Relic (Sri Dalada Maligawa). Additionally, it is ethnically diverse due to the historical religious places located not only in the city area but also in its surrounding area.	Buddhist temples, churches, and mosques		
Archaeological sites	The ancient monuments and archaeological sites create a pleasant settlement environment with scenic beauty. Additionally, they provide value due to satisfaction from preserving a historic environment; in other words, they maintain the natural heritage or cultural heritage for future generations [44]. People love to live close to archaeological sites. Land and settlement selling companies also mainly emphasize archaeological sites as a propagation technique.	Archaeological and heritage sites		

LQ: life quality, SDOSL = Survey Department of Sri Lanka, MDOSL = Meteorology Department of Sri Lanka, USGS = United States Geological Survey. * Because of different types of ethnic groups, all types of religious places were accounted.

2.5. Determination of Threshold Using the LQI

As explained in Section 2.4, information from the experts and some similar research was used to determine the threshold for each factor [9,12]. In the process, the LQ was determined by assigning five classes from the highest to the lowest life quality, as shown in Table 2. The most suitable living areas, which consisted of most of the resources for both environmental and socioeconomic aspects, was designated as the highest LQ zone (Z1). The second zone, which contained resources required for quality life but not all resources like Z1, was named the high LQ zone (Z2). The moderately habitable zone, which consisted of fundamental living factors, was denoted the moderate LQ zone (Z3), while areas with some limitations for livability were classified as the low LQ zone (Z4). The areas with more limitations and fewer resources were categorized as the lowest LQ zone (Z5). Subsequently, the reclassification method or buffer zones of multiple radii were used on each factor to make the five classes listed above (Table 2). This is the most commonly applied technique in spatial multi-attribute decision making research [4,9,12].

2.6. Pairwise Comparison and Weights Calculation

Logical decisions result from the comprehensive analysis of sets of criteria and alternatives because many factors may have influenced one decision. Hence, identification of which factors are the most important and the construction of rank order with priorities is quite a challenging task. The experts carried out the AHP by pairing two factors according to their relative preferences on a scale of 1–9 [45] by answering the question, "By how many times is the selected factor more significant/better than the other one of the pair" [46,47]. In the process, the following steps were carried out:

First: Pairwise comparison questions were prepared by using the criteria and factors presented in Figure 2 and Table 1.

Second: A group of experts with comprehensive knowledge about factors in different fields was carefully selected to avoid factor bias due to their interests (the details of the experts are mentioned in Section 2.4)

Third: Formal interviews were conducted with 52 experts to complete the AHP questions. Then, fault-free questionnaires were rectified and lined up for the next step.

Fourth: Based on the answers given by the experts, pairwise comparison matrices were prepared for criteria, sub-criteria, and factors.

Fifth: The consistency ratio was calculated, and the prioritization of factors was then conducted.

Consistency of expert judgment is a vital attribute of an AHP to ensure the reliability of the results. Hence, in the AHP framework, the consistency ratio (CR) was calculated to ensure the overall consistency of judgments. According to Thomas L. Saaty (2008) [48], the CR value uses a priority vector, which could be 0.10 (10%) or less (≤ 0.1) to present a consistency of preferences. If the CR is higher than 10%, judgments should be revised subjectively. In this study, the CR was computed using Equation (5) [49,50].

$$CR = \frac{CI}{RI} \tag{5}$$

where CR is the consistency ratio, RI is a random index introduced by Saaty in 1980 [45], and CI is the consistency index that represents a value of departure from consistency. The CI index was calculated using Equation (6) [50].

$$CI = \frac{\lambda_{max} - n}{n - 1} \tag{6}$$

where λ_{max} is the largest eigenvalue in matrix A, and n is the dimension of the matrix.

The row geometric mean method (RGMM), which is one of the most popular methods in AHP and MCDM research [50], was used to calculate priority weights, as explained by Equation (7). The priority weights of factors were multiplied by the weights of the criteria and sub-criteria within the same hierarchical level. Then, all factors were normalized and ranked, as summarized in Table 3 [49,50].

$$P_g\left(A_j\right) = \prod_{i=1}^{n} P_i\left(A_j\right)^{w_i} \tag{7}$$

where $P_g(A_j)$ presents the final comparison of criterion j, $P_i(A_j)$ is individual judgement i of criterion j, and W_i is the weight of individual judgement I. $\sum_{i=1}^{n} w_i = 1$ and n is the number of experts [51].

2.7. Computation of LQI with MCDM

The derived weights were multiplied by their respective factors, and all factors were merged into a single layer to compute the LQI in Kandy City, as per Equation (8).

$$LQI = \sum_{i=1}^{n=13} x_i w_i \tag{8}$$

where *LQI* is the life quality index, xi is factor i, and wi is the weight of factor i.

2.8. Spatial Analysis with Gradient Zones

The spatial variation in the quality of life from the city midpoint to the peripheral area was computed by the gradient analysis method. In the process, a sequence of buffers was created around the city center with a 1 km distance interval. It covered 10 buffer zones. Subsequently, the land fraction of each buffer zone was calculated by LQI classes. Finally, the spatial distribution of the the life quality and some statistical analyses were performed.

Table 2. Factor threshold and suitability classes.

LQI	Highest Livable Zone	High Livable Zone	Moderate Livable Zone	Low Livable Zone	Lowest Livable Zone
Factor threshold classes	Z1	Z2	Z3	Z4	Z5
Slope [52]	<5°	5°–11°	11°–17°	17°–31°	>31°
Geology [53]	Charnockitic Gneiss, Garnet Sillimanite Biotite Gneiss, Granite Gneiss, Undifferentiated Charnockitic Biotite Gneiss	Biotite Hornblende Garnet Gneiss	Calc Gneisses and/or granulites, Hornblende biotite migmatites	Marble	Impure quartzite and quartz schists, Quartzite
Green 1 [6,37]	Inside G1 (excluding restricted area)	<0.25 km	0.25–0.5 km	0.5–0.75 km	>0.75 km and all restricted area as explained in Table 1
Green 2 [6,37]	0.5 km	0.5–1.0 km	1.0–1.5 km	1.5–2.0 km	>2 km
Rainfall (annual) [38,54,55]	1500–1800 mm	1800–2100 mm	2100–2300 mm	2300–2600 mm	>2300 mm and < 1500 mm
LST [39,56]	<22 °C	22–24 °C	24–26 °C	26–28 °C	>28 °C
Transportation [9]	<0.5 km	0.5–1.0 km	1.0–1.5 km	1.5–2.0 km	>2 km
Main city	<2.5 km	2.5–5 km	5–7.5 km	7.5–10 km	>10 km
Healthcare [9]	<1 km	1–2 km	2–3 km	3–4 km	>4 km
Education [9]	<0.5 km	0.5–1.0 km	1.0–1.5 km	1.5–2.0 km	>2 km
Safety [9]	<1 km	1–2 km	2–3 km	3–4 km	>4 km
Religious places	<2 km	2–4 km	4–6 km	6–8 km	>8 km
Archaeological sites	<2 km	2–4 km	4–6 km	6–8 km	>8 km

3. Results

3.1. Determination of Weight and Weight Prioritization

Weights were calculated for criteria, sub-criteria, and factors using pairwise comparison matrices, as explained in Section 2.6. Subsequently, the ranks of the factors were computed to determine the importance of factors. The descriptive statistics of the calculation are summarized in Table 3. As a result, socioeconomic criteria obtained approximately double the weight of environmental criteria. From the sub-criteria, the largest weight (0.669) was observed for proximity, while the lowest was presented for the climate as 0.055. The results of the priority weights show that the highest value is presented for the transportation system, while the third priority was observed from the main city. One of the highlighted results is that both factors (main city and transportation) are categorized under the sub-criteria of proximity. Moreover, these results emphasize that proximity to transportation and the main city are the key factors in LQ in Kandy City. Transportation, main city, and school were the socioeconomic criteria represented in the top five factors, while the other two factors (slope and green 1) were environmental criteria, as shown in Table 3.

Table 3. Weights of criteria, sub-criteria, and factors concerning the priority weights and ranks.

Criteria		Sub-Criteria		Factors		Priority Weight	Rank
Category	Weight	Category	Weight	Category	Weight		
Environmental	0.333	Topography	0.655	Slope	0.833	0.18	2
				Geology	0.167	0.04	8
		Live with green	0.29	Green 1	0.750	0.07	5
				Green 2	0.250	0.02	9
		Climate	0.055	Rainfall	0.667	0.01	11
				LST	0.333	0.01	13
Socioeconomic	0.667	Proximity	0.669	Transportation	0.750	0.33	1
				Main city	0.250	0.11	3
		Public services	0.243	Healthcare	0.258	0.04	7
				Education	0.637	0.10	4
				Safety	0.105	0.02	10
		Cultural and religious	0.088	Religious places	0.833	0.05	6
				Archaeological sites	0.167	0.01	12

3.2. Spatial Distribution Pattern of Factors

The spatial distribution patterns of each factor are illustrated in Figure 3, while the descriptive statistics are tabulated in Table 4. High slope areas are dispersed in the southern part of the study area, while a low elevation area is observed in the north of the study area, as shown in Figure 3a. The Z1 and Z2 slope classes represent 68% (Table 4b) of the slope factor. There are 10 geology categories scattered over the study area. By considering the suitability for living (residential and commercial), a geology map was categorized into five classes. Among them, Z1 and Z2 jointly contributed to 172.9 km^2 (43.2%).

As illustrated in Figure 3c, more than 90% of the green 1 area comes under Z1 and Z2. Moreover, both restricted and reserved forest areas dispersed in the middle and northeast sides (Figure 3c) are considered to be Z5 areas due to the restriction to residences and construction. However, lodging in surrounding forest areas is associated with long livability due to the power of forests for improving people's mental and physical health. Buffer zones for forest areas were made, as shown in Figure 3d. For the green 1 fraction, 45.5% and 47.3% of the area was classified as Z1 and Z2, respectively. Out of 13 factors, green 2 is the only factor that has a higher portion of land (65.8%/263 km^2) classified into Z5, as shown in Table 4. Less rainfall can lead to a dry environment, which is not convenient for living, while more rainfall may cause hazards. Hence, areas with both lower and higher levels of rainfall are considered to be in Z5, although only 6.6% of the classification was due to the rainfall factor. The rest of the rainfall area was categorized into four classes (Figure 3e). A high-temperature zone was observed in the city's core area and some parts of the periphery, as shown in Figure 3f. A raised temperature in an urban zone may affect the wellbeing of urban dwellers and can affect human health [23]. Thus, these areas were considered as not comfortable for living and were rated as lower (Z5) or low livable (Z4) zones, representing 12.1% (48.3 km^2) of LST classes.

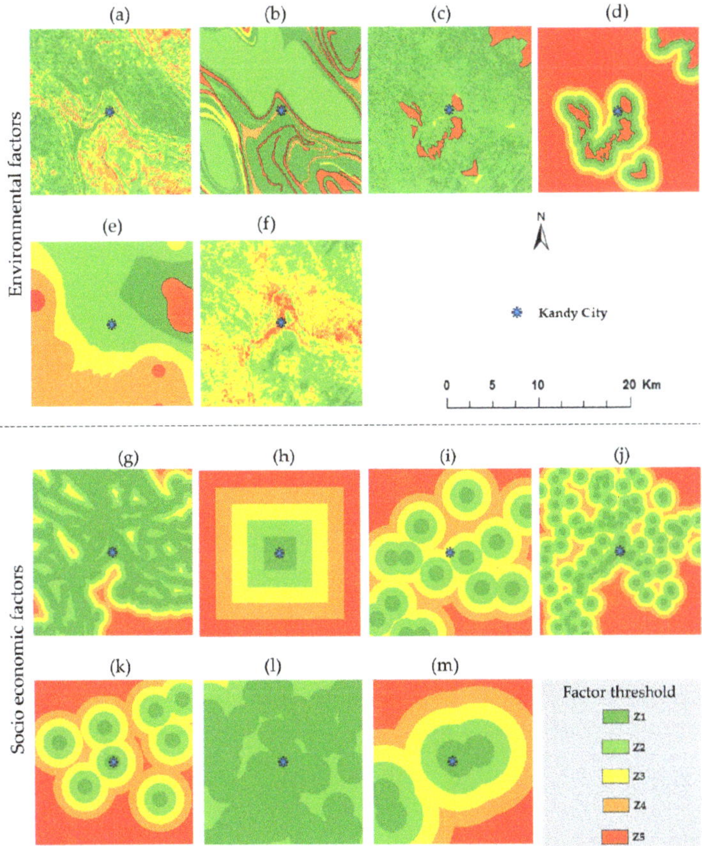

Figure 3. Factor thresholds: (**a**) slope, (**b**) geology, (**c**) green 1, (**d**) green 2, (**e**) rainfall, (**f**) LST, (**g**) transportation, (**h**) main city, (**i**) healthcare, (**j**) education, (**k**) safety, (**l**) religious places, and (**m**) archaeological sites.

Table 4. Descriptive statistics of life quality (LQ) classes by factors.

a. Area (km^2) by factors

LQI classes	Slope	Geology	Green 1	Green 2	Rainfall	LST	Transportation	Main city	Healthcare	Education	Safety	Religious places	Archaeological sites
Z1	127.8	172.9	182.2	34.5	49.0	7.3	229.3	16.0	41.2	67.3	24.7	332.8	47.1
Z2	144.1	156.0	189.1	33.5	133.2	151.6	92.8	48.0	105.9	130.5	73.6	65.9	84.0
Z3	72.8	12.4	1.9	35.2	63.1	192.7	35.4	80.0	114.1	86.5	96.2	1.4	97.2
Z4	53.4	20.7	0.0	33.7	128.5	41.1	17.4	112.0	70.4	40.3	82.3	0.0	83.8
Z5	1.9	38.0	26.9	263.0	26.2	7.2	25.1	144.0	68.4	75.4	123.2	0.0	87.8
Total	400.0	400.0	400.0	400.0	400.0	400.0	400.0	400.0	400.0	400.0	400.0	400.0	400.0

b. Area percentage by factors

LQI classes	Slope	Geology	Green 1	Green 2	Rainfall	LST	Transportation	Main city	Healthcare	Education	Safety	Religious places	Archaeological sites
Z1	31.9	43.2	45.5	8.6	12.2	1.8	57.3	4.0	10.3	16.8	6.2	83.2	11.8
Z2	36.0	39.0	47.3	8.4	33.3	37.9	23.2	12.0	26.5	32.6	18.4	16.5	21.0
Z3	18.2	3.1	0.5	8.8	15.8	48.2	8.8	20.0	28.5	21.6	24.1	0.3	24.3
Z4	13.3	5.2	0.0	8.4	32.1	10.3	4.4	28.0	17.6	10.1	20.6	0.0	21.0
Z5	0.5	9.5	6.7	65.8	6.6	1.8	6.3	36.0	17.1	18.9	30.8	0.0	22.0
Total	100	100	100	100	100	100	100	100	100	100	100	100	100

The second higher portion (57.3%) of Z1 as influenced by transportation (Figure 3g). However, access to transportation facilities was much less in the southeast part of the study area because of the mountainous topography. The classification of the main city factor gradually increased from Z1 to Z2, as illustrated by Figure 3h. When all thirteen factors were compared, the lowest land portion of Z1 was attributed to the main city factor 4% of the total area. More than 50% of the total area of healthcare was represented by Z2 and Z3, and the same scenario was also observed for the education factor, as shown in Table 4b. The south and southeast parts of the study area were mostly not covered by a safety factor, while more than 50% of places classified as having this factor belonged to Z4 and Z5, suggested that safety is currently unsatisfactory. A total of 83.2% (332.8 km^2), which is the highest land portion, of areas belonging to Z1 are represented by religious places, as revealed in Table 4b. Further, it is the only factor that is absent of any land fraction in Z4 or Z5, as illustrated in Figure 3l.

3.3. LQI in Kandy City

MCDM and the AHP were used to produce a weighted life quality index for Kandy City. Many researchers have theoretically and practically stated that the range of numerical values generated by MCDM has no scientific sense other than to explain the research area in relative terms. The values of LQI range from 1.06 to 4.56 throughout the study area. There are many classification methods powered by various kinds of software and applications. Among them, we tested six methods powered by Arc GIS version 10.6, and results were crosschecked with field observations. In addition, we discussed the results with experts. Finally, as a result of a lot of trial and error, the natural break method, which is the most frequently applied technique [12,57,58], was adopted in our research to make an LQI zonation map. Its spatial distribution is shown in Figure 4a, while the summarized results of the area and area percentage of LQI in Kandy City are presented in Figure 4b.

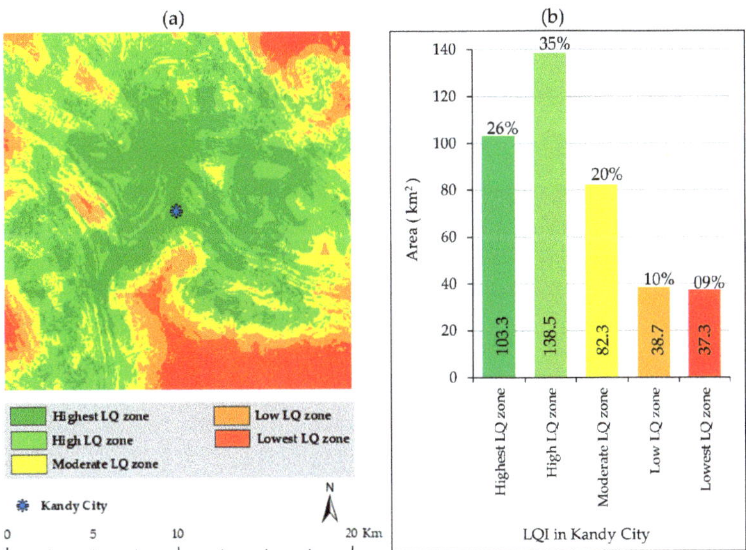

Figure 4. (**a**) Distribution of the life quality index (LQI) and (**b**) descriptive statistics of LQ index (LQI) of Kandy City.

The results show that the highest LQ zone covers 103.3 km^2, which is 26% of the total study area, as illustrated in Figure 4b. As one of the highlighted results, the highest LQ zone is somewhat skewed to the north side of the study area. A scatter distribution pattern presents a high LQ zone which is covered with the highest land fraction of 138.5 km^2 (35%). Moreover, the highest and high LQ zones jointly make up 61% or 241.8 km^2 of the land fraction, as shown in Figure 4b. The moderate, low, and lowest LQ zones cover 20%, 10%, and 9%, respectively. The lowest LQ zone is mainly found in the southeast and northeast of the study area, as illustrated by Figure 4a.

3.4. Distribution of Life Quality from the City Center

A gradient zone analysis was carried out to get a deeper understanding of the spatial distribution of life quality from the city's midpoint to the border area. LQI zones with gradients are presented in Figure 5a, while each gradient zone (from GZ_1 to GZ_{10}) is shown in the LQI zones in Figure 5b. The area of the highest LQ zone gradually declines from the city midpoint to the border area, while the low and lowest LQ zones have the opposite trend, as shown in Figure 5b. The areas of the other two zones (high and moderate IQ zones) also decline when moving away from the city center, but slight fluctuations were noticed. More than 80% of the first GZ is represented by the highest LQ zone, and the rest is covered by high and moderate LQ zones. The other two LQ zones are not represented by GZ_1, as shown in Figure 5b. At least 80% of the total extended area of GZ_2 to GZ_5 is covered by the highest and high LQ zones, and the trend gradually declines so that the last zone (GZ_{10}) covers only 33.5%.

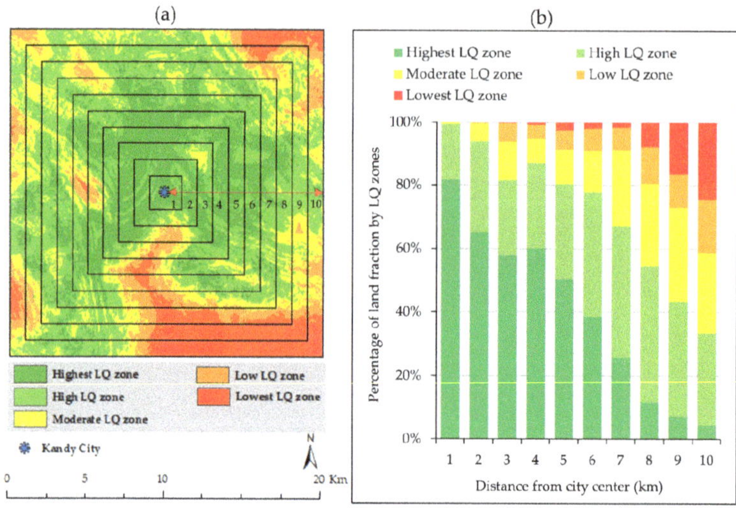

Figure 5. (**a**) Spatial distribution of the LQI with gradient zones from the city center to the peripheral area, and (**b**) area proportion by LQ zone in each gradient zone (from GZ1 to GZ10).

4. Discussion

4.1. Urbanization and Its Effect on Life Quality in Kandy City

Kandy City is a fast-growing city and essential tourist destination [18] located in the central mountain area, and it is wealthy with well-being attributes (social, economic, and environmental), which brings high livability. Additionally, it is the main city of the central province and is one of the important economic hubs in Sri Lanka. Thus, conducting this type of research is a timely vital task to identify the level of life quality of residents. In the process, we first identified the importance of Kandy City as a commercial and living place by exploring past research [19,20] and development plans [59]. However, research related to the quality of life or livability of the city was still lacking, and this was the motivation factor for selecting this research title. Both primary and secondary data were gathered from reliable and trustworthy data sources, and methodology including AHP and MCDM was adopted. A group of experts was selected in order to cover all factors, and factor bias was avoided. Finally, a weighted LQI with five zones was made. Throughout the whole process, especially in field activities, our research team took responsibility for minimizing liveware faults as much as possible.

The results of the AHP show that, among the criteria (level 1), socioeconomic factors obtained a higher weight (0.667) than environmental factors. These results indicate that socioeconomic factors influence the life quality in Kandy City more strongly than environmental factors. Proximity, which was categorized as a socioeconomic factor, obtained the highest weight of all sub-criteria (level 2). Furthermore, the highest weight (0.33) from all 13 factors was transportation (level 3). In 2019, Dissanayake et al. [5] stated that Kandy City has shown a linear city development pattern in the last twenty years. Our LULC classification results also proved this linear development pattern. Figure A1 shows that impervious surfaces are mainly distributed around the road network. We also observed this pattern through our field visits. These results indicate that the factor triggering life quality in Kandy City is the transportation network. Primarily, new settlements have been constructed in close proximity to the roads. New apartments and land fragmentation projects for new settlements have taken place around the main and minor roads. Our results and those from secondary sources reflected the ground-level scenario in the Kandy City area.

While we conducted field observations, some qualitative information related to the quality of life was also collected from residents using informal interviews. Most respondents who resided in the peripheral area stated that they wish to settle close to Kandy City due to the proximity to resources and infrastructure. The group of experts also gave the same idea. Additionally, the results of the life quality GZs show that the highest LQ zone is also distributed around the main city (Figure 5). It is approximately 5 km from the city center. By observing these facts, we can see that highly livable areas or high life quality areas are grounded by surrounding the central city area. Hence, respective government and private companies should have a plan to provide more lodging facilities surrounding the main city area, or they could plan to decentralize the resources to other areas.

4.2. The Implication of Our Results for Enhancing the Life Quality of Urban Life

Except for the green factors (forest and Kandian home garden), physical factors (geology and rainfall) are governed by nature. We cannot change these factors according to our preferences. Hence, human activities should be based on physical factors by changing socioeconomic attributes in order to reach a higher quality of life. On the other hand, we must adapt to nature by making changes to natural factors to promote a high quality of life. As an example, a green environment can be enhanced through the green belt and other applications identified by past researchers, even in Kandy City [5]. The green environment brings several benefits, and it is a suitable solution for controlling the negative effects of LST [18]. Higher LST significantly reduces the quality of life. Thus, enhancing the green environment is an essential, timely task, and green 1 was ranked in the top five most important factors, as shown in Table 3.

An area of 158.2 km^2 or 39% of the study area was classified into the lowest three LQ zones (moderate, low, and lowest), and most of these areas were located at least 7 km away from the city center (Figure 5b), especially in the northeast and southeast, as shown in Figure 4a. Further, these areas lack public services including healthcare (Figure 3i), education (Figure 3j), and transportation (Figure 3g). However, these services are fundamentally required for enhancing life quality. Hence, respective administrative bodies should take responsibility for enhancing these public services in order to sustain the quality of life in peripheral areas.

As mentioned above, Kandy City is a fast-growing city. It has a high demand for land for both new investments and settlements. Mainly, the demand surrounds the main city area. Thus, identification of the most livable areas is essential for people who are expecting to buy land in Kandy City and respective administrative bodies (government and private) who provide services for the new settlements. We believe that the results of our research can be used as proxy indicators to generate policies regarding life quality planning and future development scenarios. Regarding the implications of the research results, future city plans should be life quality-oriented, aligning with "goal 11" of the sustainable development goals (SDGs) [60].

As we have detailed, discussed, and illustrated, quality of life depends on both socioeconomic and environmental factors. Some of the factors are more important, while some of the factors are less important. However, we only investigated 13 factors that are applicable for spatial analysis. In this regard, methods applied by past researchers and experts' ideas were mainly considered. Due to the difficulty of collecting data and inapplicability to spatial analysis, human emotion factors or other preferences influenced by social caste or ethnic groups were not accounted. There may be more than 13 factors that can directly and indirectly influence the quality of life, but available data sources were limited, and some of the data sources could not be trusted. We omitted these data sources from this research. Thus, the results may be interpreted by considering these limitations.

5. Conclusions

This study attempted to develop a life quality index for Kandy City, Sri Lanka using a geospatial approach with AHP and MCDM. The decomposition of factors and construction of a hierarchical order were carried out using information from experts and past research. The AHP approach was performed through pairwise comparisons and derivation of the weight of each factor. MCDM was used to resolve the complex decision environment generated by multiple factors. Socioeconomic criteria were more found to be more necessary than environmental criteria, while proximity was the most sensitive among all sub-criteria. The transportation facility was the most significant factor and presented the highest rank on the LQI in Kandy City. The highest LQ zone was shown to surround the main city, and this was proven by GZ analysis also. The results imply that urban pressure and overpopulation can be controlled by improving socioeconomic facilities in the peripheral and rural areas. Due to the socioeconomic resource availability, people attempt to live in urban areas or close to urban areas. We conclude that the overall findings of this research can be used as guidance for the sustainability of Kandy City. Selected factors will not be a one-size-fits-all approach for any city, but the LQI model is flexible for any calibration. Hence, the constructed model can be applied to other cities to assess their life quality.

Author Contributions: The corresponding author, D.D., proposed the topic and spearheaded the data processing and analysis, as well as the writing of the manuscript. T.M., Y.M., M.R., and E.P. helped in the design, research implementation and analysis, and writing of the manuscript. All authors have read and agreed to the published version of the manuscript.

Funding: This study was supported by the Japan Society for the Promotion of Science (JSPS) through Grant-in-Aid for Scientific Research (B) 18H00763 (2018-20) and it is partly supported by the University of Tsukuba through the Data Bank Project of the Division of Policy and Planning Science Commons.

Acknowledgments: The authors are grateful to the anonymous reviewers for their helpful comments and suggestions to improve the quality of this paper.

Conflicts of Interest: The authors declare no conflict of interest.

Appendix A

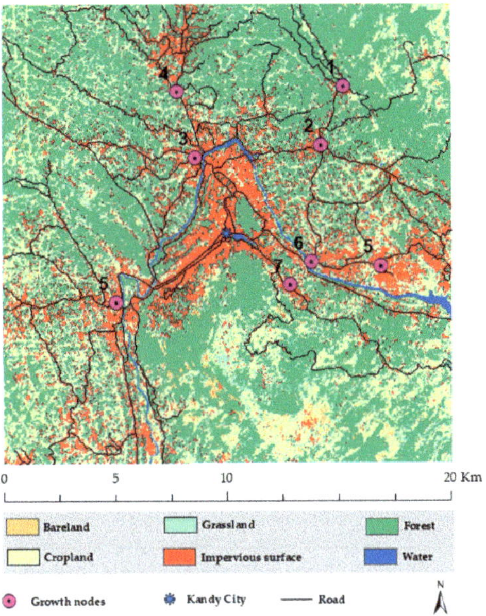

Figure A1. Land use and land cover information in 2018 in Kandy City.

References

1. United Nations Department for Economic and Social Affairs. *World Population Prospects 2019: Highlights*; United Nations Department for Economic and Social Affairs: New York, NY, USA, 2019.
2. United Nations Department for Economic and Social Affairs. *The World's Cities in 2018*; United Nations: New York, NY, USA, 2018.
3. United Nations Department of Economic and Social Affairs. *World Urbanization Prospects: The 2014 Revision, Highlights*; United Nations: New York, NY, USA, 2015.
4. Psatha, E.; Deffner, A.; Psycharis, Y. Defining the quality of urban life: Which factors should be considered? In Proceedings of the 51st ERSA Congress European Regional Science Association, Barcelona, Spain, 30 August–2 September 2011; Volume 19.
5. Dissanayake, D.; Morimoto, T.; Ranagalage, M.; Murayama, Y. Land-use/land-cover changes and their impact on surface urban heat islands: Case study of Kandy City, Sri Lanka. *Climate* **2019**, *7*, 99. [CrossRef]
6. Maas, J.; Spreeuwenberg, P.; Van Winsum-Westra, M.; Verheij, R.A.; de Vries, S.; Groenewegen, P.P. Is green space in the living environment associated with people's feelings of social safety? *Environ. Plan. A* **2009**, *41*, 1763–1777. [CrossRef]
7. Ranagalage, M.; Wang, R.; Gunarathna, M.H.J.P.; Dissanayake, D.; Murayama, Y.; Simwanda, M. Spatial forecasting of the landscape in rapidly urbanizing hill stations of south Asia: A case study of Nuwara Eliya, Sri Lanka (1996–2037). *Remote Sens.* **2019**, *11*, 1743. [CrossRef]
8. Ranagalage, M.; Murayama, Y.; Dissanayake, D.; Simwanda, M. The impacts of landscape changes on annual mean land surface temperature in the tropical mountain city of Sri Lanka: A case study of Nuwara Eliya (1996–2017). *Sustainability* **2019**, *11*, 5517. [CrossRef]
9. Onnom, W.; Tripathi, N.; Nitivattananon, V.; Ninsawat, S. Development of a liveable city index (Lci) using multi criteria geospatial modelling for medium class cities in developing countries. *Sustainability* **2018**, *10*, 520. [CrossRef]

10. Feneri, A.M.; Vagiona, D.; Karanikolas, N. Multi-criteria decision making to measure quality of life: An integrated approach for implementation in the urban area of Thessaloniki, Greece. *Appl. Res. Qual. Life* **2015**, *10*, 573–587. [CrossRef]

11. Zarghami, E.; Sharghi, A.; Olfat, M.; Salehi Kousalari, F. Using multi-criteria decision-making method (MCDM) to study quality of life variables in the design of senior residences in Iran. *Ageing Int.* **2018**, *43*, 279–296. [CrossRef]

12. Pawattana, C.; Tripathi, N.K. Analytical Hierarchical Process (AHP)—Based flood water retention planning in Thailand. *GISci. Remote Sens.* **2008**, *45*, 343–355. [CrossRef]

13. Kaklauskas, A.; Zavadskas, E.K.; Radzeviciene, A.; Ubarte, I.; Podviezko, A.; Podvezko, V.; Kuzminske, A.; Banaitis, A.; Binkyte, A.; Bucinskas, V. Quality of city life multiple criteria analysis. *Cities* **2018**, *72*, 82–93. [CrossRef]

14. Saaty, T.L.; Luis, G. *Vargas. Models, Methods, Concepts & Applications of the Analytic Hierarchy Process*; Springer: Berlin/Heidelberg, Germany, 2012; Volume 175.

15. Chandio, I.A.; Matori, A.N.B.; WanYusof, K.B.; Talpur, M.A.H.; Balogun, A.L.; Lawal, D.U. GIS-based analytic hierarchy process as a multicriteria decision analysis instrument: A review. *Arab. J. Geosci.* **2013**, *6*, 3059–3066. [CrossRef]

16. Deswal, M.; Laura, J.S. GIS based modeling using analytic hierarchy process (AHP) for optimization of landfill site selection of Rohtak city, Haryana (India). *J. Appl. Nat. Sci.* **2018**, *10*, 633–642. [CrossRef]

17. Dissanayake, D.; Morimoto, T.; Ranagalage, M. Accessing the soil erosion rate based on RUSLE model for sustainable land use management: A case study of the Kotmale watershed, Sri Lanka. *Model. Earth Syst. Environ.* **2018**, *5*, 291–306. [CrossRef]

18. Ranagalage, M.; Dissanayake, D.; Murayama, Y.; Zhang, X.; Estoque, R.C.; Perera, E.; Morimoto, T. Quantifying surface urban heat island formation in the world heritage tropical mountain city of Sri Lanka. *ISPRS Int. J. Geo-Inf.* **2018**, *7*, 341. [CrossRef]

19. Weerasundara, L.; Amarasekara, R.W.K.; Magana-Arachchi, D.N.; Ziyath, A.M.; Karunaratne, D.G.G.P.; Goonetilleke, A.; Vithanage, M. Microorganisms and heavy metals associated with atmospheric deposition in a congested urban environment of a developing country: Sri Lanka. *Sci. Total Environ.* **2017**, *584*, 803–812. [CrossRef] [PubMed]

20. Weerasundara, L.; Magana-Arachchi, D.N.; Ziyath, A.M.; Goonetilleke, A.; Vithanage, M. Health risk assessment of heavy metals in atmospheric deposition in a congested city environment in a developing country: Kandy City, Sri Lanka. *J. Environ. Manag.* **2018**, *220*, 198–206. [CrossRef] [PubMed]

21. Google Erath Engine. Image Collection Reductions. Available online: https://developers.google.com/earthengine/reducers_image_collection (accessed on 15 January 2020).

22. Weng, Q. Land use change analysis in the Zhujiang Delta of China using satellite remote sensing, GIS and stochastic modelling. *J. Environ. Manag.* **2002**, *64*, 273–284. [CrossRef]

23. Dissanayake, D.; Morimoto, T.; Murayama, Y.; Ranagalage, M. Impact of landscape structure on the variation of land surface temperature in sub-saharan region: A case study of Addis Ababa using Landsat data (1986–2016). *Sustainability* **2019**, *11*, 2257. [CrossRef]

24. Phiri, D.; Morgenroth, J. Developments in Landsat land cover classification methods: A review. *Remote Sens.* **2017**, *9*, 967. [CrossRef]

25. Erkan, U.; Gökrem, L. A new method based on pixel density in salt and pepper noise removal. *Turkish J. Electr. Eng. Comput. Sci.* **2018**, *26*, 162–171. [CrossRef]

26. Thapa, R.; Murayama, Y. Image classification techniques in mapping urban landscape: A case study of Tsukuba city using AVNIR-2 sensor data. *Tsukuba Geoenviron. Sci.* **2007**, *3*, 3–10.

27. Sakthidasan, K.; Sankaran, K.; Velmurugan Nagappan, N. Noise free image restoration using hybrid filter with adaptive genetic algorithm. *Comput. Electr. Eng.* **2016**, *54*, 382–392. [CrossRef]

28. Sobrino, J.A.; Jiménez-Muñoz, J.C.; Paolini, L. Land surface temperature retrieval from LANDSAT TM 5. *Remote Sens. Environ.* **2004**, *90*, 434–440. [CrossRef]

29. Yulianto, T.; Suripin, S.; Purnaweni, H. Zoning landslide vulnerable area according to geological structure, slopes, and landuse parameters in Trangkil Sukorejo Gunungpati Semarang City's Residental Area. In Proceedings of the 8th International Seminar on New Paradigm and Innovation on Natural Science and Its Application, Semarang, Indonesia, 26 September 2018; Volume 1227.

30. Thinh, N.X.; Vogel, R. Application of the analytic hierarchy process in the multiple criteria decision analysis of retention areas for flood risk management. *Environ. Inform. Syst. Res.* **2007**, *2007*, 675–682.

31. Perera, E.N.C.; Jayawardana, D.T.; Jayasinghe, P.; Ranagalage, M. Landslide vulnerability assessment based on entropy method: A case study from Kegalle district, Sri Lanka. *Model. Earth Syst. Environ.* **2019**, *5*, 1635–1649. [CrossRef]

32. Prakash, S.; Singh, S.K.; Jain, P.A.; Dwivedi, N.; Dwivedi, S.; Mishra, V.K.; Pundir, V.S. Importance of geology in construction and prevent the hazards. *Libr. Adv. Appl. Sci.* **2015**, *6*, 75–80.

33. Bell, F.G. *Engineering Geology and Construction*; CRC Press: Boca Raton, FL, USA, 2004.

34. Ranagalage, M. Landslide Hazards Assessment in Nuwara Eliya District in Sri Lanka. In Proceedings of the Japanese Geographical Meeting, Tsukuba, Japan, 28–30 March 2017; p. 100336.

35. Jacob, V.J.; Alles, W.S. Kandyan gardens of Sri Lanka. *Agrofor. Syst.* **1987**, *5*, 123–137. [CrossRef]

36. Perera, A.H.; Rajapakse, R.M.N. A baseline study of Kandyan forest gardens of Sri Lanka: Structure, composition and utilization. *For. Ecol. Manag.* **1991**, *45*, 269–280. [CrossRef]

37. De Vries, S.; Verheij, R.A.; Groenewegen, P.P.; Spreeuwenberg, P. Natural environments—Healthy environments? An exploratory analysis of the relationship between greenspace and health. *Environ. Plan.* **2003**, *35*, 1717–1731. [CrossRef]

38. Potuhera, O.L.; Primali, V.; Weerasinghe, A. Improving existing landslide hazard zonation map in KMC area, Sri Lanka. *Suan Sunandha Sci. Technol. J.* **2015**, *2*, 24–29.

39. World Health Organization. *Housing, Energy and Thermal Comfort: A Review of 10 Countries within the WHO European Region*; WHO Regional Office for Europe: Copenhagen, Denmark, 2007.

40. Priyankara, P.; Ranagalage, M.; Dissanayake, D.; Morimoto, T.; Murayama, Y. Spatial Process of Surface Urban Heat Island in Rapidly Growing Seoul Metropolitan Area for Sustainable Urban Planning Using Landsat Data (1996–2017). *Climate* **2019**, *7*, 110. [CrossRef]

41. Sze, N.N.; Christensen, K.M. Access to urban transportation system for individuals with disabilities. *IATSS Res.* **2017**, *41*, 66–73. [CrossRef]

42. Serag El Din, H.; Shalaby, A.; Farouh, H.E.; Elariane, S.A. Principles of urban quality of life for a neighborhood. *HBRC J.* **2013**, *9*, 86–92. [CrossRef]

43. Behere, P.B.; Das, A.; Yadav, R.; Behere, A.P. Religion and mental health. *Indian J. Psychiatry* **2013**, *55*, S187. [CrossRef] [PubMed]

44. Sharp, B.; Kerr, G. *Option and Existence Values for the Waitaki Catchment*; Ministry for the Environment Manatū Mō Te Taiao: Wellington, New Zealand, 2005.

45. Ammarapala, V.; Chinda, T.; Pongsayaporn, P.; Ratanachot, W.; Punthutaecha, K.; Janmonta, K. Cross-border shipment route selection utilizing analytic hierarchy process (AHP) method. *Songklanakarin J. Sci. Technol.* **2018**, *40*, 31–37.

46. Bozóki, S.; Fülöp, J.; Poesz, A. On pairwise comparison matrices that can be made consistent by the modification of a few elements. *Cent. Eur. J. Oper. Res.* **2011**, *19*, 157–175. [CrossRef]

47. Omamalin, B.N.; Canoy, S.R.; Rara, H.M. Differentiating total dominating sets in the join, corona and composition of graphs. *Int. J. Math. Anal.* **2014**, *8*, 1275–1284. [CrossRef]

48. Saaty, T.L. Decision making with the analytic hierarchy process. *Int. J. Serv. Sci.* **2008**, *1*, 83. [CrossRef]

49. Aguarón, J.; Teresa Escobar, M.; Moreno-Jiménez, J.; Turón, A. AHP-group decision making based on consistency. *Mathematics* **2019**, *7*, 242. [CrossRef]

50. Escobar, M.T.; Aguarón, J.; Moreno-Jiménez, J.M. A note on AHP group consistency for the row geometric mean priorization procedure. *Eur. J. Oper. Res.* **2004**, *153*, 318–322. [CrossRef]

51. Forman, E.; Peniwati, K. Aggregating individual judgments and priorities with the Analytic Hierarchy Process. *Eur. J. Oper. Res.* **1998**, *108*, 165–169. [CrossRef]

52. National Building Research Organization (NBRO). *Hazard Resilient Housing Construction Manual*; National Building Research Organisation: Colombo, Sri Lanka, 2015.

53. Kapila, D. *Handbook on Geology and Mineral Resources of Sri Lanka: A Collection of Authoritative State-of-the-Art Papers by Specialists on Some Geological and Mineralogical Aspects of Sri Lanka*; Geological Survey and Mines Bureau (GSMB): Kotte, Sri Lanka, 1995.

54. Disaster Management Centre. *Sri Lanka Post-Disaster Needs Assessment*; Ministry of Disaster Management: Colombo, Sri Lanka, 2016.

55. Japan International Cooperation Agency. *The Data Collection Survey on Road Protection against Natural Disaster (Landslide-Disaster)*; Japan Conservation Engineers & Co., Ltd.: Saitama, Japan, 2012.
56. Näyhä, S.; Rintamäki, H.; Donaldson, G.; Hassi, J.; Jousilahti, P.; Laatikainen, T.; Jaakkola, J.J.K.; Ikäheimo, T.M. Heat-related thermal sensation, comfort and symptoms in a northern population. *Eur. J. Public Health* **2014**, *24*, 620–626. [CrossRef] [PubMed]
57. Nazri Che, D.; Abu Hassan, A.; Zulkiflee Abd, L.; Rodziah, I. Application of geographical information system-based analytical hierarchy process as a tool for dengue risk assessment. *Asian Pac. J. Trop. Dis.* **2016**, *6*, 928–935.
58. Thirumalaivasan, D.; Karmegam, M.; Venugopal, K. AHP-DRASTIC: Software for specific aquifer vulnerability assessment using DRASTIC model and GIS. *Environ. Model. Softw.* **2003**, *18*, 645–656. [CrossRef]
59. Ministry of Urban Development, Water Supply and Drainage Sri Lanka. *Kandy City Region Strategic Development Plan—2030*; Urban Development Authority: Kandy, Sri Lanka, 2015.
60. United Nations (UN). Transforming Our World. In *The 2030 Agenda for Sustainable Development*; United Nations General Assembly: New York, NY, USA, 2015; pp. 1–35.

 *sustainability*

Article

Spatiotemporal Analysis of the Nonlinear Negative Relationship between Urbanization and Habitat Quality in Metropolitan Areas

Jingfeng Zhu, Ning Ding, Dehuan Li, Wei Sun, Yujing Xie and Xiangrong Wang *

Department of Environmental Science and Engineering, Fudan University, 2005 Songhu Road, Shanghai 200438, China; 17210740022@fudan.edu.cn (J.Z.); 17210740003@fudan.edu.cn (N.D.); lidehuansnow@163.com (D.L.); 16110740017@fudan.edu.cn (W.S.); xieyj@fudan.edu.cn (Y.X.)
* Correspondence: xrxrwang@fudan.edu.cn; Tel.: +86-21-3124-8988

Received: 24 December 2019; Accepted: 13 January 2020; Published: 16 January 2020

Abstract: Urbanization intensity (UI) affects habitat quality (HQ) by changing land patterns, nutrient conditions, management, etc. Therefore, there is a need for studies on the relationship between UI and HQ and quantification of separate urbanization impacts on HQ. In this study, the relationship between HQ and UI and the direct and indirect impacts of urbanization on HQ were analyzed for the Yangtze River Delta Urban Agglomeration (YRDUA) from 1995 to 2010. The results indicated that the regional relationship between HQ and UI was nonlinear and negative, with inflection points where urbanization reached 20% and 80%. Furthermore, depending on different urbanization impacts, the relationship types generally changed from a steady decrease to stable in different cities. Negative indirect impacts accelerate habitat degradation, while positive impacts partially offset habitat degradation caused by land conversion. The average offset extent was approximately 28.23%, 17.41%, 22.94%, and 16.18% in 1995, 2000, 2005, and 2010, respectively. Moreover, the dependency of urbanization impacts on human demand in different urbanization stages was also demonstrated. The increasing demand for urban land has exacerbated the threat to ecological areas, but awareness about the need to protect ecological conditions began to strengthen after the antagonistic stage of urbanization.

Keywords: urbanization; habitat quality; DMSP-OLS; spatiotemporal analysis; Yangtze River Delta Urban Agglomeration

---

1. Introduction

Since the 20th century, urbanization, including population shifting, urban expansion, economic development and so on, has been one of the most significant characteristics of human civilization [1,2]. The progress of urbanization has led to more human demand that needs to be provided for by the ecosystem services in natural ecosystems [3]. Grimm at al. found that the unprecedented rates of urban population growth over the past century have occurred on less than 3% of the global terrestrial surface, yet the impact has been global, with 78% of carbon emissions, 60% of residential water use, and 76% of wood used for industrial purposes attributed to urban areas [4]. At the same time, urbanization has brought significant land conversion from ecological spaces to urban usage, which creates seminatural or completely artificial ecosystems. In contrast to natural ecosystems, the main driving force of habitat patterns in these ecosystems is anthropogenic activity. Fertilization, irrigation, unified management, species introduction and other alterations are commonly seen in urban landscapes and have caused various impacts on habitats in terms of soil composition and nutrients [5], water, heat and carbon balance [6], species communities [7,8], atmospheric and climatic conditions [9], etc. In other words, the impact on natural ecosystems will affect the city itself. Therefore, quantifying the eco-environmental dynamic variation in the urbanization gradient and identifying the ecological impacts of urbanization provide an efficacious approach for decoupling human well-being from the consumption of natural capital [10].

In the context of urbanization, eco-environmental variation is spatially organized and dominates the composition and structure of habitats from the core urban area to the outskirts [11]. It has been widely believed that urbanization threatens biodiversity, with areas of high urbanization levels having impoverished species community composition [12–14] due to environmental stresses (e.g., poor soil quality, high pollution, and limited habitat) and the artificial conversion of ecological spaces to impervious surfaces [15]. However, the correlation trends are not completely consistent among different studies. The relationships between urbanization and the diversity of plants [16], amphibians [17] and mammals [18] have been separately analyzed, and it was found that there were no negative and even positive effects in urban-rural gradients. In addition, using non-field survey approaches, the positive influence of urbanization on environmental variation has been observed at the city, regional and national scales. For example, Meng et al. found that the overall ecological conditions showed a fluctuating increasing trend in the process of urbanization, with 20% of the area in the Yangtze River Delta Urban Agglomeration (YRDUA) being deteriorated and 40% being improved [19]. Michael et al., upon reviewing 105 studies on the effects of urbanization on the abundance of non-avian species, indicated that urbanization increases the species richness of plants, invertebrates and vertebrates by approximately 65%, 30% and 12%, respectively [20]. Jia et al. observed that approximately 90% of urban areas showed vegetation growth enhancement in the United States [21], which was also observed in 32 major cities across China [22].

The main reason for the inconsistency in environmental dynamics can be summarized in two aspects. First, simplified urbanization gradients, individual-level studies and the selection of study sites contribute to the significant differences in relationship trends. The extent to which the selected urban–rural gradient is representative of the pattern determines the sensitivity of variability in environmental dynamics. As data availability and remote sensing techniques have increased, some finer gradient approaches have been applied in the study of the abundance and community composition of species in natural and urban ecosystems [23,24]. The "Habitat Quality (HQ)" module in the "Integrated Valuation of Environmental Services and Trade-offs" (InVEST-HQ) model suite is a novel tool used for assessing HQ of habitats under anthropogenic threats. It provides a means for conducting biodiversity assessments at different scales, applied in study sites where there are multiple habitat types or a lack of species distribution data [25]. Second, the relationship between environmental variation and urbanization is also largely dependent on the selection of urbanization indicators. Commonly used indicators of urbanization intensity (UI) were urban population conversion, finance aggregation, and residential expansion [26]. However, in different urbanizing areas, the UI of each aspect shows an inconsistent growth rate. In developed countries such as the United States, finance aggregation increased without significant landscape pattern changes. In contrast, in some developing countries, such as China, the residential expansion rate significantly exceeds the speed of urban population conversion, which leads to the difference in responses to diverse UI indicators in identical study sites [27]. For instance, Peng et al. found an "inverse U" shape between ecosystem services and UI characterized by population and economy, while a negative-linear relationship existed when the UI was measured by the proportion of construction land [26]. Therefore, selecting a spatially high-resolution data source for multiple human activities is necessary.

The Defense Meteorological Satellite Program's Operational Linescan System (DMSP-OLS) stable nighttime light (NTL) data were used to detect urban lights and areas with low brightness, such as small settlements and traffic distinguished from dark rural backgrounds. This unique dataset provides a special perspective on the study of urban development and relevant human activities across various spatiotemporal scales, such as urban extent [28], urban expansion [29], urbanization [30], population density [31], socioeconomic activities [32–34] and energy consumption [35]. The spatial resolution of

1 km × 1 km and long-term monitoring periods make it possible to study the details of HQ dynamics in urban-rural gradients through a spatiotemporal perspective [36–38].

Previous studies have mentioned that there was a non-linear relationship between UI and biodiversity [20], ecosystem services supply [39], and vegetation growth [21]. The status of HQ is closely related to the diversity and abundance of species in the ecosystem, and its dynamics also affect the supply capacity of ecosystem services. Assuming a nonlinear relationship between HQ and UI, there must exist at least one inflection point, indicating where the ecological impacts of urbanization will shift from having one influence to another. Predecessors have combined UI with vegetation surface conversion [40], land use transition [41,42], and the distance from anthropogenic threats [43,44] to understand the ecological impacts of urbanization, which has been investigated for multiple HQ aspects, such as biodiversity status [7,45], ecosystem services [39] and landscape patterns [46]. Comparing and summarizing the results of these studies, which are based on field observations and remote sensing approaches, is challenging. The research scales range from vegetation, species, and ecological conditions to landscape types. The quantitative relationship between urbanization indicators is not well understood. Zhao et al. developed a general framework for the quantitative assessment of separate urbanization impacts in urban environments [22]. This framework has been used to assess urbanization impacts on vegetation growth [21] and carbon storage [47], but very few studies have investigated the direct and indirect impacts on urban habitat quality.

Urban agglomerations, consisting of densely populated, highly urbanized areas and underpopulated surrounding townships, have become a dominant form of spatial organization in urban development [48]. As one of the six world-class urban agglomerations, the Yangtze River Delta Urban Agglomeration (YRDUA) has witnessed the evolution of humans and the environment within urban ecosystems in China. The rapid urbanization of YRDUA is accompanied by indigenous landscape changes and habitat degradation [49]. However, the relationship between the HQ and UI in urban agglomerations is not clear. A systematic understanding of the ecological impacts of urbanization in metropolitan areas is necessary. Hence, using the urban–rural gradient of YRDUA as a case study, this study sought to address the following objectives:

- Identify the spatiotemporal variations in UI and HQ in YRDUA.
- Analyze the relationship between UI and HQ.
- Quantify the direct and indirect impacts of urbanization on HQ.

2. Material and Methods

2.1. Study Area and Data Source

Proposed in the Yangtze River Delta Regional Development Plan (2016–2030) and approved by the State Council of China, the Yangtze River Delta Urban Agglomeration (29°20′ W–32°34′ W, 115°46′ E–123°25′ E) is located on the largest alluvial plain in China, which is formed at the point before the Yangtze River enters the sea. It is situated in the coastal region of East China and borders the Yellow Sea and the East Sea (Figure 1). The region occupies an area of 211,700 km². A total of 49.4% of the region is mountainous (the southwestern region), and the rest is plains (the northern and northeastern regions). The YRDUA consists of 26 cities from four main basic jurisdictions, including the Shanghai municipality, 9 cities in southern Jiangsu Province, 8 cities in northern Zhejiang Province and 8 cities in parts of Anhui Province. This region has a subtropical monsoon climate, with an annual precipitation of 1371.7 mm and an annual average temperature of 15.5 °C. The Yangtze River Delta region is rich in water resources, with more than 200 lakes and dense river networks.

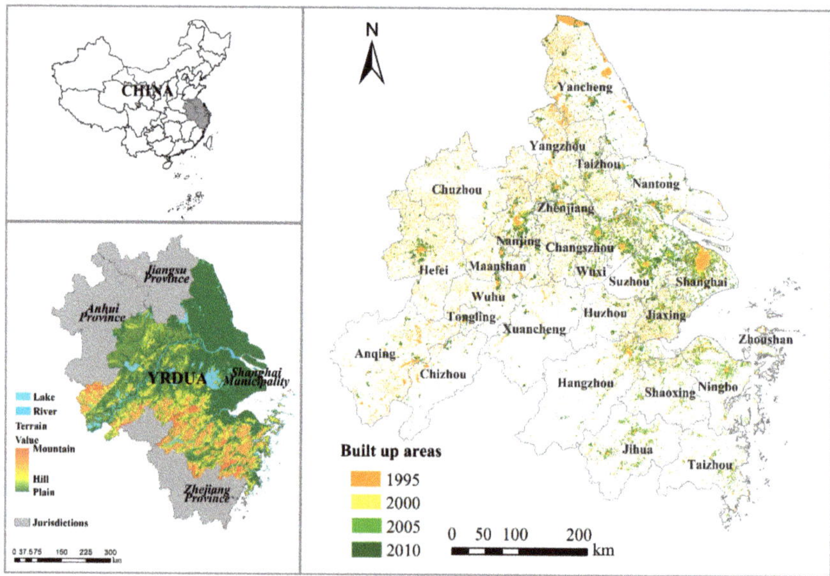

Figure 1. The Yangtze River Delta Urban Agglomeration (YRDUA) study location in China showing the topography and expansion of built-up areas during 1995–2010.

As the driving engine of China's economic development, the YRDUA has experienced an unprecedented rate of drastic and massive urbanization, which increased at an average rate of 1.41%. The population has increased from 114.3 million in 1990 to 151.0 million in 2015, with the energy consumption increasing from 114.3 million tons of coal equivalent (Mtce) to 630.4 Mtce during the same period. The region has 2.2% of the whole country's territory to support 11.0% of the total population, making it one of the most densely populated areas in China. In addition, the regional average economic growth rate of 11% is five times that of the national economic growth rate, with 18.5% ($2.07 trillion USD) of the gross domestic product in 2014 [50]. However, due to this rapid socioeconomic development and high population density, the YRDUA has had to face the challenges of habitat degradation and the construction of ecological cities. During recent decades, accelerated urbanization has exacerbated the negative impacts on local habitats, such as water eutrophication, soil erosion and natural ecosystem fragmentation. At the same time, the establishment of national parks to protect habitats, application of renewables to replace fossil fuels, disposal of pollutants to reduce environmental pollution, and other ecological-protection and restoration measures were also widely used in China to reduce the adverse impacts of urbanization [51].

Considering the availability of data sources, two spatial scales (regional and city scales) and four temporal nodes (1995, 2000, 2005, and 2010) were used to investigate the spatiotemporal impact of urbanization on HQ. Land cover data for the years 1995, 2000, 2005 and 2010 at a spatial resolution of 30 m × 30 m and land cover types were divided into 11 different categories, supplied by the Resource and Environment Data Cloud Platform (http://www.resdc.cn/data.aspx?DATAID=99). DMSP-OLS nighttime light data for the same period (F121995, F152000, F162005, and F182010) with a 1 km spatial resolution were obtained from the National Geophysical Data Center (https://www.ngdc.noaa.gov/eog/dmsp/downloadV4composites.html). The digital number (DN) of these data ranges from 0 to 63. The study was carried out on each 1 km × 1 km pixel, corresponding to 900 land units and one DN value.

2.2. Mapping Habitat Quality and Urbanization Intensity

2.2.1. Habitat Quality

As a proxy of biodiversity, habitat quality (HQ) refers to the ability of the ecosystem to provide conditions suitable for survival, reproduction and population persistence [52]. In this study, HQ was defined as the habitat status of the non-built-up part of a pixel. Due to the spatial heterogeneity in HQ, we used the InVEST-HQ, which analyzes land use/land cover (LU/LC) in conjunction with suitability and threats, to evaluate HQ throughout the study site. The model is based on the hypothesis that higher quality habitat areas can support higher native species abundance and that a reduction in HQ leads to a loss of biodiversity [44]. A half-saturation function of threats to translate habitat degradation is estimated based on the habitat quality score in InVEST-HQ:

$$Q_{obs} = H_j\left(\frac{k^2}{D^z_{xj} + k^z}\right) \tag{1}$$

where Q_{obs} is the score of HQ, H_j is the habitat suitability of land use type j, D_{xj} is the degradation score of pixel x in land cover type j, k is a half saturation coefficient (usually half of the maximum value of D_{xj}), and z is a constant to reflect the spatial heterogeneity.

$$D_{xj} = \sum_{r=1}^{R}\sum_{y=1}^{Y_r}\left(\frac{w_r}{\sum_{r=1}^{R}w_r}\right)r_y i_{rxy}\beta_x S_{jr} \tag{2}$$

where R is the number of threats ($r = 1,2,3 \ldots R$), Y_r is the set of pixels occupied by the threat r, $\left(\frac{w_r}{\sum_{r=1}^{R}w_r}\right)r_y$ evaluates the relative impact of threat r, w_r is the impact weight of r, r_y is the degradation score of threat r in pixel y, i_{rxy} is the degradation attenuation function through distance, which could be expressed as a linear or exponential function of distance from threats to habitats, β_x is the legal reachability of pixel x, which defaults to 1 in this study, and S_{jr} is the related sensitivity of each habitat type j to each threat source r. The values used as input elements for the HQ model are reported in Table S1, and the maps of habitat types and magnitude of threats in 2010 in YRDUA are shown in Figure S1. To reduce the occurrence of accidental regional errors, the HQ layer was aggregated and resampled from 30 m × 30 m to 1 km × 1 km pixels. Each output pixel contained the mean value calculated by the input pixels around that pixel.

2.2.2. Urbanization Intensity

Urbanization intensity (UI) reflects multiple aspects of urbanization, including population, industrial structure and regional space. Three recognized indicators were used in the study to calculate the traditional urbanization level index (ULI), such as the proportion of urban population, the proportion of secondary and tertiary industries, and the proportion of built-up area, which are representative of population urbanization, economic urbanization, and spatial urbanization, respectively.

$$ULI = \sum_{i}^{3}(w_i \times U_i) \tag{3}$$

U_i is the three factors of traditional urbanization level evaluation, and w_i is the weight of factors. This study considers that these three factors have equal influence on urbanization level, so they were given equal weight.

The average intensity of nighttime light in DMSP-OLS embodies the comprehensive responses of the interaction among these factors [33]. Following Chen et al. and Yang et al., the UI of an urban pixel is defined as the night light compositive index (NLCI or β) within a pixel from the DMSP-OLS stable light data, ranging from 0 to 1 [53,54].

$$\beta = \text{NLCI} = p_1 \times \text{NLII} + p_2 \times \text{NLAI} \tag{4}$$

$$\text{NLII} = \begin{cases} \dfrac{\sum_{i=t}^{63}(DN_i \times n_i)}{N_t \times 63} \times 100\% & Nt \neq 0 \\ 0 & Nt = 0 \end{cases} \tag{5}$$

$$\text{NLAI} = \dfrac{Nt}{Count} \times 100\% \tag{6}$$

The average nighttime light intensity index (NLII) and nighttime light area index (NLAI) were defined and calculated using the following indicators, which are linearly weighted to calculate NLCI. p_1 and p_2 are the weight coefficients, ranging from 0 to 1. t is the DN threshold of illuminated pixels, ranging from 0 to 63 [54]. DN_i is the original DN value of pixel i, n_i is the number of pixels with DN value i, N_t is the number of pixels with a DN value greater than or equal to threshold t, and $Count$ is the total number of regional pixels.

The value of p_1 and p_2 has a great influence on the results of the NCLI index. In order to improve the accuracy of the estimation model, a coefficient matrix with a combination of NLII and NLAI weights was adopted with a step size of 0.1 (Table 1).

Table 1. Weight combination of nighttime light intensity index (NLII) and nighttime light area index (NLAI).

Number	p_1	p_2	Weight Combination
1	0	1	NLAI
2	0.1	0.9	0.1 NLII + 0.9 NLAI
3	0.2	0.8	0.2 NLII + 0.8 NLAI
4	0.3	0.7	0.3 NLII + 0.7 NLAI
5	0.4	0.6	0.4 NLII + 0.6 NLAI
6	0.5	0.5	0.5 NLII + 0.5 NLAI
7	0.6	0.4	0.6 NLII + 0.4 NLAI
8	0.7	0.3	0.7 NLII + 0.3 NLAI
9	0.8	0.2	0.8 NLII + 0.2 NLAI
10	0.9	0.1	0.9 NLII + 0.1 NLAI
11	1	0	NLII

To reflect the actual situation as authentically as possible, Pearson correlation coefficients between the NCLI and the ULI under different weight combinations were calculated, using the sample data of 26 cities in YRDUA in 1995, 2000, 2005, and 2010, respectively. The optimal weight combination for each period was selected based on the criterion of "maximum correlation coefficient". It is interesting to note that under the selection of the best correlation coefficients ($R_{1995} = 0.727$, $R_{2000} = 0.731$, $R_{2005} = 0.714$, $R_{2010} = 0.786$), the combination of weights for the four time periods was consistent ($p_1 = 0.7$, $p_2 = 0.3$).

2.3. Analyzing the Relationship between UI and HQ

Due to the extensive amount of raw data, we first calculated the mean value of habitat quality (Q_{mean}) for all the UI pixels at intervals of 0.015 in YRDUA using R 3.4.2 software. Then, polynomial regression was applied to clarify the relationship between HQ and UI. This approach ignores the spatial heterogeneity of pixels and the development direction of the city. Since a lower polynomial order (order <3) could not faithfully characterize the tendency of the scatters, and using higher orders resulted in a trivial difference in the fitting effect because of the addition of outliers, cubic polynomial regression was used to identify the $Q_{mean} \sim \beta$ relationship in each city from 1995 to 2010. The descending rate (Q') was calculated to find the threat level of the corresponding UI on HQ in each pixel.

$$Q' = \frac{dQ_{mean}}{d\beta} \tag{7}$$

2.4. Quantifying the Urbanization Impacts on HQ

Urbanization has different impacts on habitats, including vegetation replacement, green infrastructure development, the occurrence of heat islands, and changes due to agricultural management. Zhao et al. proposed a framework that can identify the direct and indirect effects of urbanization on the net primary productivity (NPP) [22]. To quantify the impact of urbanization on HQ, we have improved the original framework of Zhao et al. (Figure S2).

The total impact of urbanization on HQ was decomposed into direct and indirect impacts. The direct impacts were defined as the variation in HQ in a pixel due to the replacement of the ecological land units by urbanized cover, either partially or completely. The indirect impacts were defined as the variation in HQ in a pixel due to other anthropogenic factors.

Conceptually, the HQ of one surface pixel is the remainder of the local background HQ under the overall impact of urbanization:

$$Q_{obs} = (1 - \omega)Q_b \tag{8}$$

where Q_{obs} is the HQ value of the surface pixel, ω is the overall impact of urbanization on HQ, and Q_b is the value of the background habitat quality before urbanization or in fully vegetated areas. There are two ways to determine Q_b. One method uses the mean or median HQ value of all the fully vegetated pixels for each city. The other approach uses the intercept of the regression between Q_{mean} and β. The high correlation coefficient ($R = 0.62{\sim}0.98$) for all the cities and years indicated a significant correlation between HQ and UI and the suitability of the background HQ measurement.

The direct variation in HQ caused by land conversion was expressed as a zero-impact line, referring to the condition in which urbanization has no indirect impact on habitats and is determined by two factors corresponding to local background HQ and the state of UI in a pixel. The indirect variation in HQ is the difference between the overall impact and the direct impacts.

$$Q_d = Q_b - \beta Q_b \tag{9}$$

$$Q_{id} = Q_{obs} - Q_d \tag{10}$$

The Q_d and Q_{id} are the HQ values under direct and indirect urbanization impacts, respectively. β is the nighttime light composite index NLCI, used to represent the UI value.

In addition, the ratio of the indirect urbanization impact (ω_d) to the direct urbanization impact (ω_i) is defined as the relative contribution coefficient τ:

$$\tau = \frac{\omega_i}{\omega_d} \times 100\% \tag{11}$$

The value of the relative contribution coefficient represents the extent to which the effect of other anthropogenic factors on the remaining ecological patches can offset (if τ is positive) or exacerbate (if τ is negative) the habitat degradation caused by the direct replacement of the original vegetation cover with fully urbanized surfaces.

The direct and indirect impacts of urbanization on HQ were calculated as follows:

$$\begin{cases} \omega_d = \frac{\Delta Q_d}{\Delta Q} = \frac{Q_b - Q_{zi}}{Q_b - Q_{obs}} \times 100\% \\ \omega_i = \frac{\Delta Q_i}{\Delta Q} = \frac{Q_{zi} - Q_{obs}}{Q_b - Q_{obs}} \times 100\% \end{cases} \tag{12}$$

where ω_d and ω_i are the proportions of direct and indirect impacts, respectively, and ΔQ_d, ΔQ_i and ΔQ are the direct HQ change (i.e., $Q_b - Q_{zi}$), indirect HQ change (i.e., $Q_{zi} - Q_{obs}$) and total HQ change (i.e., $Q_b - Q_{obs}$), respectively.

3. Results

3.1. The HQ and UI Spatiotemporal Variations in YRDUA

3.1.1. The Spatial and Temporal Changes in Habitat Quality

Habitat quality and urbanization intensity maps have different spatial patterns across YRDUA (Figure 2). The HQ varied from 0 to 1 for the four time periods analyzed. With two clearly isolated high-value areas corresponding to Huangshan Mountain and the Taihu Lake, the HQ values were generally higher along the southwestern mountains in Zhejiang Province and decreased in the northeastern plains in Jiangsu Province. Furthermore, the lowest HQ values were concentrated around the major urban areas of Shanghai, Suzhou, Nanjing and Wuxi, which have the highest population densities. Among the different land use types, forestland was the main provider of HQ in YRDUA, followed by cultivated land and water bodies. HQ was lower in urban areas than it was in the other land cover types. This is mainly due to the small amount of ecological land resulting from the large quantities of construction land, leading to a decline in the total habitat quality.

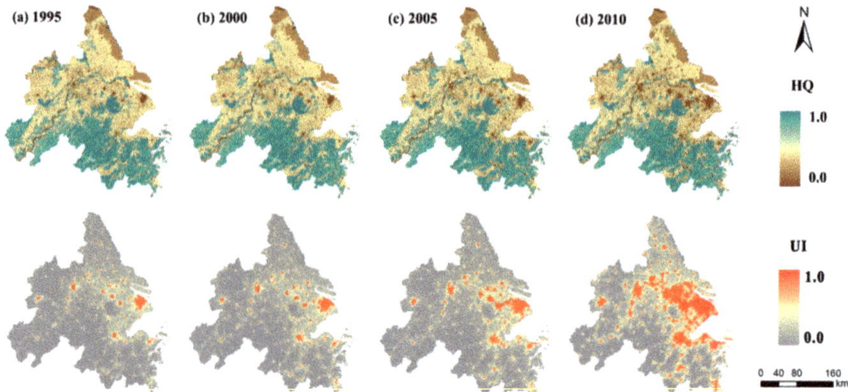

Figure 2. Spatial patterns of habitat quality and urbanization intensity in different years in YRDUA. (a) 1995, (b) 2000, (c) 2005, and (d) 2010.

With a decreasing rate of 4.95% (Table 2), the average values of regional HQ were 0.586, 0.581, 0.572 and 0.557 in 1995, 2000, 2005 and 2010, respectively. HQ decreased not only on a regional scale, but also in 23 cities of YRDUA, the largest of which was Shanghai (−18.83%), followed by Suzhou (−17.19%). HQ increased slightly only for Chuzhou, Xuancheng and Anqing, by 0.98%, 0.31% and 0.12%, respectively (Figure 3). In terms of jurisdictions, the change in HQ in Suzhou was the most substantial (−0.098), while the lowest amount of change was found in Chuzhou (0.005). The average HQ was highest in Hangzhou (0.79) and lowest in Shanghai (0.27) in 2010.

Table 2. Habitat quality (HQ) and urbanization intensity (UI) for cities in YRDUA during 1995–2010.

	HQ			UI		
	Max	**Min**	**Average**	**Max**	**Min**	**Average**
1995	0.803 (Hangzhou)	0.337 (Shanghai)	0.586	0.413 (Shanghai)	0.010 (Chizhou)	0.092
2000	0.800 (Hangzhou)	0.335 (Shanghai)	0.581	0.456 (Shanghai)	0.015 (Chizhou)	0.109
2005	0.791 (Hangzhou)	0.301 (Shanghai)	0.572	0.527 (Shanghai)	0.023 (Chizhou)	0.134
2010	0.785 (Hangzhou)	0.274 (Shanghai)	0.557	0.723 (Shanghai)	0.056 (Chizhou)	0.256
Value change	0.005 (Chuzhou)	−0.098 (Suzhou)	−0.029	0.456 (Suzhou)	0.045 (Anqing)	0.164
Change ratio	0.98% (Chuzhou)	−18.83% (Shanghai)	−4.95%	439.59% (Chizhou)	74.96% (Shanghai)	177.56%

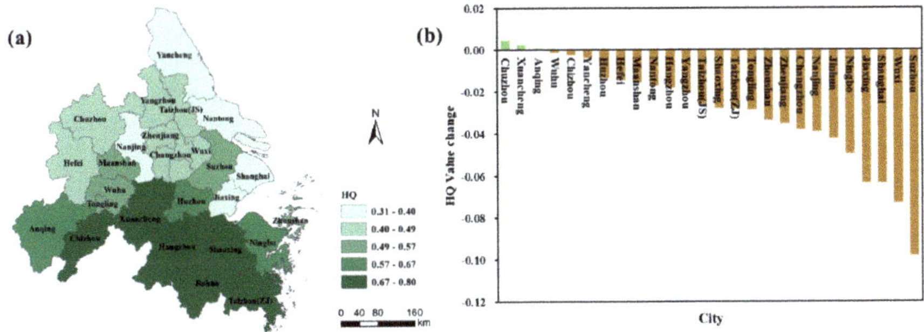

Figure 3. (**a**) Annual average value of HQ and (**b**) the variation in HQ in YRDUA from 1995 to 2010.

3.1.2. The Spatial and Temporal Changes in Urbanization Intensity

The spatial distribution pattern of UI in YRDUA was opposite to that of HQ (Figure 2). UI was the highest along the Yangtze River and Hangzhou Bay, especially in the Yangtze River Estuary (Shanghai, Suzhou and Wuxi), where urbanization gradually decreased from the center to the outskirts. As an important ecological barrier and drinking water source for YRDUA, a variety of ecological protection projects were implemented in the southwestern region; therefore, the UI was relatively low compared with that in other places. With a rate of increase of 177.56%, the average value of regional UI was 0.092, 0.109, 0.134, and 0.256 in 1995, 2000, 2005 and 2010, respectively. In 2010, the average value of UI was highest in Shanghai (0.723), followed by Xuancheng (0.062), Anqing (0.063) and Chizhou (0.056) (Figure 4). Suzhou, which had the largest decrease in HQ, was also the most prominent city for the increase in urbanization (0.456), while Anqing had the smallest decrease (0.045).

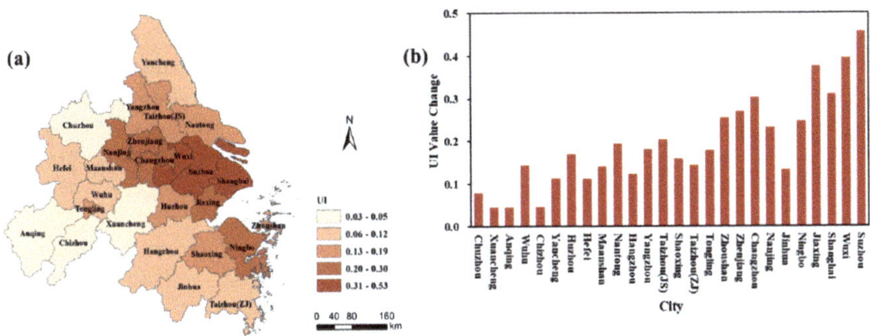

Figure 4. (**a**) Annual average value of UI and (**b**) the variation in UI in YRDUA from 1995 to 2010.

3.2. The Relationship between HQ and UI

3.2.1. Regional Scale

To identify the threat level of UI to HQ in each pixel, the spatial pattern of the descending rate (Q') between UI and HQ was analyzed among all the periods (Figure 5). Overall, HQ was negatively correlated with UI, i.e., the regression coefficients in all the regional pixels were less than zero. The threat level from the urban center to the outskirts is a "reverse U" trend, which is enhanced after the decrease. The minimum negative urbanization impact on HQ, which was mainly distributed in the surrounding suburban area, covered nearly 50% of the urbanization range ($\beta = 0.2 \sim 0.7$), while Q_{obs} declined significantly with increasing UI in the urban core districts (Shanghai, Suzhou, Wuxi, Changzhou, Nanjing, Hangzhou and Hefei) and unfrequented areas. Similar to the mountains surrounding the urban center, along the sides of the ridge, the impact of urbanization on HQ gradually increased. In the temporal gradients, each threat level was continuously pushed outward and closer together until they merged into a larger multicenter band, such as the Hangzhou Bay Belt, the Yangtze River Estuary and the Yangtze River Belt. This was consistent with the result of the polycentric megaregion evolution model in previous studies [49].

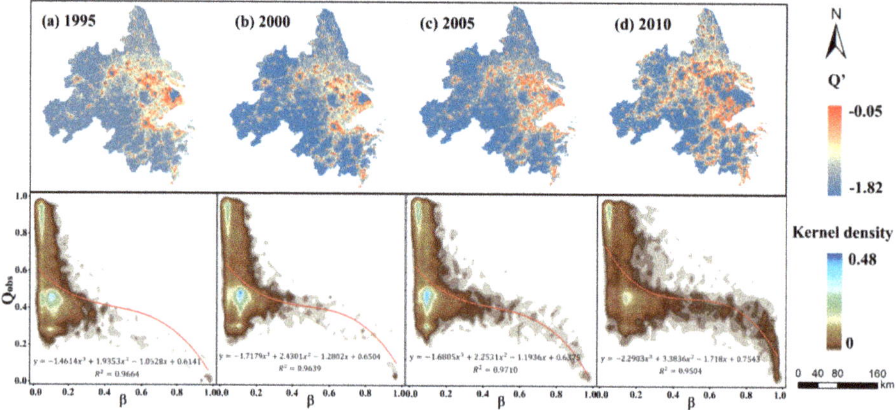

Figure 5. Spatial pattern of Q' (threat level of UI to HQ) and kernel density distribution of Q_{obs} in each UI gradient in 1995–2010 in YRDUA. The red curve is the cubic polynomial fit of Q_{mean} to UI. (**a**) 1995, (**b**) 2000, (**c**) 2005, and (**d**) 2010.

Based on the kernel density distribution of HQ, the Q_{mean} had a distinct nonlinear relationship with UI ($p < 0.001$) along the urban-rural gradients of YRDUA (Figure 5). There were two inflection points in the relationship between HQ and UI. In the position where the urbanization was 20%, the response of HQ to UI changed from a steady decrease to stable. However, when urbanization reached 80%, HQ went from stable back to a steady decrease. In the early period, only some pixels with low Q_{obs} gradually evolved to a higher UI. Over time, the urbanized level of areas with a high value of Q_{obs} started to increase, and the medium value of Q_{obs} appeared in all the urbanization gradients, even in extremely highly urbanized areas. The transformation in the relationship indicated that more natural areas were affected by urbanization and that the habitat quality in urban areas was improved in the process of urbanization.

3.2.2. City Scale

The relationships between UI and HQ across cities in YRDUA in 2010 are shown in Figure 6, and the results for other years in Figures S2–S4. Generally, HQ decreased with the enhancement of urbanization in each city, because the proportion of vegetation types in the grids was gradually replaced by built-up areas along the range of urbanization. Comparing the relationship curves of various cities, 15% of cities, such as Chizhou, Wuxi, Ningbo, and Suzhou, were linearly and negatively correlated. Seventy percent of cities, such as Tongling, Nanjing, and Hangzhou, have the same nonlinear relationship as that of the regional scale. Nevertheless, there were some exceptions, such as Nantong, Shanghai, Yancheng and Jiaxing, where HQ was not significantly negatively correlated with UI. The HQ maintained a relatively consistent value in areas of moderate urbanization. Different forms of the relationships were largely related to the local development orientation and urbanization level. The background value of each city may be associated with the local natural background and climatic conditions.

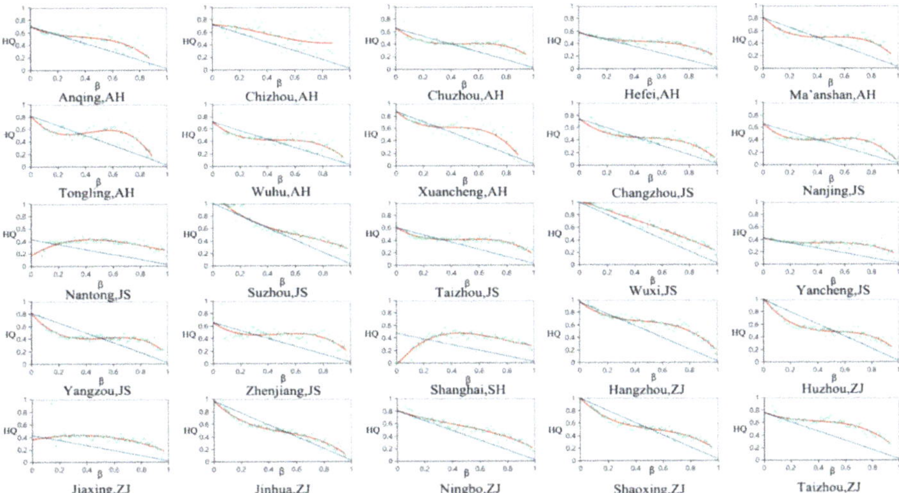

Figure 6. The relationship between UI and HQ across 26 cities in YRDUA in 2010. The red line represents the polynomial fitting curve of Q_{obs} in each scatter diagram, and the straight blue line represents the zero-impact line. The HQ responses to UI for 1995, 2000 and 2005 are shown in Figures S3–S5.

3.3. The Direct and Indirect Impacts of Urbanization on HQ

3.3.1. Regional Urbanization Impacts on HQ

The preliminary findings showed that Q_{obs} and Q_d, which showed nonlinear and linear variation, respectively, declined along the UI gradient across cities (Figure 7a,b). Some UI gradients were empty (i.e., β = 0.0317 and 0.0476) because the DN values of 2 and 3 were not found in the DMSP-OLS NTL data for YRDUA. To study the reason for the difference between the nonlinear relationship and the linear relationship, the part of the HQ variation under indirect impacts was separately extracted and studied (Figure 7c). The indirect impact of HQ (Q_{id}) (i.e., below zero) was mainly concentrated in the early stage of urbanization, and the Q_{id} in moderate-high UI was above the zero-impact line, with a maximum value of β = 0.8. As shown in Figure 7d, the relative contribution coefficient (τ) was negative at values less than β = 0.4 and tended to stabilize after growth beyond zero, which indicated that urbanization in the primary stage has a negative indirect impact on local HQ, but at a relatively high urbanization level, it gradually turned into a positive impact. Conceptually, a negative indirect impact will accelerate habitat degradation, while a positive indirect impact can partially offset the habitat degradation caused by land conversion. The average offset extent was approximately 28.23%, 17.41%, 22.94%, and 16.18% in 1995, 2000, 2005, and 2010, respectively. Notably, the variation in τ was more significant in areas of lower urbanization intensity because the observed HQs were higher than the background HQ value in pixels with relatively low urban intensity, especially for highly ecologically sensitive areas.

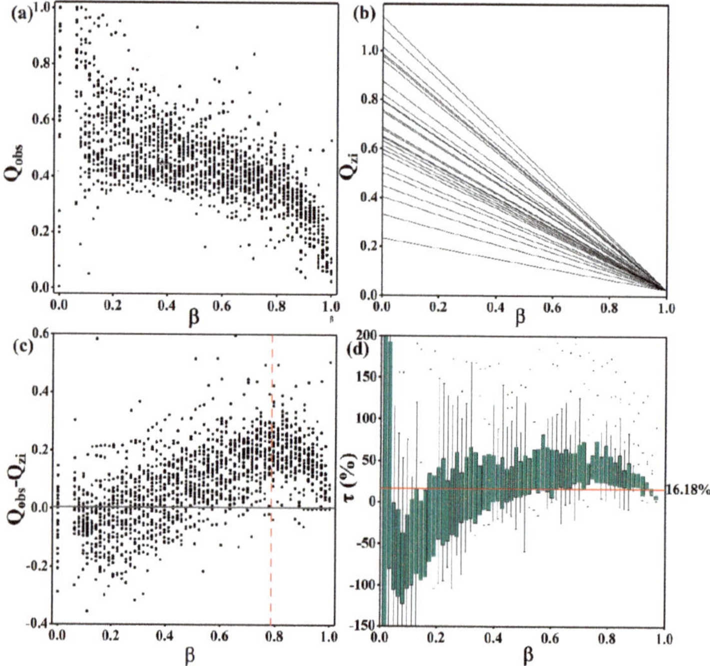

Figure 7. The relationship between UI and HQ for YRDUA in 2010: (**a**) the observed HQ along the UI gradients (Q_{obs}); (**b**) the direct HQ value change (Q_d); (**c**) the indirect HQ change (Q_{id}), in which the indirect growth of HQ peaked at approximately 0.8, indicated by the red dotted line; and (**d**) the relative contribution coefficient (τ). The boxplot for τ in each box shows 25% and 75% (the black points are outliers, the medians for each box are marked by the orange points, and the mean of the medians is 16.18%, as shown by the red lines).

3.3.2. Urbanization Impacts on HQ in Cities

It is worth mentioning that the HQ value of almost all the pixels in the areas with relatively high urbanization intensity was above the zero-impact line for 26 cities, which indicated the positive ecological impact of urbanization in the urban environment. The median of the observed habitat quality (Q_{mean}), the background habitat quality index (Q_b) and the mean of the medians of the relative contribution coefficient (τ) in all the urban pixels for 26 cities in YRDUA for the period of 1995 to 2010 are shown in Tables S2–S5. For example, the median observed HQ among all the pixels of Ningbo was 0.53, and the Q_b was 0.80, which offset 22.21% of the loss of habitat quality through vegetation cover reduction in 2010. The observed HQ in some pixels, which was even larger than the background value, caused a higher offset of HQ, such as 54.34% in Yancheng and 53.98% in Hefei in 2010.

4. Discussion

4.1. Nonlinear Relationship between Habitat Quality and Urbanization Intensity

Based on land use [55], urban populations [26] and impervious surfaces [56], the overall ecological urbanization impact on HQ was negative. Nighttime light data serve as the composite response to the interaction of these factors and were used to characterize the overall urbanization intensity [57].

We found that the nonlinear negative relationship between HQ and UI changed from a steady decrease to stable and then back to a steady decrease, with inflection points where urbanization reaches 20% and 80%. The transformation in the relationship indicated that more natural areas were affected by urbanization, and the habitat quality in urban areas was improved in the process of urbanization, which was consistent with the results of previous research. The ecological conditions in 20% of YRDUA deteriorated, and the ecological conditions in 40% increased from 1995 to 2010 [19]. The improvement in habitat quality in urban areas has also been testified in accelerating vegetation growth [21,22,58] and enhancing species richness [18,20] in urban environments compared to those in rural equivalents.

Habitat quality is used as a surrogate to assess the status of extant biodiversity under human activities [59]. Different dynamics of HQ were largely dominated by local ecological factors and complex land use patterns (Figure 8). In woodland areas, the ecosystem structures were more complicated to support the survival and reproduction of relatively diverse species. In contrast, under intensive human intervention, cultivated land represents poor habitat, hosting relatively few species, which is equivalent to urban areas [60]. In rural areas of YRDUA ($\beta < 0.2$), land use was bifurcated into woodland and cropland, which established completely different ecological backgrounds. We also found some cities, such as Changzhou, Nanjing, Nantong, Taizhou, Yancheng, Wuxi, Zhenjiang, Shanghai and Jiaxing, where the HQ dynamics were relatively consistent in moderate or even higher UI. Cropland (53–85%) accounted for the majority of the land cover in these cities and had stable relationships, while woodland accounted for less than 10%, leading to landscape homogenization from the city center to the outskirts ($\beta = 0.2\sim0.8$), with a relatively consistent habitat quality. At an extremely high urbanization level ($\beta > 0.8$), urban land use was mainly converted from cultivated land to urban land, resulting in a drastic reduction in habitat quality. The same relationship type has also been proposed in corresponding studies of plant diversity [61], birds [43] and amphibian abundance [17].

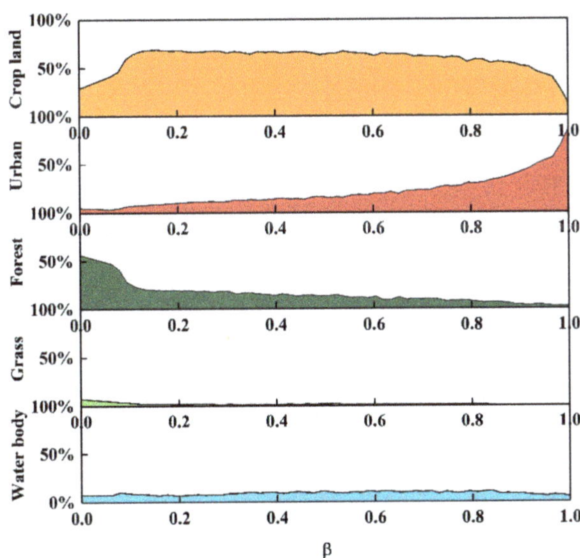

Figure 8. The proportion of each land cover type with UI gradients in YRDUA in 2010.

4.2. The Necessity of Distinguishing Urbanization Impacts on Habitat Quality

Some existing studies focused on overall impact rather than distinguishing the urbanization effect into direct impact and indirect impacts [22,62,63]. Although the habitat in YRDUA degenerated with urbanization growth, the HQ of the urban area increased during the urbanization process (Figure 7).

The reduction in HQ was the result of ecological land being occupied by constructed land, while the remaining habitats in the urban areas retained relatively good ecological conditions. The impact of urbanization on HQ is difficult to fully explain with the replacement of ecological land, which was also related to human demands in different urbanization stages [46]. Therefore, analyzing and quantifying the direct and indirect impacts of urbanization on HQ dynamics is necessary.

The cities in YRDUA have varying urbanization stages, and the impact of urbanization on HQ was also different. Based on the variations in HQ, UI and τ, the cities can be divided into four clusters (Figure 9). (I) A total of 11.5% of the cities, including Jiaxing, Wuhu and Maanshan, were in the first category, which was characterized by relatively low HQ values and an exacerbated rate of habitat degradation as a result of urbanization. (II) A total of 42.3% of the cities, including Nantong, Zhoushan, and Huzhou, were in the second category, which had relatively low HQ values and habitat degradation. (III) A total of 34.6% of the cities, including Wuxi and Suzhou, were in the third category, which had a relatively high HQ. In this category, the habitat degradation was offset by urbanization. (IV) Approximately 11.5% of the cities, including Shanghai, Nanjing and Zhenjiang, were in the fourth category, which had slightly improved HQ values. Urbanization still exacerbated habitat degradation in this category.

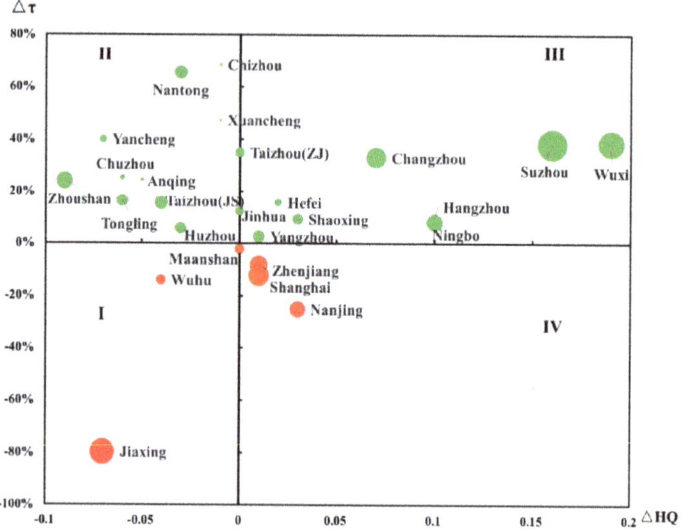

Figure 9. The variation in the median observed habitat quality (Q_{mean}), urbanization intensity (β) and the offset of habitat degradation (τ) of 26 cities in YRDUA from 1995 to 2010.

The diversity of the demands created by human beings in different urbanization stages impacts the approach used to manage landscapes [46,64]. Based on this study, the indirect urbanization impact on habitat quality dynamics was categorized into four stages: the damage stage (I), antagonistic stage (II), coordination stage (III), and degenerative stage (IV). The indirect urbanization impact was distinguished as negative impacts and positive impacts (Figure 10). In the damage stage, the negative impact on habitats increased significantly at low urbanization levels due to patch fragmentation, a shortage in the food supply, and a weakening of ecosystem resistance and resilience. Resource exploitation (e.g., deforestation and overkilling) also destroyed the ecological conditions in the available habitat. In the second stage, rapid urbanization accelerated the occupation of ecological land. At the same time, some ecological protection and restoration works were applied to match the urgent demand for blue-green ecological space (e.g., green space and aquatic landscapes) and corresponding ecosystem services. In the coordination stage, with awareness about human ecological protection and continuous investment in ecological restoration and management, some key ecological corridors and functions had been repaired, and the positive ecological impacts caused by urbanization had gradually offset the negative impacts [64]. In the degenerative stage, the extremely high proportion of built-up areas provided very limited space for species to thrive, which increased the hazard exposure of species to the external environment. The positive ecological utility generated by manual management had a threshold and could not be increased without limitation. The negative impact from the surrounding environment increased persistently, resulting in a certain degree of reduction in comprehensive utility.

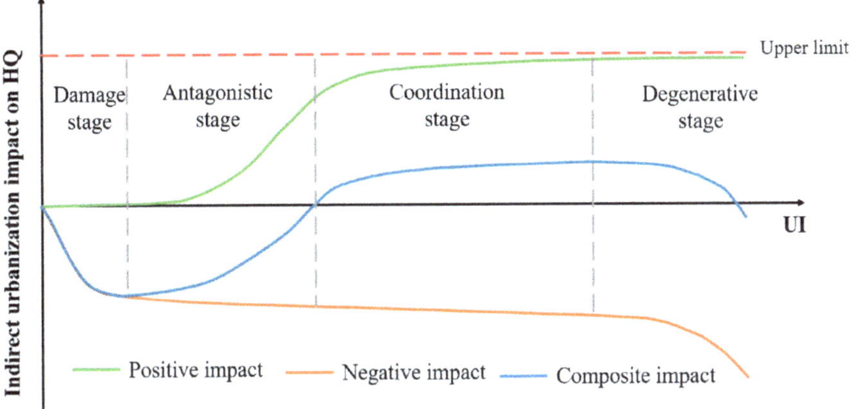

Figure 10. The urbanization stages and their indirect impacts on HQ dynamics. The dotted red line represents the upper limit of the positive impacts.

4.3. Limitations and Future Directions

Although the application of the InVEST model to regional biodiversity conservation has proven to be effective, the assessment of HQ has limitations related to two factors: the accuracy of the land cover data and the impact threshold of the threat factors. In the estimation of HQ in YRDUA, 30 m resolution land cover data were used and separated into 17 categories, and the variation in habitat types was not sensitive at some lower scales with small patches. This may obscure the complexity of the habitat, especially in Zhoushan, a city of multiple islands. Moreover, the isolated habitat fragments present in urban areas often maintain different plant and insect communities, but they are most likely only visited by individuals of the same widespread avian and mammal species. In our study, we used a relatively small impact threshold (i_{rxy}) for each threat factor used to calculate HQ, which is more appropriate for species such as plants, invertebrates, and small mammals, which have a range of survival activities less than or equal to this value. In reality, the assessment of large-scale HQ often contains uncertainty in its results due to the lack of corresponding observation data. Moreover, the quality of the data is related to the issues that need to be addressed in the research. Although the remote sensing approach to assessing HQ may simplify the understanding of complex ecological processes, our study aims to analyze the prevalent dynamics of habitats in urbanization gradients to achieve the vision of broad protection of biodiversity in rapid urbanization. It involves the protection of not one species or community but multiple positive elements in a habitat (e.g., habitat quality, ecological carrying capacity, and biodiversity) and the identification and control of threats that may cause damage to habitats.

The urbanization characterized by NTL data could also create uncertainty. Urbanization can be represented by economic aggregation and residential expansion, which is also related to the variation in the NTL data. However, the quantitative relationship between nighttime light data and various urbanization indicators is not well understood [27]. The variation in NTL data may reflect the spatial and aggregation statuses of economies and populations in YRDUA, although it was not consistent with the actual urbanization at the pixel level [65]. The 1 km spatial resolution DMSP-OLS dataset spans two decades (1992–2013), which leads to limited timeliness and spatial precision in this study. The coarse resolution and blooming effect of DMSP-OLS NTL data require future improvements in the accuracy of extracting urbanization information. The NPP-VIIRS NTL data collected on the Suomi satellite have relatively high resolution and no pixel oversaturation; therefore, they could be used for a higher quality evaluation of urbanization.

In general, although there is some uncertainty in the data sources, from a macro perspective, our study is still valuable for the exploration of the ecological impact of different urbanization stages on terrestrial ecosystems. Therefore, to achieve a more in-depth picture of the ecological impacts of various urbanization stages on habitats, multisource, higher-resolution datasets (e.g., Quickbird images and NPP-VIIRS NTL data) will be used, and field observation experiments will be added to lessen the uncertainty of assessment results from remote sensing approaches.

5. Conclusions

In this study, the relationship between HQ and UI and the direct and indirect impacts of urbanization on HQ were delineated and analyzed for YRDUA from 1995 to 2010 through the application of remote sensing data. The results indicated that urbanization might lead to habitat degradation, while awareness about protecting ecological conditions began to increase after the antagonistic stage of urbanization. The main conclusions can be summarized as follows:

- The YRDUA underwent rapid urbanization from 1995 to 2010, intensifying urban expansion and human activities. The vast majority of urban expansion was concentrated in the Hangzhou Bay Belt, the Yangtze River Estuary and the Yangtze River Belt, accompanied by a large proportion of habitat degradation.
- The overall dynamic of HQ was generally nonlinear and negative along the urbanization gradient, whereas the nonlinear negative relationship between HQ and UI changed from a steady decrease to stable and then back to a steady decrease, with inflection points where urbanization reached 20% and 80%. The transformation in the relationship indicated that more natural areas were affected by urbanization and that the habitat quality in urban areas was improved in the process of urbanization.
- With an improved conceptual framework, the difference between linear and nonlinear relationships depends on the indirect urbanization impact. Negative indirect impacts will accelerate habitat degradation, while positive impacts can partially offset habitat degradation caused by land conversion. The average offset extent was approximately 28.23%, 17.41%, 22.94%, and 16.18% in 1995, 2000, 2005, and 2010, respectively. Nearly 76.9% of the cities showed positive indirect impacts, and 55% of them showed improved habitat quality.

Supplementary Materials: The following are available online at http://www.mdpi.com/2071-1050/12/2/669/s1.

Author Contributions: Conceptualization, J.Z. and N.D.; methodology, J.Z., N.D. and Y.X.; software, J.Z. and D.L.; validation, J.Z., and D.L.; formal analysis, J.Z.; resources, W.S.; data curation, J.Z. and D.L.; writing—original draft preparation, J.Z.; writing—review and editing, J.Z., W.S. and Y.X.; visualization, J.Z.; supervision, X.W. and Y.X.; project administration, X.W.; funding acquisition, X.W. All authors have read and agreed to the published version of the manuscript.

Funding: This research was funded by the National Key Research and Development Program of China (Grant No. 2016YFC0502700) and the National Social Science Major Foundation of China (Grant No. 14ZDB140).

Conflicts of Interest: The authors declare no conflict of interest.

References

1. Li, C.; Li, J.X.; Wu, J.G. Quantifying the speed, growth modes, and landscape pattern changes of urbanization: A hierarchical patch dynamics approach. *Landsc. Ecol.* **2013**, *28*, 1875–1888. [CrossRef]
2. Liu, Z.F.; He, C.Y.; Zhou, Y.Y.; Wu, J.G. How much of the world's land has been urbanized, really? A hierarchical framework for avoiding confusion. *Landsc. Ecol.* **2014**, *29*, 763–771. [CrossRef]
3. Huang, G.L.; Zhou, W.Q.; Cadenasso, M.L. Is everyone hot in the city? Spatial pattern of land surface temperatures, land cover and neighborhood socioeconomic characteristics in Baltimore, MD. *J. Environ. Manag.* **2011**, *92*, 1753–1759. [CrossRef]
4. Grimm, N.B.; Faeth, S.H.; Golubiewski, N.E.; Redman, C.L.; Wu, J.G.; Bai, X.M.; Briggs, J.M. Global change and the ecology of cities. *Science* **2008**, *319*, 756–760. [CrossRef]
5. Islam, K.R.; Weil, R.R. Land use effects on soil quality in a tropical forest ecosystem of Bangladesh. *Agric. Ecosyst. Environ.* **2000**, *79*, 9–16. [CrossRef]
6. Zhou, W.Q.; Qian, Y.G.; Li, X.M.; Li, W.F.; Han, L.J. Relationships between land cover and the surface urban heat island: Seasonal variability and effects of spatial and thematic resolution of land cover data on predicting land surface temperatures. *Landsc. Ecol.* **2014**, *29*, 153–167. [CrossRef]
7. He, C.Y.; Liu, Z.F.; Tian, J.; Ma, Q. Urban expansion dynamics and natural habitat loss in China: A multiscale landscape perspective. *Glob. Chang. Biol.* **2014**, *20*, 2886–2902. [CrossRef]
8. Pickett, S.T.A.; Cadenasso, M.L.; Grove, J.M.; Boone, C.G.; Groffman, P.M.; Irwin, E.; Kaushal, S.S.; Marshall, V.; McGrath, B.P.; Nilon, C.H.; et al. Urban ecological systems: Scientific foundations and a decade of progress. *J. Environ. Manag.* **2011**, *92*, 331–362. [CrossRef] [PubMed]
9. Carreiro, M.M.; Tripler, C.E. Forest remnants along urban-rural gradients: Examining their potential for global change research. *Ecosystems* **2005**, *8*, 568–582. [CrossRef]
10. Youngsteadt, E.; Dale, A.G.; Terando, A.J.; Dunn, R.R.; Frank, S.D. Do cities simulate climate change? A comparison of herbivore response to urban and global warming. *Glob. Chang. Biol.* **2015**, *21*, 97–105. [CrossRef]
11. McDonnell, M.J.; Pickett, S.T.A. Ecosystem structure and function along urban rural gradients—An unexploited opportunity for ecology. *Ecology* **1990**, *71*, 1232–1237. [CrossRef]
12. Saunders, M.I.; Atkinson, S.; Klein, C.J.; Weber, T.; Possingham, H.P. Increased sediment loads cause non-linear decreases in seagrass suitable habitat extent. *PLoS ONE* **2017**, *12*, 21. [CrossRef] [PubMed]
13. Isbell, F.; Gonzalez, A.; Loreau, M.; Cowles, J.; Diaz, S.; Hector, A.; Mace, G.M.; Wardle, D.A.; O'Connor, M.I.; Duffy, J.E.; et al. Linking the influence and dependence of people on biodiversity across scales. *Nature* **2017**, *546*, 65–72. [CrossRef] [PubMed]
14. Mimura, M.; Yahara, T.; Faith, D.P.; Vazquez-Dominguez, E.; Colautti, R.I.; Araki, H.; Javadi, F.; Nunez-Farfan, J.; Mori, A.S.; Zhou, S.L.; et al. Understanding and monitoring the consequences of human impacts on intraspecific variation. *Evolut. Appl.* **2017**, *10*, 121–139. [CrossRef] [PubMed]
15. Aronson, M.F.J.; La Sorte, F.A.; Nilon, C.H.; Katti, M.; Goddard, M.A.; Lepczyk, C.A.; Warren, P.S.; Williams, N.S.G.; Cilliers, S.; Clarkson, B.; et al. A global analysis of the impacts of urbanization on bird and plant diversity reveals key anthropogenic drivers. *Proc. R. Soc. B Biol. Sci.* **2014**, *281*, 8. [CrossRef]
16. Pregitzer, C.C.; Charlop-Powers, S.; Bibbo, S.; Forgione, H.M.; Gunther, B.; Hallett, R.A.; Bradford, M.A. A city-scale assessment reveals that native forest types and overstory species dominate New York City forests. *Ecol. Appl.* **2019**, *29*, 12. [CrossRef]
17. Jennette, M.A.; Snodgrass, J.W.; Forester, D.C. Variation in age, body size, and reproductive traits among urban and rural amphibian populations. *Urban Ecosyst.* **2019**, *22*, 137–147. [CrossRef]
18. Thomas, J.P.; Jung, T.S. Life in a northern town: Rural villages in the boreal forest are islands of habitat for an endangered bat. *Ecosphere* **2019**, *10*, 15. [CrossRef]
19. Meng, Z.; Liu, M.; She, Q.; Yang, F.; Long, L.; Peng, X.; Han, J.; Xiang, W. Spatiotemporal Characteristics of Ecological Conditions and Its Response to Natural Conditions and Human Activities during 1990(–)2010 in the Yangtze River Delta, China. *Int. J. Environ. Res. Public Health* **2018**, *15*, 2910. [CrossRef]
20. McKinney, M.L. Effects of urbanization on species richness: A review of plants and animals. *Urban Ecosyst.* **2008**, *11*, 161–176. [CrossRef]
21. Jia, W.X.; Zhao, S.Q.; Liu, S.G. Vegetation growth enhancement in urban environments of the Conterminous United States. *Glob. Chang. Biol.* **2018**, *24*, 4084–4094. [CrossRef]

22. Zhao, S.Q.; Liu, S.G.; Zhou, D.C. Prevalent vegetation growth enhancement in urban environment. *Proc. Natl. Acad. Sci. USA* **2016**, *113*, 6313–6318. [CrossRef] [PubMed]

23. Dobbs, C.; Nitschke, C.; Kendal, D. Assessing the drivers shaping global patterns of urban vegetation landscape structure. *Sci. Total Environ.* **2017**, *592*, 171–177. [CrossRef] [PubMed]

24. Ossola, A.; Hopton, M.E. Climate differentiates forest structure across a residential macrosystem. *Sci. Total Environ.* **2018**, *639*, 1164–1174. [CrossRef] [PubMed]

25. Polasky, S.; Nelson, E.; Pennington, D.; Johnson, K.A. The Impact of Land-Use Change on Ecosystem Services, Biodiversity and Returns to Landowners: A Case Study in the State of Minnesota. *Environ. Resour. Econ.* **2011**, *48*, 219–242. [CrossRef]

26. Peng, J.; Tian, L.; Liu, Y.X.; Zhao, M.Y.; Hu, Y.N.; Wu, J.S. Ecosystem services response to urbanization in metropolitan areas: Thresholds identification. *Sci. Total Environ.* **2017**, *607*, 706–714. [CrossRef]

27. Elvidge, C.D.; Sutton, P.C.; Baugh, K.E.; Ziskin, D.C.; Anderson, S. National Trends in Satellite Observed Lighting: 1992–2009. In *AGU Fall Meeting Abstracts*; AGU: Washington, DC, USA, 2011.

28. Shi, K.F.; Huang, C.; Yu, B.L.; Yin, B.; Huang, Y.X.; Wu, J.P. Evaluation of NPP-VIIRS night-time light composite data for extracting built-up urban areas. *Remote Sens. Lett.* **2014**, *5*, 358–366. [CrossRef]

29. Liu, Z.F.; He, C.Y.; Zhang, Q.F.; Huang, Q.X.; Yang, Y. Extracting the dynamics of urban expansion in China using DMSP-OLS nighttime light data from 1992 to 2008. *Landsc. Urban Plan.* **2012**, *106*, 62–72. [CrossRef]

30. Zhou, Y.Y.; Smith, S.J.; Elvidge, C.D.; Zhao, K.G.; Thomson, A.; Imhoff, M. A cluster-based method to map urban area from DMSP/OLS nightlights. *Remote Sens. Environ.* **2014**, *147*, 173–185. [CrossRef]

31. Gallo, K.P.; Tarpley, J.D.; McNab, A.L.; Karl, T.R. Assessment of urban heat islands—A satellite perspective. *Atmos. Res.* **1995**, *37*, 37–43. [CrossRef]

32. Propastin, P.; Kappas, M. Assessing Satellite-Observed Nighttime Lights for Monitoring Socioeconomic Parameters in the Republic of Kazakhstan. *GISci. Remote Sens.* **2012**, *49*, 538–557. [CrossRef]

33. Fan, J.F.; Ma, T.; Zhou, C.H.; Zhou, Y.K.; Xu, T. Comparative Estimation of Urban Development in China's Cities Using Socioeconomic and DMSP/OLS Night Light Data. *Remote Sens.* **2014**, *6*, 7840–7856. [CrossRef]

34. Wu, J.S.; Wang, Z.; Li, W.F.; Peng, J. Exploring factors affecting the relationship between light consumption and GDP based on DMSP/OLS nighttime satellite imagery. *Remote Sens. Environ.* **2013**, *134*, 111–119. [CrossRef]

35. Chand, T.R.K.; Badarinath, K.V.S.; Elvidge, C.D.; Tuttle, B.T. Spatial characterization of electrical power consumption patterns over India using temporal DMSP-OLS night-time satellite data. *Int. J. Remote Sens.* **2009**, *30*, 647–661. [CrossRef]

36. Davies, T.W.; Bennie, J.; Inger, R.; De Ibarra, N.H.; Gaston, K.J. Artificial light pollution: Are shifting spectral signatures changing the balance of species interactions? *Glob. Chang. Biol.* **2013**, *19*, 1417–1423. [CrossRef]

37. Milesi, C.; Elvidge, C.D.; Nemani, R.R.; Running, S.W. Assessing the impact of urban land development on net primary productivity in the southeastern United States. *Remote Sens. Environ.* **2003**, *86*, 401–410. [CrossRef]

38. Imhoff, M.L.; Tucker, C.J.; Lawrence, W.T.; Stutzer, D.C. The use of multisource satellite and geospatial data to study the effect of urbanization on primary productivity in the United States. *IEEE Trans. Geosci. Remote Sens.* **2000**, *38*, 2549–2556.

39. Wang, J.L.; Zhou, W.Q.; Pickett, S.T.A.; Yu, W.J.; Li, W.F. A multiscale analysis of urbanization effects on ecosystem services supply in an urban megaregion. *Sci. Total Environ.* **2019**, *662*, 824–833. [CrossRef]

40. Maes, J.; Paracchini, M.L.; Zulian, G.; Dunbar, M.B.; Alkemade, R. Synergies and trade-offs between ecosystem service supply, biodiversity, and habitat conservation status in Europe. *Biol. Conserv.* **2012**, *155*, 1–12. [CrossRef]

41. Li, B.; Zhang, W.; Shu, X.X.; Pei, E.L.; Yuan, X.; Wang, T.H.; Wang, Z.H. Influence of breeding habitat characteristics and landscape heterogeneity on anuran species richness and abundance in urban parks of Shanghai, China. *Urban For. Urban Green.* **2018**, *32*, 56–63. [CrossRef]

42. Sun, X.; Crittenden, J.C.; Li, F.; Lu, Z.M.; Dou, X.L. Urban expansion simulation and the spatio-temporal changes of ecosystem services, a case study in Atlanta Metropolitan area, USA. *Sci. Total Environ.* **2018**, *622*, 974–987. [CrossRef] [PubMed]

43. Blair, R.B.; Johnson, E.M. Suburban habitats and their role for birds in the urban-rural habitat network: Points of local invasion and extinction? *Landsc. Ecol.* **2008**, *23*, 1157–1169. [CrossRef]

44. Terrado, M.; Sabater, S.; Chaplin-Kramer, B.; Mandle, L.; Ziv, G.; Acuna, V. Model development for the assessment of terrestrial and aquatic habitat quality in conservation planning. *Sci. Total Environ.* **2016**, *540*, 63–70. [CrossRef] [PubMed]

45. Sun, C.Z.; Zhen, L.; Wang, C.; Yan, B.Y.; Cao, X.C.; Wu, R.Z. Impacts of ecological restoration and human activities on habitat of overwintering migratory birds in the wetland of Poyang Lake, Jiangxi Province, China. *J. Mount. Sci.* **2015**, *12*, 1302–1314. [CrossRef]

46. Yi, L.; Chen, J.; Jin, Z.; Quan, Y.; Han, P.; Guan, S.; Jiang, X. Impacts of human activities on coastal ecological environment during the rapid urbanization process in Shenzhen, China. *Ocean Coast. Manag.* **2018**, *154*, 121–132. [CrossRef]

47. Guan, X.B.; Shen, H.F.; Li, X.H.; Gan, W.X.; Zhang, L.P. A long-term and comprehensive assessment of the urbanization-induced impacts on vegetation net primary productivity. *Sci. Total Environ.* **2019**, *669*, 342–352. [CrossRef]

48. Cidell, J. Concentration and decentralization: The new geography of freight distribution in US metropolitan areas. *J. Transp. Geogr.* **2010**, *18*, 363–371. [CrossRef]

49. Feng, Y.H.; Wu, S.F.; Wu, P.X.; Su, S.L.; Weng, M.; Bian, M. Spatiotemporal characterization of megaregional poly-centrality: Evidence for new urban hypotheses and implications for polycentric policies. *Land Use Policy* **2018**, *77*, 712–731. [CrossRef]

50. *Yangtze River Delta Yearbook (YRDY)*; Hohai University Press: Shanghai, China, 2015.

51. Zhang, D.; Qu, L.P.; Zhang, J.H. Ecological security pattern construction method based on the perspective of ecological supply and demand: A case study of Yangtze River Delta. *Acta Ecol. Sin.* **2019**, *39*, 13.

52. Sharp, R.; Tallis, H.T.; Ricketts, T.; Guerry, A.D.; Wood, S.A.; Chaplin-Kramer, R.; Nelson, E.; Ennaanay, D.; Wolny, S.; Olwero, N.; et al. *VEST 3.3.1 User's Guide*; The Natural Capital Project: Stanford, CA, USA, 2016.

53. Chen, J.; Zhuo, L.; Shi, P.; Ichinose, T. A study of the urbanization process in China based on DMSP/OLS data: Development of a light index for urbanization level estimation. *J. Remote Sens.* **2003**, *7*, 168–175, 241.

54. Yang, M.; Wang, S.X.; Zhou, Y.; Wang, L.T. A Method of Urbanization Level Estimation Using DMSP/OLS Imagery. *Remote Sens. Inf.* **2011**, *26*, 100–106.

55. Sallustio, L.; De Toni, A.; Strollo, A.; Di Febbraro, M.; Gissi, E.; Casella, L.; Geneletti, D.; Munafo, M.; Vizzarri, M.; Marchetti, M. Assessing habitat quality in relation to the spatial distribution of protected areas in Italy. *J. Environ. Manag.* **2017**, *201*, 129–137. [CrossRef] [PubMed]

56. Helms, B.S.; Feminella, J.W.; Pan, S. Detection of biotic responses to urbanization using fish assemblages from small streams of western Georgia, USA. *Urban Ecosyst.* **2005**, *8*, 39–57. [CrossRef]

57. Li, X.; Zhou, Y. Urban mapping using DMSP/OLS stable night-time light: A review. *Int. J. Remote Sens.* **2017**, *38*, 6030–6046. [CrossRef]

58. Pretzsch, H.; Biber, P.; Uhl, E.; Dahlhausen, J.; Schutze, G.; Perkins, D.; Rotzer, T.; Caldentey, J.; Koike, T.; van Con, T.; et al. Climate change accelerates growth of urban trees in metropolises worldwide. *Sci. Rep.* **2017**, *7*, 10. [CrossRef] [PubMed]

59. Stephens, P.A.; Pettorelli, N.; Barlow, J.; Whittingham, M.J.; Cadotte, M.W. Management by proxy? The use of indices in applied ecology. *J.Appl. Ecol.* **2015**, *52*, 1–6. [CrossRef]

60. Gamez-Virues, S.; Perovic, D.J.; Gossner, M.M.; Borsching, C.; Bluthgen, N.; de Jong, H.; Simons, N.K.; Klein, A.M.; Krauss, J.; Maier, G.; et al. Landscape simplification filters species traits and drives biotic homogenization. *Nat. Commun.* **2015**, *6*, 8. [CrossRef]

61. Yan, Z.G.; Teng, M.J.; He, W.; Liu, A.Q.; Li, Y.R.; Wang, P.C. Impervious surface area is a key predictor for urban plant diversity in a city undergone rapid urbanization. *Sci. Total Environ.* **2019**, *650*, 335–342. [CrossRef]

62. Xie, W.X.; Huang, Q.X.; He, C.Y.; Zhao, X. Projecting the impacts of urban expansion on simultaneous losses of ecosystem services: A case study in Beijing, China. *Ecol. Indic.* **2018**, *84*, 183–193. [CrossRef]

63. Li, Y.; Zhao, S.Q.; Zhao, K.; Xie, P.; Fang, J.Y. Land-cover changes in an urban lake watershed in a mega-city, Central China. *Environ. Monit. Assess.* **2006**, *115*, 349–359. [CrossRef]

64. Peng, J.; Shen, H.; Wu, W.H.; Liu, Y.X.; Wang, Y.L. Net primary productivity (NPP) dynamics and associated urbanization driving forces in metropolitan areas: A case study in Beijing City, China. *Landsc. Ecol.* **2016**, *31*, 1077–1092. [CrossRef]
65. Zhou, C.; Wang, S.; Wang, J. Examining the influences of urbanization on carbon dioxide emissions in the Yangtze River Delta, China: Kuznets curve relationship. *Sci. Total Environ.* **2019**, *675*, 472–482. [CrossRef] [PubMed]

 *sustainability*

Article

Dynamic Monitoring and Analysis of Ecological Quality of Pingtan Comprehensive Experimental Zone, a New Type of Sea Island City, Based on RSEI

Xiaole Wen, Yanli Ming, Yonggang Gao * and Xinyu Hu

College of Environment and Resources, Fuzhou University, Fuzhou 350108, China; wenxl@fzu.edu.cn (X.W.); mingyanli@foxmail.com (Y.M.); 386311895@163.com (X.H.)
* Correspondence: yggao@fzu.edu.cn; Tel.: +86-136-6506-9980

Received: 18 November 2019; Accepted: 14 December 2019; Published: 18 December 2019

Abstract: Islands face increasingly prominent environmental problems with rapid urbanization. Hence, timely and objective monitoring and evaluation of island ecology is of great significance. This study took the Pingtan Comprehensive Experimental Zone (PZ) in the east sea of Fujian Province of China as the research object. Based on remote sensing technology, four Landsat images from 2007 to 2017 and the remote sensing ecological index (RSEI) were used to explore the ecological status and space–time change. The results showed that from 2007 to 2011, the average RSEI decreased from 0.519 to 0.506, indicating that the ecological quality generally showed a slight downward trend, mainly due to large-scale development brought by the construction; by 2014, although the ecology of the original area improved, the overall ecology was still declining with 0.502 mean RSEI mainly because of large-scale reclamation projects; by 2017, the average RSEI rebounded to 0.523, which was attributed to the fact that ecological construction and protection were emphasized in the construction of PZ, especially in reclamation areas. In conclusion, the increase of large area bare soil will lead to the decline of regional ecology, but the implementation of scientific ecological planning is conducive to ecological restoration and construction.

Keywords: RSEI; remote sensing; ecological status; dynamic motoring; Pingtan Island

1. Introduction

Islands, an important part of the marine system, have the special geographical position and resource superiority. Recently, with the rapid growth of population size and economy, the rapid development of islands has triggered a series of ecological and environmental problems, such as environmental degradation, coastline erosion, biodiversity degradation, etc [1–3]. As an important ecological function reservoir, the island is the carrier of human habitation and the fulcrum for the protection and utilization of the ocean. Unlike land, the island ecosystem has its own unique vulnerability, which is difficult to recover once disturbed or destroyed [4]. Therefore, timely and accurate monitoring of island ecology is of great significance for regional sustainable development.

Currently, there are many methods for ecological environment monitoring and evaluation. One of them, remote sensing spatial information technology, has been widely used in the field of ecological environment due to its advantages of rapid, real-time and large-scale monitoring [5–8], making up for the deficiency of traditional semi-quantitative ecological monitoring and evaluation methods, and providing an effective research method for regional ecological evaluation. Many scholars have made use of various remote sensing indices to study the ecological evaluation of different habitats, such as cities, rivers and forests, but most of them are limited to the monitoring and evaluation of a single piece of remote sensing information, such as using the vegetation index to evaluate forestry ecology [9], using the building index and surface temperature to evaluate the urban heat island [10] or using the water body index to extract river information and then evaluate the water environment [11]. However, in the actual ecosystem, the impact of a single ecological factor on the ecosystem is far less than the comprehensive effect of multiple factors, and the quality of the ecological environment is controlled by multiple ecological factors [12]. Hence, it is very important to explore the changing rules of ecological factors and their synergistic relationship and a comprehensive indicator that can couple these multiple factors is needed to carry out comprehensive evaluations of ecological environments from the perspective of ecological systems. Mozuderer et al. [13] used the normalised difference water index (NDWI), the modified normalised difference water index (MNDWI), the normalised difference pond index (NDPI), the normalised difference vegetation index (NDVI) and field data to classify Ramsar wetland Deepor Beel and then evaluate the ecosystem. Hazaymeh et al. [14] selected six commonly used indicators in agricultural drought, and integrated the three least relevant indicators selected by principal component analysis (PCA) to generate conditions for four agricultural drought categories. The data validation results showed that the method has certain applicability to monitoring of the agriculture drought conditions in semi-arid areas with moderately high spatial and time resolution images. Yanchuang et al. [15] found that the remotely sensed reflectivity is related to multifunctionality by studying the relationship between six albedo metrics and two VIs (normalized difference vegetation index (NDVI) and enhanced vegetation index (EVI)), and multifunctionality has been related to the alternative states in global drylands, indicating that albedo may monitor changes in dryland ecosystem functioning. In addition, there are also many studies on island ecosystems, most of which were based on ecological vulnerability research. Rodgers et al. [16] applied the integrated landscape indexes to assess the ecological health of Hawaiian Islands' coral reefs and associated waters. Farhan et al. [17] selected seven indicators to evaluate the ecological vulnerability of the Sabribu Islands in Indonesia. Mukherjee et al. [18] used the normalized difference water index (NDWI), the normalized difference vegetation index (NDVI), the visible and shortwave drought index (VSDI), the normalized multiband drought index (NMDI), the moisture stress index (MSI), land surface temperature (LST), etc., to explore the impact of climate change on the natural and socio-economic vulnerability of Mousuni Island.

There are still some shortcomings in the above results: Some of them affect promotion and use due to the cumbersome evaluation factors and the difficulty in obtaining the required evaluation indicators, while some reduce the credibility of evaluation result mainly because of the subjective manual setting. The remote sensing-based ecological index (RSEI) proposed by Xu [19] is an ecological evaluation index based on remote sensing information, which integrates multiple indicators dominated by natural factors and reflects the most intuitive ecological environment. It can make up for some of the shortcomings of the existing research methods, achieve objective and quantitative evaluation of regional ecological environment conditions, analyze and visualize the spatial and temporal evolution of an ecological environment and has verified its reliability in multiple regions [20–24].

As the fifth largest island in China and the largest island in Fujian, Pingtan Island is the nearest place to Taiwan in mainland China. In view of its special geographical location, Fujian Province officially approved the establishment of PZ in 2010 in accordance with the instructions of the state council on accelerating the construction of the economic zone on the west coast of the straits, and thus started a large-scale development boom of sea–island cities. As an island, the ecological environment of Pingtan Island was originally very fragile. Under large-scale construction, whether its ecological environment will further degrade has become a hot spot of public concern. Therefore, it is necessary to conduct measurements of the ecological changes triggered by the construction of the PZ, and to provide a scientific decision-making basis for the next stage of development and construction.

Currently, there are few studies on the construction of sea–island cities and the ecological environment changes induced by them, while there are few reports on the evaluation of the ecological status of the construction of PZ, which is extremely unfavorable for the scientific guidance of the ecological protection in the construction of PZ. In view of this, this paper uses the remote sensing-based ecological index (RSEI) evaluation method to conduct a rapid and objective evaluation of the ecological status of the PZ in the early-middle stage (2007–2017), analyzes the characteristics and causes of the spatio-temporal changes of the ecological status of Pingtan Island during the period of rapid development over the last 10 years and provides a scientific decision-making basis for protecting the ecological environment in the further construction of PZ.

2. Materials and Methods

2.1. Study Area

PZ is located in Pingtan County in the eastern waters of Fujian Province. Pingtan County consists of 126 islands and nearly a thousand reefs, including the main island Haitan Island (also known as Pingtan Island, 25°16′–25°44′ N 119°32′–120°10′ E), which is the largest island in Fujian and the fifth largest island in China with an area of 267 km^2, and this study uses this island as a study area (Figure 1). The terrain of the island is higher in the north and south, mostly with the rolling hills and low mountains, and is lower in the middle with the marine plain. The area has short summers and long winters, is warm and humid and has an average annual temperature of 19.6 °C and an annual precipitation of 1161.4 mm.

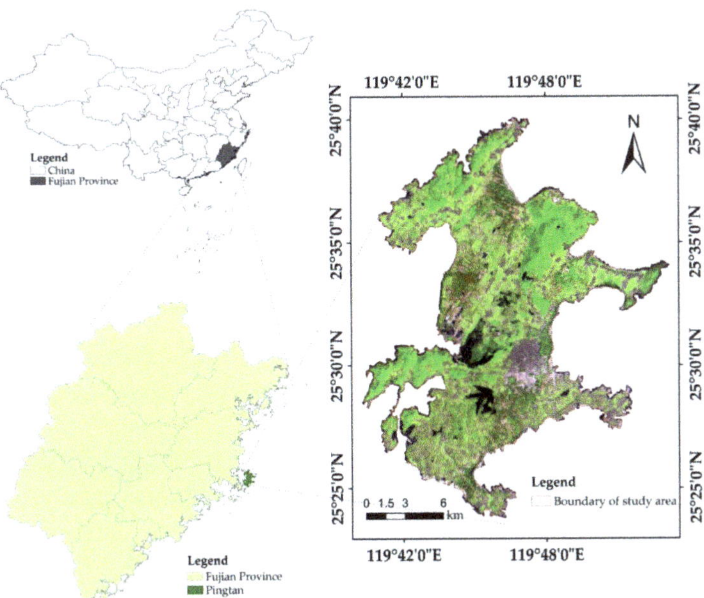

Figure 1. Location of the study area.

2.2. Data Resources and Image Pre-Processing

This study selected Landsat TM and OLI/TIRS images on 2007-09-14, 2011-09-09, 2014-09-01 and 2017-09-25, and the four images are close to each other in time, with a maximum difference of about one month (Table 1). Therefore, the vegetation has a similar growth status, which ensures the comparability of the ecological research results.

The TM images were co-registered to the OLI/TIRS using second-order polynomial and nearest-neighbour re-sampling, and the root mean square error (RMSE) of registration was less than 0.5 pixels. Then radiometric correction was carried out to convert the digital number (DN) values into planetary TOA reflectance. For TM images, the IACM atmospheric correction model of Chander et al. [25] and Chavez [26] was used. For OLI images, the algorithms provided in the Landsat 8 Data

Users Handbook posted by the USGS website were used to reduce the differences in terrain, light and atmosphere of the images with different phases [27].

Table 1. Landsat images used for analysis.

Satellite	Acquisition Date
Landsat TM	2007-09-14
	2011-09-09
Landsat OLI/TIRS	2014-09-01
	2017-09-25

2.3. Methodology

The remote sensing-based ecological index (RSEI) selects four factors related to human survival, such as greenness, wetness, heal and dryness, as ecological evaluation factors among many natural factors reflecting ecological quality, and quickly extracts representative images from remote sensing images through remote sensing information enhancement technology. The information of the indicators, namely, the normalized difference vegetation index (NDVI) [28], wetness component of the tasseled cap transformation (TCT) [29], the normalized difference impervious surface index (NDBSI) [30] and the land surface temperature (LST) [31] represent greenness, wetness, dryness and heat, respectively.

2.3.1. The Greenness Index

The normalized difference vegetation index (NDVI), one of the most widely used vegetation indexes, is a vegetation index that has demonstrated its usefulness in many ecological studies [32,33]. Thus, NDVI can be selected as the greenness index in this paper, and its formula is:

$$NDVI = (NIR - Red)/(NIR+Red) \tag{1}$$

where NIR and Red are the amounts of near-infrared and red light, respectively.

2.3.2. The Wetness Index

Tasseled cap wetness (TCW) has demonstrated its sensitivity to the moisture conditions of vegetation, water and soil [34]. Thus, the TCW is used as the wetness index here, and it can be estimated for Landsat TM and Landsat 8 using Equations (2) [29] and (3) [35], respectively.

$$TCW_{TM} = 0.0315\rho_{Blue} + 0.2012\rho_{Green} + 0.3102\rho_{Red} + 0.1594\rho_{NIR} - 0.6806\rho_{SWIR1} - 0.6109\rho_{SWIR2} \tag{2}$$

$$TCW_8 = 0.1511\rho_{Blue} + 0.1973\rho_{Green} + 0.3283\rho_{Red} + 0.3407\rho_{NIR} - 0.7117\rho_{SWIR1} - 0.4559\rho_{SWIR2} \tag{3}$$

where the ρ_i is the reflectance of each band in the TM and OLI sensors, respectively.

2.3.3. The Dryness Index

Soil drying, mainly because of construction land and bare soil, will cause serious harm to a regional ecological environment. Thus, this paper chose the normalized difference impervious surface index (NDBSI), combined with the index-based built-up index (IBI) and soil index (SI), and they are calculated as follows [36,37]:

$$NDBSI = (IBI + SI)/2 \tag{4}$$

$$IBI = \{2\rho_{SWIR1}/(\rho_{SWIR1} + \rho_{NIR}) - [\rho_{NIR}/(\rho_{NIR} + \rho_{Red}) + \rho_{Green}/(\rho_{Green} + \rho_{SWIR1})]\}/\{2\rho_{SWIR1}/(\rho_{SWIR1} + \rho_{NIR}) + [\rho_{NIR}/(\rho_{NIR} + \rho_{Red}) + \rho_{Green}/(\rho_{Green} + \rho_{SWIR1})]\} \tag{5}$$

$$SI = [(\rho_{SWIR1}+\rho_{Red}) - (\rho_{NIR}+\rho_{Blue})]/[(\rho_{SWIR1}+\rho_{Red})+(\rho_{NIR}+\rho_{Blue})] \tag{6}$$

where the ρ_i is the reflectance of each band in the TM and OLI sensors, respectively.

2.3.4. The Heat Index

The heat index is represented by the land surface temperature (LST). While there are two thermal infrared bands in the Landsat 8 TIRS sensor, this paper only chose band 10 to retrieve the LST because of the radiation calibration problem of band 11 [38]. For TM, of course, we still got the LST based on band 6 [39]. First, the digital number (DN) values was converted into the spectral radiance at the sensors aperture (L_λ), expressed as:

$$L_\lambda = G_{rescale} \times Q_{cal} + B_{rescale} \tag{7}$$

where G_{scale} is the band-specific rescaling gain factor, Q_{cal} is quantized calibrated pixel value (DN), and $B_{rescale}$ is the band-specific rescaling bias factor. The at-sensor spectral radiance (L_λ) can be converted into effective at-sensor brightness temperature using the following formula:

$$T_\lambda = K2/\ln(K1/L_\lambda + 1) \tag{8}$$

where T_λ is the effective at-sensor brightness temperature, and K1 (607.76 W/(m^2·sr·μm) for TM band 6 and 774.89 W/(m^2·sr·μm) for TIR band 10) and K2 (260.56K for TM band 6 and 1321.08K for landsat 8 band 10) are the calibration constant 1 and 2, respectively. Finally, T_λ was used to obtain the LST, and the formula is expressed as:

$$LST = T_\lambda/[1 + (\lambda T/\rho) \ln\varepsilon] \tag{9}$$

where λ is the wavelength of the emitted radiance (11.435 μm for TM band 6 and 10.9 μm for TIR band 10); ρ is a contant (1.438 × 10^{-2} mK); ε is the surface emissivity, which can be estimated by NDVI using Sobrino's model [40].

2.3.5. Construction of Remote Sensing-Based Ecological Index (RSEI)

The four component indicators obtained above are coupled by principal component analysis (PCA), and the first principal component (PC1) was used to build the RSEI, the biggest advantage of which is that the weight of the integrated indexes is not determined artificially, but is determined automatically and objectively according to the contribution of each indicator to PC1. Thus, the possible deviation of the result caused by the weight setting that varies from person to person and from method to method is avoided in the calculation, which greatly improves the objectivity and credibility of the result. Due to the uneven dimensions of the above four factors, if these factors are directly used to calculate PCA, the weight of each index will be unbalanced. Therefore, the above factors are supposed to be normalized to convert each index value into a dimensionless value within the range of 0–1 before PCA is calculated. The general normalization formula of each index is:

$$NI_i = (I_i - I_{min})/(I_{max} - I_{min}) \tag{10}$$

where NIi is a normalized index value; Ii is the value of the index in pixel I; Imax is the maximum value of this index; Imin is the minimum value of this index.

After the normalization of the four indicators, PC1 was calculated with the help of the principal component analysis module in ENVI software. In order to make the large value represent the good ecology, 1 can be used to minus PC1 to get the initial ecological index RESI0

$$RSEI_0 = 1 - PC1 \tag{11}$$

Moreover, in order to the obtain comparable RSEI values in different time and space, RSEI values should be normalized as well:

$$RSEI = (RSEI_0 - RSEI_{0\_min})/(RSEI_{0\_max} - RSE_{I0\_min}) \tag{12}$$

2.3.6. Calculation of RSEI

First, four ecological indexes of four year images were obtained, and then they were normalized and combined into a new image. In addition, in order not to let the water affect the results of principal component analysis, the modified normalized water index (MNDWI) was used to mask the water. Finally, principal component transformation was performed on the newly synthesized image to obtain RSEI images of different years (Figure 2).

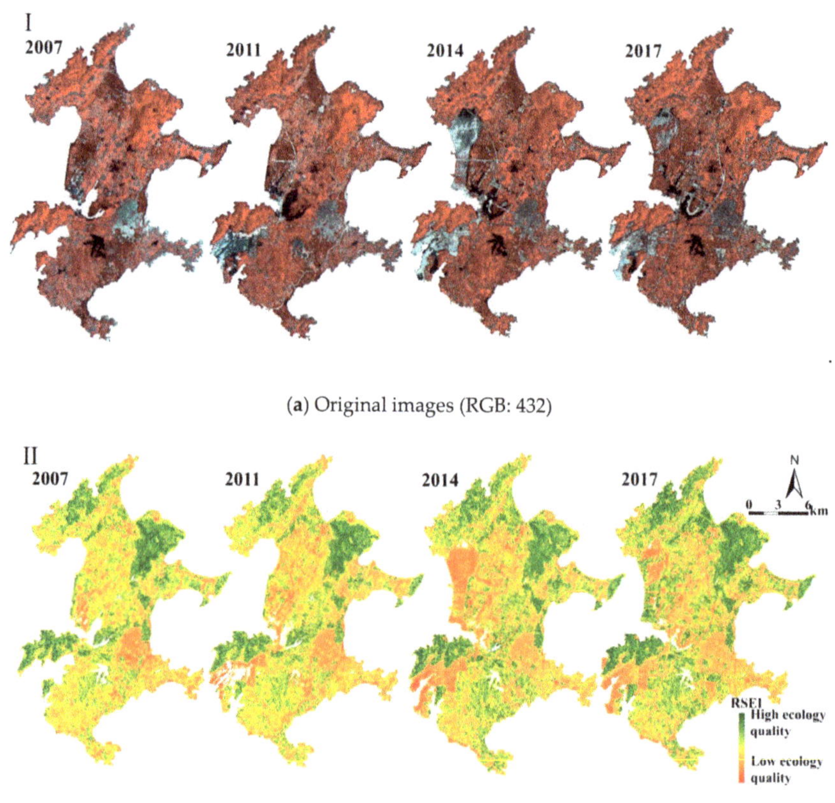

(**a**) Original images (RGB: 432)

(**b**) Corresponding remote sensing-based ecological index (RSEI) images

Figure 2. Original and corresponding remote sensing-based ecological index (RSEI) images of Pingtan Island.

3. Results

3.1. Pingtan Island Ecological Status

The PCA was used to obtain the load values of PC1 for the four indicators in the study area in 2007, 2011, 2014 and 2017. Then, the mean values of the four indicators and RSEI in each year and the magnitude of their changes were calculated, as shown in Table 2.

Table 2. Statistics of four indicators and RSEI in different years.

Index	2007		Rate 07–11 (%)	2011		Rate 11–14 (%)	2014		Rate 14–17 (%)	2017	
	Mean	PC1 Load		Mean	PC1 Load		Mean	PC1 Load		Mean	PC1 Load
NDVI	0.628	−0.474	−8.92	0.572	−0.485	6.64	0.610	−0.496	1.97	0.622	−0.497
WET	0.757	−0.509	−5.02	0.719	−0.509	0.83	0.725	−0.504	−6.76	0.676	−0.506
NDBSI	0.536	0.541	1.66	0.550	0.541	−4.18	0.527	0.535	4.93	0.553	0.537
LST	0.448	0.473	23.44	0.553	0.461	−17.72	0.455	0.462	−4.62	0.434	0.457
RSEI	0.519		−2.50	0.506		−0.79	0.502		4.18	0.523	

Table 2 demonstrates that, during the study period, the ecological quality of the PZ first decreased and then increased. The average RSEI decreased from 0.519 in 2007 to 0.506 in 2011, with a decrease of 2.50%; and then continued to decrease to 0.502 in 2014, with a decrease of 0.79%, and finally rose to 0.523 in 2017, with an increase of 4.18%.

At the initial stage of the construction of the PZ, that is, from 2007 to 2011, the average value of greenness and humidity, which had a positive effect on the ecological status, decreased by 8.92% and 5.02%, respectively. Moreover, the dryness and heat, which had a negative effect on the ecological status, increased by 1.66% and 23.44%, respectively. Thus, the overall ecological situation had a downward trend. In the middle period of construction of Pingtan Island, that is, from 2011 to 2014, the average green value favorable to the ecological situation increased by 6.64%, the humidity increased by 0.83%, the dryness unfavorable to the ecological situation decreased by 4.18% and the heat decreased by 17.72%, but, as the reclamation area has more bare land, the overall ecology declined. From 2014 to 2017, the ecologically beneficial greenness increased by 1.97%, the humidity decreased by 6.76%, the ecologically unfavorable dryness increased by 4.93% and the heat decreased by 4.62%. On the whole, the negative effects of reduced humidity and increased dryness were less than the positive effects of the increased greenness and decreased heat on the ecology, and the ecological situation showed an improvement trend.

In order to further explore the changes of local ecological conditions in the study area, the normalized RSEI ecological index was divided into five grades according to the numerical interval of 0.2. The values from low to high represent ecologically poor, fair, moderate, good and excellent, respectively [19,20,23], and the area and proportion of each ecological level in each year was calculated (Table 3).

Table 3. Area and percentage change of each RSEI level in different years.

RSEI Level	2007		2011		2014		2017	
	Area (km²)	Pct. (%)	Area (km²)	Pct. (%)	Area (km²)	Pct. (%)	Area (km²)	Pct. (%)
Level 1: Poor (0.0~0.2)	6.24	2.5	9.41	3.7	24.99	9.1	11.63	4.2
Level 2: Fair (0.2~0.4)	62.17	25.3	77.97	30.6	74.99	27.2	81.60	29.4
Level 3: Moderate (0.4~0.6)	101.15	41.2	93.78	36.8	79.12	28.7	82.11	29.5
Level 4: Good (0.6~0.8)	55.42	22.6	53.74	21.1	65.78	23.8	66.14	23.8
Level 5: Excellent (0.8~1.0)	20.33	8.3	19.78	7.8	31.06	11.3	36.47	13.1
Total	245.31	100	254.68	100	275.93	100	277.95	100

Table 3 shows that, in 2007, the overall ecological quality of PZ was mainly at the moderate level (Level 3), with an area ratio of 41.2%. The total areas of the ecologically good and excellent levels (Levels 4 and 5) were almost equivalent to the total areas of the ecologically poor and fair (Levels 1 and 2), which were 30.9% and 27.8%, respectively. By 2011, the proportion of the area with excellent, good and moderate levels decreased by 10.7%, 6.6% and 6.0%, respectively, while the ecologically poor and fair levels increased by 48.0% and 20.9% respectively, indicating that the overall ecological condition of the PZ was degraded at the beginning of construction. By 2014, the ecological status of the PZ had undergone further changes, mainly with ecologically excellent and good areas beginning to increase slowly, increasing to 11.3% and 23.8%, respectively, while the ecologically fair and medium areas decreased to 30.2% and 29.1%, respectively, indicating that the ecological environment problems have been paid attention to during the construction process, and the ecological status has been improved a bit. However, it is worth noting that the proportion of ecologically poor areas greatly increased to 9.1% due to the impact of large-scale reclamation, which was at the time mainly in the form of bare soil and bare sand; hence, the study area still shows a decline trend of overall ecological quality. By 2017, with the further advancement of construction, the proportion of ecologically excellent areas continued to rise, with an increase of 25.0%, ecologically poor areas decreased sharply, with a drop of 51.8%, and ecologically fair, moderate and good areas increased slightly, rising to 29.4%, 29.5% and 23.8%, respectively, indicating that the ecological status of the study area greatly improved. Thus, it can be seen that between 2007 and 2017, the ecological status of the study area was changing, with the overall ecology of the area showing a trend of first degradation and then improvement.

Figure 3 shows the changes in spatial distribution of the graded RSEI, with green, light green, yellow, orange and red representing the excellent, good, moderate, fair and poor ecological levels, respectively. In 2011, the ecologically fair and poor (orange–red) areas significantly increased with wider distributions, and the area with moderate ecology was significantly reduced mainly due to degradation of the environment in the central and northern regions compared with 2007, and there were also large areas in the southwest with fair and poor ecological levels due to land reclamation. By 2014, the ecology of the southeast, south and south-central areas improved compared with 2011, but there were many new areas in the north-central and southwest due to reclamation, most of which were in a state of poor ecology, which lowered the overall ecological level in 2014. By 2017, ecologically excellent area (green) in the north had further expanded, and the bare soil that appeared in the southwest and north-central areas due to reclamation was covered by vegetation; thus, the overall ecological situation improved.

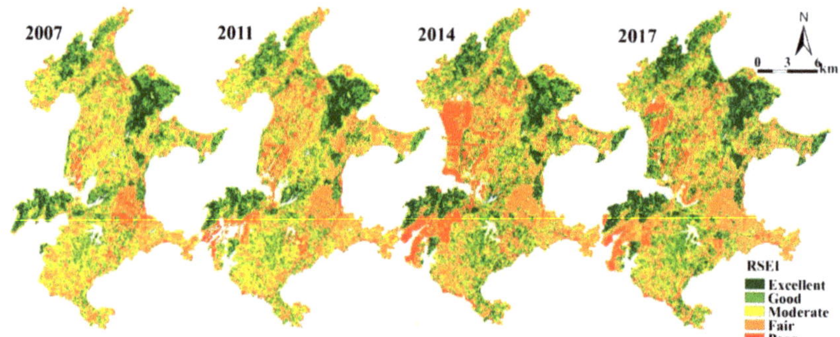

Figure 3. Five-leveled RSEI images of Pingtan Island in different years.

3.2. Dynamic Changes in the Ecological Status of Pingtan Island

In order to reveal the dynamic changes of the ecological status of the PZ and the impact of reclamation on the ecology in different time periods in detail, we divided the study area into two parts: The original area and the reclamation area. For the original area, the image difference method was used to obtain its ecological dynamic change information from 2007 to 2017 (Figure 4), and an area ratio of the corresponding changes was counted (Table 4). For the reclamation area, according to the image resolution and the actual land use situation of the study area, four land use types, namely water, vegetation, build-up land and unused land, were selected to reclassify the reclamation area by random forest classification, and then their ecological changes through land use changes were analyzed (Table 5).

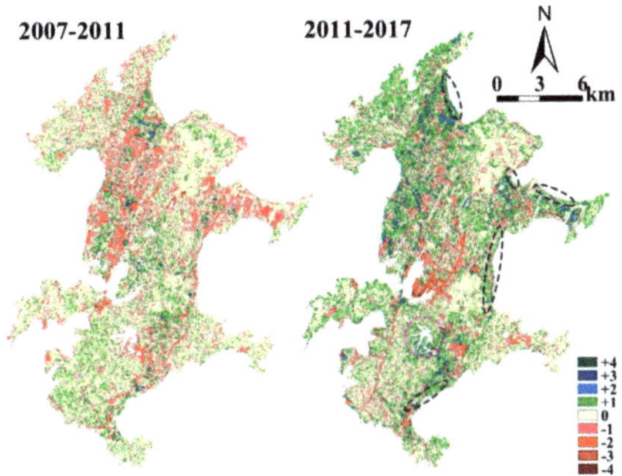

Figure 4. Change images of RSEI between 2007 and 2017.

Table 4. Percentage changes of ecological levels in Pingtan Island during 2007–2017.

Ecological Change Type	Grade Differential	2007–2011		2011–2017	
		Area Proportion (%)	Increase/Decrease in Total (%)	Area Proportion (%)	Increase/Decrease in Total (%)
Improved	+4	0.001	15.077	0.009	32.190
	+3	0.067		0.319	
	+2	0.615		3.868	
	+1	14.394		27.994	
Unchanged	0	64.940	–	54.355	–
Degraded	−1	17.325	19.983	10.879	13.455
	−2	2.413		2.060	
	−3	0.243		0.457	
	−4	0.002		0.059	

Table 5. Percentage changes of land use type part 1 (**a**) and part 2 (**b**) in the land-reclamation area of Pingtan Island during 2007–2017.

(a)

Land Use Type	2007		2011		2014		2017	
	Area (km²)	Pct. (%)	Area (km²)	Pct. (%)	Area (km²)	Pct. (%)	Area (km²)	Pct. (%)
Water	8.95	89.48	2.30	22.63	0.61	6.01	0.23	2.26
Vegetation	0.16	1.64	0.13	1.27	0.31	3.04	1.25	12.50
Build-up land	0	0	0	0	2.62	25.95	4.08	40.67
Unused land	1.05	10.52	7.73	76.10	6.57	65.00	4.47	44.57
Total	10.17	100	10.16	100	10.12	100	10.03	100

(b)

Land Use Type	2007		2011		2014		2017	
	Area (km²)	Pct. (%)	Area (km²)	Pct. (%)	Area (km²)	Pct. (%)	Area (km²)	Pct. (%)
Water	13.78	89.84	8.83	57.38	1.92	12.50	1.66	10.82
Vegetation	1.12	7.11	0.33	2.12	2.38	15.45	7.88	51.19
Unused land	0.48	3.05	6.23	40.50	11.09	72.04	5.85	37.99
Total	15.39	100	15.39	100	15.39	100	15.39	100

As seen in Table 4 and Figure 2, the ecological changes in the original area of PZ during 2007–2017 were relatively small, and were mostly based on ecological improvement level 1 and decline level

1. From 2007 to 2011, most of the original areas remained unchanged, which accounted for 64.940% of the total area, mainly distributed in the central, southern and northeastern corners. In addition, the degraded ecological area accounted for 19.983%, of which the western and eastern areas were larger, while the ecological improvement area accounted for only 15.077%, indicating that the degree of ecological degradation in the research area exceeded the degree of ecological improvement. From 2011 to 2017, the area with improved ecology accounted for 32.190%, and the green tone that characterizes ecological improvement was mainly distributed in the northwest, the five major outlets (black dotted oval in Figure 4) and the periphery of the 36-foot lake (purple dotted oval in Figure 4). The area with declining ecology accounted for 13.455%, and was mainly distributed in the surrounding areas of the central urban main road. This indicated that the overall ecological change was stable and rising.

Table 5 and Figure 4 show that the land use types in the reclamation area varied greatly during 2007–2017. In 2007, reclamation had not yet begun and was mostly in the form of water. By 2011, with the progress of reclamation projects, the water area was greatly reduced, and the vegetation was also reduced slightly. Meanwhile, the area of unused land increased rapidly, with an 11.9 times increase in the north-central region (Figure 5a) and a 6.3 times increase in the southwest region (Figure 5b). After that, the changes in the two regions were different. For the north-central region (Figure 5a), large-scale afforestation was carried out during the continuous blowing of sand so that the area of vegetation reached 51.19% by 2017. For the southwest corner (Figure 5b), in 2014, the water continued to decrease, and some unused land began to transform into construction land. In 2017, the reclamation was basically over, and the unused land was gradually covered by a large number of buildings and vegetation. Overall, the ecology of the reclamation area dropped sharply first and then rose slowly.

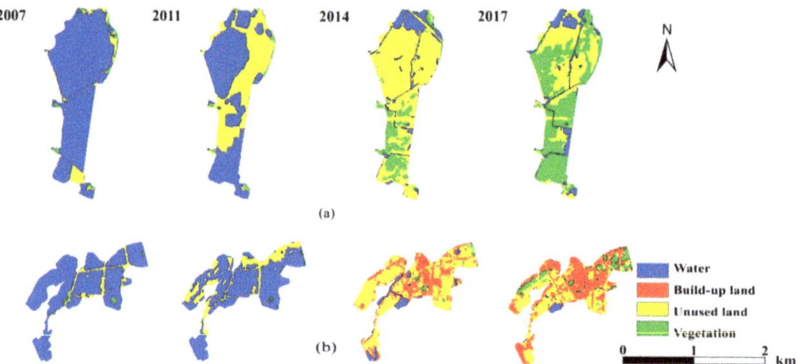

Figure 5. Land use classification of part 1 (**a**) and part 2 (**b**) in reclamation areas.

3.3. Ecological Effects of Overall Planning of Pingtan Island

According to the master plan map of PZ [41] (Figure 5b), the development of the comprehensive experimental zone has been mainly concentrated in the central and western regions, excluding the southwest, north and northeast regions. The southwest corner and northeast corner of Pingtan Island are mostly mountainous areas with high vegetation coverage. Moreover, they have been less affected by human activities so far, so the ecological condition is good and remains basically unchanged. Besides, the main constructions in the north, south and east are afforestation on sand wasteland, forestland afforestation, green village construction and restoration of the basic forest belts. From the perspective of the spatial distribution of RSEI dynamic changes, these areas were the main distribution areas with the unchanged or improved ecological grade.

The ecologically degraded areas were mainly distributed in newly developed construction areas in the western, southwest and central new districts that had been greatly affected by human development activities (Figure 6a). They almost correspond to the locations of several development zones, shown in Figure 6b (dotted oval), that were affected by ecological damage caused by human activities.

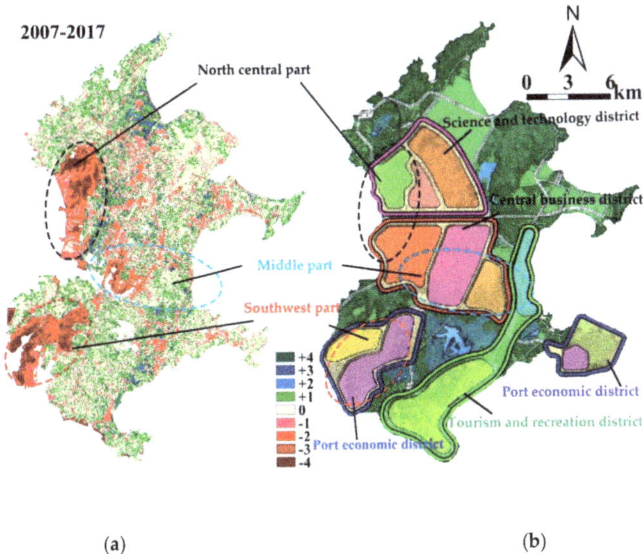

Figure 6. The contrast of the two images: (**a**) Change images of RSEI between 2007 and 2017; (**b**) master plan map of Pingtan Comprehensive Experimental Zone (PZ) (2010–2030).

Among them, the central area (blue dotted oval in Figure 6a) includes township areas such as Zhujing, Yucheng, Tancheng, etc. According to the overall planning of the experimental area and the construction target of the recent plan (2010–2015), the central part will be the main functional area of the future experimental area, and Zhuji, Yucheng and its adjacent areas have been designed as the first batch of development zones, which will be built into a multi-functional central business district. From 2007 to 2017, the area was changed from an agro-ecological landscape to an urban eco-landscape. Moreover, the land use type has changed greatly. For example, some major road networks have been newly formed, occupying a large amount of farmland and forest land; and large-scale areas of bare construction land have replaced the original green land, thus reducing the ecological status of the area.

The dramatic changes in the ecological conditions in the southwest and north-central parts were mainly due to the ecological impact of large-scale sand-blown land reclamation. According to the master plan map of PZ, the planned area within the entire experimental area is about 30 km^2. The sea area in the southwest corner of the main island (the black dotted oval in Figure 6a) will make full use of the advantages of sea and air transportation to build a modern port economic and trade zone with important influence. Moreover, the north-central sea (red dotted oval in Figure 6a) is planned to be a scientific and technology district. In 2007, the two parts were still large areas of water. By 2014, there were large areas of bare land within the two areas (Figure 2), and the area of the reclamation had reached 30.62 km^2. The land use type had been transformed from the good ecological status of seawater into building bare land, and the newly added bare land in a short period of time will inevitably have had a certain adverse impact on the local ecological environment. Therefore, it showed a deep red in the 2007–2017 ecological change map.

4. Discussion

The vulnerability of Pingtan Island's ecological environment is the result of the dual factors of natural conditions and human activities. Human development activities are the main reason for the recent ecological changes on the island. In the initial stage of construction, the overall ecological level of PZ declined mainly because the tract development of the experimental area brought about the destruction of the previous vegetation-based natural ecology, which made the originally fragile island ecology face greater pressure; in the middle and late stages of construction, the overall ecological level of PZ began to slowly recover, except for the reclamation area, mainly due to the promotion of ecological agriculture, urban greening and other measures and the construction of five major vents near the coastline, green corridors and green corridors around the island after the planning and construction of the experimental area. According to actual statistics, from 2011 to the end of 2016, more than 33 km^2 of afforestation were completed in the experimental area, and the forest coverage increased from 29% to 35.82% [42]. Therefore, in the later construction of the experimental area, the contents of the PZ Environmental Master Plan (2011–2020) should be implemented, the coordinated development should be paid attention to and the green city should be built.

The reclamation area is a new land area in which the PZ uses sand-blown reclamation to protect farmland and save land resources. However, due to the greater intensity, the land use type has changed drastically, and the early stage is mostly in the form of bare soil and bare sand. The existence and destruction of the ecological environment of the land and offshore waters led to a downward trend in the overall ecology in 2014. By 2017, some areas had completed greening and construction. Among them, the total area of the newly built shelter forest of the north-central Xingfuyang group was over 16 km^2 [43]. The ecological situation has been improved, and the overall ecological level has risen, but most of the ecological conditions are still in poor and fair grades. Among them, the main land use type of the Jinjingwan group in the southwest has been transformed from a water body to bare land. While nearly half of the north-central Xingfuyang group is covered by vegetation, half of the area is still bare soil, and most of the vegetation is still in a young state, contributing less to ecology. As the reclamation areas are all located in coastal areas, and Pingtan Island, one of the strong wind regions in the country, rages with wind and sand all year round, vegetation plays an important role in wind and sand fixation. Therefore, in the later construction, the scientific planning of the reclamation area should be carried out, the proportion of the planned area of construction land and green space should be coordinated and the ecological quality should be improved.

5. Conclusions

The remote sensing ecological index (RSEI) was used to analyze the ecological status and temporal and spatial trends and their causes. From 2007 to 2011, the ecological quality of PZ showed a slight decline mainly due to the large area development in the early stage of construction. From 2011 to 2014, the overall ecology of PZ was still in a declining state due to the large-scale reclamation project, but the ecology of the original area was improved. By 2017, the overall ecology had improved, which was attributed to the emphasis on ecological construction and protection during the construction of PZ and the active implementation of ecological measures. To sum up, the increase of bare soil in large areas of buildings will lead to regional ecological decline, but the implementation of scientific ecological planning is conducive to ecological recovery and construction. Therefore, in the construction process, in order to improve the overall quality, it is necessary to strengthen greening construction in the newly-built areas of the county, especially planting trees, which can not only prevent wind and sand, but also prevent soil erosion; for the old city, local renovation and reconstruction should be carried out; the original green areas with good ecology are supposed to continue to be repaired and protected to stabilize and enhance their ecological advantages; the reclaimed area needs a lot of afforestation.

Author Contributions: X.W. designed the research; Y.M. performed research and analyzed the date; Y.G. and X.H. collected and processed the data; X.W. and Y.M. wrote and revised the paper. The final manuscript is the result of the collaborative effort of all authors. All authors have read and agreed to the published version of the manuscript.

Funding: This research was funded by Nature Science Foundation of Fujian, China, grant number 2014J01156 and Key Technologies Program of Fujian Educational Committee, grant number JA15044.

Acknowledgments: The authors sincerely thank the editors and anonymous reviewers for their kindly view and constructive suggestions.

Conflicts of Interest: The authors declare no conflict of interest.

References

1. Cao, W.; Li, R.; Chi, X.; Chen, N.; Zhang, H.; Zhang, F. Island urbanization and its ecological consequences: A case study in the Zhoushan Island, East China. *Ecol. Indic.* **2017**, *76*, 1–14. [CrossRef]
2. Kerr, J.T.; Ostrovsky, M. From space to species: Ecological applications for remote sensing. *Trends Ecol. Evol.* **2003**, *18*, 299–305. [CrossRef]
3. Katovai, E.; Burley, A.L.; Mayfield, M.M. Understory plant species and functional diversity in the degraded wet tropical forests of Kolombangara Island, Solomon Islands. *Biol. Conserv.* **2012**, *145*, 214–224. [CrossRef]
4. Yuan, C.; Shi, H.S.; Sun, J.K.; Zhen, G.; Ma, D.M. Evaluation on Island Resources and Environment Carrying Capacity under the Background of Urbanization. *J. Nat. Resour.* **2017**, *32*, 1374–1384.
5. Boyd, D.S.; Foody, G.M. An overview of recent remote sensing and GIS based research in ecological informatics. *Ecol. Inf.* **2011**, *6*, 25–36. [CrossRef]
6. Roerink, G.J.; Danes, M.H.G.I. Quantification of Ecological Changes by Remote Sensing. *Geophys. Res.* **2010**, *12*, 2010–4643.
7. Willis, K.S. Remote sensing change detection for ecological monitoring in United States protected areas. *Biol. Conserv.* **2015**, *182*, 233–242. [CrossRef]
8. Crétaux, J.-F.; Jelinski, W.; Calmant, S.; Kouraev, A.; Vuglinski, V.; Berge-Nguyen, M.; Gennero, M.-C.; Nino, F.; Abarca Del Rio, R.; Cazenave, A.; et al. SOLS: A lake database to monitor in the Near Real Time water level and storage variations from remote sensing data. *Adv. Space Res.* **2011**, *47*, 1497–1507. [CrossRef]
9. Ochoa-Gaona, S.; Kampichler, C.; Jong, B.H.J.D.; Hernández, S.; Geissen, V.; Huerta, E. A multi-criterion index for the evaluation of local tropical forest conditions in Mexico. *Forest Ecol. Manag.* **2010**, *260*, 618–627. [CrossRef]
10. Xu, H.Q. A new index-based built-up index (IBI) and its eco-environmental significance. *Remote Sens. Technol. Appl.* **2007**, *22*, 301–308.
11. Wen, X.L.; Xu, H.Q. Remote sensing analysis of impact of Fuzhou City expansion on water quality of lower Minjiang River, China. *Sci. Geogr. Sin.* **2010**, *30*, 624–629.
12. Suter, G.W.; Norton, S.B.; Cormier, S.M. A methodology for inferring the causes of observed impairments in aquatic ecosystems. *Environ. Toxicolo. Chem.* **2002**, *21*, 1101–1111. [CrossRef]
13. Mozumder, C.; Tripathi, N.K.; Tipdecho, T. Ecosystem evaluation (1989–2012) of Ramsar wetland Deepor Beel using satellite-derived indices. *Environ. Monit. Assess* **2014**, *186*, 7909–7927. [CrossRef] [PubMed]
14. Hazaymeh, K.; Hassan, Q.K. A remote sensing-based agricultural drought indicator and its implementation over a semi-arid region, Jordan. *J. Arid Land.* **2017**, *3*, 4–15. [CrossRef]
15. Yanchuang, Z.; Xinyuan, W.; Novillo, C.J.; Patricia, A.F.; René, V.J.; Fernando, T.M. Albedo estimated from remote sensing correlates with ecosystem multifunctionality in global drylands. *J. Arid Environ.* **2018**, *157*, 116–123.
16. Rodgers, K.S.; Kido, M.H.; Jokiel, P.L.; Eric, K.B. Use of Integrated Landscape Indicators to Evaluate the Health of Linked Watersheds and Coral Reef Environments in the Hawaiian Islands. *Environ. Manag.* **2012**, *50*, 21–30. [CrossRef] [PubMed]
17. Farhan, A.R.; Lim, S. Vulnerability assessment of ecological conditions in Seribu Islands, Indonesia. *Ocean Coast. Manag.* **2012**, *65*, 1–14. [CrossRef]
18. Mukherjee, N.; Siddique, G. Climate change and vulnerability assessment in Mousuni Island: South 24 Parganas District. *Spat. Inf. Res.* **2018**, *26*, 163–174. [CrossRef]
19. Xu, H.Q. A remote sensing index for assessment of regional ecological changes. *China Environ. Sci.* **2013**, *33*, 889–897. (In Chinese)
20. Hu, X.S.; Xu, H.Q. A new remote sensing index for assessing the spatial heterogeneity in urban ecological quality: A case from Fuzhou City, China. *Ecol. Indic.* **2018**, *89*, 11–21. [CrossRef]
21. Xu, H.Q.; Wang, M.Y.; Shi, T.T.; Guan, H.D.; Fang, G.Y.; Lin, Z.L. Prediction of ecological effects of potential population and impervious surface increases using a remote sensing based ecological index (RSEI). *Ecol. Indic.* **2018**, *93*, 730–740. [CrossRef]
22. Bai, X.Y.; Du, P.J.; Guo, S.C.; Zhang, P.; Lin, C.; Tang, P.F.; Zhang, C. Monitoring Land Cover Change and Disturbance of the Mount Wutai World Cultural Landscape Heritage Protected Area, Based on Remote Sensing Time-Series Images from 1987 to 2018. *Remote Sens.* **2019**, *11*, 1332. [CrossRef]

23. Yue, H.; Liu, Y.; Li, Y.; Lu, Y. Eco-environmental quality assessment in China's 35 major cities based on remote sensing ecological index. *IEEE Access* **2019**, *7*, 51295–51311. [CrossRef]
24. Shan, W.; Jin, X.B.; Ren, J.; Wang, Y.C.; Xu, Z.G.; Fan, Y.T.; Gu, Z.M.; Hong, C.Q.; Lin, J.H.; Zhou, Y.K. Ecological environment quality assessment based on remote sensing data for land consolidation. *J. Clean. Prod.* **2019**, *239*, 118126. [CrossRef]
25. Chander, G.; Markham, B.L.; Helder, D.L.l. Summary of current radiometric calibration coefficients for Landsat MSS, TM, ETM+, and EO-1 ALI sensors. *Remote Sens. Environ.* **2009**, *113*, 893–903. [CrossRef]
26. Chavez, P.S., Jr. Image-based atmospheric corrections: Revisited and revised. *Photogramm. Eng. Remote Sens.* **1996**, *62*, 1025–1036.
27. Xu, H.Q. Image-based normalization technique used for Landsat TM/ETM+imagery. *Geomatics Inf. Sci. Wuhan Univ.* **2007**, *32*, 62–66. (In Chinese)
28. Rouse, J.W.; Haas, R.H.; Schell, J.A.; Deering, D.W. Monitoring vegetation systems in the Great Plains with ERTS. In *ERTS Symposium Conference*; NASASP-351: Greenbelt, MD, USA, 1973; pp. 309–317.
29. Eric, P.C. A TM Tasseled Cap equivalent transformation for reflectance factor data. *Remote Sens. Environ.* **1985**, *17*, 301–306.
30. Xu, H.Q. Analysis of Impervious Surface and its Impact on Urban Heat Environment using the Normalized Difference Impervious Surface Index (NDISI). *Photogramm. Eng. Remote Sens.* **2010**, *76*, 557–565. [CrossRef]
31. Jimenez-Munoz, J.C.; Cristobal, J.; Sobrino, J.A.; Sòria, G.; Ninyerola, M.; Pons, X. Revision of the Single-Channel Algorithm for Land Surface Temperature Retrieval from Landsat Thermal-Infrared Data. *IEEE Trans. Geosci. Remote Sens.* **2009**, *47*, 339–349. [CrossRef]
32. Boelman, N.T.; Stieglitz, M.; Rueth, H.M.; Sommerkon, M.; Grifin, K.L.; Shaver, G.R.; Gamon, J.A. Response of NDVI, Biomass, and Ecosystem Gas Exchange to Long-Term Warming and Fertilization in Wet Sedge Tundra. *Oecologia* **2003**, *135*, 414–421. [CrossRef] [PubMed]
33. Verbesselt, J.; Jonsson, P.; Lhermitte, S.; Aardt, J.V.; Coppin, P. Evaluating satellite and climate data-derived indices as fire risk indicators in savanna ecosystems. *IEEE Trans. Geosci. Remote Sens.* **2006**, *44*, 1622–1632. [CrossRef]
34. Jin, S.; Sader, S.A. Comparison of time series tasseled cap wetness and the normalized difference moisture index in detecting forest disturbances. *Remote Sens. Environ.* **2005**, *94*, 364–372. [CrossRef]
35. Baig, M.H.A.; Zhang, L.; Shuai, T.; Tong, Q.X. Derivation of a tasselled cap transformation based on Landsat 8 at-satellite reflectance. *Remote Sens. Lett.* **2014**, *5*, 423–431. [CrossRef]
36. Xu, H.Q. A new index for delineating built-up land features in satellite imagery. *Int. J. Remote Sens.* **2008**, *29*, 4269–4276. [CrossRef]
37. Essa, W.; Verbeiren, B.; Van der Kwast, J.; Van de Voorde, T.; Batelaan, O. Evaluation of the DisTrad thermal sharpening methodology for urban areas. *Int. J. Appl. Earth Obs. Geoinf.* **2012**, *19*, 163–172. [CrossRef]
38. Julia, B.; John, S.; Simon, H.; Raqueno, N.G.; Markham, B.L.; Radocinski, R.G. Landsat-8 Thermal Infrared Sensor (TIRS) Vicarious Radiometric Calibration. *Remote Sens.* **2014**, *6*, 11607–11626.
39. Nichol, J. Remote Sensing of Urban Heat Islands by Day and Night. *Photogramm. Eng. Remote Sens.* **2005**, *71*, 613–622. [CrossRef]
40. Sobrino, J.A.; Jimenez-Munoz, J.C.; Paolini, L. Land surface retrieval from LANDSAT TM5. *Remote Sens. Environ.* **2004**, *90*, 434–440. [CrossRef]
41. Fujian Urban and Rural Planning and Design Institute. Available online: http://www.fjplan.org/chgnr.asp?id=225 (accessed on 10 September 2019).
42. PTNET. Available online: http://en.ptnet.cn/index.html (accessed on 10 December 2019).
43. Sohu. Available online: http://www.sohu.com/a/167112662_263546 (accessed on 10 December 2019).

 *sustainability*

Article

Role of Urban Public Space and the Surrounding Environment in Promoting Sustainable Development from the Lens of Social Media

Thuy Van T. Nguyen [1,2], Haoying Han [1,* and Noman Sahito [1,3,*] 

[1]   Institute of Urban and Rural Planning Theories and Technologies, College of Civil Engineering and Architecture, Zhejiang University, Hangzhou 310058, China; khut_khit@yahoo.com
[2]   Department of Architecture, Danang Architecture University, Da Nang 550000, Vietnam
[3]   Department of City & Regional Planning, Mehran University of Engineering and Technology, Jamshoro 76062, Pakistan
*   Correspondence: hanhaoying@zju.edu.cn (H.H.); noman_sahito@yahoo.com (N.S.)

Received: 5 October 2019; Accepted: 25 October 2019; Published: 27 October 2019

Abstract: The development of mobile social network has shown the power to change the ways people gather and communicate in urban public spaces (UPSs). In this study, we utilized a check-in database collected from Instagram in 2016 and 2017 in two central districts of Ho Chi Minh City (HCMC) to analyze the city dynamics and activities over the course of the day. By quantifying the popularity of contemporary UPSs, a comprehensive study was conducted on many attraction features spreading over the two central districts. Pearson's correlation was used to explore the proximity and attractiveness associated with the surrounding environment of three types of UPSs. The results show that the lifetime of UPSs is very stable during weekdays and weekends. Within that, commercial UPSs are proved to play a dominant role in urban dynamics. This paper's finding is at odds with the urban planning stereotype that public facilities often help people to get around. In the case of HCMC, it has proved the opposite: people are attracted to urban public spaces even though there are not many cultural and social specialties there. The results will contribute to enhancing the predictability of each UPS on socio-economic performance and to understanding of the role of urban facilities in urban sustainability.

Keywords: urban public space; environment; check-in data; social media platform; point of interest

1. Introduction

A city is a result of modern human civilization, where people live and work in an extensive and well-developed urban system. However, in the 21st century, cities have undergone a unique period of rapid economic and technological growth. In a case of Southeast Asia, Dick and Rimmer [1] mentioned Asian exceptionalism. Their arguments suggest that the motion of Southeast Asia's new urbanization would be the avoidance of social discomfort. Urban geography of the region features in the movements between public and private space, which can take place from diverse dimensions: center to periphery, periphery to periphery, center to center, or periphery to center [2]. Along with modernization and globalization, urban public space, which is a traditional connotation in urban planning, now faces an inevitable transition in the new era. Urban public spaces (UPSs) indeed have acquired a renewed visibility in urban planning discourse as an essential ingredient of urban sustainability. Therefore, the major difficulty for authorities is to create and sustain public goods in a manner that distributes proportionally health, well-being, viability, and economic benefits. Hence, investigating the UPSs use and its correlation helps us understand how it can bring out different socio-economic benefits.

In defining public space, it is essential to consider the meaning of the term "public". Madanipour [3] suggests that "the word public originates from the Latin and refers to people, indicating a relationship to both society and the state". This suggests that "public" may be any entity, regardless of whether tangible or not, that relates to people and is shared by and open to them in a community as a whole. The concern here is space as the physical entity that is linked to the term "public". This provides a basic understanding of public space as the space that concerns people and may be interpreted "as the space open to people as a whole" [3]. These ideas are echoed in the various definitions of public space.

Public space plays an important role in sustaining the public realm. There is a renewed interest in public space with a growing belief that while modern societies no longer depend on the town square or the piazza for basic needs, good public space is required for the social and psychological health of modern communities. New public spaces are emerging around the world and old public space typologies are being retrofitted to contemporary needs. Good public space is responsive, democratic and meaningful. However, few comprehensive instruments exist to measure the quality of public space [4–7].

According to Carr, et al. [4], a true urban public space should be responsive, democratic and meaningful. Responsiveness of a public space requires that it should serve the needs of its users such as comfort, relaxation, active and passive engagement and discovery. Democratic public space, which should be accessible to all groups, is that which protects the rights of its users.

Over the last two decades, UPSs have been the subject of considerable debate. There are theories and beliefs about the knowledge of urban design and, in particular, public places, which have always focused on the role of space characteristics in the production of urban spaces [8]. Some physical and social factors of environment can have impact on the selection and utilization of UPS in developed countries [9]. Many recent reviews have studied relevance between proximity and use of UPS at an intense volume of physical activity [10], and the vital elements and their intricate interaction influence the use of UPS such as sociability [8], behavior of individuals and groups in small UPSs [11], design elements in public spaces [12,13] accessibility [14–16] and the surrounding environmental factors [17–19]. Specifically, the presence of facilities is thought to become the most essential element reflecting UPS, in particularly urban green zones [20,21]. Nevertheless, these reports were mostly in developed countries, which had social and economic constraints according to diverse contexts.

Brown indicated that sustainability should construct the balance between the natural environment and artificial configurations [22]. The built environment is an essential element of sustainable society [19]. Focusing on the typical urban spaces in the city, such as: open spaces, commercial space, etc... This paper aims to study the relationship between the surrounding environment and UPSs utilization in low- and middle-income countries. Nowadays, due to the restricted space, unsafety, noise and air contamination of previous public spaces [23], along with an emergence of financially-driven urban redevelopment processes, commercial-related public spaces (such as restaurants, cinemas, and shopping malls) have become the new civic centers and social hubs in contemporary cities [24,25] This new phenomenon is posing crucial questions: how do the people experience these urban public spaces in the city? Vice versa, how surrounding contexts, have influence on people's choice of gathering could also become a good research question. Answering these two issues may lead to a more insightful understanding of spatial-temporal needs for the daily life of local communities.

Often conventional methods for large scale data-collecting are labor-intensive, hence too costly; which turns out to be a drawback in the research practice on people's spatial and temporal activities. Recently, however, fast-advanced technology has given way to many practical innovations, especially in the field of the Internet. It is also worth noting a new development of geographic science [26], which is demonstrated by a huge number of geographic attachments of "geo-tagged" data providing both the latitude and longitude coordinates [27–29]. In spite of this promising potential for new studies, there are some specific constraints of the exploration of social media data, namely issues of sampling, situation-related unreliability, and also deprivation of theoretical firmness [30]. Nevertheless, after being carefully refined, the check-in database has an advantage to answer different questions over traditional data, which is often combined and outdated [29]. Point-of-interest (POIs) and check-in data in social media, a refined dataset obtaining more information, are becoming new focal subjects in urban study, which allow us to examine how new communities rise and form sociocultural norms in the online world [31,32]. Instagram is regarded as a socially visual-locative means, which can contribute as a participatory sensing system [33,34]. By using Instagram data to record urban dynamics, it is found that spatiotemporal models are correlated with routine activities of citizens. Consequently, Instagram has the capability of identifying the places of cultural activities [27,35].

In this paper, with consideration to the spatial-temporal dimensions and activity categories for analysis, we aim to give the detailed descriptions of citizens' activities. Relying on the Instagram platform, the research intends to collect people's check-in data in Instagram, then keep track of and analyze relevant information of UPS. UPS is framed to be the organic combination of the place itself, and a destination for people to stay and carry out activities, which results in a particular geographic space and other surrounding environmental, infrastructure, and relevant conditions.

We conducted research in Ho Chi Minh City (HCMC) in Vietnam. HCMC is the biggest city in Vietnam. From a literature review, we found that there are few studies on urban public spaces in HCMC and no study is present using people's check-in data in Instagram methods. Therefore, in this study, with help social media check in data POIs we have three research questions: what are the social characteristics of urban public spaces? Which are most preferred/visited urban public spaces throughout the day? Why are hotspots located at different times during weekdays as well as weekends and are causes of attractiveness in urban public spaces?

This paper follows this order: Section 2 will go through related literature review on this field; Section 3 will demonstrate the methodology and dataset in detail; Section 4 will present our results; finally, Section 5 comes to conclusions with suggestions for further studies.

2. Literature Review

2.1. Urban Public Space (UPS)

In the last 20 years, public spaces have adapted to people's changing behavior, which consequently needs a renewed observation on urban planning discourse. Conventional wisdom holds that public spaces, playing a role as a vital element for urban sustainability, are now becoming indispensable keys to the formation of inclusive communities and more specifically, enhancing public culture and cultural diversity [36]. The term "public space" here also refers to more commercial-related public sites of consumption (such as café shops), because these spaces usually perform the same social role as a space for sociality and recreation. In short, public spaces allow people to "assemble and socialize away from home and work", where Oldenburg observed and coined the connotation "Third place" [37]. Nguyen et al. added the book-store café as a "third place", which emphasized that these are places where people can gather for casual occasions, yet this important sociality can be without excessive social or personal obligations [38]. However, public spaces recently have not always purely served the public. Low [39] pointed out the relation of the privatization and commercialization of public spaces by corporate or commercial interests, claiming that "during the past 20 years, privatization of UPSs has been significantly increasing, concerning with the closing, redesign, and regulating of public parks, the building of malls, departments or plazas, the opening of central business districts that have certain control over local streets and parks, and the transfer of public air rights for the commercial buildings ostensibly open to the public". This trend may simply raise awareness to broaden the definition of public spaces; specifically, to incorporate some of the new forms of commercial-related public spaces that have been thriving in urban society. Later, Banerjee [40] suggested that urban planners should keep close attention to broader notions of public life, instead of merely physical public spaces, given the new reality that many public activities exist in private spaces, as they are "not just in incorporate theme parks, but also in small businesses such as coffee shops, bookstores … ". Moreover, there is also a growing interest in re-establishing the correspondence between public space and urban mobility.

Empirical studies have shown that a warm welcome for attractive public spaces, which also comprise mobility hubs (like stations, airports, bus stops), as well as the use of commercial-related public spaces [19] (for instance, department stores and commercial complexes) add more key challenges for an innovative design of viable mobility systems and a goal for sustainable and livable cities.

2.2. Geo-Tagged Social Media-Related Study

A strong rise of social media platforms along with an outburst of smartphones have helped many generations, especially youth, to share their daily routines and activities on the Internet, also known as the online community, which means to create their digital footprint in urban areas. Undoubtedly, this newly emerged community along with its digital data turn out to be precious sources for academic urban behavior research. Moreover, it has recently been recognized that data collection of social media can help provide reliable geographic information in replacement of conventional data-collection methods [41].

Currently, geo-tagged social media and Global Positioning System (GPS) tracking data have been claimed to further understand human mobility patterns as well as spatiotemporal models [42,43]. Although there are several platforms that have more users globally, Twitter is so far the most popular platform to extract geographic data [41]. For instance, Tsou [44] demonstrated an academic method for recording and investigating the spatial content of Twitter which helps enable analysis of social events from a spatial temporal perspective. Studying social media "check-in" pattern hence contribute a more profound rationale of urban dynamics [45]. In 2012, Frias-Martinez [46] demonstrated another framework to identify land use structure by geo-tagged tweets in Manhattan, London, and Madrid, claiming that these data can provide a compatible statistical input for local governments. Having said that, researchers trying to rely on the geo-tagged data might be concerned with social media platforms that do not simply show the citizens' movement patterns. Different platforms might have different preferences, such as Twitter for updating news and Instagram for everyday photos and videos [47], or levels of posting immediacy [48]. Foursquare and Instagram both demonstrate almost identical urban features (population), and also show distinctions: Foursquare distributes better user routing data, while Instagram illustrate better cultural behaviors of users [35].

Instagram users are believed to be picky about showcasing their lifestyles and personal images that they feel suited for acquaintances or followers [49]. This implies that they choose to represent the city and their places with a highly curated content. By connecting with other users in cyberspace and tagging the same locations, Instagram users create communities on both online and offline worlds. Consequently, by demonstrating places where the association takes place, we can explore and examine how communities really rise and develop at the two interfaces and then create social and cultural contexts. This provides an impressively innovative approach as before such data could only be collected and processed by surveys or observation settings, but now we can access social media to collect much more data at a much shorter time, and to conduct a comprehensive investigation of how citizens interact and assemble with each other.

3. Methods

3.1. Analysis of Spatial Distribution Patterns

In order to detect where check-in posts are, a grid mapping is virtualized by 100-meter × 100-meter grid-size squares. The purpose and number of visiting inside each square is also examined. This helps generate activity spatial distribution maps by pointing out the frequently-hit places as well as their main functioning. Citizens' activities are very diversified at different periods of time in one day, weekdays, and between weekdays and weekends [50]. When check-in mapping shows contrasts for multiple activity groups, this suggests a high effect of urban areas on people's visiting choices.

3.2. Indicators of UPSs' Visitation Rates

To evaluate how the surrounding built environment influences the UPS, we examined two design indicators: 1) UPSs' check-in times 2) the surrounding built environment. For UPSs' check-in times, we counted the check-in times which stipulate the attractiveness of the particular space towards the people. We hypothesized that UPSs themselves have their own surrounding advantages, which include the inside facilities and its characteristic, such as, area of a place, the environment, the interior decoration, etc. At the level of neighborhood, we evaluate transportation access and other surrounding attractive places, and we assumed that one place would attract or accommodate more visitors. The neighborhood characteristics are computed using a kernel density estimation (KDE) procedure [51].

KDE is a method that produce representations of local density estimates from two-dimensional point perspective, hence becoming an effective tool for analyzing projected hotspots, estimating intensity and visualizing the distribution of points by spreading a radius by a kernel function with defined bandwidth [52,53]. To Schabenberger and Gotway [54], the KDE can be calculated as Equation (1)

$$f_{(s)} = \sum_{i=1}^{n} \frac{1}{h^2} k\left(\frac{d_{is}}{h}\right) \tag{1}$$

where $f_{(s)}$ is the KDE function at the location, h is the bandwidth, d_{is} refers to the distance from point i to s, and k the classic Gaussian kernel function.

3.3. Correlation Coefficient Modelling

To evaluate how the surrounding built environment has influence on UPS visitation, we examined two design indicators: 1) UPSs' check-in times Dp and 2) the surrounding built environment Dk as below:

The UPSs' check-in times: The UPSs's check-in times is counted as the check-in times which stipulates the attractiveness of the UPS towards the people.

The surrounding built environment D_k: Assuming that the attractiveness of each POI is not only relied on its own endowed advantage, but it is also remarkably influenced by the surrounding environment, such as locations, traffic conditions, and many other external factors. In this study, we assumed that the surrounding environment include all neighborhood accessibility variables at the neighborhood level, such as public transportation (bus stops), public facilities (parks and shopping malls), and other commercial spaces. The KDE then examines the point distribution of adjacent domains (with h bandwidth). The choice of h bandwidth strongly affects the resulting surfaces. As the study was mostly concerned with the fine-scale of the city, we defined the neighborhood of each UPS as a 400-meter buffer zone. The reason is that previous studies suggested 400 meters would be a primary maximum walking distance based on 5 to 15-minute maximum walking times in transit design guides [55]. Thus, we chose 400 meters as the value of h bandwidth in the KDE calculation.

Next, we use a geographic information system (GIS) to collect and sort out data. Assuming that the D_p and D_k are in different units, the standardization stage was processed to put different variables on the same scale. This stage allows to compare scores between different types of variables.

In order to indicate whether the choice of people of any space related to the surrounding facilities or not, Pearson's correlation (Pearson's R) was used as a statistical measurement that helped to calculate the strength of the relationship between the advantages of external factors and the UPSs visitation. The value range varied from −1.0 to 1.0. A correlation of −1.0 showed a perfect negative correlation, meanwhile, a correlation of 1.0 showed a perfect positive correlation [56]. The correlation coefficient is formulated as below:

$$R = \frac{\left(\sum_{i=1}^{n} D_p D_k\right) - \left(\sum_{i=1}^{n} D_p\right)\left(\sum_{i=1}^{n} D_k\right)}{\sqrt{\left[\sum_{i=1}^{n} D_p^2 - \left(\sum_{i=1}^{n} D_p\right)^2\right]\left[\sum_{i=1}^{n} D_k^2 - \left(\sum_{i=1}^{n} D_k\right)^2\right]}} \tag{2}$$

4. Materials

4.1. Case Study

HCMC is famous as the business and financial center of Vietnam, as well as a popular tourist destination. With a population over nine million, HCMC nowadays has become a host for a large number of energetic people in the nation, particularly the young. Especially, with the dynamics of ongoing urban sprawl, agglomeration has resulted in HCMC being the first megacity in Vietnam [57]. This study was based on Instagram check-in data within two districts (district 1 and district 3) in the center of HCMC. The analysis unit for studying urban activities is considered different based on the borders of municipal administration units, for example, neighborhood and traffic zones, because the latter are too broad to tell apart the features of urban activities and assure sufficient accuracy. Therefore, the analysis unit was a 100-meter × 100-meter grid. This grid unit serves the calculation of hour-based check-in numbers and coordinate the output with POI data.

4.2. Data Collection

Points of Interest (POI) are particular location of points, for example, shopping malls, office, restaurant, open spaces, which provide a tool for location-based services. The POI data was chosen because: (1) POI data has good adaptability for scaling problems, (2) by people's interaction, POIs help show their personal biases and also a place's social roles (3) statistical gridding of POI data is much finer, hence providing more comprehensive information [58,59].

Street networks are collected from OpenStreetMap (OSM). OSM is a team-worked web-mapping project that collects geospatial data in online maps which often offers country downloads [60]. In our study, we obtained all the locations of public transportations from the Google website.

A check-in database is a set of geo-tagged data which can be used to record people's movements at one time and place. Hence, a large-scale set of check-in data can help us extend our perspective on people's actual activities in terms of space and time [61]. Launched in 2010, Instagram has been the worldwide social network platform for photo and video sharing [62]. As a result, Instagram, which can be regarded as socially visual-locative means, can contribute as a participatory sensing system [33]. This study uses data crawled from the Instagram Application Program Interface (API) platform over two successive years (from January 2016 to December 2017). Only geo-tagged posts from two central districts were selected. Besides the crawling coordinates, each Instagram user also carries information about the user ID, the link of each post, check-in time, and check-in locations. In total, we downloaded 142,620 posts within district 1 and district 3 of HCMC.

Identification of activity categories: Geo-tagged social media platform is valued for the capability to identify activity sets. When queried in the location-based service provider, this link provides information about the categories based on the visited venue. After that, we manually verified and, when needed, merged the place names. This research paper classifies all of the obtained POIs into 7 types based on type of the visited locations as well as according to Mathew Carmona's classification of UPSs [63], namely Open Space, Food and Drink, Life Service, Art and Entertainment, Shopping, Residence and Office; Education (Table 1). Also, Figure 1 demonstrates the POIs in each category. Then, the mixing POIs, including open space-oriented POIs (OS-POI), commercial-oriented POIs (C-POI) and other-oriented POIs (O-POIs) were categorized and assessed. A number of each OS-POIs, C-POIs, O-POIs were calculated in the designed grid. Here, OS-POIs comprise open public spaces such as park, square, etc. In contrast, C-POI includes food and average, shopping services, life services, art and entertainment. O-POI consists of residence, office and education.

However, check-in data has several constraints, namely sampling bias or location inconsistencies. Therefore, some standards were applied to refine the check-in data [64]: (1) POIs not covered in study area were removed; one or two-time check-in POIs were taken out; (3) user accounts having only one check-in were considered invalid; and (4) in 30-minute blocks, users who checked in more than once at the same place were counted as one record. After filtering, 66.055 check-in records are kept, with 344

POIs (Figure 1, Figure 2). Next, GIS software (ArcGIS 10.4) received all the data, and made a layer of x-y coordinated points, then each point was illustrated (Figure 3).

Table 1. Activity category classification.

Activity Category	Type of Visited Location
Open Space	Park, Square, Playground, Streets, etc.
Food & Drink	Coffee shop, Tea house, Restaurant, Pizza, Pub, Nightclub, Bar.
Life Service	Gym, Spa, Nail, Salon, etc.
Art & Entertainment	Entertainment, Theater, Concert Hall, Event Space, etc.
Shopping	Supermarket, Plaza, Pharmacy, Mall, Boutique, Bookstore, etc.
Residence and Office	Residential building, Office, etc.
Education	School, College, University, etc.

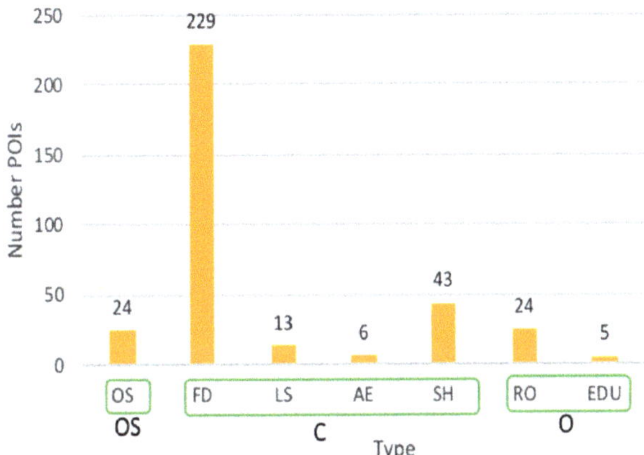

Figure 1. Total number of points of interest (POIs) in each category. Note: OS: Open space; FD: Food and Drink; LS: Life Service; AE: Art and Entertainment; SH: Shopping; RO: Residence and Office; EDU: Education; C: Commercial-oriented POI; OS: Open space-oriented POI; O: Other. Source: Authors.

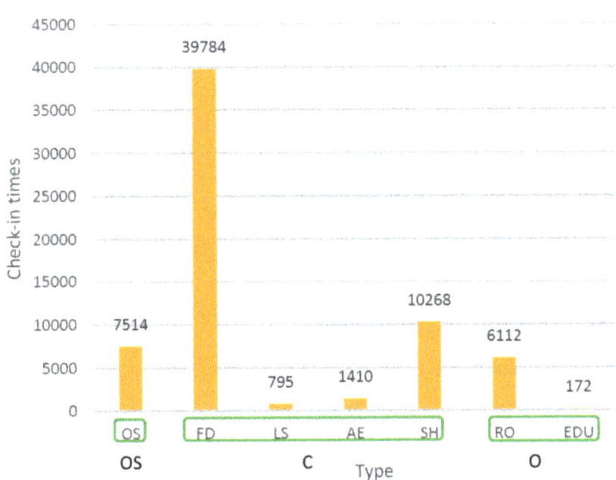

Figure 2. Total number of check-ins in each category. Source: Authors. **Note:** OS: Open space; FD: Food and Drink; LS: Life Service; AE: Art and Entertainment; SH: Shopping; RO: Residence and Office; EDU: Education; C: Commercial-oriented POI; OS: Open space-oriented POI; O: Other

Figure 3. Spatial grid and check-in data in the study area. Source: Authors.

Normalization of data and calculation of the number of check-ins during period time of day. Figure 4 shows the one-hour time intervals of Instagram posts frequency within the two districts throughout 24 hours. The frequency of Instagram posts fluctuates considerably as social media is used the most during the evening time. The peak time occurred at 9pm (6386 posts), whereas this figure fell to 1530 posts at 8am. Subsequently, the data was aggregated into 5 time slots to facilitate analysis.

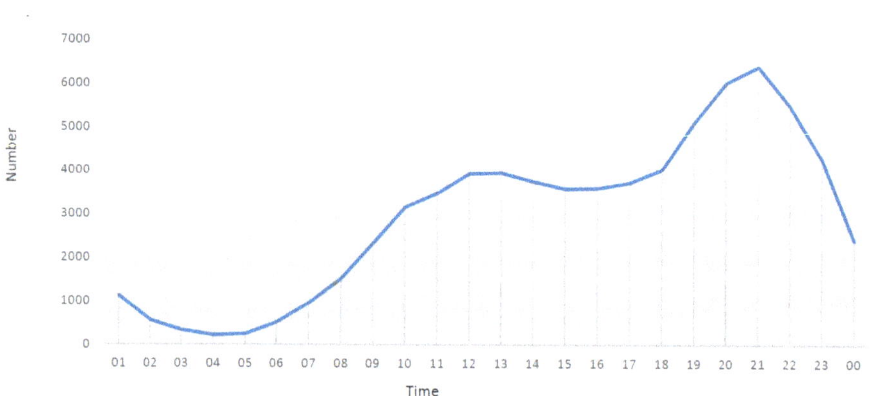

Figure 4. Temporal check-in densities for different activity categories.

5. Data Analysis

5.1. Spatial Distribution of the Most Popular Spaces

Figure 5 shows the spatial distribution of different activities in district 1 and district 3 of HCMC. For this study, urban activities are divided into specific timeslot with a division of per day, per weekdays, and between weekdays and weekends [50].

We counted the total number of check-ins for different activities within 24 hours in the 100-meter × 100-meter grid during weekdays and weekends. Subsequently, the Instagram activity and land use were linked and compared. By performing descriptive analysis based on the crawling data of geo-tagged Instagram posts in accordance with the predominant area in two central districts, Figures 5 and 6 show that for most activities related to C_POI spaces, the clear radiance becoming bolder from the morning to the evening, from weekdays to weekends, except for OS_POI areas (which had a pretty fair radiance) and O_POI areas (which showed more activity during weekdays rather than weekends). In HCMC, by an outnumber of Instagram check-in posts, commercial spaces take over a dominant role in citizens' lifestyle compared to open spaces, regardless of weekdays or weekends (as shown by Figure 5(a1,a2,b1,b2)). However, there are contrasting trends of two types of spaces: open spaces have fewer activities during weekends along with the shrinking size (Figure 5(a1,a2)); meanwhile, commercial spaces gain attractiveness in more diverse areas, particularly the periphery of research scope (Figure 5(b1,b2)). Therefore, in terms of spatial distributions, a visible contradiction can be detected among public spaces, commercial spaces and other spaces, with a growing tendency of leisure activities in commercial spaces; and the areas of public activity in open spaces, which were much more active during the daytime, meanwhile other spaces are not remarkable.

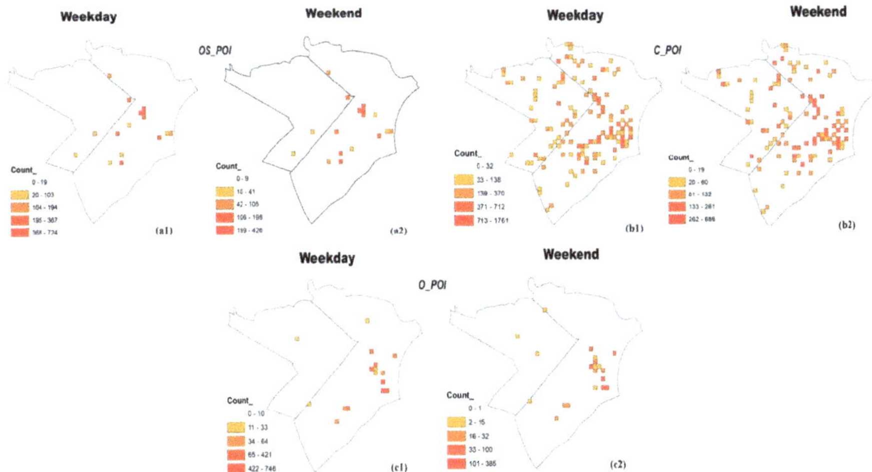

Figure 5. Spatial distributions of different activity categories.

5.2. Temporal Mobility Patterns

To display the temporal frequency of the two central districts' mobility, we examined the distribution of visits for activity purposes throughout different time courses of the day (see Figure 6). We also investigated the weekly regularity of these visits. Impressively, the number of commercial spaces have the most regular and highest visiting frequency compared with open spaces and other spaces (approximately 10 times higher). During the day, each space has two distinctive peaks: 12:00 and 20:00, noted that nightlife's activities are two times higher. There is also a fairly stable distribution among three investigated spaces, however, we can observe that weekends fairly host more activities than weekdays, regardless of space categories.

5.3. Kernel Density Estimation of Temporal–Spatial Distribution of the Popular Spaces

We conducted a kernel density analysis to find the density of each cell for all activities in different categories in 2-hour intervals. This Kernel Density Estimation (KDE) analysis helps comprehensive research on the vibrant transformation of activities in terms of both space and time. Figure 6 below shows the kernel density estimation results for district 1 and district 3 in HCMC, which point out the pattern of citizens' activities.

As for the spatial allocation, the central area at Ben Nghe ward presents obvious high frequencies and densities, especially the main agglomeration area which is bordered by the most crowded streets: Le Thanh Ton street, Nam Ky Khoi Nghia street, Ham Nghi street and Le Loi street; besides, there is also a high agglomeration near Notre-Dame Cathedral Basilica of Saigon. From the temporal perspective over space, people's activity frequencies are relatively high at 8:00–10:00; 12:00–14:00 and 18:00–20:00, which are associated with the eating time (breakfast, lunch and dinner) as well as the leisure nighttime for hanging out.

As shown in Figure 7, it is noteworthy that the permanently dense hotspots are also the most bustling and lively locations of HCMC in terms of individual density. Besides, stability both in space and time is also observed in the city center.

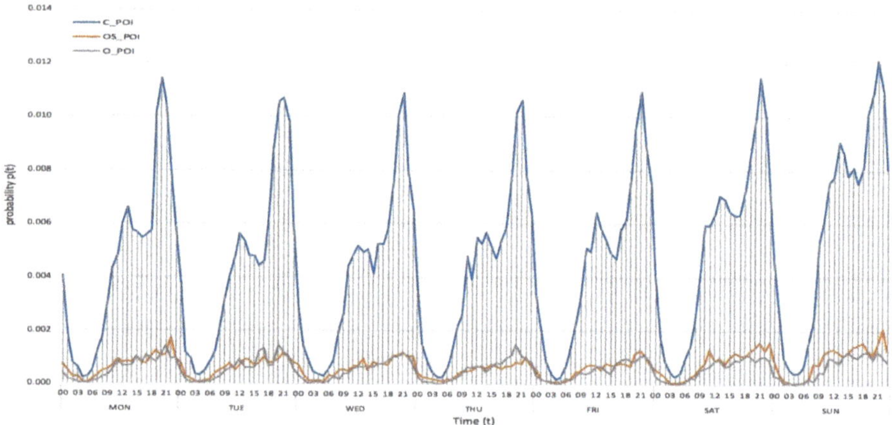

Figure 6. The average daily temporal trend (24 h) of Instagram posts in the study area.

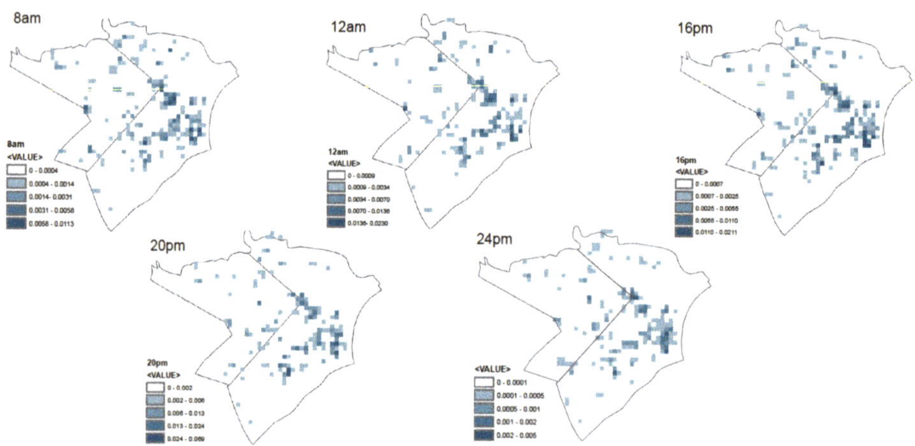

Figure 7. Kernel density estimation results of the check-in densities at different times.

5.4. Correlation Coefficient

Based on Equation (2) above, we examined a correlation coefficient of check-in numbers and kernel density estimation as shown in Figure 8. In the same group, the correlation coefficient can help facilitate the relation of scores among different measures. Accordingly, an intense direct relationship is considered as a high value (advancing toward +1.00), values around 0.50 are moderate, whereas a loose relationship are values below 0.30. In contrast, an intense inverted relationship is considered as a negative value (advancing toward –1.00), and values about 0.00 mean little or no relationship [56].

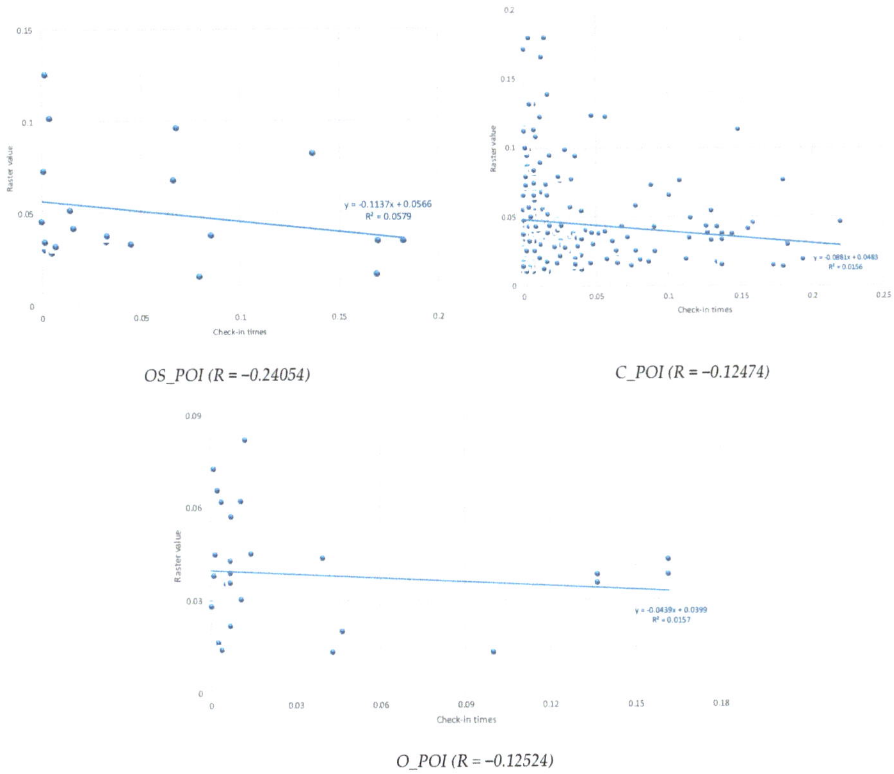

Figure 8. Pearson's corelation coefficient.

As shown in Figure 8, the R = –0.24, POI of Open Spaces revealed a relatively weak negative relationship between check-in data and the surrounding facilities. Meanwhile, with R = –0.12, both Other Spaces and Commercial Spaces showed little or negligible relationship. This result, surprisingly, implies that surrounding built environment has almost no influence on the attractiveness of one urban space.

6. Discussion and Conclusions

An assistant of big data for urbanization issues is more widely accepted due to the fast development of technology and the expansion of many social media platforms. Social media data is quite time-sensitive, a special characteristic, which can reflect status in real time. More importantly, it is a kind of public participation, which make it cheap and easy to collect. In this research paper, we used data from Instagram, which is one of the most popular social networks for Vietnamese citizens, to analyze their spatiotemporal patterns. Instagram provides a great tool and network for users to showcase where they are, when they come and what they do. These data can, therefore, illustrate urban activities and evolution over space and time.

We proposed a framework that integrates the spatial-temporal distribution for different activity categories, the urban public spaces' check-in time and UPSs surrounding built environment. This paper aims to concentrate on the correlation between the distribution of active Instagram posts regarding UPSs in the different periods of time in a day, days of week, and between weekdays and weekends, offering an initial understanding of urban dynamics, and pointing out the UPSs which gain or lose attractiveness in each time slot over the course of the day. Subsequently, we analyzed the spatial distribution of popular space in 2-hour time slots. The results show that it is noteworthy that the permanently dense hotspots are also the most bustling and lively locations of HCMC in terms of individual density. Besides, stability both in space and time is also observed in the central city.

Finally, with what is revealed by new sources of delocalized data, an inevitable direction for further studies will be to bridge the existing knowledge about the influence of the environment on the attractiveness of urban space. This Instagram-based approach has enabled us to recognize visually that commercial-related public spaces play a dominant role in the urban dynamics of HCMC. Commercial spaces like retail areas are believed to be the most active in the evening and at the weekends, whereas open spaces host more activities in the morning on weekdays. This paper's finding is at odds with the urban planning stereotype that the surrounding environment such as public transportation (bus stop), public facilities (parks and shopping malls), and other commercial spaces, often help appeal people to get around. In the case of HCMC, it has proved the opposite: people are attracted to UPSs even though there are not many cultural and social specialties there.

This paper's research consolidates one assumption from our urban observation: different categories of urban spaces have different spatiotemporal patterns. Taking advantage of this newly emerging source of social network data makes it feasible to measure, analyze, model and predict. Consequently, this real-time spatiotemporal distribution can be a good estimation for further urban planning as the outcomes help to seek more insight into urban vibrancy in relation to diverse functional spaces. This is desirable because by understanding the link between activities and land use, governments and related organizations can predict future trend patterns of citizens' activity as well as new urban sustainable developments.

However, this research has some limitations. The approach in this paper overcomes the purely descriptive limitation by using KDE analysis in order to examine the attractiveness of different spaces in the two populous central districts on Instagram activity according to time slot. It is obvious that there are some limitations with Instagram data because it cannot represent the entire population of the city; therefore, we can use other social media platforms for future comprehensive research along with the Instagram data. If possible, researchers should obtain data from many social platforms within the whole city to create more objective and comprehensive understanding ho w vitality in relation to urban design. Other indicators for measuring facilities in urban surroundings as proposed by Prem Chhetri et al. [65] e.g, aesthetic, amenity, and social interaction should also be included in the model to gain more deeper understanding. Last but not least, in addition to the correlation coefficient applied in this paper, researchers could also try spatial regression model or ordinary least squares to look at the comparative relationship between each surrounding urban environment indicator with people's choice at any particular place.

7. Data Availability Statement

Some or all data, models, or code generated or used during the study are available from the corresponding author by request:

List of data generated or used during the study:

1. Point of interest data of Ho Chi Minh City collected from the Google map in 2017.
2. Check-in database collected from Instagram in 2016 and 2017 in Ho Chi Minh City.
3. Street map of Ho Chi Minh City collected in Open Street Map.

Author Contributions: Conceptualization, H.H., T.V.T.N. and N.S.; Methodology: T.V.T.N.; Validation: H.H. and T.V.T.N.; Formal Analysis: T.V.T.N.; Investigation: H.H. and T.V.T.N.; Resource: H.H. and T.V.T.N.; Data Curation: T.V.T.N.; Writing-Original Draft Preparation: T.V.T.N.; Writing-Review and Editing: H.H., T.V.T.N. and N.H.; Supervision: H.H.; Funding Acquisition: H.H.

Funding: This research is funded by the National Natural Science Foundation of China, Project No.: 51778560.

Acknowledgments: We thank Xianfan Shu for his helpful advices and suggestions in ArcGIS analysis. We are grateful to Tram Ngoc Lam for her research assistance.

Conflicts of Interest: The authors declare no conflict of interest.

References

1. Dick, H.; Rimmer, P. Beyond the Third World City: The New Urban Geography of South-east Asia. *Urban Stud.* **1998**, *35*, 2303–2321. [CrossRef]
2. Shatkin, G. The City and the Bottom Line: Urban Megaprojects and the Privatization of Planning in Southeast Asia. *Environ. Plan. A Econ. Space* **2008**, *40*, 383–401. [CrossRef]
3. Madaniour, A. Why are the design and developement of public spaces significant for cities? *Environ. Plan. B Plan. Des.* **1999**, *26*, 879–891. [CrossRef]
4. Carr, S.; Francis, M.; Rivlin, L.G.; Stone, A.M. *Public Space*; Cambridge University Press: Cambridge, UK, 1992.
5. Afacan, Y. Achieving Inclusion in Public Spaces: A Shopping Mall Case Study. In *Designing Inclusive Systems*; Springer: Berlin/Heidelberg, Germany, 2012; pp. 85–92.
6. Akkar Ercan, Z.M. Public spaces of post-industrial cities and their changing roles. *METU J. Fac. Archit.* **2007**, *24*, 115–137.
7. Chang, H.S.; Liao, C.H. Exploring an intergrated method for measuring the relative spatial equity im public facilities in the context of urban parks. *Cities* **2011**, *28*, 361–371. [CrossRef]
8. Kazemi, A.V.; Dousti, F.; Behzadfar, M. A New Reading of Sociable Public Spaces: The Nexus between Urban Design and Microsociology. *Armanshaltr Archit. Urban Dev.* **2018**, *11*, 39–49.
9. Rung, A.L.B.; Mowen, A.; Cohen, D.A. The significance of parks to physical activity and public health: A conceptual model. *Am. J. Prev. Med.* **2005**, *28*, 159–168. [CrossRef]
10. Ding, D.; Sallis, J.F.; Kerr, J.; Lee, S.; Rosenberg, D.E. Neighborhood environment and physical activities among youth. *American Journal of Preventive Medicine.* **2014**, *41*, 442–455. [CrossRef]
11. Ghavampour, E. The Contribution of Natural Design Elements to the Sustained Use of Public Space in a City Centre. Ph.D. Thesis, Victoria University of Wellington, Wellington, New Zeland, 2014.
12. Aspinall, P. On environmental preference: applying conjoint analysis to visiting parks and buying houses. In *Innovative Approaches to Research Landscape and Health: Open Space: People Space 2*; Routledge: Abingdon, UK, 2010.
13. Kaplan, R. The role of nature in the urban context. In *Environment and Behaviroural Studies*; Altman, I., Christensen, K., Eds.; Pelnum Press: New York, NY, USA, 1983.
14. Koohsari, M.J.; Kaczynski, A.T.; Giles-Corti, B.; Karakiewicz, J.A. Effects of access to public open spaces on walking: Is proximity enough? *Landsc. Urban Plan.* **2013**, *117*, 92–99. [CrossRef]
15. Schipperijn, J.; Stigsdotter, U.K.; Randrup, T.B.; Troelsen, J. Influences on the use of urban green space – A case study in Odense, Denmark. *Urban For. Urban Green.* **2010**, *9*, 25–32. [CrossRef]
16. Tzoulas, K.; Korpela, K.; Yli-Pelkonen, V.; Kazmierczak, A.; Niemela, J.; James, P. Promoting ecosystem and human health in urban areas using green infrastrucutre: A literature review. *Landsc. Urban Plan.* **2007**, *81*, 167–168. [CrossRef]
17. Chen, Y.; Liu, T.; Xie, X.; Marusic, B.G. What attracts people to visit community open spaces? A case study of the overseas Chinese town community in Shanzhen, China. *Environ. Res. Public Health* **2016**, *13*, 644. [CrossRef] [PubMed]
18. Coutts, C. Multiple Case Studies of the Influence of Land-Use Type on the Distribution of Uses along Urban River Greenways. *J. Urban Plan. Dev.* **2009**, *135*, 31–38. [CrossRef]
19. Han, H.; Sahito, N.; Nguyen, T.V.T.; Hwang, J.; Asif, M.; Nguyen, T.T. Exploring the Features of Sustainable Urban Form and the Factors that Provoke Shoppers towards Shopping Malls. *Sustainability* **2019**, *11*, 4798. [CrossRef]
20. Kaczynski, A.T.; Potwarka, L.R.; Smale, B.J.A.; Havitz, M.E. Association of Parkland Proximity with Neighborhood and Park-based Physical Activity: Variations by Gender and Age. *Leis. Sci.* **2009**, *31*, 174–191. [CrossRef]

21. Sugiyama, T.; Gunn, L.D.; Christian, H.; Francis, J.; Foster, S.; Hooper, P.; Owen, N.; Giles-Corti, B. Quality of Public Open Spaces and Recreational Walking. *Am. J. Public Health* **2015**, *105*, 2490–2495. [CrossRef]
22. Brown, G. Mapping spatial attributes in survey research for natural resource managerment: Methods and applications. *Soc. Nat. Res.* **2005**, *18*, 17–39. [CrossRef]
23. Gehl, J.; Richard, L.R. *Cities for People*; Island Press: Washington, DC, USA, 2013.
24. Abaza, M. Shopping Malls, Consumer Culture and the Reshaping of Public Space in Egypt. *Theory Cult. Soc.* **2001**, *18*, 97–122. [CrossRef]
25. Stillerman, J.; Salcedo, R. Transposing the Urban to the Mall: Routes, Relationships, and Resistance in Two Santiago, Chile, Shopping Centers. *J. Contemp. Ethnigr.* **2012**, *41*, 309–336. [CrossRef]
26. Sui, D.; Goodchild, M. The convergence of GIS and Social media: Challenges for GISscience. *Int. J. Geogr. Inf. Sci.* **2011**, *25*, 1737–1748. [CrossRef]
27. Croitoru, A.; Wayant, N.; Crooks, A.; Radzikowski, J.; Stefanidis, A. Linking cyber and physical spaces through community detection and clustering in social media feeds. *Comput. Environ. Urban Syst.* **2014**, *53*, 47–64. [CrossRef]
28. Lin, J.; Cromley, R.G. Evaluating geo-located Twitter data as a control layer for areal interpolation of population. *Appl. Geogr.* **2015**, *58*, 41–47. [CrossRef]
29. Shelton, T.; Poorthuis, A.; Graham, M.; Zook, M. Mapping the data shadows of Hurricane Sandy: Uncovering the cocial soatial dimensions of big data. *Geoforum* **2014**, *52*, 167–179. [CrossRef]
30. Boyd, D.; Crawford, K. Critical questions for big data: provocations for a culture, technological, and scholarly phenomenon. *Inf. Commun. Soc.* **2012**, *15*, 662–679. [CrossRef]
31. Martí, P.; Serrano-Estrada, L.; Nolasco-Cirugeda, A. Using locative social media and urban cartographies to identify and locate successful urban plazas. *Cities* **2017**, *64*, 66–78. [CrossRef]
32. Zhai, S.; Xu, X.; Yang, L.; Zhou, M.; Zhang, L.; Qiu, B. Mapping the popurarity of urban restaurants using social media data. *Appl. Geogr.* **2015**, *63*, 113–120. [CrossRef]
33. Lane, N.D.; Eisenman, S.B.; Miluzzo, E.; Compbell, A.T. Urban sensing systems. In Proceedings of the 9th Workshop on Mobile Computing Systems and Applications—Hot Mobile, Napa Valley, CA, USA, 25–26 February 2008.
34. Noulas, A. *Human Urban Mobility in Location-Based Social Networks: Analysis, Models and Applications*; University of Cambridge: Cambridge, UK, 2013.
35. Silva, T.H.; De Melo, P.O.S.V.; Almeida, J.M.; Salles, J.; Loureiro, A.A.F. A comparison of Foursquare and Instagram to the study of city dynamics and urban social behavior. In Proceedings of the 2nd ACM SIGKDD International Workshop, Chicago, IL, USA, 11 August 2013.
36. Setha, L.; Taplin, D.; Scheld, S. *Rethinking Urban Parks: Public Space and Cultural Diversity*; University of Texas Press: Austin, TX, USA, 2005.
37. Oldenburg, R. *The Great Good Place*; Marlowe: New York, NY, USA, 1999.
38. Nguyen, T.V.T.; Han, H.; Sahito Nand Lam, N.T. The Bookstore-Café: Emergence of a New Lifestyle as a Third Place in Hangzhou, China. *Space Cult.* **2019**, *22*, 216–233. [CrossRef]
39. Low, S.; Smith, N. *The Politics of Public Space*; Routledge: New York, NY, USA, 2006.
40. Banerjee, T. The Future of Public Space:Beyond Invented Streets and Reinvented Places. *J. Am. Plan. Assoc.* **2001**, *67*, 9–24. [CrossRef]
41. Stock, K. Mining location from social media: A systematic review. *Comput. Environ. Urban Syst.* **2018**, *71*, 209–240. [CrossRef]
42. Song, C.; Koren, T.; Wang, P.; Barabasi, A.-L. Modelling the scaling properties of human mobility. *Nat. Phys.* **2010**, *6*, 818–823. [CrossRef]
43. Xiao, G.; Juan, Z.; Zhang, C. Travel mode detection based on GPS track data and Bayesian networks. *Comput. Environ. Urban Syst.* **2015**, *54*, 14–22. [CrossRef]
44. Tsou, M.-H.; Leitner, M. Visualization of social media: seeing a mirage or a message? *Cartogr. Geogr. Inf. Sci.* **2013**, *40*, 55–60. [CrossRef]
45. Preotiuc-Peotro, D.; Cohn, T. Mining user behaviours: A case study of check-in patterns in location based social networks. In Proceedings of the 5th Annual ACM Science Conference, Paris, France, 2–4 May 2013.
46. Frias-Martinez, V.; Soto, V.; Hohwald, H.; Frias-Martinez, E. Characterizing Urban Landscapes Using Geolocated Tweets. In Proceedings of the 2012 International Conference on Privacy, Security, Risk and Trust

and 2012 International Confernece on Social Computing, Amsterdam, The Netherlands, 3–5 September 2012; pp. 239–248.

47. Xia, C.; Hu, J.; Zhu, Y.; Naaman, M. What Is New in Our City? A Framework for Event Extraction Using Social Media Posts. In Proceedings of the Computer Vision—ECCV 2012, Firenze, Italy, 7–13 October 2012; Springer: Berlin/Heidelberg, Germany, 2015; Volume 9077, pp. 16–32.

48. Hyvärinen, O.; Saltikoff, E. Social Media as a Source of Meteorological Observations. *Mon. Weather Rev.* **2010**, *138*, 3175–3184. [CrossRef]

49. Boy, J.D.; Uitermark, J. How to Study the City on Instagram. *PLoS ONE* **2016**, *11*, e0158161. [CrossRef]

50. Candia, J.; González, M.C.; Wang, P.; Schoenharl, T.; Madey, G.; Barabasi, A.-L. Uncovering individual and collective human dynamics from mobile phone records. *J. Phys. A Math. Theor.* **2008**, *41*, 224015. [CrossRef]

51. O'Sullivan, D.; Unwin, D. *Geographic Information Analysis*; John Wiley & Sons: Hoboken, NJ, USA, 2003.

52. Agterberg, F.P. Interactive Spatial Data Analysis. *Comput. Geosci.* **1996**, *22*, 953–954. [CrossRef]

53. Silverman, B.W. *Density Estimation for Statistics and Data Analysis*; Chapman and Hall: London, UK, 1999.

54. Schabenberger, O.; Gotway, C.A. *Statistical Methods for Spatial Data Analysis*; Chapman and Hall: London, UK, 2017.

55. Miyake, K.K.; Maroko, A.R.; Grady, K.L.; Maantay, J.A.; Arno, P.S. Not Just a Walk in the Park: Methodological Improvements for Determining Environmental Justice Implications of Park Access in New York City for the Promotion of Physical Activity. *Cities Environ.* **2010**, *3*, 1–17. [CrossRef]

56. Correlation Coefficient: Simple Definition, Formula, Easy Steps. 2019. Available online: http://www.statisticshowto.datasciencecentral.com/probability-and-statistics/correlation-coefficient-formula/ (accessed on 10 September 2019).

57. Waibel, M. *Ho Chi Minh Mega City*; Regio Spectra Verlag: Berlin, Germany, 2013.

58. Long, Y.; Huang, C.C. Does block size matter? The impact of urban design on economic vitality for Chinese cities. *Environ. Plan. B Urban Anal. City Sci.* **2019**, *46*, 406–422. [CrossRef]

59. Donahue, M.L.; Keeler, B.L.; Wood, S.A.; Fisher, D.M.; Hamstead, Z.A.; McPhearson, T. Using social media to understand drivers of urban park visitation in the Twin Cities. *Landsc. Urban Plan.* **2018**, *175*, 1–10. [CrossRef]

60. Curran, K.; Fisher, G.; Crumlish, J. OpenStreetMap. *Int. J. Interact. Commun. Syst. Technol.* **2012**, *2*, 69–78. [CrossRef]

61. Shen, Y.; Karimi, K. Urban function connectivity: Characterisation of functional urban streets with social media check-in data. *Cities* **2016**, *55*, 9–21. [CrossRef]

62. Duggan, M. Mobile Messaging and Social Media. 2015. Available online: https://www.pewinternet.org/2015/08/19/mobile-messaging-and-social-media-2015/ (accessed on 19 August 2015).

63. Carmona, M. Contemporary public space, part two: Classsification. *J. Urban Des.* **2010**, *15*, 123–148. [CrossRef]

64. Wu, L.; Zhi, Y.; Sui, Z.; Liu, Y. Intra-Urban Human Mobility and Activity Transition: Evidence from Social Media Check-In Data. *PLoS ONE* **2014**, *9*, e97010. [CrossRef] [PubMed]

65. Prem, C.; Robert, S.; John, W. Modelling the Factors of Neighbourhood Attractiveness Reflected in Residential Location Decision Choices. *Stud. Reg. Sci.* **2006**, *36*, 393–417.

sustainability

Article

Comparison on Multi-Scale Urban Expansion Derived from Nightlight Imagery between China and India

Liang Zhou [1,2,3,4], Qinke Sun [1,3,4,*], Xuewei Dang [1,3,4] and Shaohua Wang [5,*]

1   Faculty of Geomatics, Lanzhou Jiaotong University, Lanzhou 730070, China
2   State Key Laboratory of Resources and Environmental Information System, Institute of Geographic Sciences and Natural Resources Research, CAS, Beijing 100101, China
3   National-Local Joint Engineering Research Center of Technologies and Applications for National Geographic State Monitoring, Lanzhou 730070, China
4   Gansu Provincial Engineering Laboratory for National Geographic State Monitoring, Lanzhou 730070, China
5   School of Geographical Sciences and Urban Planning, Arizona State University, Tempe, AZ 85287, USA
*   Correspondence: 0217708@stu.lzjtu.edu.cn (Q.S.); swang344@asu.edu (S.W.)

Received: 2 July 2019; Accepted: 16 August 2019; Published: 20 August 2019

Abstract: "The Dragon and the Elephant" between China and India is an important manifestation of global multipolarization in the 21st century. As engines of global economic growth, the two rising powers have followed similar courses of development but possess important differences in modes of development and urban development, which have attracted the widespread attention of scholars. From a geospatial perspective, and based on continuous annual night light data (Defense Meteorological Satellite Program-Operational Linescan System, DMSP-OLS) from 1992 to 2012, this paper conducts a multi-scale comparative analysis of urban development differences between China and India by employing various approaches such as the Gini coefficient, Getis–Ord Gi* index, and the Urban Expansion Intensity Index (UEII). The results show that: (1) The urban land space of the two countries expand rapidly, with the average annual expansion rate of China and India being 5.24% and 3.85%, respectively. The urban land expansion rate in China is 1.36 times faster than that in India. Resource-typed towns in arid northwest China and the resource-typed towns in central India have developed rapidly in recent years. (2) The unbalanced development in India is more prominent than in China; and the regional and provincial development imbalances in China are shrinking, while India's imbalances are improving slowly and its regional differences are gradually widening. (3) The spatial pattern of land use in both countries shows significant coastal and inland differences. The difference between the east, the central regions, and the west is the main spatial pattern of China's regional development, while the difference between the north and the south is the spatial pattern of India's regional development. (4) There are obvious differences in the expansion intensity of core cities between the two countries. From 1997 to 2007, the expansion intensity of core cities in China was relatively higher than that in India, while that in India was relatively higher than that in China from 2007 to 2012.

Keywords: spatial patterns; spatial differences; DMSP-OLS; China; India

---

1. Introduction

China and India are two rising world powers and are often referred to as "the driving force of Asia in global change" [1]. To promote the rapid development of the economy, economic reforms and liberalization measures were implemented in China and India in 1978 and 1991, respectively. Since the implementation of reform and opening-up policy of China, the total GDP of China has increased from 0.15 trillion dollars in 1978 to 12.24 trillion dollars in 2017, with its world ranking having risen nine places during that period. Since the implementation of economic reforms in India, the total GDP has increased from 0.26 trillion dollars in 1991 to 2.6 trillion dollars in 2017, and its world ranking has risen by 10 places during that period. China and India have quickly developed into the world's second and sixth largest economies by virtue of their successive accession to the World Trade Organization (WTO). However, India currently has a low urbanization rate, which increased from 25.72% in 1991 to 33.60% in 2017 [2], while China saw its urbanization rate increased from 17.92% in 1978 to 58.52% in 2017 [3]. The urbanization rate of the two countries differs by 24.92% with a population of about 350 million. Similar developmental histories and different developmental paths and models of China and India make their competition and development relations widely concerned.

In the 1980s, scholars from the United States, Japan, and the United Kingdom conducted comparative studies on the economic development of China and India. Later, some scholars proposed the concept of "The Dragon and the Elephant" to describe their competitive relationship. The international discussion on "The Dragon and the Elephant" has also become a constant topic [4,5]. Most of the literature compared the sustainability of competition and development of the two countries from economic and political perspectives. For example, Mukherjee et al. [6] analyzed the differences in rural industrial development between China and India from the perspective of their inherent political systems. Hölscher et al. [7] compared developmental differences between China and India in terms of economic structural characteristics. Nigam studied the evolution of the Chinese and Indian economies and their impact from an economic perspective [8]. Compared with China's, India is regarded as the world's largest democracy, with a political system that is more in line with the Western concept of democracy. Therefore, the advantages and disadvantages of the developmental paths of China and India and the comparison of development models have aroused great interest among scholars [9,10]. However, in addition to the existing differences between China and India in terms of economics and politics, both have vast swathes of land with their own developmental characteristics, and regional spatial development differences are also therefore significant. Due to limitations in efficient spatial information acquisition and processing, relevant empirical research is difficult to fully develop. Some scholars have attempted to use statistical data to study the regional differences between Chinese and Indian urban systems [11]. However, traditional spatial statistics lack sufficient spatial information, and the spatial and temporal features and differences of regions are not detailed enough. Moreover, high-resolution remote sensing image data is costly and difficult to handle and adapt to large-scale and long-term spatial differences [12,13]. Nighttime remote sensing images provide a new means of gathering data for regional spatial difference research. The object of traditional remote sensing is the object of surface, while the object of nighttime light remote sensing is human. Research shows that nighttime light data can reflect the intensity of human activities. The total amount of regional lighting is positively correlated with GDP and population [14], which can be used to estimate regional GDP and population [15]. At the same time, nighttime light data can be used to extract urban built-up areas and study the spatial and temporal evolution of cities and towns [16]. Interestingly, night light remote sensing satellite imaging began in earnest in 1992, the second year of economic reform in India. Nighttime lighting data still provides complete and powerful support for the comparative study of the development of China and India many years after the economic reform. At present, there are more comparative studies on nighttime lighting data between the world [17], regions [18,19], China and Russia [20], China and America [21,22], as well as among different regions within China or India [23–26]. However, there are few comparative studies on the systems of the two countries with respect to one another.

Considering the special role, status, and similar development histories of China and India in the world, this paper carries out analysis at the regional, provincial (state), prefecture, and county (district) multi-level spatial scales. To understand the developmental differences between China and India, it is necessary to quantify their spatial and temporal evolution accurately. In the past, research has focused on economic and political differences but lacked a geospatial perspective to study differences in the spatial development between the two countries. In this paper, long-term sequence Defense

Meteorological Satellite Program-Operational Linescan System (DMSP/OLS) nighttime remote sensing images are selected as the basic data, and multiple methods—such as the Gini coefficient, Getis–Ord Gi* index, and extended intensity index—are used to conduct a multi-scale comparative analysis of urban development differences between China and India, and reveals the evolution of the law and the spatial pattern of urban development in the two countries.

2. Materials and Methods

2.1. Study Area

This article takes China and India as the research areas. China is situated in the Asia-Pacific core region combining eastern Asia to the west coast of the Pacific Ocean, which ranks third in the world of a land area about 9.6 million km^2 and is also the most populous country in the world. India is the largest country in the South Asian subcontinent, which ranks seventh in the world in land area, with an area of 2.98 million km^2 about one-third of China's land area and is the second most populous country after China. On a regional scale, China is divided into seven regions, and India is divided into six regions (Figure 1). At present, there are 35 administrative divisions in India, of which the state of Telangana was separated from Andhra Pradesh in 2014. In 2012, there were 34 first-level administrative divisions in India, including 27 states (Pradesh) and 7 union territories; 640 second-level administrative divisions, namely district units; and 3894 third-level administrative divisions, namely tehsil units, also known as towns in India. At the end of 2010, there were 34 first-level administrative divisions in China, including 23 provinces (including Taiwan), 4 municipalities, 5 autonomous regions, and 2 special administrative regions (Hong Kong and Macao). The number of second-level administrative division was 333, including prefecture-level cities, districts, and autonomous prefectures. The third-level administrative divisions include 2856 county-level units (including municipal districts and county-level cities). Different levels of administrative divisions are taken as research units in this paper (excluding Hong Kong, Macao, and Taiwan of China) to conduct an in-depth analysis of differences in regional development between China and India on a spatial scale.

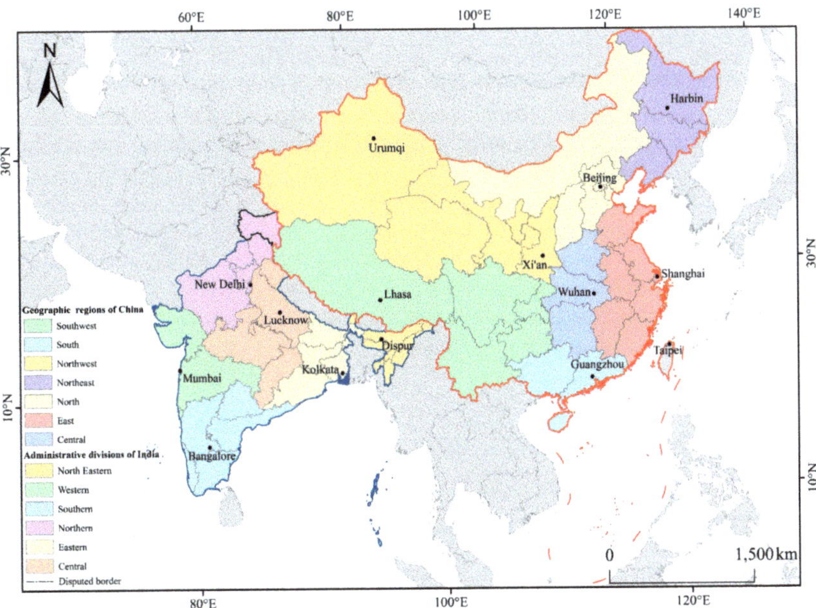

Figure 1. The location of the study area.

2.2. Data Sourcing and Preprocessing

At present, nighttime light (NTL) remote sensing data, obtained by the US Air Force DMSP-OLS sensor in the 1970s, have the longest time series and are the most widely used [27]. The data were released by the National Environmental Satellite Data and Information Service (NESDIS) (http://ngdc.noaa.gov/eog/index.html), part of the National Oceanic and Atmospheric Administration (NOAA). DMSP-OLS was originally used to monitor nighttime cloud volume and was later found to be able to detect light emitted by towns at night [28], which led to extensive research and the application of nighttime light remote sensing. At the same time, NTL brightness index reflects the activity intensity of the urban spatial economy, which is very suitable for large-scale urban spatial expansion research. The DMSP/OLS used in this study is the fourth edition of global nighttime lighting data. Six satellites were used to collect NTL data, among which two satellites were in service at the same year (1994, 1997–2007), and a total of 34 images were collected. DMSP/OLS non-radiation-calibrated nighttime light image includes three image types: global cloudless observation frequency image data, global average light image data, and global stable light image data. Globally stable light images include long-lasting light sources in cities, towns, and other places, and have eliminated the effects of accidental noise such as moonlit clouds and oil and gas combustion from them, which is useful for analyzing light in cities and towns. The night light image has a latitude ranging from 65° S to 75° N and a longitude ranging from 180° E to 180° W, with a spatial resolution of about 1 km and a digital number (DN) ranging from 0–63. The area where the DN value is 0 is an unlit area. The larger the DN value of the cell in the image, the larger the light intensity value of the area.

Since OLS sensors mounted on different DMSP satellites do not perform on-board calibration and mutual correction of image data, the data with a long time series between different sensors in different years is not continuous and comparable. In order to minimize the inter-annual error caused by different sensors and environmental factors, it is necessary to preprocess the original night light data. In this paper, the quadratic regression model proposed by Elvidge et al. [29] is used to conduct mutual calibration of the NTL image data from 1992 to 2012. The regression model is

$$DN_C = C_0 + C_1 \times DN + C_2 \times DN^2 \tag{1}$$

where C_0, C_1, C_2 are regression coefficients provided in Elvidge et al. [29], DN is the original pixel value of DMSP-OLS, and DN_C is the corrected pixel value. Elvidge et al. determine that F121999 has the brightest average values, so all other nighttime light image is intercalibrated to match its range. For our choice of an invariant area for the regression, since both China and India have a wide land area as the constant target area, we followed the method of Elvidge et al. global-scale regression in using Sicily.

The mutually calibrated images are cropped to the spatial extent of China and India, re-projected onto an Albert's equal-area projection, and resampled to 1 km. Since there are two satellites in service in some years, the stable light data of the same year were averaged as the pixel values of that year. For ease of study, the cells of DN < 3 were reclassified as 0, and the cells of the DN > 63 after intercalibration were reclassified to 63. Next, lights were summed for each region and province in China and India, generating total lights per region and province for every year from 1992 to 2012.

2.3. Methodology

2.3.1. Gini Coefficient

The Gini coefficient is a relative indicator that can generally reflect the degree of difference in the overall or partial distribution [30]. Its value ranges from 0 to 1, the value below 0.2 means 'absolute average', and above 0.5 means 'disparity'. This paper uses the Gini coefficient to study the balance of light at the regional and provincial levels in China and India. An increase of the Gini coefficient corresponds to an increase in inequality, indicating the spatial imbalance of national development.

Conversely, a decrease corresponds to an increase in the degree of equality, indicating that the country's development is spatially balanced.

2.3.2. Coefficient of Variation

The coefficient of variation is used to measure the difference in regional development at different scales and described the temporal and spatial dynamics of regional development at different scales in China and India. This coefficient has been widely used in geospatial difference research [31].

$$C_v = \frac{1}{\bar{x}} \sqrt{\sum_{i=1}^{n} (x_i - \bar{x})^2 / (n-1)} \tag{2}$$

where C_v is the coefficient of variation, n is the total number of cells corresponding to a certain scale, x_i is the average DN value of the ith unit, and $\bar{x}$ is the average of x_i.

2.3.3. Hot Spot Analysis

The Getis–Ord Gi* index is mainly used to determine the hot spots and cold spots of geographical phenomena [32]. This paper uses the Getis–Ord Gi* index to analyze the cold spots and hot spots at different scales in China and India, revealing the heterogeneity of the spatial pattern of development in China and India. The formula is

$$G_i^*(d) = \sum_{j=1}^{n} W_{ij}X_j - \bar{X}\sum_{j=1}^{n} W_{ij} / S \sqrt{(n\sum_{j=1}^{n} W_{ij}^2 - (\sum_{j=1}^{n} W_{ij})^2)/(n-1)} \tag{3}$$

where X_j is the attribute value of element j; W_{ij} is the space weight matrix; space is adjacent to 1, space is not adjacent to 0; n is the total number of elements; $\bar{X}$ is the mean; and S is the standard deviation. From the obtained P value and Z score, it can be found that the high value or low value element spatially generates a clustering position. If the P value is significant, the Z score is positive and the value is large, indicating that the high value clustering of the region is tight. If the Z score is negative and the value is small, the low values of the region are clustered. The tighter the class, the more prominent the cold spot. If the Z score is close to zero, there is no obvious spatial clustering.

2.3.4. Urban Expansion Intensity Index

In order to compare the strength of national core city expansion in different research periods, this paper uses the urban expansion intensity index to represent the spatial distribution of urban expansion for those periods [33]. The formula is

$$UEII = \frac{\left[[UDN_{t+i} - UDN_t] \times \frac{100}{i}\right]}{SDN} \tag{4}$$

UEII represents the expansion strength of the urban built-up area in the spatial range between t and $t+i$ years; UDN_{t+i} and UDN_t represent the area of the urban built-up area in $t+i$ and t, respectively; and *SDN* represents the spatial extent of the total area inside.

2.3.5. Extraction of Urban Built-Up Areas

Identifying the threshold of light which can represent the urban spatial range is very important for analyzing urban spatial expansion. Based on the urban spatial boundary extracted from high-resolution remote sensing image data, Henderson [34] selected core cities, such as San Francisco and Beijing, to determine the threshold of urban spatial boundary, extracted from night light data, and found that the overall precision rate of the Beijing urban area reaches 92.80% when DN exceeds 30. Ma et al. [35] divided the area of human activity into five categories using the division method of a local scale quadratic curve, and the area with a DN greater than 32 represented the suburb connected with the edge of the central city. Therefore, in order to make the urban area extracted by nighttime light data more realistic, this paper synthetically considers the best experiential segmentation threshold selected by predecessors, and on this basis, sets a dynamic threshold for each city's light data in different years. Then the threshold is constantly changed to match the data of a reference urban built-up area, and finally, the local optimal threshold is determined for each city in different years. After determining the area of urban built-up area by the local optimal threshold, the urban expansion of the city in i years was calculated as the area of urban built-up area in $t + i$ year minus the area of urban built-up area in t year.

3. Results

3.1. Spatial and Temporal Evolution Characteristics of National Scale Differences

3.1.1. Temporal Evolution Characteristics

As emerging powers, China and India have similar development processes, whose total lights have a similar evolutionary trend over time. In 1992, total lights in China were slightly higher than those in India. Under the continuous influence of economic reform and globalization strategies, India began to inflow a large amount of foreign capital, and its total lights briefly surpassed China's for the first time in 1996. In the following years, total lights in India closely followed China. However, after the accession to the WTO in 2001, China has rapidly integrated with the international community and the global economy, and the total lights in China changed greatly. As a result, the gap between the two countries is extending. India was affected by the third India-Pakistan war in 1999, during which heavy military expenditure seriously hindered India's economic development, and India's total lights gradually decreased from 2000 but began to develop rapidly in 2007 (Figure 2). It is worth noting that the average annual growth rates of China and India in 1992–2012 were about 5.24% and 3.85%, respectively, with China's growth being 1.36-times that of India.

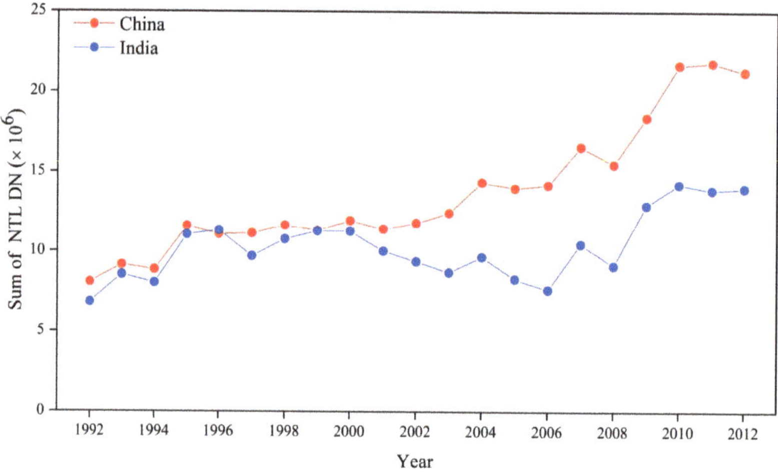

Figure 2. Annual total lights from 1992 to 2012 for China and India after intercalibration.

Descriptive statistics are used to calculate and analyze the changes in various types of light intensity on a national scale from 1992 to 2012. The DN values are classified according to the classification rules established by Henderson [34] combined with the light brightness values (Table 1), where DN = 0 represents "no illumination area", DN = 4–5 and DN = 6–10 represent "dim area", DN = 11–20 and DN = 21–62 represent "bright illumination area", and DN = 63 represents "light saturation area". Studies have shown that from 1992 to 2012, light in China and India both showed a trend of range extension and brightness increasing across the country. The light percentage of DN = 0 in China decreased from 92% in 1992 to 82% in 2012, while that in India decreased from 71% in 1992 to 50% in 2012, which further confirms the extensive diffusion of human beings over the earth since the early 1990s [36]. In particular, the sum of the proportions of three categories of pixel (DN = 4–5, DN = 6–10, and DN = 11–20) in India increased from 27 to 47%, along with the percentage of DN = 0 dropped from 71 to 50%, making the light distribution of India more uniform at low brightness levels. The categories of pixel values with the largest growth rate in China were DN = 21–62 and light saturation value. However, because of the dense population in southeastern China and the sparse population in northwestern China, the percentage of DN = 0 only decreased from 92 to 82%. As a result, the overall coverage of lights in India was wider than that in China, but the range of high brightness values in India was lower than in China (Figure 3).

Table 1. Descriptive statistics for lights in China and India

Attribute	India		China	
	1992	**2012**	**1992**	**2012**
DN = 0 (%)	70.82	49.59	91.95	82.31
DN = 4–5 (%)	19.15	26.12	3.89	5.86
DN = 6–10 (%)	5.97	15.91	2.05	6.80
DN = 11–20 (%)	2.50	5.41	1.14	2.24
DN = 21–61[a]/60[b] (%)	1.48	2.66	0.94	2.67
DN = 62[a]/61[b] (%)	0.08	0.31	0.02	0.12
Sum of all lights (DN)	6819899	13987576	8056700	21246952

[a] Top-code for intercalibrated composite in 1992. [b] Top-code for intercalibrated composite in 2012.

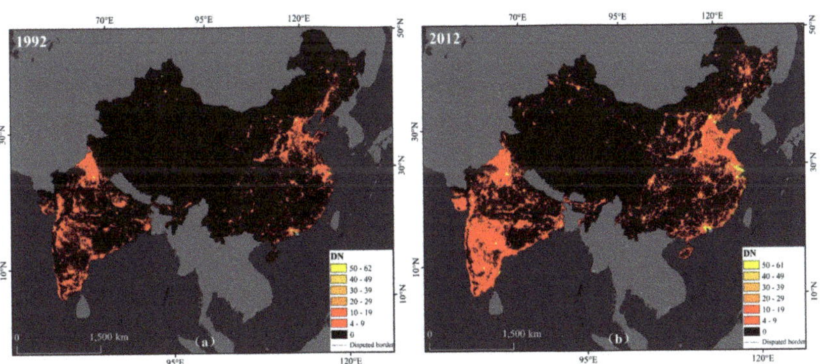

Figure 3. (**a**) Intercalibrated DMSP-OLS 1992 image over China and India. (**b**) Intercalibrated DMSP-OLS 2012 image over China and India.

3.1.2. Spatial Evolution Characteristics

In order to reflect the geographical significance of the spatial light variation between China and India over the past 20 years. we visualize the differences in lights between the first, middle, and final years of the 1992–2012 time period, the F101992, F152002, and F182012 composites were combined into a single three-band raster in ArcGIS 10.5. This resulted in the tritemporal map of lights for China and India. In Figure 4, 1992 values were input into the blue channel, 2002 into the green channel, and 2012 into the red channel, with the raster displayed using a standard deviation stretch (*n* = 2.5). In the result of composite, white indicates a stable area having a relatively constant brightness every year. Blue–purple represents a degraded area, that is, an area where the brightness is decreasing. Red represents the areas was brightest in 2012, suggesting recent growth. Green indicates the area is slow-growing in light, indicating that the lights in this region grew at the beginning of the 21st century, but that the growth rate was slow. As shown in Figure 4, the most obvious white areas in China are concentrated in the Beijing-Tianjin-Hebei, Yangtze River Delta, and Pearl River Delta urban agglomerations. The Yangtze River Delta urban agglomeration, with Shanghai as the core, is surrounded by red pixels, indicating that the cities near the Yangtze River Delta have grown more recently. The Beijing-Tianjin-Hebei urban agglomeration with Beijing as the core is surrounded by red and green pixels, indicating that the cities around Beijing-Tianjin-Hebei began to grow rapidly in the early 21st century. Although the Pearl River Delta urban agglomeration is one of the fastest growing regions in the world, it appears to be relatively white, possibly due to the limitations of DMSP-OLS sensors that cannot detect further light enhancement beyond the saturation level. In addition, the most obvious red areas in China were Hohhot and Ordos in Inner Mongolia and Urumqi in Xinjiang. Although located in the arid northwest, these newly growing cities have been able to grow rapidly because of their rich resources [37]. China's blue–purple regions, namely recession-type regions, are concentrated in cities such as Shenyang City in Liaoning Province and Changchun City in Jilin Province, which belong to the core cities of the old industrial bases in Northeast China, but due to China's reform and opening up policy and the southward shift of the economic center, a "Rust Belt", similar to that of Detroit in the United States, has appeared in northeast China.

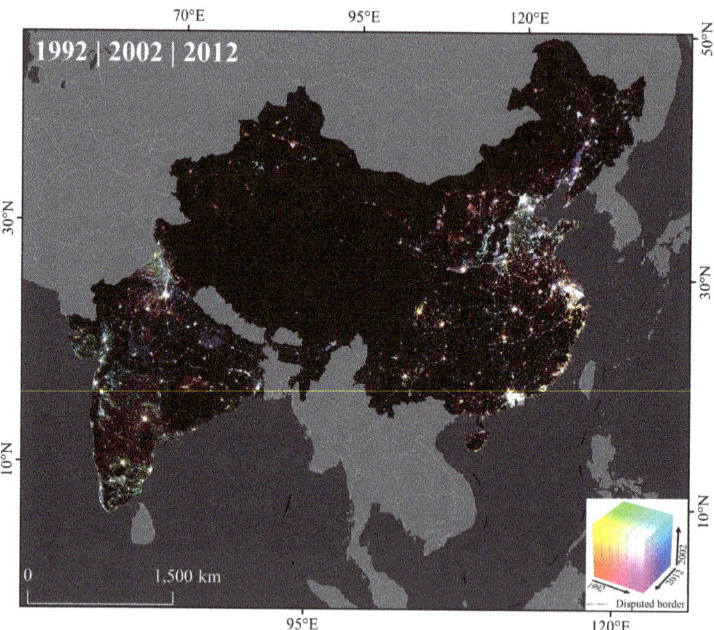

Figure 4. Tritemporal map composite showing F10 1992, F15 2002, and F18 2012 images over China and India.

The stable development zone of light in India is concentrated in big cities such as New Delhi, Mumbai, and Bangalore. The developments on both sides of New Delhi show an obvious difference, and the west side is red–green and the east side is blue–purple, indicating that the towns on the west side grow faster while the growth of towns on the east side is gradually declining. In addition, the difference in development between northern and southern India is obvious. The southern areas are mainly green, while red and purple predominate in northern areas. In the early 21st century, the southern area began to develop rapidly while development in the northern area lagged behind. There are more green areas in Himachal Pradesh and Punjab, due to the influence of the National Rural Employment Guarantee Act (NREGA) implemented by the Indian government in the early 21st century to address unemployment and poverty [38]. It is worth noting that India's newly growing cities are clustered in cities such as Ragar, in the north of Chhattisgarh and are growing rapidly on the basis of regional resources, while recession regions are located in cities such as Varanasi in Uttar Pradesh. Due to the large populations of these cities and the influence of traditional religions, a "Rust Belt" similar to that of Detroit in the United States has appeared.

3.2. Evolution Analyses of Regional and Provincial Scale Differences

3.2.1. Regional Differential Evolution Analysis

The "World Inequality Report", written by Thomas Piketty et al. [39], analyzed the global state of income inequality and highlighted that the degree of inequality was reduced to some extent because of the rise of China and India. Developing economies, led by China and India, have succeeded in closing the gap between poorer and more highly developed economies. The analysis of regional scale shows that the development levels of different regions of China and India shows increase tends (Figure 5). The northeastern areas in both China and India show different developing trends from other regions. Northeastern India maintains a flattening growth rate while Northeastern China shows high overall volatility. Forming an overall evolutionary trend of different regions in China and India, the development curve is divided into different stages, and it of China is divided into three stages, and India is divided into four stages. China experienced a period of rapid growth from 1992 to 1996, because China launched economic reforms under the leadership of Deng Xiaoping in 1992, and its economy maintained a sustained high growth rate. Between 1997 and 2001, China experienced a slowdown in development rate affected by the Asian financial turmoil in 1997–1998, though its economy was not directly affected due to the implementation of more cautious financial policies. The "Western Development" strategy was implemented to reduce regional inequality in 1999, and the "Rise of Central China" and "Northeast Revitalization" strategies were proposed in the following years. 2002–2012 was a sustained growth period for the development of China, and China's accession to the WTO in 2001 and the acceleration of opening-up have brought China's economic take-off to the forefront. China has rapidly integrated with the international community and the global economy and has gradually become the developing country that attracts the most foreign investment. Due to the impact of the subprime mortgage crisis in 2008, however, its growth rate declined. China adopted a series of protective measures to stimulate economic growth and took the lead in obviating the economic crisis, and in 2010, China became the second largest economy in the world. The development of India from 1992–2012 is divided into four stages: 1992–1996 was a period of rapid growth which witnessed the substantial positive effects of major economic reforms implemented in 1991 and have put the economy on a fast track of rapid growth during 1992 to 1996. The period from 1997 to 2004 was a stable stage because, due to the exchange crisis and the Asian financial turmoil in 1997, economic development did not improve significantly. The period between 2004 and 2008 was a stage of volatile decline. The impact of the subprime mortgage crisis in 2008 and the country's fragile economic base forced the economic growth rate to decline rapidly. In contrast, 2009–2012 was a period of rapid growth, and India entered the top 10 economies in the world in 2010.

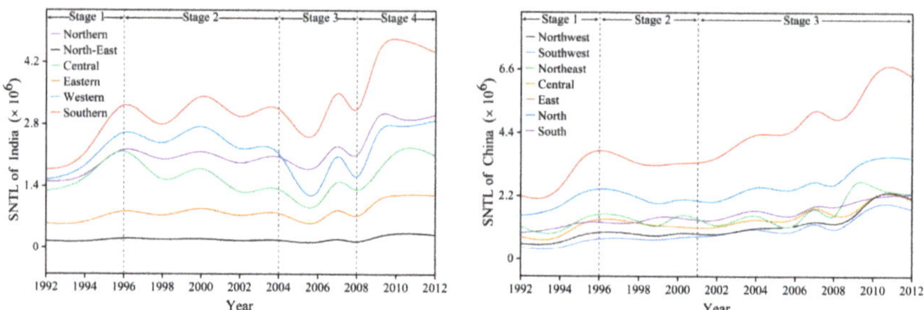

Figure 5. Time series of total lights plotted for each region in India and China.

3.2.2. Provincial Differential Evolution Analysis

In order to compare the development of the lights of the provinces (states) of China and India from 1992 to 2012, the natural logarithm of the lights was used to compare the provinces with large differences (Figure 6), and the curves in the figure were smoothed to highlight the overall varying trend. The overall lights in China and India show a growing trend, the difference between them was the variation rate of light. The change of lights in most provinces in China show a high rate, while the variation rate in most states in India tends to be flat and lower than that in various provinces in China. In 1992, the provinces with lower lights in China caught up with those with higher light, and some provinces in western and northern China, such as Tibet and Qinghai, experienced the fastest rate of variation in lights. These underdeveloped northwestern provinces have grown faster than developed regions such as Beijing and Shanghai. What is noteworthy is that Hainan Province, the southernmost island-province in China, has a slower rate of variation in light compared with that of the northwestern region of China, but it also shows a rapid growth trend, which can be attributed to slow development in the past and the recent strong government support for tourism [40]. In addition, western China is also rich in energy resources, such as oil and gas. At the beginning of the 21st century, the Chinese government applied energy development as part of the regional development strategy in northwest China [41]. In India, the fastest light variation rate occurs in the states of the northeast and west such as Sikkim, Goa, and Dadra-Nagar Haveli, which are growing faster than developed states such as Chandigarh, Kerala, and Maharashtra. It is worth noting that Himachal Pradesh, a state in the northwestern part of the Himalayas, has about 90% of its population living in rural areas, and its light variation rate shows a high growth trend. Its rapid growth of lights may be attributed to many plans formulated by the Indian government to combat poverty, unemployment, and inequality. A slower increase rate of light occurred in Uttar Pradesh in northern India, where the 'Rust Belt' is located, and similar with conditions in Jilin and Liaoning provinces in China.

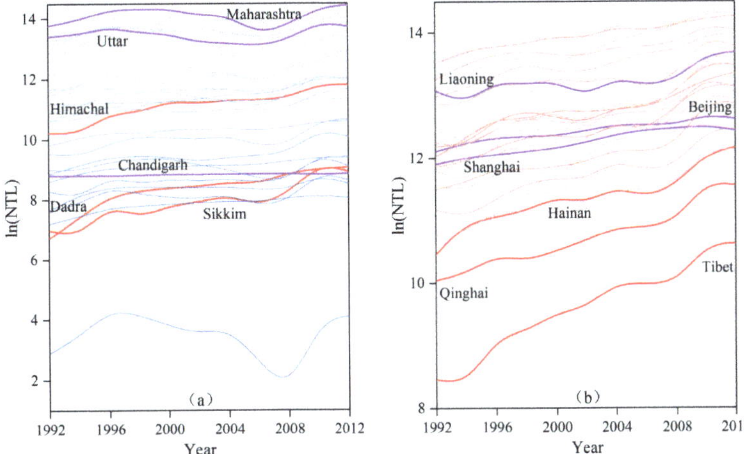

Figure 6. Time series of total lights plotted for each province in (**a**) India and (**b**) China.

3.2.3. Regional and Provincial Difference Measures

In order to analyze the developmental differences between China and India at the regional and provincial scales, and to explore the differences in evolution law, the Gini coefficient and coefficient of variation were used. The Gini coefficient is one of the indicators in the United Nations Commission on Sustainable Development (CSD) sustainable development evaluation index system. The Gini coefficient of lights is mostly used for comparison at the national level [42,43]. Here we combine the coefficient of variation to study development status at the regional and provincial levels in China and India. (1) Regional scale: The coefficient of variation among regions in China is larger than that in India, but the Gini coefficient of China is smaller than that in India (Figure 7a). The difference in development among different regions in China shows a slight decreasing trend. Both the Gini coefficient and the C_v rise at first and then decline. The Gini coefficient increases from 0.29 in 1992 to 0.32 in 1994 and then decreases to 0.24 in 2012. The C_v increases from 0.74 in 1992 to 0.84 in 2006 and then decreases to 0.76 in 2012. The differences between the six regions in India are still relatively obvious, and the regional development differences generally increase. The Gini coefficient and C_v rise first and then decline, showing an overall increasing pattern, increasing from 0.28 and 0.44 in 1992 to 0.32 and 0.49 in 2012, respectively, reaching a maximum of 0.37 and 0.62 in 2003. (2) Provincial scale: The Gini coefficient and coefficient of variation of China are smaller than those of India, and the regional difference in India is more obvious than that in China. The evolution of differences between provinces in China shows a decreasing trend on the whole (Figure 7b). The Gini coefficient rises first and then declines, increasing from 0.44 in 1992 to 0.46 in 1994 and then decreasing to 0.38 in 2012, while C_v shows a continuous decline, decreasing from 1.71 in 1992 to 1.23 in 2012. The evolutionary trend of the differences among the state in India is similar to that in China; that is, variations in the Gini coefficient and in C_v are similar, showing a slow and general decreasing trend. India's Gini coefficient increases from 0.64 in 1992 to 0.66 in 1999 and then falls to 0.63 in 2012, while C_v decreases from 2.16 in 1992 to 1.56 in 2012. In summary, the Gini coefficient and coefficient of variation of China and India differ in the evolution curves at different scales, and the years with the highest value are also different. The development difference between China and India is more obvious at the provincial scale than at the regional scale. This indicates that the scales have an important impact on development differences. Within the same research time series, even if the same indicators and measurement methods are adopted, the statistical results will greatly depend on the specific scale.

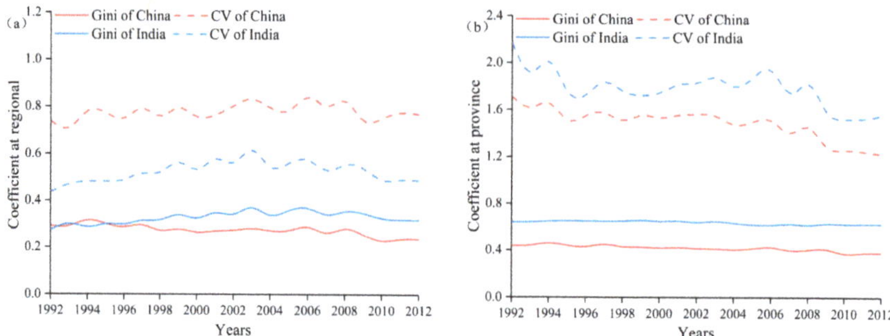

Figure 7. Gini coefficient and C_v at (**a**) regional level and at (**b**) province level in China and India.

3.3. Spatial Pattern Analyses of Multi-Scale Regional Differences

Here, we further analyze spatial correlation characteristics of regional development and spatial heterogeneity characteristics at different scales and reveal the mechanisms of regional development of China and India. The Getis–Ord Gi* index is used to analyze at the provincial, prefecture, and county level scales and to express them as hierarchical symbols. First, on the provincial scale, the pattern of hot spots in China has changed little, and cold spots have clustered further, while the hot and cold spots in India are relatively stable (Figure 8a,b). India's hot spots are always located in Punjab, Haryana, Himachal Pradesh, and New Delhi in northern India, while its cold spots are concentrated in the northeastern states of Meghalaya and Assam. The hot spots in China have evolved from Beijing-Tianjin-Hebei and Jiangsu to Beijing-Tianjin-Hebei, and Zhejiang and Jiangsu. The cold spots are further concentrated in the northwest and southwest of China. The results at the prefecture level scale show that China's hot spots are converging, while its cold spots are shrinking from east to west (Figure 8c,d). Cold-spot distribution in India is distinct and the number of cold spots in the north is increasing. In 1992, China's hot spots were mainly distributed in the core cities of the Beijing-Tianjin, Dalian, Jinan, Nanjing, Shanghai, and Guangzhou in southern China. In 2012, the hot spots were further distributed in the Beijing-Tianjin-Hebei, Yangtze River Delta urban agglomerations. The number of hot spots in the Yangtze River Delta region increased, and the hot spots of Jinan and Dalian evolved into sub-hot spots, while Zhengzhou City in Henan Province evolved into a new hot spot. The pattern of cold spot areas in China was relatively stable, mostly distributed in the central and western parts of the country, and the number of cold spots also gradually decreased from east to west. In 1992, India's prefecture-level hot spots were mainly distributed in Chennai in Tamil Nadu, Pune in Maharashtra, Mumbai, Punjab, and cities near New Delhi. In 2012, hot spots further expanded on the basis of 1992, especially those near southern Tamil Nadu [44]. The distribution of cold spots in India was relatively stable, mostly distributed in central and northeastern India, and the quantitative change shows an increasing trend in the north. The results of the county level Getis–Ord Gi* values indicate that the spatial patterns and evolutionary characteristics of hot and cold spots are more detailed (Figure 8e,f). The differences between coastal and inland areas in China have become increasingly diverse, and hot spots have spread widely in the eastern coastal areas, while the number of cold spots has increased inland. India's north–south gap have gradually widened, cold spots have been concentrated in northern India, and hot spots are widely distributed in the south and near the capital New Delhi. In 1992, China's hot spots were distributed in most parts of the Northern China, Eastern China, and Southern China. In 2012, the hot spots spread further and were widely distributed in the eastern coastal areas, and some parts of Xinjiang evolved into sub-hot spots. The number of cold spot areas increased, spreading from southwestern China to parts of southwestern, northwestern, and northeastern China. The transition zone between cold and hot spots is also more obvious. India's hot spots expanded further in areas such as Maharashtra, Kerala, Tamil Nadu, Punjab, and New Delhi, in which the number of hot spots near the southern state of Tamil Nadu increased most significantly, while the number of sub-hot spots was more concentrated in southern India. The cold spot area gradually shifts from the original north–south distribution to the north of India, and the distribution pattern change is obvious.

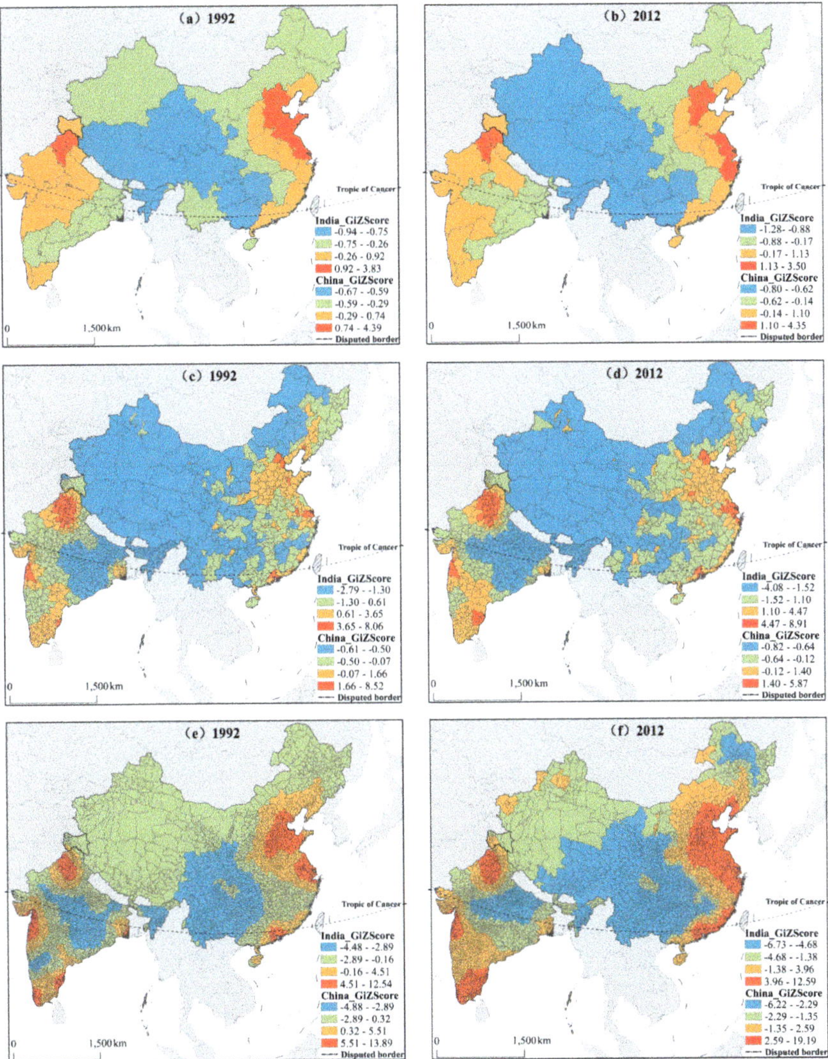

Figure 8. Distribution of Getis–Ord Gi* values in different levels in China and India: (**a**,**b**) at the province level; (**c**,**d**) at the prefecture level; (**e**,**f**) at the county level.

In general, China's hot and sub-hot spots at the provincial scale are distributed in the eastern coastal areas, and the cold spots are concentrated in the inland areas [45]. The hot spots in India are concentrated in the north and the cold spots in the northeast. At the prefecture-level scale, China's hot and sub-hot spots are concentrated in the eastern coastal areas dominated by the Beijing-Tianjin-Hebei, Yangtze River Delta, and Pearl River Delta urban agglomerations. Cold and sub-cold areas are concentrated in inland areas and gradually decrease from coastal areas to inland. The north–south distribution of India's cold hot spots is obvious, and the hot and sub-hot spots are mostly concentrated in the south and near the capital, New Delhi. The number of cold spots in the north is increasing. At the county level, China's hot and sub-hot spots are widely distributed in the eastern coastal areas, and the cold spots are concentrated in the inland areas and are increasing. India's hot and sub-hot spots are concentrated in the south and in the capital New Delhi, while the cold spots are more concentrated in Northern India.

3.4. Comparative Analyses of National Core Cities

Cities are areas with a high population density and have gathered a large number of innovative resources such as capital, technology, and manpower. Core cities with different functions have important influences throughout the country in the fields of politics, economy, and science and technology, and their development level is not only related to the consolidation and promotion of the their own status but also to the overall economic and social development of the country, as well as the international competitiveness and status of the country, thus core cities are also the hot areas in geography research [46]. Urban expansion is a process of spatial outward expansion of cities, especially municipal districts and built-up areas [47]. This paper selects cities that can represent the country's international division of labor, cooperation, and competition from the political, economic, scientific, and technological fields—such as Beijing-Delhi, Shanghai-Mumbai, and Shenzhen-Bangalore—to compare the spatial development differences between China and India. Nighttime lights data are used to simulate urban space expansion and compare urban expansion intensity.

After processing nighttime light data, a time-space expansion map of the core cities of China and India was obtained (Figure 9). Based on a comparative analysis of the core cities of the two countries in terms of an expansion mode and expansion intensity (Table 2), it can be found that China's core cities are all in the multi-core expansion mode, while India is dominated by the single-core expansion mode. The expansion model of Beijing and New Delhi are quite different. Beijing has a multi-core expansion model and developed in a coordinated way with the main city as the core and a multi-sub-core in the periphery, with the expansion strength, maintained at around 0.20 and at its maximum in 1997–2002, reaching 0.27. New Delhi has a single-core expansion model. The light area based on the main city area was gradually expanded, with the expansion strength reached a maximum of 0.45 in 2007–2012, while the expansion intensity was small in 1997–2007, all below 0.05. The expansion model of Shanghai and Mumbai are similar to those of Beijing and New Delhi, respectively. Shanghai has a multi-core expansion model and its expansion intensity has remained at a high level. The expansion intensity of Shanghai reached 0.75 in 2002–2007, higher than that of the core cities of China and India in the same period. Mumbai is not only the financial capital of India, but also the world's largest slum. From the perspective of the expansion mode of the light, Mumbai has less external expansion, and the expansion mainly based on the original foundation and the expansion strength is low, with the maximum expansion intensity being 0.24, which is at the lower level of the expansion intensity of the two countries. As the national science and technology innovation centers, the expansion modes of Shenzhen and Bangalore are dominated by the filling mode. Shenzhen centers on internal and external filling modes and is developing to the east. During 1992–1997, the expansion intensity reached the maximum value of 0.52. Bangalore was dominated by external filling modes, following an 'urban sprawl' model, whose expansion intensity has remained at a high level and reached 0.58 during 2007–2012, the maximum value in the core cities of China and India during the same period. In summary, the core cities of China and India have different expansion models. China's core cities all have multi-core expansion models, while India is characterized by a single-core expansion model. During 1997–2007, China's core urban expansion intensity remained at a relatively high level and was at a relatively low level in 2007–2012. India, however, had a lower level of expansion intensity in its core cities in 1997–2007 and had a high level in 2007–2012.

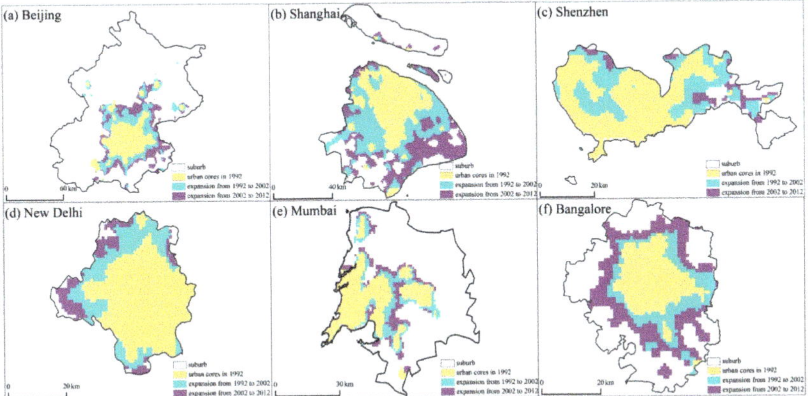

Figure 9. Changes in the lighting of core cities of China and India.

Table 2. Core cities' expansion intensity index of China and India in 1992–2012.

City	1992–1997	1997–2002	2002–2007	2007–2012
Beijing[p]	0.2589	0.2704	0.1929	0.2561
Delhi[p]	0.2365	0.0292	0.0053	0.4465
Shanghai[e]	0.5494	0.4164	0.7453	0.0882
Mumbai[e]	0.2433	0.0777	0.0259	0.2489
Shenzhen[s]	0.5159	0.1720	0.2337	0.0816
Bangalore[s]	0.3062	0.2694	0.3468	0.5756

p, *e*, and *s* represent the political, economic, and scientific centers of China and India, respectively.

4. Discussion

Nighttime light data can more closely reflect spatial information than traditional statistical data. Therefore, the use of light data instead of statistical data to analyze the temporal and spatial dynamics of Chinese and Indian towns is more objective, simple, and has a high inter-annual comparability. However, DMSP-OLS lighting satellites still have low resolution and supersaturation problems, which affect the identification of urban growth boundaries and, hence, research accuracy to some extent. In this study, we combined the empirical thresholds and the local-optimized threshold methods to detect the core cities' urban expansion of China and India during 1992–2012. This combined method not only identified the urban extents from both the large urban areas and small urban patches, but also effectively prevented some fragments of the suburban areas from being included in the extracted urban area for small- and medium-sized cities. In order to assess NTL-derived urban areas by this method, we compared the maps with the extracted urban areas from finer-resolution using The Atlas of Urban Expansion (TAUE) data which have been widely used to extract urban area in previous studies [48,49]. Because TAUE data has a higher resolution than NTL images (30 m vs. 1 km), it has been extensively used for assessing NTL-derived urban extents. Compared with the extracted urban areas from TAUE data, urban areas extracted using corrected NTL data well represented the spatial patterns in the study area. The average overall accuracy (OA) of urban expansion of Chinese and Indian cities in 2012 was 85.79%, and the average Kappa was 0.61. Our results are promising at the city level (Figure 10).

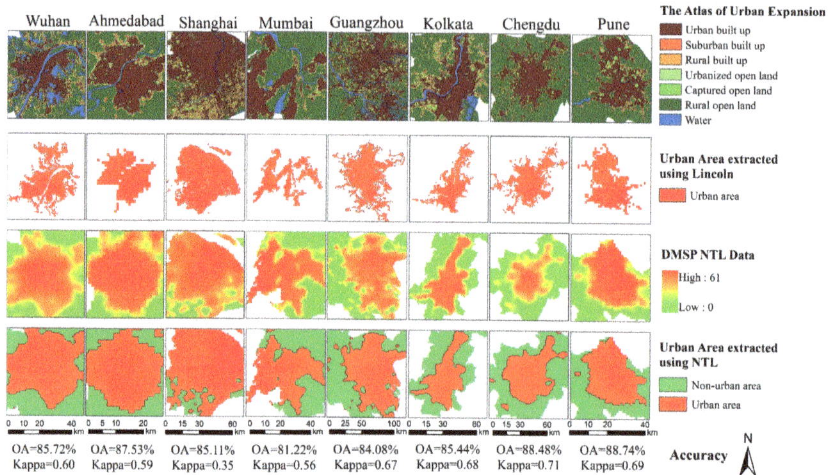

Figure 10. Accuracy assessment of extracted urban areas from nighttime light (NTL) data through comparison with the classified maps of The Atlas of Urban Expansion in 2012.

The urban areas extracted from TAUE data or NTL data still have many uncertainties, the main reasons for these can be categorized as: (1) the difference between impervious surfaces and NTL-based urban definitions, and (2) the oversaturation effects and coarser spatial resolution of DMSP-OLS NTL data, as well as radiation correction problems. In addition, although TAUE data are an urban area data extracted from Landsat Thematic Mapper (TM) data and proven to be feasible and acceptable, the available images still contain uncertainties due to weather effects. Meanwhile, data preprocessing inevitably leads to a loss in image fidelity. Therefore, in the future, DMSP-OLS data can be combined with a new generation of the Suomi National Polar Partnership-Visible Infrared Imaging Radiometer Suite (NPP-VIIRS) data [50] with a higher radiation resolution to improve the accuracy of town simulations. However, since nighttime lighting data from different satellites cannot be directly compared, research on strengthening data combinations should be explored in future work.

5. Conclusions

As the reform and opening-up policy enters its fourth decade, cities in China and India have developed rapidly. Based on the spatial study of night light data, it is found that the fastest growing areas in both countries are concentrated in the core cities: the Beijing-Tianjin-Hebei, Yangtze River Delta, and Pearl River Delta urban agglomerations in China; and New Delhi, Mumbai, and Bangalore in India. Cities in the resource-based regions of the two countries have developed rapidly recently, represented by Hohhot, Inner Mongolia and Urumqi, Xinjiang, in China; and Raigarh, Chhattisgarh in India.

The differences of urban development between China and India are manifested in the spatial differences in inland and coastal areas. The spatial pattern of regional development in China is mainly an east–central–west difference, while the spatial pattern of regional development in India is dominated by the north–south differences. The imbalance at the regional and provincial levels is more prominent both in China and India, and the developments imbalance among states in India is even more marked. Imbalanced development is one of the predicaments of SDGs. The Chinese government has made great achievements in reducing regional inequality. Although India has formulated a series of anti-poverty measures, the effect is not as significant as China.

Author Contributions: L.Z. and Q.S. conceived and designed the experiments; Q.S. and X.D. carried out the method; Q.S. performed the analysis and wrote the paper; L.Z. and S.W. reviewed and edited the manuscript.

Funding: This research was funded by National Natural Science Foundation of China (grant no. 41701173), The Strategic Priority Research Program of the Chinese Academy of Sciences (grant no. XDA19040502), Science Foundation for the Excellent Youth Scholars of Ministry of Education of China (grant no.17YJCZH268), LZJTU EP (grant no. 201806).

Acknowledgments: The authors would like to express their appreciation for the anonymous reviewers and journal editor whose comments have helped to improve the overall quality of this paper.

Conflicts of Interest: The authors declare no conflict of interest.

References

1. Kaplinsky, R.; Messner, D. Introduction: The Impact of Asian Drivers on the Developing World. *World Dev.* **2008**, *36*, 197–209. [CrossRef]
2. Tripathi, S. Do Economic Reforms Promote Urbanization in India? *Asia-Pac. J. Reg. Sci.* **2018**, 1–28. [CrossRef]
3. Wan, G. Introduction to the Special Section on "Urbanization in China". *China Econ. Rev.* **2018**, *49*, 141–142. [CrossRef]
4. Yuan, J. The dragon and the elephant: Chinese-Indian relations in the 21st century. *Washington Quarterly* **2007**, *30*, 131–144. [CrossRef]
5. Korukonda, A.R.; Carrillo, G.; Bathala, C.; Afza, M. The Dragon and the Elephant: A Comparative Study of Financial Systems, Commerce, and Commonwealth in India and China. *ICFAI J. Int. Bus.* **2007**, *2*, 7–20.
6. Mukherjee, A.; Zhang, X. Rural Industralization in China and India: Role of Policies and Institutions. *World Dev.* **2007**, *35*, 1621–1634. [CrossRef]
7. Hölscher, J.; Marelli, E.; Signorelli, M. China and India in the global economy. *Econ. Syst.* **2010**, *34*, 212–217. [CrossRef]
8. Nigam, N. China versus India: Emerging Giants in the World Economy. In *The China Business Model*; Paulet, E., Rowley, C., Eds.; Chandos Publishing: Cambridge, MA, USA, 2017; pp. 215–249.
9. Kan, K.; Wang, Y. Comparing China and India: A factor accumulation perspective. *J. Comp. Econ.* **2013**, *41*, 879–894. [CrossRef]
10. Jianglin, Z. China and India: A Comparative Study of Economic Development Stage. *South Asian Stud.* **2011**, *2*, 49–68.
11. Jiaming, L.; Yu, Y.; Jie, F.; Fengjun, J.; Wenzhong, Z.; Shenghe, L.; Bojie, F. Comparative research on regional differences in urbanization and spatial evolution of urban systems between China and India. *Acta Geogr. Sin.* **2017**, *72*, 986–1000.
12. Zhang, J.; Zhou, Q.; Shen, X.; Li, Y. Cloud Detection in High-Resolution Remote Sensing Images Using Multi-features of Ground Objects. *J. Geovis. Spat. Anal.* **2019**, *3*, 14. [CrossRef]
13. Frick, A.; Tervooren, S. A Framework for the Long-term Monitoring of Urban Green Volume Based on Multi-temporal and Multi-sensoral Remote Sensing Data. *J. Geovis. Spat. Anal.* **2019**, *3*, 6. [CrossRef]
14. Nordhaus, W.; Chen, X. A sharper image? Estimates of the precision of nighttime lights as a proxy for economic statistics. *J. Econ. Geogr.* **2014**, *15*, 217–246. [CrossRef]
15. Ghosh, T.; Powell, L.R.; Elvidge, D.C.; Baugh, E.K.; Sutton, C.P.; Anderson, S. Shedding light on the global distribution of economic activity. *Open Geogr. J.* **2010**, *3*, 148–161.
16. Yi, K.; Tani, H.; Li, Q.; Zhang, J.; Guo, M.; Bao, Y.; Wang, X.; Li, J. Mapping and Evaluating the Urbanization Process in Northeast China Using DMSP/OLS Nighttime Light Data. *Sensors* **2014**, *14*, 3207–3226. [CrossRef] [PubMed]
17. Elvidge, C.D.; Sutton, P.C.; Ghosh, T.; Tuttle, B.T.; Baugh, K.E.; Bhaduri, B.; Bright, E. A global poverty map derived from satellite data. *Comput. Geosci.* **2009**, *35*, 1652–1660. [CrossRef]
18. Stathakis, D.; Tselios, V.; Faraslis, I. Urbanization in European regions based on night lights. *Remote. Sens. Appl. Soc. Environ.* **2015**, *2*, 26–34. [CrossRef]
19. Small, C.; Elvidge, C.D. Night on Earth: Mapping decadal changes of anthropogenic night light in Asia. *Int. J. Appl. Earth Obs. Geoinf.* **2013**, *22*, 40–52. [CrossRef]
20. Bennett, M.M.; Smith, L.C. Using multitemporal night-time lights data to compare regional development in Russia and China, 1992–2012. *Int. J. Remote. Sens.* **2017**, *38*, 5962–5991. [CrossRef]
21. Weipan, X.; Xun, L.; Haohui, C. A comparative research on the rank-size distribution of cities in China and the United States based on urban nighttime light data. *Prog. Geogr.* **2018**, *37*, 385–396.

22. Kuang, W.; Chi, W.; Lu, D.; Dou, Y. A comparative analysis of megacity expansions in China and the U.S.: Patterns, rates and driving forces. *Landsc. Urban Plan.* **2014**, *132*, 121–135. [CrossRef]
23. Zhou, Y.; Ma, T.; Zhou, C.; Xu, T. Nighttime Light Derived Assessment of Regional Inequality of Socioeconomic Development in China. *Remote. Sens.* **2015**, *7*, 1242–1262. [CrossRef]
24. Zhang, Q.; Su, S. Determinants of urban expansion and their relative importance: A comparative analysis of 30 major metropolitans in China. *Habitat Int.* **2016**, *58*, 89–107. [CrossRef]
25. Ghosh, S.; Das, A. Exploring the lateral expansion dynamics of four metropolitan cities of India using DMSP/OLS night time image. *Spat. Inf. Res.* **2017**, *25*, 779–789. [CrossRef]
26. Pandey, B.; Joshi, P.; Seto, K.C. Monitoring urbanization dynamics in India using DMSP/OLS night time lights and SPOT-VGT data. *Int. J. Appl. Earth Obs. Geoinf.* **2013**, *23*, 49–61. [CrossRef]
27. Elvidge, C.D.; Cinzano, P.; Pettit, D.R.; Arvesen, J.; Sutton, P.; Small, C.; Nemani, R.; Longcore, T.; Rich, C.; Safran, J.; et al. The Nightsat mission concept. *Int. J. Remote. Sens.* **2007**, *28*, 2645–2670. [CrossRef]
28. Croft, T.A. Nighttime Images of the Earth from Space. *Sci. Am.* **1978**, *239*, 86–98. [CrossRef]
29. Elvidge, C.D.; Hsu, F.C.; Baugh, K.E.; Ghosh, T. National trends in satellite-observed lighting. *Glob. Urban Monit. Assess. Through Earth Obs.* **2014**, *23*, 97–118.
30. Domeij, D.; Flodén, M. Inequality trends in Sweden 1978–2004. *Rev. Econ. Dyn.* **2010**, *13*, 179–208. [CrossRef]
31. Champernowne, D.G.; Cowell, F.A. *Economic inequality and income distribution*; Cambridge University Press: London, UK, 1998.
32. Mitchel, A. *The ESRI Guide to GIS Analysis, Volume 2: Spartial Measurements and Statistics*; ESRI Press: Redlands, CA, USA, 2005.
33. Zou, Y.; Peng, H.; Liu, G.; Yang, K.; Xie, Y.; Weng, Q. Monitoring Urban Clusters Expansion in the Middle Reaches of the Yangtze River, China, Using Time-Series Nighttime Light Images. *Remote Sens.* **2017**, *9*, 1007. [CrossRef]
34. Henderson, J.V.; Storeygard, A.; Weil, D.N. Measuring Economic Growth From Outer Space. *Am. Econ. Rev.* **2012**, *102*, 994–1028. [CrossRef] [PubMed]
35. Ma, T.; Zhou, Y.; Zhou, C.; Haynie, S.; Pei, T.; Xu, T. Night-time light derived estimation of spatio-temporal characteristics of urbanization dynamics using DMSP/OLS satellite data. *Remote Sens. Environ.* **2015**, *158*, 453–464. [CrossRef]
36. Geldmann, J.; Joppa, L.N.; Burgess, N.D. Mapping Change in Human Pressure Globally on Land and within Protected Areas. *Conserv. Boil.* **2014**, *28*, 1604–1616. [CrossRef] [PubMed]
37. Zhou, L.; Zhou, C.; Yang, F.; Che, L.; Wang, B.; Sun, D. Spatio-temporal evolution and the influencing factors of PM2.5 in China between 2000 and 2015. *J. Geogr. Sci.* **2019**, *29*, 253–270. [CrossRef]
38. Ahuja, U.R.; Tyagi, D.; Chauhan, S.; Chaudhary, K.R. Impact of MGNREGA on rural employment and migration: A study in agriculturally-backward and agriculturally-advanced districts of Haryana. *Agric. Econ. Res. Rev.* **2011**, *24*, 495–502.
39. Alvaredo, F.; Chancel, L.; Piketty, T.; Saez, E.; Zucman, G. *World inequality report 2018*; Belknap Press of Harvard University Press: Cambridge, MA, USA, 2018.
40. Stone, M.; Wall, G. Ecotourism and Community Development: Case Studies from Hainan, China. *Environ. Manag.* **2004**, *33*, 12–24. [CrossRef] [PubMed]
41. Woodworth, M.D. Disposable Ordos: The making of an energy resource frontier in western China. *Geoforum* **2017**, *78*, 133–140. [CrossRef]
42. Li, X.; Ge, L.; Chen, X. Detecting Zimbabwe's Decadal Economic Decline Using Nighttime Light Imagery. *Remote. Sens.* **2013**, *5*, 4551–4570. [CrossRef]
43. Cauwels, P.; Pestalozzi, N.; Sornette, D. Dynamics and spatial distribution of global nighttime lights. *EPJ Data Sci.* **2014**, *3*, 2. [CrossRef]
44. Desmet, K.; Ghani, E.; O'Connell, S.; Rossi-Hansberg, E. The spatial development of India. *J. Reg. Sci.* **2015**, *55*, 10–30. [CrossRef]
45. Lu, Z.; Zuoquan, Z.; Wei, W. The spatial pattern of economy in coastal area of China. *Econ. Geogr.* **2014**, *34*, 14–19.
46. Liwei, W.; Chanchun, F. Spatial expansion pattern and its driving dynamics of Beijing-Tianjin-Hebei metropolitan region: Based on nighttime light data. *Acta Geogr. Sin.* **2016**, *71*, 2155–2169.

47. Angel, S.; Sheppard, S.; Civco, D.L.; Buckley, R.; Chabaeva, A.; Gitlin, L.; Kraley, A.; Parent, J.; Perlin, M. *The Dynamics of Global Urban Expansion*; Transport and Urban Development Department, The Wold Bank: Washington, DC, USA, 2005.

48. Xu, G.; Dong, T.; Cobbinah, P.B.; Jiao, L.; Sumari, N.S.; Chai, B.; Liu, Y. Urban expansion and form changes across African cities with a global outlook: Spatiotemporal analysis of urban land densities. *J. Clean. Prod.* **2019**, *224*, 802–810. [CrossRef]

49. Dong, T.; Jiao, L.; Xu, G.; Yang, L.; Liu, J. Towards sustainability? Analyzing changing urban form patterns in the United States, Europe, and China. *Sci. Total Environ.* **2019**, *671*, 632–643. [CrossRef] [PubMed]

50. Elvidge, C.D.; Baugh, K.E.; Zhizhin, M.; Hsu, F.C. Why VIIRS data are superior to DMSP for mapping nighttime lights. *Proc. Asia-Pac. Adv. Netw.* **2013**, *35*, 62–69. [CrossRef]

sustainability

Article

Impact of COVID-19 Induced Lockdown on Environmental Quality in Four Indian Megacities Using Landsat 8 OLI and TIRS-Derived Data and Mamdani Fuzzy Logic Modelling Approach

Sasanka Ghosh [1], Arijit Das [2], Tusar Kanti Hembram [2], Sunil Saha [2], Biswajeet Pradhan [3,*] and Abdullah M. Alamri [4]

1   Department of Geography, Kazi Nazrul University, Asansol 713340, West Bengal, India; sasankaghoshsg@gmail.com
2   Department of Geography, University of Gour Banga, Malda 732103, West Bengal, India; Arijit3333@gmail.com (A.D.); tusarpurulia1991@gmail.com (T.K.H.); sunilgeo.88@gmail.com (S.S.)
3   Centre for Advanced Modeling and Geospatial Information Systems (GIS), Faculty of Engineering and Information Technology, University of Technology Sydney, Sydney, NSW 2007, Australia
4   Department of Geology & Geophysics, College of Science, King Saud University, P.O. Box 2455, Riyadh 11451, Saudi Arabia; amsamri@ksu.edu.sa
*   Correspondence: Biswajeet.Pradhan@uts.edu.au

Received: 10 June 2020; Accepted: 3 July 2020; Published: 7 July 2020

Abstract: The deadly COVID-19 virus has caused a global pandemic health emergency. This COVID-19 has spread its arms to 200 countries globally and the megacities of the world were particularly affected with a large number of infections and deaths, which is still increasing day by day. On the other hand, the outbreak of COVID-19 has greatly impacted the global environment to regain its health. This study takes four megacities (Mumbai, Delhi, Kolkata, and Chennai) of India for a comprehensive assessment of the dynamicity of environmental quality resulting from the COVID-19 induced lockdown situation. An environmental quality index was formulated using remotely sensed biophysical parameters like Particulate Matters PM_{10} concentration, Land Surface Temperature (LST), Normalized Different Moisture Index (NDMI), Normalized Difference Vegetation Index (NDVI), and Normalized Difference Water Index (NDWI). Fuzzy-AHP, which is a Multi-Criteria Decision-Making process, has been utilized to derive the weight of the indicators and aggregation. The results showing that COVID-19 induced lockdown in the form of restrictions on human and vehicular movements and decreasing economic activities has improved the overall quality of the environment in the selected Indian cities for a short time span. Overall, the results indicate that lockdown is not only capable of controlling COVID-19 spread, but also helpful in minimizing environmental degradation. The findings of this study can be utilized for assessing and analyzing the impacts of COVID-19 induced lockdown situation on the overall environmental quality of other megacities of the world.

Keywords: COVID-19 pandemic; spatiotemporal analysis; environmental quality; PM_{10} concentration; GIS; remote sensing

---

1. Introduction

Environmental deterioration has emerged as a rising alarm in urban centres around the globe, exclusively in developing nations. The unrestrained urban expansion and subsequent industrialization, along with atypical population-boost, have magnetized environmental degradation as a burning concern for city planning [1]. After the liberalization in the 1990s, India has been incessantly experiencing economic advancement, the rapid growth of the urban centres, uncontrolled infrastructural improvement, and industrial expansion [2]. According to the Central Pollution Control Board of India (CPCB), the major five constituents of air pollution are particulate matters (PM), nitrogen oxide, sulpher dioxide (SO_2), carbon monoxide (CO), and ozone (O_3) and, among them, the most prevalent threats for human health are PM_{10} and $PM_{2.5}$. Since the ongoing decade, Indian metro cities have placed within the top 20 utmost polluted cities in the world exceeding the standard of specified air quality index (AQI) by the CPCB India and the World Health Organization (WHO) [3,4]. During 2015, in India, nearly one million lives were lost due to the high concentration of ambient PM [5].

Simultaneously, the environmental degradation vulnerability in the four megacities of India (Delhi, Kolkata, Mumbai, and Chennai) has increased, causing human health-related problems. According to the report (2016) of the World Health Organization (WHO) on the environmental performance index, Delhi is one of the most polluted cities of the world [6]. The revised environmental monitoring databases (from the year 2011 to 2016) among the largest megacities of 100 countries have listed Delhi on high rank in terms of PM_{10} concentration [7]. A typical death due to the heat wave in Delhi reveals the importance of Land Surface Temperature (LST) as articulated by Mohan and Kandya [8]. The concentration of PM_{10} in the city of Kolkata has exceeded the National Ambient Air Quality Standards (NAAQS) [9]. Changing patterns of the thermal environment, along with uncontrolled urbanization, are transforming vegetative lands into dry environments [10]. The satellite-based estimation of LST reveals a warming trend, i.e., average LST values range from 27.36 °C to 30.025 °C (from the year 1989–2006) and in 2010, it was 33.023 °C. Besides this, urban expansion has also reflected the impression on the alterations of other biophysical factors [10]. A similar trend in the ecological setup of the cities of Mumbai and Chennai can be found in studies of Kumar et al. (2016), Sundaram (2011), Partheeban et al. (2020), and Sathyakumar et al. (2020) [11–14].

Updated temporal and spatial information regarding environmental quality in the megacities of India could be the prerequisite to formulate new strategies to endorse further development. Urban Environmental Quality (UEQ) assessment depends on multidimensional aspects (spatial, physical, social, and economic traits of the city milieu) of the environment, which makes it more complex; this complex nature demands a comprehensive understanding of the environmental degradation process and its driving forces [15]. In this regard, one concerning issue is to determine the effective indicators of UEQ which may not only facilitate better mapping than can assist in city planning and sustainable growth as proficient management techniques. The remote-sensing-data-based Urban Environmental Quality (UEQ) assessment mainly provides information about land surface temperature, atmospheric elements such as air moisture and suspended particulate matters, vegetation status and biomass quantity, land use/land cover, and water [15]. Vegetation in the urban areas is a key variable to determine UEQ in several ways, i.e., freshening air through consuming CO_2, filtering sunlight and water, sieving pollutants, cooling heat influx, modifying local climatic setting, providing protection to animals and amusement field to society [16–20]. Land surface temperature (LST) is another foremost indicator of urban climate—controlling directly the comfortability and healthiness of urban environment, which is connected to the concept of urban heat island [15,21–23]. Various studies have regarded LST as an aspect of UEQ [24,25]. An exceeded standard amount of suspended particulate matters (PM_{10} and $PM_{2.5}$) in the air is a high threat for public respiratory systems produced from industries, vehicles, dust, and residential energy [5,26]. Keeping in mind the importance of all these parameters for assessing Urban Environmental Quality Index (UEQI), this study considered five main elements which are mentioned in the literature, namely, PM_{10} concentration, LST, NDVI, NDWI and NDMI. Although NDVI, NDWI, and NDMI are more or less stable in nature, aside from the undisturbed condition of the natural environment due to imposition of the lockdown phenomenon, these parameters may also change slightly, and, above all, these parameters are very important for assessing environmental quality [15,16,24,25].

COVID-19 is an extremely transmittable virus and was firstly originated in the Wuhan city of Central China during December 2019. Further, there have been outbreaks all over the world, especially in various nations of Asia, Europe, and the United States of America. Up to 9 May 2020, globally 3,822,382 COVID-19 positive cases have been confirmed and 263,658 people have died according to WHO. In India, to date, 59,881 people have been infected and 1990 deaths by COVID-19 have been recorded (https://www.covid19india.org/). To break the infection chain in India, a nationwide lockdown was enforced from 24 March 2020 till 14 April 2020, and later it was extended a 2nd time till 17 May 2020. This lockdown was also imposed on all industrial sectors (excluding emergency service sectors) and on the mass transportation systems. Hence, the consequence of the biophysical environment has shown a drastic change. Sharma et al. (2020) reported that pollution status has been radically reduced across the 88 cities in India due to lockdown upshot as per the data provided by CPCB [27]. Following this, Mahato et al. (2020) suggested that short-term lockdown can be assumed to be an operative substitute scheme to be executed to control air pollution to preserve standard environmental quality [1].

Although some research has been done on the effect of the COVID-19 pandemic on ecology in various cities of India, very little or no studies could be found regarding the impact of the COVID-19 induced lockdown situation on the overall bio-physical Environmental Quality (EQ) of Indian cities. Researches have shown a more specific interest in air quality index (AQI) assessment based on station-wise monitored secondary nonspatial data. The present research is an effort to trace the effect of the COVID-19 pandemic on the alteration of the physical environment using satellite-image-derived PM_{10} concentration, LST, NDVI, NDWI, and NDMI of the pre-lockdown period and during the lockdown period in four main big cities of India.

2. Study Area: Four Megacities of India

Four big cities of India, i.e., Mumbai, Delhi, Kolkata, and Chennai, were considered for assessing the impact of the COVID-19 induced lockdown situation on environmental quality. These four cities are strategically located in four different parts of India, playing a big role in the Indian economy (Figure 1). National Capital Territory (NCT) Delhi is the biggest Urban Agglomeration (UA) of India and the second most populated megacity of the world, with a population of 30.29 million [1] and with a population density of 11,297 people/sq. km. On the other hand, Mumbai is the ninth most populous megacity of the world, with a total population of 20.41 million. Mumbai is also ranked as one of the most polluted urban areas, having a 2.33 million vehicle count [11]. Kolkata has a total population of 14.85 million, which makes it the 16th largest megacity of the world, located in the eastern part of India. Chennai obtained the position of 30th most populated megacity of the world, with a population of 10.97 million. Due to intense emissions from vehicles, garbage, and biomass scorching, building construction, and natural space annihilation, the cities' ecological setups are under severe threat, and the environmental condition is gradually deteriorating for all four big cities, which are currently under investigation [8,28,29]. This increasing threat to the natural environment warrants an urgent assessment of the environmental quality during this dark time of the COVID-19 epidemic.

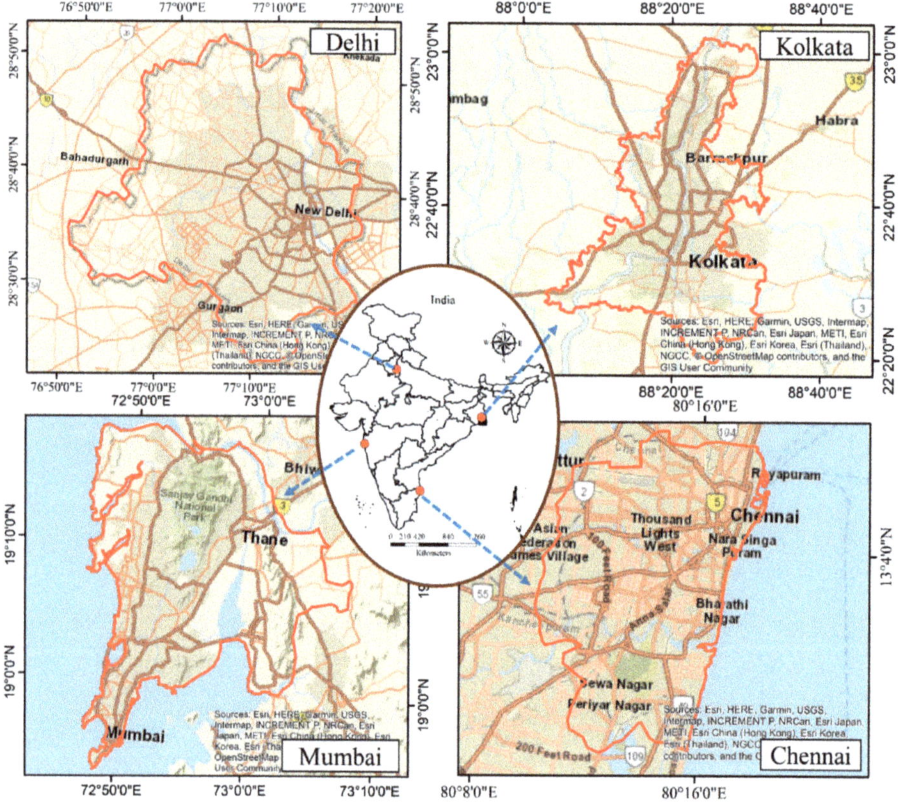

Figure 1. Location of the study areas.

3. Materials

Satellite Data

This research was performed using the satellite imageries which were downloaded from https://landsat.usgs.gov/ (Table 1). Landsat 8 OLI (Operational Land Imager) and TIRS (Thermal Infrared Sensor) images covering the selected megacities for both pre-lockdown and during the lockdown period were collected for further exploration. Landsat 8 satellites with two sensors, i.e., OLI (Operational Land Imager) and TIRS (Thermal Infrared Sensor), are comprised of nine spectral and two thermal bands. Spectral bands have a spatial resolution of 30 m, except the band 8 (panchromatic band of 15 m spatial resolution) and TIRS bands are attained at 100 m spatial resolution, though resampled into 30 m for the end user (Table 2).

Table 1. Corresponding dates, image ID, and other properties of satellite images for each of the megacities.

Study Area		Date	Sensor	Image ID	% of Cloud	Remarks
Mumbai	Same time of pre-lockdown, 2019	10\04\2019		LC08_L1TP_148047_20190410_20190422_01_T1	0.72	No cloud cover on the study area
	Pre-lockdown, 2020	11\03\2020		LC08_L1TP_148047_20200311_20200325_01_T1	5.77	7.29% on land but very less on the study area
	During lockdown, 2020	12\04\2020		LC08_L1TP_148047_20200412_20200422_01_T1	1.62	No cloud cover on the study area
Delhi	Same time of pre-lockdown, 2019	28\04\2019	LANDSAT 8 OLI	LC08_L1TP_146040_20190428_20190508_01_T1	0	No cloud cover on the study area
	Pre-lockdown, 2020	13\03\2020		LC08_L1TP_146040_20200313_20200325_01_T1	34.4	No cloud cover on the study area
	During lockdown, 2020	29\03\2020		LC08_L1TP_146040_20200329_20200409_01_T1	0.03	No cloud cover on the study area
Kolkata	Same time of pre-lockdown, 2019	20\04\2019		LC08_L1TP_138044_20190420_20190507_01_T1	18.64	No cloud cover on the study area
	Pre-lockdown, 2020	21\03\2020		LC08_L1TP_138044_20200321_20200326_01_T1	32.57	No cloud cover on the study area
	During lockdown, 2020	06\04\2020		LC08_L1TP_138044_20200406_20200410_01_T1	0.58	No cloud cover on the study area
Chennai	Same time of pre-lockdown, 2019	31\03\2019		LC08_L1TP_142051_20190331_20190404_01_T1	2.46	No cloud cover on the study area
	Pre-lockdown, 2020	17\03\2020		LC08_L1TP_142051_20200317_20200326_01_T1	3.54	No cloud cover on the study area
	During lockdown, 2020	02\04\2020		LC08_L1TP_142051_20200402_20200410_01_T1	0.28	No cloud cover on the study area

Table 2. Band details of the satellite imagery used in this study.

| Band Description | | |
Band	Spectral Resolution (μm)	Spatial Resolution (m)
Band 1-Coastal/Aerosol	0.433 to 0.453	30
Band 2-Visible blue (BLUE)	0.450 to 0.515	30
Band 3-Visible green (GREEN)	0.525 to 0.600	30
Band 4-Visible red (RED)	0.630 to 0.680	30
Band 5-Near-infrared (NIR)	0.845 to 0.885	30
Band 6-Short wavelength infrared1 (SWIR1)	1.56 to 1.66	30
Band 7-Short wavelength infrared2 (SWIR2)	2.10 to 2.30	30
Band 8-Panchromatic	0.50 to 0.68	15
Band 9-Cirrus	1.36 to 1.39	30
Band 10-Thermal Infrared (TIRS) 1	10.3 to 11.3	100 * (30)
Band 11-Thermal Infrared (TIRS) 2	11.5 to 12.5	100 * (30)

* represents the original spatial resolution of the bands 10 and 11 of TIRS 1 and 2.

4. Methodology

The present research was completed using the steps mentioned in Figure 2.

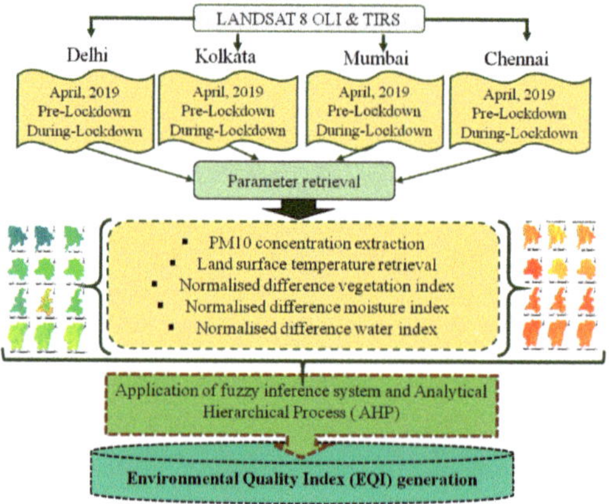

Figure 2. Workflow diagram of the present research.

4.1. PM_{10} Concentration

The concentration of PM_{10} is an important element for air quality, but these important air quality parameter data are not available sufficiently for Indian cities. Lesser number of observation points are available which can be used for interpolation to get a spatial variation of PM_{10} concentration with a smaller spatial unit. This problem of PM_{10} concentration assessment can be minimized through the integration of satellite-derived images and ground observation data. The entire process of PM_{10} concentration extraction is discussed below.

4.1.1. Conversion from DN Value to Top of the Atmosphere (TOA) Reflectance

The digital number (DN) value of a satellite band is the converted representation of actual reflectance from the surface. This converted reflectance value should be transformed into actual reflectance value for measuring the actual amount of reflectance from the land surface. The DN value for band 1 to 4 of Landsat 8 is converted into Top of Atmosphere value using Equation (1).

$$p\lambda' = M\rho\, Qcal + A\rho \tag{1}$$

where

$p\lambda'$ = Top of Atmosphere planetary reflectance, without solar angle correction,

$M\rho$ = Multiplicative rescaling factor for a specific band (REFLECTANCE_MULT_BAND_x, where x indicates band number),

$A\rho$ = Additive rescaling factor for a specific band (REFLECTANCE_ADD_BAND_x, where x indicates band number),

$Qcal$ = Converted pixel values (DN value).

4.1.2. Sun Angle Correction

TOA reflectance is a solar angle uncorrected product which may have some deviated value due to changes in solar angle; that is why solar angle correction is important. This solar angle correction of TOA data is performed using Equation (2).

$$p\lambda = \frac{p\lambda'}{cos(\Theta SZ)} = \frac{p\lambda'}{sin(\Theta SE)} \tag{2}$$

$\rho\lambda$ = Solar angle corrected Top of Atmosphere planetary reflectance.

$p\lambda'$ = Without solar angle corrected Top of Atmosphere planetary reflectance.

ΘSE = Sun elevation angle (Local) (SUN_ELEVATION data found in the metadata section).

ΘSZ = Solar zenith angle (Local); [$\Theta SZ = 90° - \Theta SE$].

More accurate reflection calculation depends on per-pixel solar angle data, which are not correctly available with Landsat 8 image metadata, which is substantiated with the scene centre solar angle value [30].

4.1.3. Regression Analysis and PM_{10} Concentration Calculation

Correlation among ground-measured PM_{10} concentration and difference between TOA reflectance and sun angle corrected TOA reflection represents the amount of PM_{10} concentration for a pixel. Least-root-mean-square error (RMSE) value and higher R^2 value among all possible combinations of the band between band 1 to band 4 of Landsat 8 satellite image were selected to obtain multiplicative and additive coefficient value, which is used to calculate actual PM_{10} concentration value for each pixel with the help of sparsely distributed PM_{10} concentration observation points located within the studied four megacity areas.

4.2. LST Retrieval

The steps to extract the LST from the thermal band of the satellite images are as follows:

4.2.1. Digital Number to Spectral Radiance Conversion

Following the law of electromagnetic radiation of the objects (each object radiates Electromagnetic Radiation (EMR), as each has temperature more than absolute zero (K)), at-sensor radiance can be achieved through the conversion of recognized signals by the Thermal Infrared Sensor (TIRS) and Enhanced Thematic Mapper (ETM+). Spectral radiance ($L\lambda$) was computed by the formula given below [31]:

$$L\lambda = \text{"gain"} * QCAL + \text{"offset"} \tag{3}$$

where gain = radiance slope/DN transformation function; DN signifies digital number of a specific pixel; bias is the radiance intercept/DN transformation function. It may also express as:

$$L\lambda = LMIN\lambda + [(LMAX\lambda - LMIN\lambda) / (QCALMAX - QCALMIN) * QCAL] \tag{4}$$

where $QCAL$ = DN of pixels; $QCALMAX$ = 255; $QCALMIN$ = 0; $LMIN\lambda$ = spectral radiance for thermal band at DN = 0, and $LMAX\lambda$ = spectral radiance for thermal band at DN = 255. Replacement of the corresponding values in Equation (3) provides a simpler Equation (4):

$$L\lambda = (0.037059 \times DN) + 3.2 \tag{5}$$

4.2.2. Spectral Radiance to At-Satellite Brightness Temperatures

Based on the category of land cover, rectifications for emissivity (e) were done for the radiant temperatures. Generally, vegetation areas have assigned with 0.95 and non-vegetation areas have assigned with a value of 0.92, following Nichol [32]. Emissivity-corrected LST was figured ensuing Artis and Carnahan [33].

$$T_B = \frac{K_2}{\ln\left(\frac{k_1}{L_\lambda} + 1\right)} \tag{6}$$

where L_λ is Spectral Radiance in $W \cdot m^{-2} \cdot sr^{-1} \cdot \mu m^{-1}$ and K_1 and K_2 are two constants of two pre-launch calibrations.

4.2.3. LST Estimation

The values of temperature attained from the above computations are denoted to a black body. Hence, it requires rectification of spectral emissivity (ε). It can be performed rendering to the type of land cover or by accomplishing corresponding emissivity values from the NDVI values for respective pixels [34]. The method of emissivity-rectified LST formulation can be stated as the equation below [33]:

$$LST = T_B / [1 + \{(\lambda \times T_B / \rho) \times ln\epsilon\}] \tag{7}$$

Here, LST = Land Surface Temperature in Kelvin, T_B = At-sensor brightness temperature, λ = TOA reflectance, $ln\varepsilon$ = Emissivity.

$$\text{Land surface emissivity } (\epsilon) = 0.004 \times P_v + 0.986 \tag{8}$$

where P_v is the proportion of vegetation which can be set up as:

$$P_v = \left(\frac{NDVI_{j_r} - NDVI_{min}}{NDVI_{max} - NDVI_{min}}\right)^2 \tag{9}$$

4.2.4. Kelvin to Degree Celsius Conversion

To simplify the conception, the measuring unit of these derived LSTs was transformed from a Kelvin scale to degree Celsius scale, applying the formula 0 °C equal to 273.15 K.

4.3. Biophysical Indices Extraction

To assess the instant changes of Environmental Quality Index (EQI) in the present framework of research, the three most relevant biophysical parameters have been included, such as Normalised Difference Moisture Index (NDMI), Normalised Difference Vegetation Index (NDVI) and Normalised Difference Water Index (NDWI) [15]. NDWI designates the vegetation water content and the status of vegetation can appositely represent by combining it with NDVI [35]. This indicator is the combination of reflectance of green plants, dry and plant absent areas, and soil parts through two near-infrared (NIR) bands [36]. It is computed using Equation (10).

$$NDWI = (NIR - SWIR) / (NIR + SWIR) \tag{10}$$

where index values near to the 1 signify water bodies, to −1 signify dry land and close to 0 denotes moderately humid lands [37].

NDVI is the most widely considered variable to extract vegetation information, i.e., health status and spatial distribution, etc., in a region that was derived following Equation (11) [38].

$$NDVI = (NIR - RED)/(NIR + RED) \tag{11}$$

where NIR and RED are the near-infrared band and red band respectively. The value ranges from +1 to −1, signifying vegetated zone to bare lands, and value near to 0 epitomizes grasslands [39].

NDMI was used to estimate the humidity contents of various landscape features such as soils, rocks, and plants. Eco-environmental exposure could be estimated using NDMI [39]. This was computed following Jin and Sader [40].

$$NDMI = (NIR - IR)/(NIR + IR) \tag{12}$$

where IR is the infrared. A high amount of humidity designated by NDMI values > 0.1 and values near to −1 signifies a low humidity level.

4.4. Fuzzy Inference System (FIS)

FIS is a proficient methodical approach to draw inferences using a set of input variables [41]. Fuzzy logic has great acceptance as a multi-criterion-based decision-making process has been adopted in various fields [42,43]. A precise result can be obtained using this method, because it assigns different membership grades for definite pixels and maximizes the membership degrees to attain destiny [43,44].

The fuzzy Mamdani system involves human's empirical knowledge to make an inference according to the input variable [45]. This approach is operated through fuzzification of the input variable layers, rule assessment, fuzzy end inference, and defuzzification [46,47]. To achieve the goal, FIS need the following setup:

Fuzzy Model Setup

The framework of this model for this research was built upon the following two heads:

(a) Fuzzy membership function was applied to five Environmental Quality Variables (EQV) considering their influence on urban environment.
(b) Positioning Control Point (CP) in order to find out the most influential range of the five EQVs to produce Environmental Quality Index (EQI).

Membership function to each EQV was dispensed to produce a normalized fuzzy layer of each raw variable, and CP for each was also specified according to the distributional characteristics of them. This two-phase operation ensured the conversion of each EQV layer into a 0–1 value range of pixels to make the data unidimensional, which is the most acceptable approach to overlay different layers with dissimilar units [41]. Fuzzy membership with monotonically decreasing function allotted to those EQVs which influence level gradually decreases following sigmoidal curve from CP 'a' to 'b' or CP 'c' to 'd' concerning Environmental Quality Index (EQI) determination (Table 3). Whereas, monotonically increasing membership function signifies the vice versa aiming the same goal (Table 3). Membership function and shape of EQVs and corresponding CP has been enlisted in Table 3.

$$\mu = \frac{1}{\left\{1 + \left(\frac{x-c_1}{c_2-c_1}\right)^2\right\}} \tag{13}$$

where μ is the standardized score of an EQV, x signifies raw score, and c_1 and c_2 are base and end value of control point, respectively. To optimize the decision space, c_1 and c_2 were used.

$$\mu_d(x_n) = exp\big((x_n - v_n)^2 / 2S_n^2\big) \tag{14}$$

Table 3. Control point, fuzzy membership function and shape determination for fuzzy transformation of the Environmental Quality Value (EQVs).

EQFs	Control Points				Fuzzy Membership Function	Fuzzy Membership Shape
	a	b	c	d		
PM$_{10}$ concentration	1	281			Monotonically increasing	Sigmoidal
LST	7	50			Monotonically increasing	Sigmoidal
NDVI			−0.13	0.65	Monotonically Decreasing	Sigmoidal
NDWI			−0.57	0.56	Monotonically Decreasing	Sigmoidal
NDMI			−0.65	0.55	Monotonically Decreasing	Sigmoidal

Equation (14) is used to set fuzzy rules, i.e., for a fuzzy set, resultant of all the inputs of S rules, fuzzy set $\mu_e(S)$, where *s* fit in '*S*' with different decisions of agent *X*, which can be inscribed as the equation below:

$$\mu_{E(X)} = \oplus_1^s \mu_{e_{(s)}} \tag{15}$$

In the present research, monotonically increasing fuzzy membership function with sigmoid pattern distribution curve was assigned for the parameters PM$_{10}$ concentration in the air and land surface temperature as the degradation risk of EQI increases with the increasing value of these two parameters within the specified CP range (Table 3). Similarly, for the NDVI, NDWI, and NDMI, the fuzzy membership shape was assigned following the same, but membership function type was followed as monotonically decreasing which means that, as the value of these indices increases, the probability of environmental degradation decreases.

4.5. Fuzzy-Analytical Hierarchical Process (AHP)

For modeling, computation of the weights of each parameter is essential to produce the goal. AHP, in this regard, is an efficient method to obtain weights of the inducing parameters to make decisions in complex problems [48]. The method involves making a comparison among the input parameters based on their relative importance to target decision through forming a pairwise comparison matrix [48,49], which enable to compute the parameter weights [47,50]. The stepwise process of the AHP method includes a breakdown of the complex structure of the problem into different alternatives and criteria, i.e., EQVs in here, assignment of preference values to the alternatives or criteria according to the scale of importance considering their association with the goal, i.e., EQI in this research [48,51], and computation of the weights following Satty [48]. The corresponding parameter weights are computed by dividing the cell values by values of the column sum of the pairwise comparison matrix. The importance scale for the parameter in respect of EQI was determined and assigned following the earlier studies, the appearance of the study area, and experts' opinions involved in this field [52,53]. Further progression was enabled through calibration of the pairwise comparison matrix and validated by the rule of Consistency Ratio (CR), which reflects the trustworthiness and applicability of the analysis performed. A value of <0.1 of CR suggests the reliability of the analysis. The pairwise comparison matrix, consistency ratio, and relative weights of the parameters in the present analysis are enlisted in Table 4. Pairwise comparison matrix and CR can be expressed using the equations below.

$$A = \begin{bmatrix} a_{11} & a_{12} & a_{13} & a_{1n} \\ a_{21} & a_{22} & a_{23} & a_{2n} \\ \cdots & \cdots & a_{ij} & \cdots \\ a_{n1} & a_{n2} & a_{n3} & a_{nn} \end{bmatrix} \tag{16}$$

where

$$a_{ij} = \frac{Wi}{Wj} = \frac{\text{weight for attribute i}}{\text{weight for attribute j}} \tag{17}$$

Table 4. Pairwise comparison matrix for considered EQVs to derive relative weight using the Fuzzy-Analytical Hierarchical Process (AHP) method.

	PM_{10}	LST	NDVI	NDWI	NDMI
PM_{10}	1				
LST	0.33	1			
NDVI	0.20	0.20	1		
NDWI	0.14	0.14	0.33	1	
NDMI	0.11	0.11	0.20	0.33	1
Consistency Ratio is 0.07					

The computation of *CR* can be expressed as

$$CR = CI/RI \tag{18}$$

where *CI* and *RI* are the consistency index and random index, respectively, and *CI* can be expressed as the equation below:

$$CI = \lambda Max - n/n - 1 \tag{19}$$

where λMax = largest eigenvalue, and *n* = order of the matrix.

In this research, fuzzy-AHP was the process of a fuzzy behavioural structure where the fuzzy-transformed layer of the EQVs was weighted (Table 5) by the AHP method with respect to their influence to estimate dynamics status of environmental quality status in three different phases.

Table 5. Relative weight for each EQV generated through the AHP method for using in the weighted linear combination overlay method to generate the final vulnerable area map.

EQVs	Weightage
PM_{10} concentration	0.492
Land Surface Temperature	0.3136
NDVI	0.1093
NDWI	0.0553
NDMI	0.0298

4.6. EQI Generation

After assigning the weights to the fuzzy-converted layers of EQVs, to produce the EQI maps of four megacities of India, the layers were integrated using the Linear Combination (LCM) overlay technique as each parameter was unidirectional and unit-less. The LCM was performed following Equation (20).

$$LCM = \frac{1}{n} \times \sum_{i=1}^{n} D_i \times w_i \tag{20}$$

where *n* is the number of parameters considered, w_i is the weight of *i* parameter, and *D* is the decisive EQV.

5. Results

5.1. Changing Patterns of PM_{10} Concentration

PM_{10} concentration is one of the determinants of air quality as well as environmental quality. Mainly, industrial activities, motor vehicles, and construction works are the sources of PM_{10} concentration in the air which may destroy human comforts and creates lots of respiratory diseases. This is not only the direct human problem, but it also harms environmental conditions indirectly in the form of increasing air and land surface temperature. The lockdown situation has successfully restricted the spread of the COVID-19 virus in the study areas. These restrictions play a positive role in recovering the past situation of PM_{10} concentration by reducing PM_{10} supply into the atmosphere. Figure 3 indicates a clear declination of PM_{10} concentration from 2019—the same month for which lockdown data are collected to 2020 lockdown, representing the same period, although the pre-lockdown phase shows an increasing amount of PM_{10} concentration compared to 2019, which is a natural trend of increasing this air quality parameter. This situation is applicable for Mumbai and Chennai, whereas a different situation is observed for Delhi and Kolkata. The PM_{10} concentration map of Delhi shows that the pre-lockdown phase indicated a decreased amount of PM_{10} concentration, but lockdown time indicated a higher decrease in PM_{10} concentration than in 2019 and the pre-lockdown situation. This different result has taken place as a result of the early shutdown of the industrial sector and restriction on human movement before imposing lockdown in India. Kolkata is showing an exceptional pattern of PM_{10} concentration compared to the rest of the study sites, because the pre-lockdown phase showing minimum PM_{10} concentration than 2019 and lockdown phase, although the lockdown phase representing lower PM_{10} concentration than 2019 same month. This is maybe the same reason as in Delhi, which is discussed earlier.

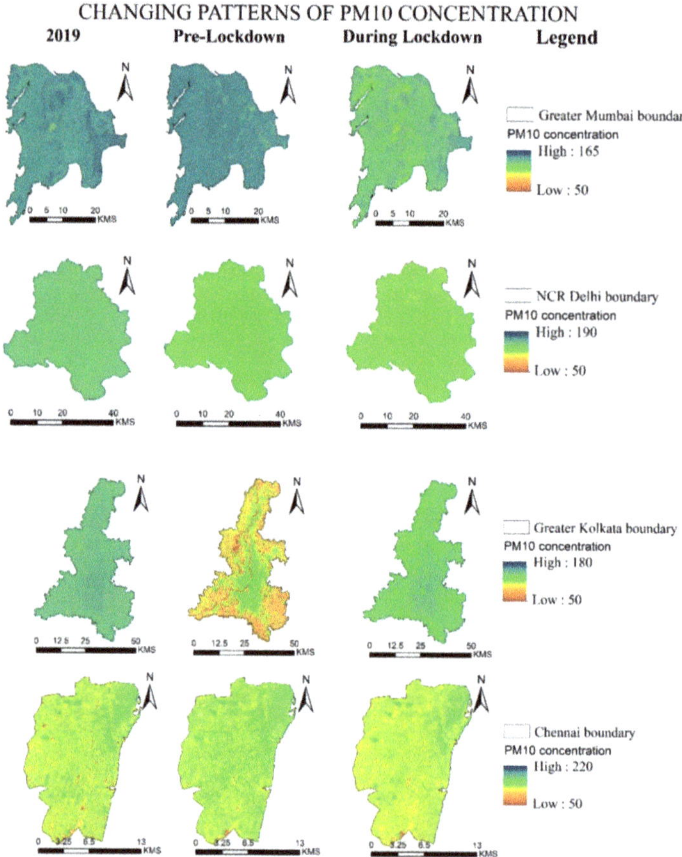

Figure 3. Changing pattern of PM_{10} concentration in four megacities of India during the same season in 2019 of the pre-lockdown 2020; pre-lockdown, 2020 and during the lockdown, 2020.

5.2. *Changing Patterns of LST*

LST is also an important parameter for environmental health, which depends on a large number of atmospheric and physical factors. The month of the study, cloudy condition, Land Use/Land Cover (LULC) patterns, and many more factors govern the patterns of LST for an area. To avoid the problems of different seasons and months, a similar month satellite image was collected with less than 10% cloud coverage. On the other hand, LULC pattern has become constant from pre-lockdown to lockdown time, although LULC patterns from 2019 to 2020 lockdown time have changed as a natural process. For this reason, LULC is not considered for assessing environmental quality.

Land Surface Temperature indicates an increasing trend over time due to LULC change and other global phenomena. LST maps indicate that 2019 and lockdown time is higher than pre-lockdown, which is a natural phenomenon due to the changes in study months; 2019 and lockdown time data were collected for the last week of March and the first two weeks of April, which generally represents higher temperature months than pre-lockdown data collected during the period of February last week to March second week. Here in India, February and March represent the transition period between the winter and summer season, whereas March last to April represents the summer season, which indicates a general trend of temperature increase. Mumbai, Delhi, and Kolkata are showing a higher temperature of 48 °C to 50 °C for 2019, pre-lockdown, and lockdown time, whereas Chennai is showing a lower high temperature of 42 °C (Figure 4). On the other hand, both of these four study areas are showing a similar trend of LST patterns; same time of lockdown in 2019 and lockdown time showing a higher temperature than pre-lockdown time.

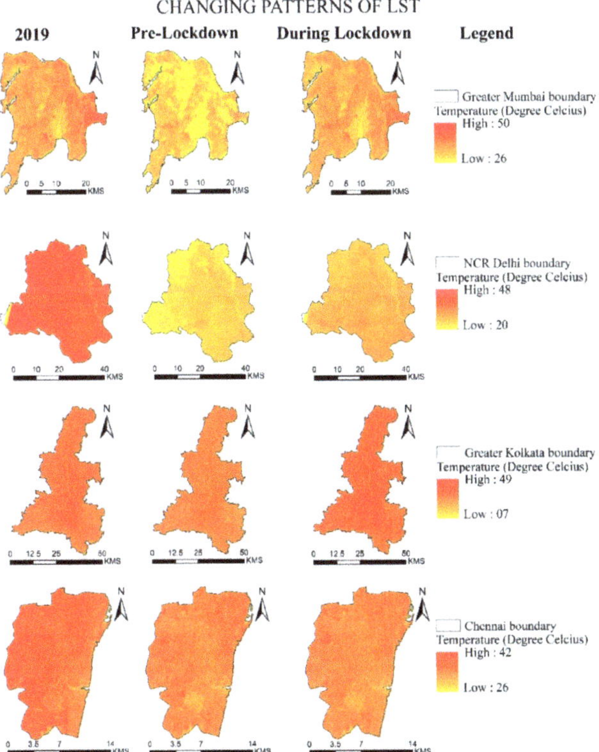

Figure 4. Changing pattern of land surface temperature in four megacities of India during the same season in 2019 of the pre-lockdown 2020, pre-lockdown, 2020, and during the lockdown, 2020.

5.3. Changing Patterns of Biophysical Indices

Bio-physical indices are calculated from remotely sensed satellite images to quantify the changing nature of some environmental parameters which may determine the quality of the environment for an area. Previous studies have used various bio-physical indices to assess the nature of environmental conditions and their changing nature such as NDVI, NDWI, NDBI, and NDMI. For this study, three biophysical indices were calculated from satellite images for each phase for all the study areas to assess the changing patterns of bio-physical environmental quality for the study time. Normalized Difference Vegetation Index (NDVI) represents the greenness of an area and positive greenness indicates a healthy environmental condition. The greenness of the study areas has increased from 2019 and pre-lockdown condition to lockdown condition due to the restriction of human movement and undisturbed condition of the existing natural vegetation. On the other hand, the emergence of newly vegetated areas in previously open spaces results in a high level of greenness during this time. This situation is also applicable to all megacities, which is a positive indication from the perspective of positive environmental health. Normalized Difference Water Index (NDWI) is also a positive indication of environmental health. Patterns of NDWI in the study areas are ranging between −0.55 and 0.32 for Mumbai, −0.55 and 0.54 in Delhi, −0.56 and 0.31 in Kolkata and −0.53 and 0.32 in Chennai, all of these ranges indicating that amounts of wetness in the study areas are not so high for any time period of this study (Figure 5). The presence of the water body or volume of wetness is important for maintaining a healthy environment. The overall analysis found that, for Mumbai, Delhi, and Chennai, NDWI value has increased during pre-lockdown time than in 2019 and lockdown time, but exceptionally Kolkata is showing a different condition with higher NDWI value during lockdown time (Figure 6).

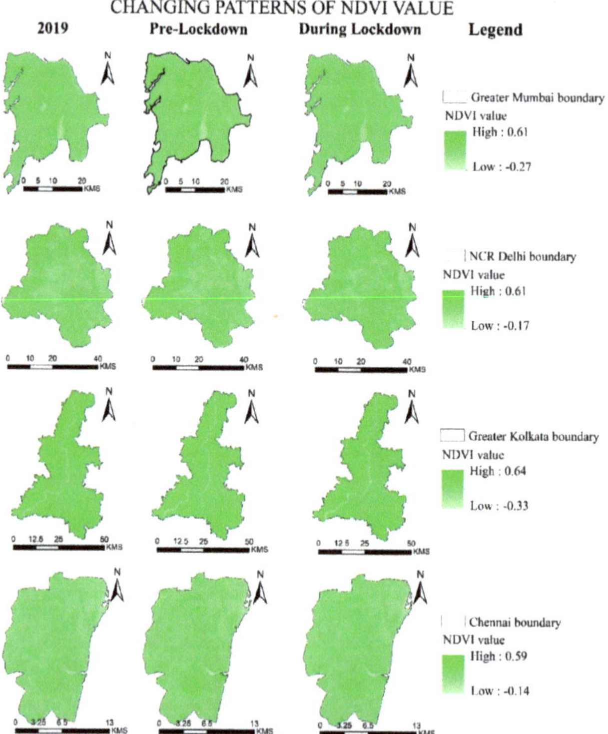

Figure 5. Changing pattern of NDVI in four megacities of India during the same season in 2019 of the pre-lockdown 2020, pre-lockdown, 2020 and during the lockdown, 2020.

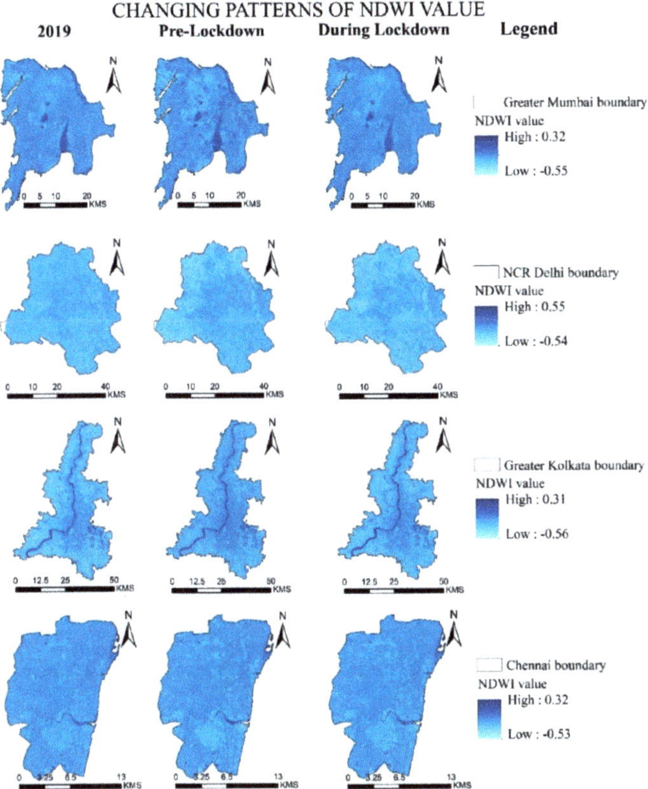

Figure 6. Changing pattern of NDWI in four megacities of India during the same season in 2019 of the pre-lockdown 2020, pre-lockdown, 2020 and during the lockdown, 2020.

Normalized Difference Moisture Index (NDMI) represents the amount of moisture content in the surface, which also determines the environmental condition of an area. Less human intervention results in an increase in NDMI value as human intervention alters the nature of moisture availability. NDMI index values for the study areas are indicating that NDMI value has increased during lockdown time than in 2019 and pre-lockdown time for Mumbai, Kolkata, and Chennai, where Delhi has to gain more moisture condition in pre-lockdown time than in 2019 and lockdown time (Figure 7). This may be due to the atmospheric condition of the Delhi area during the internment of satellite data.

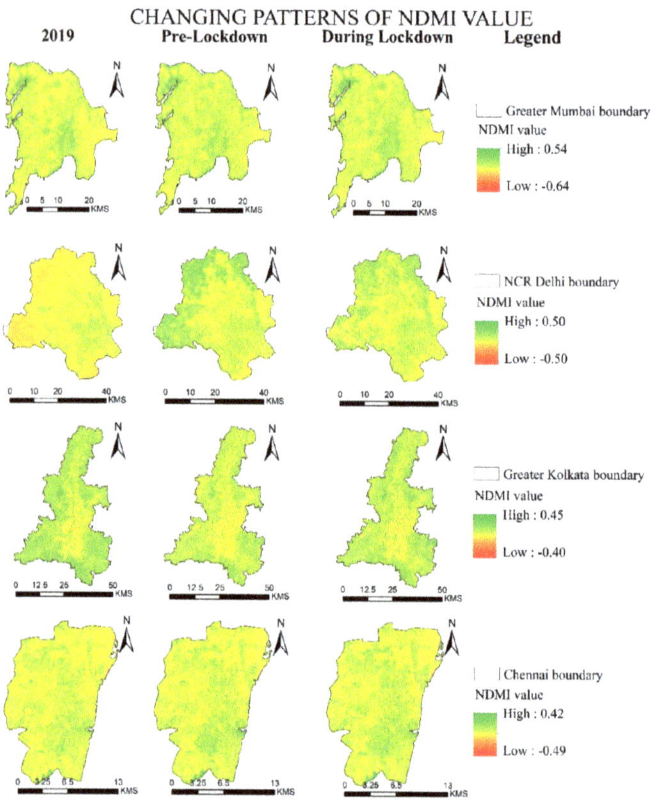

Figure 7. Changing pattern of NDMI in four megacities of India during the same season in 2019 of the pre-lockdown 2020, pre-lockdown, 2020 and during the lockdown, 2020.

5.4. Impact of COVID-19 Lockdown on Environmental Quality

Published articles on COVID-19 impact the environment, indicating that different environmental parameters, especially air quality parameters, have changed positively due to the lockdown situation in the whole world. India is not an exception, because India also started lockdown from 24 March 2020 for the whole country. A positive effect on air quality and some aspects of the physical environment is already notified by some researchers, but none of the studies assessed overall environmental quality for any part of the country. Indian megacities are the worst COVID-19 affected areas that were strictly following the lockdown conditions from the beginning, which resulted in the improved environmental quality of the studied megacities. It is found from this study that environmental quality has improved from the same month of 2019 to the same lockdown month. The general trend of environmental quality of the studied cities is degraded from its previous years due to the increase of human intervention [10]. According to this trend, the lockdown time of 2020 should have a higher environmental degradation than the same month of 2019, but this study shows that environmental quality has improved in 2020 compared with the same month of 2019, which has shown a positive indication from the perspective of environmental quality. This positive indication is a hope of light in the darkness of the COVID-19 situation. Figure 8 indicates that environmental quality has improved from 2019 to 2020 for Mumbai, Delhi, and Chennai, although pre-lockdown time shows a better environmental quality for all three study areas except Chennai, due to the change in the month, which represents low-temperature months compared with lockdown month. This stated result is also visible from Figure 9, which shows the changing patterns of physical environmental quality in five classes, divided using Jenkin's Natural Break classification system with similar class range values to make it comparable between time and study sites. Areal change of different environmental quality index classes show that very low degraded environmental quality area is increased from 10.22% in 2019 to 38.45% in 2020 for Mumbai with the same month data for both years. This fact is also true for the other three study sites of Delhi, and Chennai with a changing value from 2019 to 2020, as 1.50% to 71.03% and 58.38% to 59.95%, respectively (Table 6). On the other hand, Kolkata is indicating an increasing low environmental degraded area from 52.16% in 2019 to 35.75% in lockdown time (Figure 10). Similarly, low environmental degradation areas have also increased in the lockdown period (percentage) in Mumbai, Delhi, Kolkata, and Chennai, respectively, from 2019 to 2020 (Table 6). The very high environmental quality degradation area has decreased for three studied megacities, except for Kolkata, which is also the same for high environmental quality degradation areas for three megacities, except for Chennai and Kolkata. The overall environmental quality degradation index has improved from 2019 to lockdown time, which is a positive indication for the whole world in the era of urbanization and environmental change.

Table 6. Pattern of environmental quality degradation in four megacities of India.

Study Area	Date	Very Low	Low	Medium	High	Very High
Mumbai	2019	10.22759	19.97495	35.34755	21.62421	12.8257
	Pre-lockdown	46.19468	32.35018	18.56851	2.787521	0.099114
	Lockdown	38.44749	31.63997	22.3983	5.484552	2.029686
Delhi	2019	1.502448	1.08621	8.639542	30.67348	58.09832
	Pre-lockdown	94.39192	5.593551	0.0135	0.0004	0.000629
	Lockdown	71.02876	27.36464	1.565073	0.040099	0.00143
Kolkata	2019	52.16253	25.94085	16.33086	4.490782	1.074978
	Pre-lockdown	55.86787	37.05701	7.02363	0.019139	0.032355
	Lockdown	35.74941	31.45045	20.39915	8.486124	3.914865
Chennai	2019	58.38214	28.33292	12.44535	0.463332	0.376257
	Pre-lockdown	44.20491	37.07574	16.24046	1.987608	0.491288
	Lockdown	59.94629	31.48092	7.667207	0.740598	0.164985

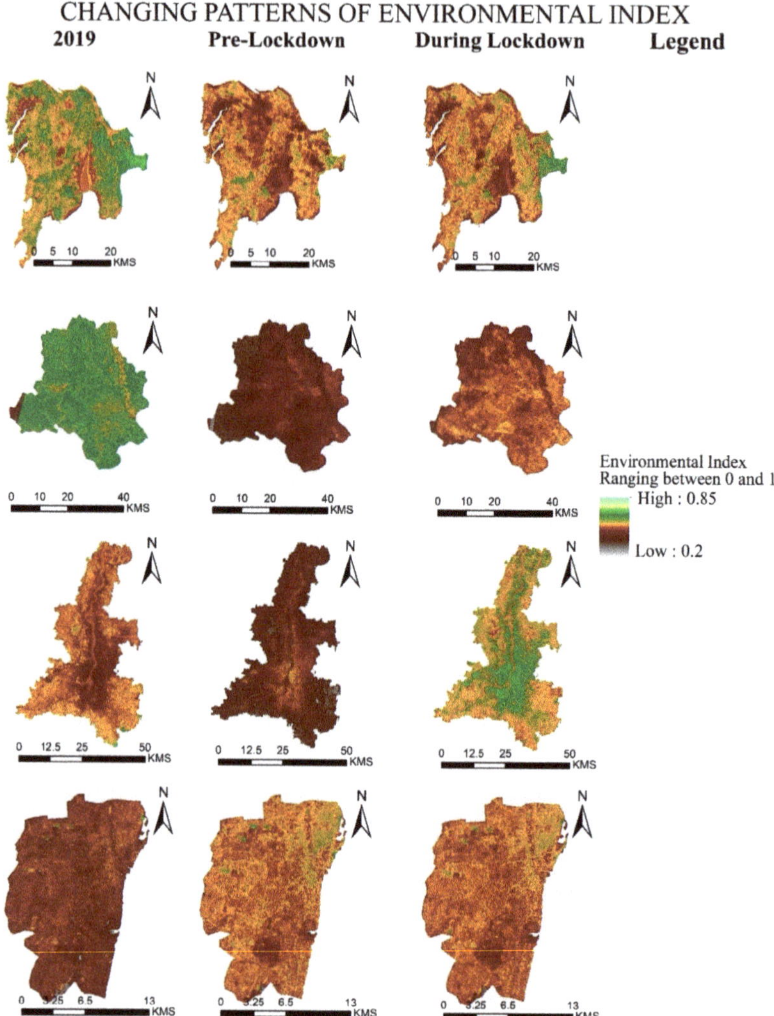

Figure 8. Changing pattern of the environmental quality index in four megacities of India during the same season in 2019 of the pre-lockdown 2020, pre-lockdown, 2020 and during the lockdown, 2020.

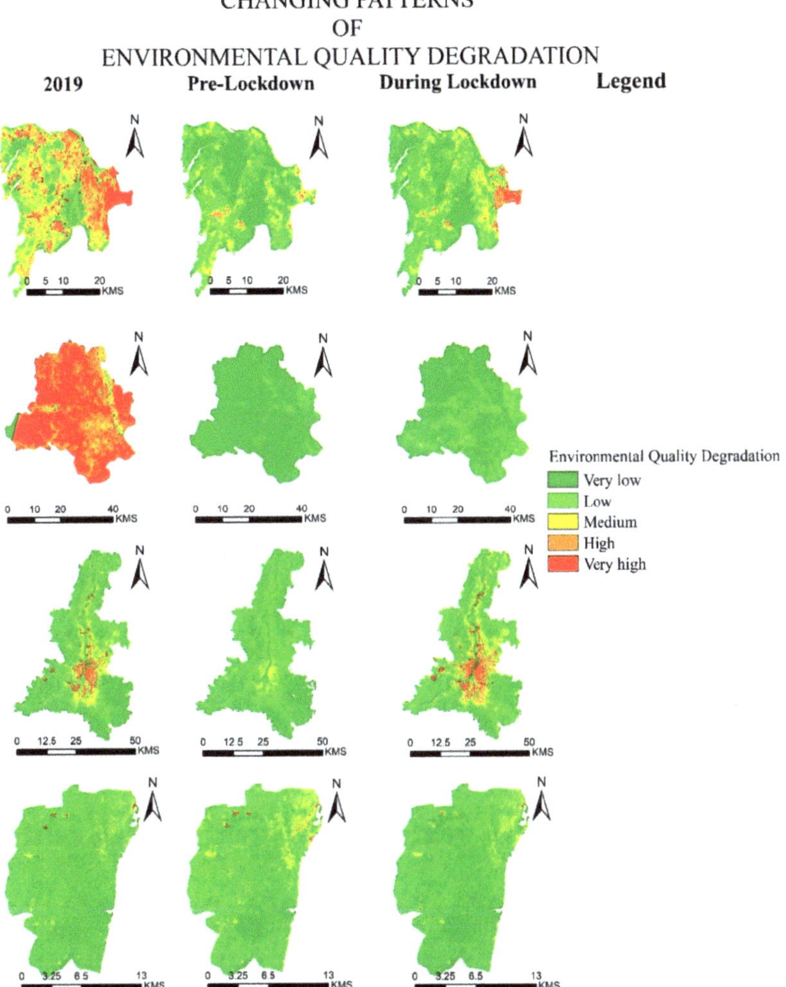

Figure 9. Spatial distribution of environmental quality vulnerability in four megacities of India during the same season in 2019 of the pre-lockdown 2020, pre-lockdown, 2020 and during the lockdown, 2020.

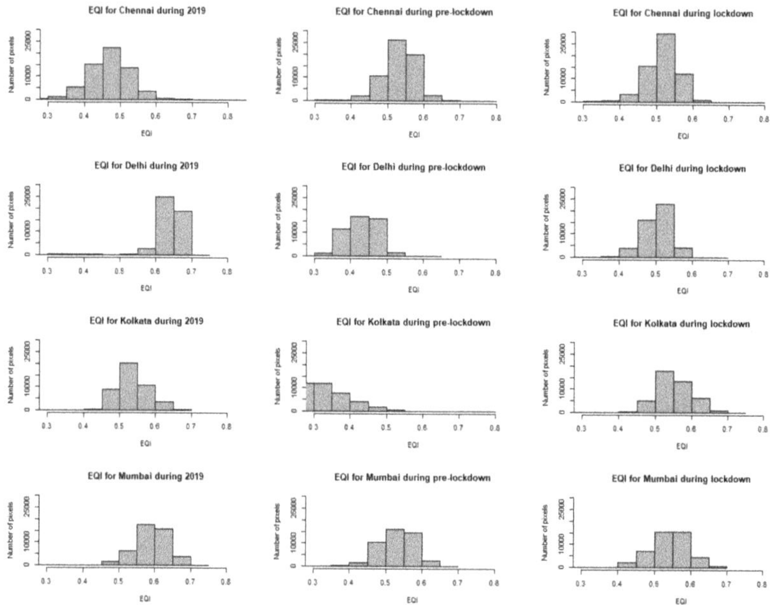

Figure 10. Changing pattern of Environmental Quality Index (EQI) in four megacities of India during the same season in 2019 of the lockdown 2020, pre-lockdown, 2020 and during the lockdown, 2020.

6. Discussion

Environmental quality for the world has changed negatively for the last few decades due to rapid population growth and increased human intervention in the natural environment; more specifically, physical environmental parameters have changed rapidly such as air quality, water quality, land surface temperature, and many more. The COVID-19 pandemic imposes a positive indication of the recovery of environmental degradation through improving the physical environmental parameters [1]. Restriction on human movement, industrial production, and vehicle movement results in the reduction of PM_{10} concentration in the atmosphere; on the other hand, less anthropogenic heat flux in the near-surface atmosphere resulting from human movement has also reduced LST, which helps to minimize the natural increase of LST from its previous year [54]. Not only these two parameters, but some biophysical parameters such as NDVI, NDWI, and NDMI have also changed due to uninterrupted or untouched condition of the natural environment.

The result indicates that overall environmental quality of all the study sites has improved during lockdown time, although the better environmental quality is found during the pre-lockdown time, which is the combination of two factors: (1) imposition of restriction rules in megacities of India before declaring formal lockdown for the whole country, (2) seasonal variation due to changes in data collection months. Most of the satellite images for pre-lockdown time were collected from February and the first half of March, which is the transition time between winter and summer season, but lockdown satellite images were collected mostly from April, except for Kolkata, which was collected for 29 March, which represents hotter days than pre-lockdown time.

National Capital Region (NCR) Delhi has repeatedly come into headlines of a reputed Indian newspaper for degraded air quality during the last 3 to 4 months [2], and other studied megacities are also in the same direction of degraded air quality, but COVID-19 lockdown on these megacities has shown a drastic change in the air quality, especially in PM_{10} concentration. LST has also been showing a similar increasing trend for the last few years, which is also reduced in this time for a lesser amount of vehicle and human exposure to the environment. Inter-megacity comparison of environmental quality indicates that Chennai has better environmental quality than the other three megacities under this study because of its locational advantage of being a coastal city, which generally reduces the amount of LST for the whole year compared to landlocked megacities.

In NCR, Delhi PM_{10} concentration was less in pre-lockdown time. It could be due to several reasons such as: (i) imposition of different regulation, i.e., allowing odd- and even-number vehicles on the road on alternative days, (ii) continuous water spray on roads for reducing the supply of particulate matters, etc., for minimizing air pollution through reducing particulate matter concentration in the air. These regulations play a positive role in minimizing air pollution and overall environmental quality of the studied cities. In the case of Kolkata, the less temperate condition may play an important role in minimizing PM_{10} concentration in the air.

Overall, this study found that the environmental quality of the studied megacities has improved from its previous years; in some instances, it is also improved from pre-lockdown time. This positive indication is generating some hope for people to get a better and recovered environmental condition, irrespective of this deadliest pandemic situation.

7. Conclusions

Changing environmental quality of the world has gained a positive move toward sustainable environment-friendly conditions due to the imposition of the lockdown owing to the COVID-19 pandemic. Environmental parameters such as PM_{10} concentration, NDWI, and NDMI have changed positively during this lockdown time, which is already assessed by some conducted and published studies. To the best of our knowledge, this is the first study that tries to present the changes in the environmental conditions due to the imposition of lockdown comprehensively owing to COVID-19 by devising an environmental quality index. The study found that overall environmental quality has improved from its previous year (same months) for all four studied megacities, although according to the general trend of environmental quality degradation, 2020 must have the worst environmental condition than the environmental condition of 2019. The environmental quality index map indicates an improved environmental condition, which is a sign of anticipation for the environmentalist in this crisis moment. This study has tried to obtain an overall view of the environmental condition, but more detailed investigation is still needed for a better understanding of the environmental response to the COVID-19 pandemic imposed lockdown situation and assessment of UQI after the lockdown situation is also important for understanding the environmental impact of COVID-19 specifically. However, the lockdown situation is still existing in India; hence, after the lockdown, a more detailed investigation on these aspects will help to maintain the better quality of environment.

Author Contributions: Methodology; formal analysis, S.G., A.D., T.K.H., and S.S.; S.G. and S.S., investigation; writing—original draft preparation, S.G., A.D., and T.K.H., S.S.; writing—review and editing, B.P. and S.S., and S.G. performed the experiments, wrote the manuscript; S.S. wrote the manuscript and analyzed the data; B.P. supervised, edited, restructured, and professionally optimized the manuscript; B.P. and A.M.A. arranged the funding acquisition. All authors have read and agreed to the published version of the manuscript.

Funding: This research is supported by the Centre for Advanced Modelling and Geospatial Information Systems (CAMGIS), Faculty of Engineering and IT, the University of Technology Sydney (UTS). This research was also supported by Researchers Supporting Project number RSP-2020/14, King Saud University, Riyadh, Saudi Arabia.

Conflicts of Interest: The authors declare no conflict of interest.

References

1. Mahato, S.; Pal, S.; Ghosh, K.G. Effect of lockdown amid COVID-19 pandemic on air quality of the megacity Delhi, India. *Sci. Total Environ.* **2020**, *730*, 139086. [CrossRef] [PubMed]
2. Sharma, S.; Mathur, S. Analyzing the Patterns of Delhi's Air Pollution. In *Advances in Data Sciences, Security and Applications*; Springer: Singapore, 2020; pp. 33–44.

3. Mukherjee, A.; Agrawal, M. Air pollutant levels are 12 times higher than guidelines in Varanasi, India. Sources and transfer. *Environ. Chem. Lett.* **2018**, *16*, 1009–1016. [CrossRef]

4. Garaga, R.; Sahu, S.K.; Kota, S.H. A review of air quality modeling studies in India: Local and regional scale. *Curr. Pollut. Rep.* **2018**, *4*, 59–73. [CrossRef]

5. Guo, H.; Kota, S.H.; Sahu, S.K.; Hu, J.; Ying, Q.; Gao, A.; Zhang, H. Source apportionment of PM2. 5 in North India using source-oriented air quality models. *Environ. Pollut.* **2017**, *231*, 426–436. [CrossRef] [PubMed]

6. World Health Organization. Ambient Air Pollution: A Global Assessment of Exposure and Burden of Disease. 2016. Available online: http://who.int/phe/publications/airpollution-globalassessment/en/ (accessed on 20 May 2020).

7. Polk, H.S. *State of Global Air 2019: A Special Report on Global Exposure to Air Pollution and Its Disease Burden*; Health Effects Institute: Boston, MA, USA, 2019.

8. Mohan, M.; Kandya, A. Impact of urbanization and land-use/land-cover change on diurnal temperature range: A case study of tropical urban airshed of India using remote sensing data. *Sci. Total Environ.* **2015**, *506*, 453–465. [CrossRef] [PubMed]

9. Haque, M.; Singh, R.B. Air pollution and human health in Kolkata, India: A case study. *Climate* **2017**, *5*, 77. [CrossRef]

10. Sharma, R.; Chakraborty, A.; Joshi, K. Geospatial quantification and analysis of environmental changes in urbanizing city of Kolkata (India). *Environ. Monit. Assess.* **2015**, *187*, 4206. [CrossRef]

11. Kumar, A.; Gupta, I.; Brandt, J.; Kumar, R.; Dikshit, A.K.; Patil, R.S. Air quality mapping using GIS and economic evaluation of health impact for Mumbai city, India. *J. Air Waste Manag. Assoc.* **2016**, *66*, 470–481. [CrossRef]

12. Sundaram, A.M. Urban green-cover and the environmental performance of Chennai city. *Environ. Dev. Sustain.* **2011**, *13*, 107–119. [CrossRef]

13. Partheeban, P.; Raju, H.P.; Hemamalini, R.R.; Shanthini, B. Real-Time Vehicular Air Quality Monitoring Using Sensing Technology for Chennai. In *Transportation Research*; Springer: Singapore, 2020; pp. 19–28.

14. Sathyakumar, V.; Ramsankaran, R.; Bardhan, R. Geospatial approach for assessing spatiotemporal dynamics of urban green space distribution among neighbourhoods: A demonstration in Mumbai. *Urban For. Urban Green.* **2020**, *48*, 126585. [CrossRef]

15. Liang, B.; Weng, Q. Assessing urban environmental quality change of Indianapolis, United States, by the remote sensing and GIS integration. *IEEE J. Sel. Top. Appl. Earth Obs. Remote Sens.* **2010**, *4*, 43–55. [CrossRef]

16. Akbari, H.; Rosenfeld, A.H.; Taha, H. Summer heat islands, urban trees, and white surfaces. *ASHRAE Trans.* **1990**, *96*, 1381–1388.

17. Akbari, S.; Rose, H.L.S.; Taha, H. Analyzing the land cover of an urban environment using high-resolution orthophotos. *Landsc. Urban Plan.* **2003**, *63*, 1–14. [CrossRef]

18. De Vries, S.; Verheij, R.A.; Groenewegen, P.P.; Spreeuwenberg, P. Spreeuwenberg, Natural environments— Healthy environments? An exploratory analysis of relationship between green space and Health. *Environ. Plan. A* **2003**, *35*, 1717–1731. [CrossRef]

19. Nichol, J.; Wong, M.S. Modeling urban environmental quality in a tropical city. *Landsc. Urban Plan.* **2005**, *73*, 49–58. [CrossRef]

20. Heynen, N. Green urban political ecologies: Toward a better understanding of inner-city environmental change. *Environ. Plan A* **2006**, *38*, 499–516. [CrossRef]

21. Weng, Q.; Lu, D.; Schubring, J. Estimation of land surface temperature–vegetation abundance relationship for urban heat island studies. *Remote Sens. Environ.* **2004**, *89*, 467–483. [CrossRef]

22. Weng, Q.; Liu, H.; Liang, B.; Lu, D. The spatial variations of urban land surface temperatures: Pertinent factors, zoning effect, and seasonal variability. *IEEE J. Sel. Top. Appl. Earth Obs. Remote Sens.* **2008**, *1*, 154–166. [CrossRef]

23. Rajasekar, U.; Weng, Q. Spatio-temporal modelling and analysis of urban heat islands by using Landsat TM and ETM+ imagery. *Int. J. Remote Sens.* **2009**, *30*, 3531–3548. [CrossRef]

24. Li, G.; Weng, Q. 15 Integration of Remote Sensing and Census Data for Assessing Urban Quality of Life: Model Development. In *Urban Remote Sensing*; Weng, Q., Quattrochi, D.A., Eds.; CRC Press: Boca Raton, FL, USA, 2006; p. 311.

25. Li, G.; Weng, Q. Measuring the quality of life in city of Indianapolis by integration of remote sensing and census data. *Int. J. Remote Sens.* **2007**, *28*, 249–267. [CrossRef]

26. Zhang, X.; Wang, C.; Li, E.; Xu, C. Assessment model of ecoenvironmental vulnerability based on improved entropy weight method. *Sci. World J.* **2014**, *2014*, 797814. [CrossRef]

27. Sharma, S.; Zhang, M.; Gao, J.; Zhang, H.; Kota, S.H. Effect of restricted emissions during COVID-19 on air quality in India. *Sci. Total Environ.* **2020**, *728*, 138878. [CrossRef]

28. Mohan, M.; Dagar, L.; Gurjar, B.R. Preparation and validation of gridded emission inventory of criteria air pollutants and identification of emission hotspots for megacity Delhi. *Environ. Monit. Assess.* **2007**, *130*, 323–339. [CrossRef] [PubMed]

29. Saraswat, I.; Mishra, R.K.; Kumar, A. Estimation of PM10 concentration from Landsat 8 OLI satellite imagery over Delhi, India. *Remote Sens. Appl. Soc. Environ.* **2017**, *8*, 251–257. [CrossRef]

30. Landsat Missions: Using the USGS Landsat 8 Product. Available online: https://landsat.usgs.gov/using-usgs-landsat-8-product (accessed on 15 December 2016).

31. Landsat Project Science Office. *Landsat 7 Science Data User's Handbook*; Goddard Space Flight Center, NASA, 2002. Available online: http://ltpwww.gsfc.nasa.gov/IAS/handbook/handbook_toc.html (accessed on 20 May 2020).

32. Nichol, J.E. A GIS-based approach to microclimate monitoring in Singapore's high-rise housing estates. *Photogramm. Eng. Remote Sens.* **1994**, *60*, 1225–1232.

33. Artis, D.A.; Carnahan, W.H. Survey of emissivity variability in thermography of urban areas. *Remote Sens. Environ.* **1982**, *12*, 313–329. [CrossRef]

34. Snyder, W.C.; Wan, Z.; Zhang, Y.; Feng, Y.Z. Classification-based emissivity for land surface temperature measurement from space. *Int. J. Remote Sens.* **1998**, *19*, 2753–2774. [CrossRef]

35. Chen, X.L.; Zhao, H.M.; Li, X.; Yin, Z.Y. Remote sensing image-based analysis of the relationship between urban heat island and land use/cover changes. *Remote Sens. Environ.* **2006**, *104*, 133–146. [CrossRef]

36. Jackson, T.J.; Chen, D.; Cosh, M.; Li, F.; Anderson, M.; Walthall, C.; Doriaswamy, P.; Hunt, E.R. Vegetation water content mapping using Landsat data derived normalized difference water index for corn and soybeans. *Remote Sens. Environ.* **2004**, *92*, 475–482. [CrossRef]

37. Gao, B.C. NDWI—A normalized difference water index for remote sensing of vegetation liquid water from space. *Remote Sens. Environ.* **1996**, *58*, 257–266. [CrossRef]

38. Purevdorj, T.S.; Tateishi, R.; Ishiyama, T.; Honda, Y. Relationships between percent vegetation cover and vegetation indices. *Int. J. Remote Sens.* **1998**, *19*, 3519–3535. [CrossRef]

39. Nguyen, A.K.; Liou, Y.A.; Li, M.H.; Tran, T.A. Zoning eco-environmental vulnerability for environmental management and protection. *Ecol. Indic.* **2016**, *69*, 100–117. [CrossRef]

40. Jin, S.; Sader, S.A. Comparison of time series tasseled cap wetness and the normalized difference moisture index in detecting forest disturbances. *Remote Sens. Environ.* **2005**, *94*, 364–372. [CrossRef]

41. Ghosh, S.; Das, A. Urban expansion induced vulnerability assessment of East Kolkata Wetland using Fuzzy MCDM method. *Remote Sens. Appl. Soc. Environ.* **2019**, *13*, 191–203. [CrossRef]

42. Keshavarzi, A.; Sarmadian, F.; Heidari, A.; Omid, M. Land suitability evaluation using fuzzy continuous classification (a case study: Ziaran region). *Mod. Appl. Sci.* **2010**, *4*, 72. [CrossRef]

43. Hembram, T.K.; Saha, S. Prioritization of sub-watsheds for soil erosion based on morphometric attributes using fuzzy AHP and compound factor in Jainti River basin, Jharkhand, Eastern India. *Environ. Dev. Sustain.* **2018**, *6*, 1–28. [CrossRef]

44. Ahmed, R.; Sajjad, H.; Husain, I. Morphometric parameters-based prioritization of sub-watersheds using fuzzy analytical hierarchy process: A case study of lower Barpani Watershed, India. *Nat. Resour. Res.* **2017**, *27*, 67–75. [CrossRef]

45. Mamdani, E.H. Application of fuzzy logic to approximate reasoning using linguistic synthesis. *IEEE Trans. Comput.* **1977**, *12*, 1182–1191. [CrossRef]

46. Bojadziev, G. *Fuzzy Logic for Business, Finance, and Management*; World Scientific: Singapore, 2007; Volume 23.

47. Sarkar, S.; Parihar, S.M.; Dutta, A. Fuzzy risk assessment modelling of East Kolkata Wetland Area: A remote sensing and GIS based approach. *Environ. Model. Softw.* **2016**, *75*, 105–118. [CrossRef]

48. Saaty, T.L. A scaling method for priorities in hierarchical structures. *J. Math. Psychol.* **1977**, *15*, 234–281. [CrossRef]

49. Saaty, T.L. *The Analytical Hierarchy Process*; McGraw Hill: New York, NY, USA, 1980; p. 350.

50. Saha, S. Groundwater potential mapping using analytical hierarchical process: A study on Md. Bazar Block of Birbhum District, West Bengal. *Spat. Inf. Res.* **2017**, *25*, 615–626. [CrossRef]

51. Satty, T.L.; Vargas, L.G. Models, methods, concepts and applications of the analytic hierarchy process. *Int. Ser. Oper. Res. Manag. Sci.* **2001**, *34*, 1–352.
52. Mondal, B.; Dolui, G.; Pramanik, M.; Maity, S.; Biswas, S.S.; Pal, R. Urban expansion and wetland shrinkage estimation using a GIS-based model in the East Kolkata Wetland, India. *Ecol. Indic.* **2017**, *83*, 62–73. [CrossRef]
53. Richardson, C.; Amankwatia, K. GIS-based analytic hierarchy process approach to watershed vulnerability in Bernalillo County, New Mexico. *J. Hydrol. Eng.* **2018**, *23*, 04018010. [CrossRef]
54. Muhammad, S.; Long, X.; Salman, M. COVID-19 pandemic and environmental pollution: A blessing in disguise? *Sci. Total Environ.* **2020**, *20*, 138820. [CrossRef]

MDPI

St. Alban-Anlage 66

4052 Basel

Switzerland

Tel. +41 61 683 77 34

Fax +41 61 302 89 18

www.mdpi.com

Sustainability Editorial Office

E-mail: sustainability@mdpi.com

www.mdpi.com/journal/sustainability